AF342326

Nonferrous Extractive Metallurgy

NONFERROUS EXTRACTIVE METALLURGY

C. B. Gill

Lafayette College

A Wiley-Interscience Publication

JOHN WILEY & SONS, New York • Chichester • Brisbane • Toronto

Copyright © 1980 by John Wiley & Sons, Inc.

All rights reserved. Published simultaneously in Canada.

Reproduction or translation of any part of this work
beyond that permitted by Sections 107 or 108 of the
1976 United States Copyright Act without the permission
of the copyright owner is unlawful. Requests for
permission or further information should be addressed to
the Permissions Department, John Wiley & Sons, Inc.

Library of Congress Cataloging in Publication Data:

Gill, Charles Burroughs, 1921–
  Nonferrous extractive metallurgy.

  "A Wiley-Interscience publication."
  Includes index.
  1. Nonferrous metals—Metallurgy. I. Title.
TN758.G49   669   79-28696
ISBN 0-471-05980-3

Printed in the United States of America

10 9 8 7 6 5 4 3 2 1

Preface

Because the extractive metallurgy of the major nonferrous metals has developed rapidly in recent years, much of the literature covering this field has become outdated. This book attempts to make up for this shortcoming by presenting the practices and processes in current use for extracting and refining the major nonferrous metals.

The extractive processes are considered in three ways. First, they are discussed in a very broad, general sense to see what the whole category of treatment accomplishes. Then the different types of processes within the general category are briefly examined. Finally, the exact processes are considered in detail for the specific metal compound or metal being treated by that process. The last approach is not merely a compilation of generalities but provides detailed information on each process that is important for the best understanding of the operation and appreciation of its scope.

Similarities in the treatment steps of the metals covered are brought out to the greatest possible degree. If the general processing operations, of which there are not too many, are clearly described and understood, then their combinations can be put together fairly simply to give the specific methods of treatment for a large number of metals.

For ease in combining and comparing the metals of a similar grouping and to emphasize their common general type of processing, the text is divided into four main sections on extractive treatments. These are pyrometallurgical processing and hydrometallurgical processing for the major metals, which have been further divided into two subgroups of nonreactive and reactive metals. The nonreactive metals considered are copper, nickel, lead, cobalt, gold, and silver, all metals that are processed essentially rather simply in contact with air. The reactive metals covered, namely zinc, uranium, aluminum, titanium, and magnesium, are all metals that require more complicated processing out of contact with air, under vacuum or in an inert gas atmosphere.

Worldwide treatments of the metals are covered to give their overall processing as completely as possible, for metallurgical technology has advanced by varying degrees and in different selective channels in various countries. There are some types of processing that have been phased out and replaced by newer methods in some countries while remaining the major operating methods in other countries. Conversely, some technological innovations developed and well established in one country may not yet have found wide acceptance outside of the country where they were developed.

An additional section considers such environmental factors as the pollution of air and water, and the treatment of product gases, which are increasingly becoming major concerns in metal processing. This section covers the various types of dust collectors and gas-cleaning equipment, heat recovery methods, and water cleaners that are used and also considers the specific treatments dealing with operational pollution for those metals where the problems are the most serious.

Major references, included with each chapter, are also intended to supply further extensive reference source material if required. This makes it unnecessary to list references of only marginal value, which detract from the more important sources that have been used.

C. B. Gill

Easton, Pennsylvania
March 1980

Contents

3. Reactive Metals Pyrometallurgical Treatments   205

4. Reactive Metals Hydrometallurgical Treatments   257

Nonferrous Extractive Metallurgy

INTRODUCTION

Nonferrous Extractive Metallurgy
Current Practices, Treatments, and Processes

The processes of metal extraction have been developed to recover and produce refined metals of high purity from ores and concentrates.

Most metals are found in nature as compounds and come in rather complex mixtures of sulfides, oxides, carbonates, and silicates. Occasionally a mineral is found as a rather pure, high-grade ore that does not require a great deal of pretreatment; but for the most part ores are low grade and must be concentrated by mineral beneficiation to separate and free them from waste rock prior to the extraction steps that produce the metal itself. This mineral beneficiation generally consists of crushing and grinding the ore to a sufficiently small size to liberate the mineral crystals from the matrix of waste rock and then collecting these mineral values by various combinations of specific gravity, magnetic permeability, and froth flotation methods. Part of the rock is discarded as valueless material, while the concentrate of the mineral compound goes on to extraction to produce the metal by either pyrometallurgy or hydrometallurgy treatment.

Pyrometallurgy involves operations where furnace treatments at high temperatures are used to separate the metal values from the considerable amount of waste rock still present in the beneficiated concentrate. This waste product is most often removed as slag, which is usually discarded. Hydrometallurgical extraction is the leaching in an aqueous solvent of the metal value out of an ore or concentrate, leaving behind the waste material as an insoluble residue, also to be discarded.

Usually the metal-bearing compound is not in a form from which the metal may be easily or economically removed by a simple process, and it is first necessary to convert the metal compound by a preliminary chemical treatment into a form that can be more easily processed. To this end sulfides are converted to oxides, sulfates, and chlorides; oxides are converted to sulfates and chlorides; and carbonates are converted to oxides.

The selection of the unit processes to be used for the actual extraction and refining should be such that the processes will be integrated in one overall pattern. There is then a connection between the different processes, with the product from the first unit process becoming the feed material to the second unit process, and so on, throughout the complete series of individual treatments.

Frequently a metal may be produced from an ore or a concentrate by more than one set of processing steps, with the choice between the different sets of processes depending on several conditions. Certain types of metal compounds in ores lend themselves to much easier extraction by one method than by another; for example, oxides and sulfates are readily dissolved and easily soluble in leach solutions, while sulfides are only soluble with difficulty. Another qualifying condition is the grade of purity required, which can be better from one set of extraction sets than from another. Zinc produced by pyrometallurgical retorting or by hydrometallurgical leaching, both from zinc sulfide flotation concentrates, illustrates this difference. With these, the zinc metal from retorting still contains some impurities of lead, arsenic, and iron but is quite adequate in purity for galvanizing or making brass. On the other hand, hydrometallurgical zinc precipitated electrolytically from a leach solution is of very high purity and is the preferred zinc to be used for the manufacture of certain zinc alloys.

The recovery of a particular impurity that has a considerable by-product value is also a factor in metal processing. An an example, copper refining methods can be considered, where fire refining removes many of the impurities but does not allow for recovery of the precious metals, while electrolytic refining does provide for collection of the precious metals as a by-product. Consequently a blister copper containing little or no precious

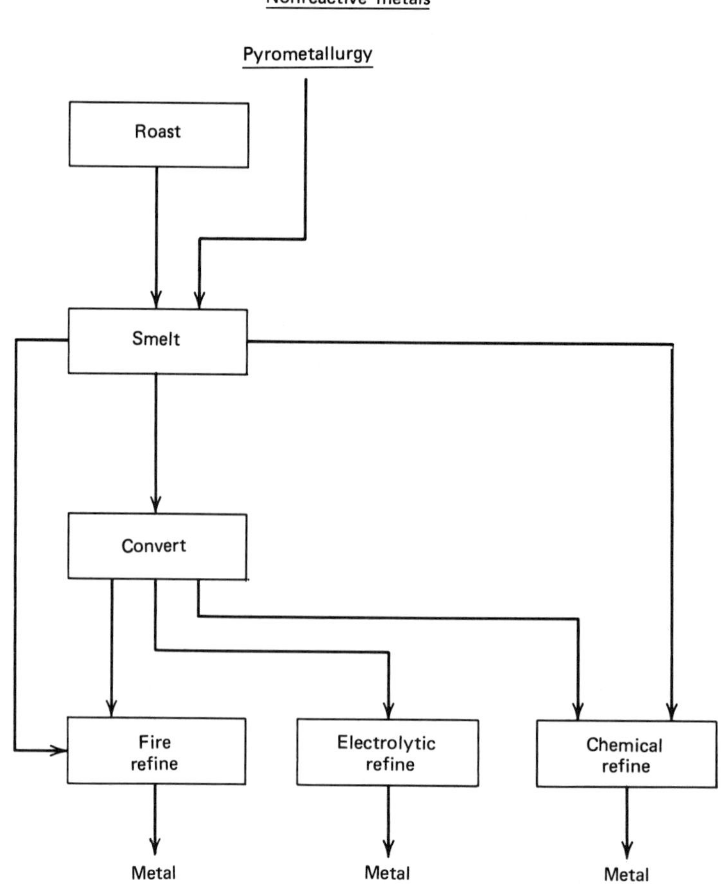

Nonreactive metals
Hydrometallurgy
Roast
Concentrate
Leach
Electrolytic precipitate
Chemical precipitate
Chemical refine
Electrolytic refine
Metal
Metal
Metal

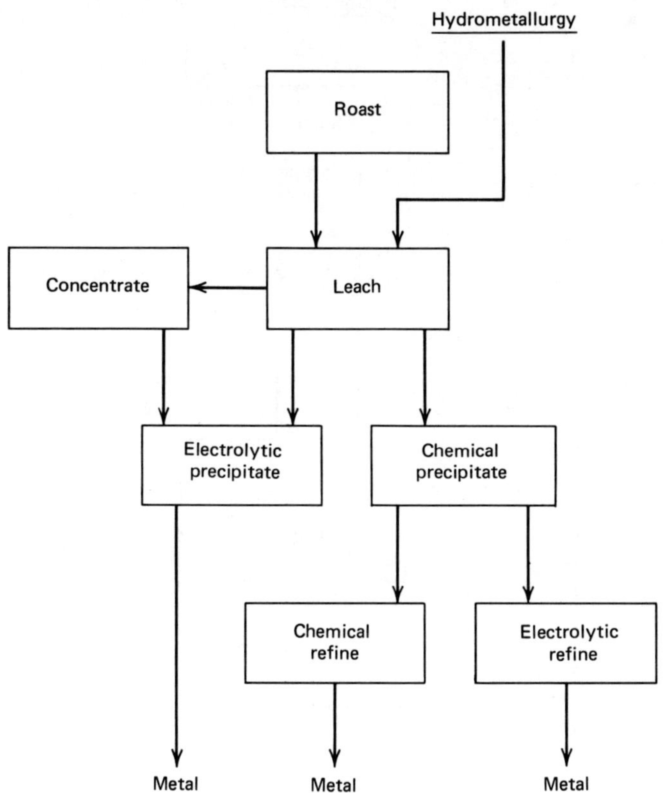

Reactive metals
Pyrometallurgy
Roast
Smelt
Fire refine
Electrolytic refine
Metal
Metal
Metal

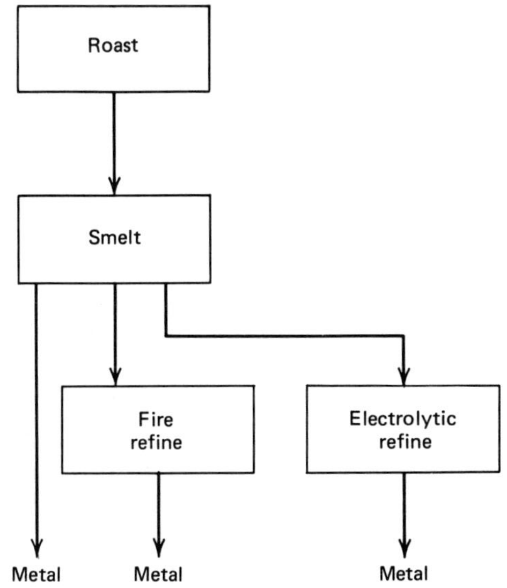

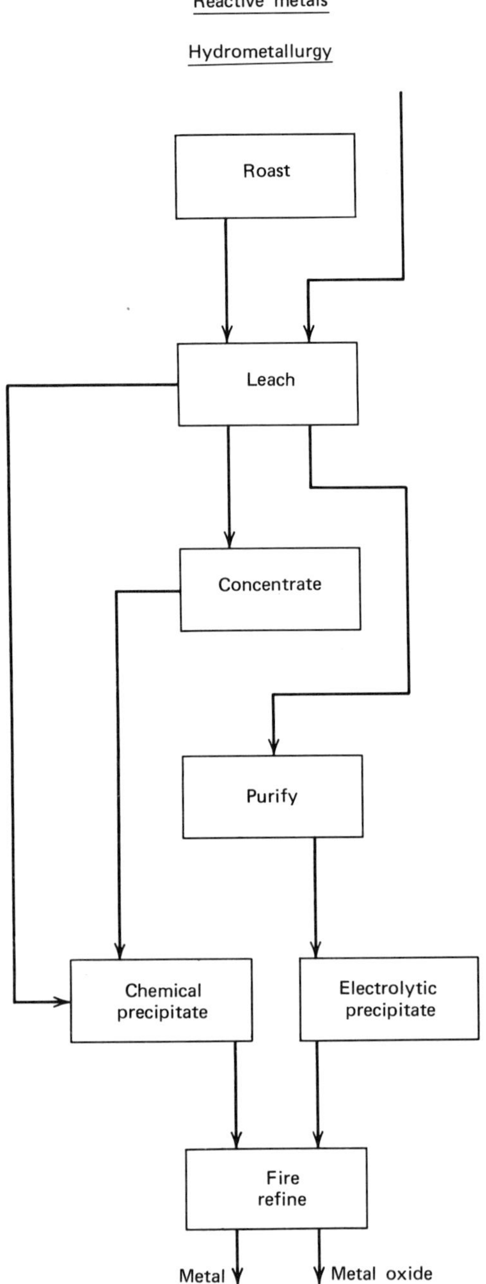

metals can be adequately refined by less expensive fire refining, while one with sufficient precious metals to make more costly processing profitable should be electrolytically refined.

The extractive treatments for the major nonferrous metals have many general features in common, both in their pyrometallurgical and hydrometallurgical processings. These can be subdivided to differentiate between the specific treatments for the nonreactive and reactive metals. The nonreactive metals are copper, nickel, lead, cobalt, gold, and silver; and the reactive metals are aluminum, titanium, magnesium, zinc, and uranium. The nonreactive metals can be processed simply in an air atmosphere without any problem of oxidation occurring during reactions when metal oxides are reduced by carbon or metal values are precipitated from aqueous leach solutions. On the other hand, the reactive metals frequently require either special equipment or processing to keep them excluded from an air atmosphere with its associated oxidation; and because there are complications during oxide reduction with carbon or it is impossible to electrowin from an aqueous solution, they often need sealed furnaces that have a vacuum or inert gas atmosphere.

To cover systematically the processing for each grouping of the nonreactive and reactive metals, they can be looked at in three different ways: first, in a broad sense as to what the particular unit process is accomplishing in the overall treatment of that group of metals; secondly, by the several different methods in general terms that are applicable for each particular unit process; and finally, by the specific unit process in detail that is carried out on a particular metal. In this format both the broad similarities and the different ramifications of the same unit process when applied to certain specific metals are taken into consideration, with emphasis on the unit process approach to the extraction and refining of the major nonferrous metals.

To accomplish this comparison of the various unit processes, they have been grouped into two main categories, namely, nonreactive and reactive metals, with a further subdivision of pyrometallurgy and hydrometallurgy treatments within each main category.

These groupings will be covered in the following order:

Nonreactive Metals—Copper, Nickel, Lead, Cobalt, Gold, Silver.

1. Pyrometallurgy treatments—roast, smelt, convert, refine.
2. Hydrometallurgy treatments—roast, leach, precipitate, refine.

Reactive Metals—Aluminum, Titanium, Magnesium, Zinc, Uranium.

1. Pyrometallurgy treatments—roast, smelt, refine.
2. Hydrometallurgy treatments—roast, leach, concentrate, purify, precipitate, refine.

1. Nonreactive Metals, Pyrometallurgical Treatments

Pyrometallurgical processes are both the oldest and most frequently used methods of metal extraction and purification. The commonest metals treated in this way include copper, nickel, lead, and cobalt, with the processes used being a combination of roasting, smelting, converting, fire refining, electrolytic refining, and chemical refining.

Roasting, the first process, is used to change the metallic compounds into forms more easily treated by the operations to follow and also to remove some easily volatile impurities in the gas stream. Smelting and converting melt the metallic compounds and form new compounds in the liquid state, which can be separated into layers of heavy metallic values and lighter slag from the waste rock. The metallic values, now with the greater part of the initial impurities removed, are further purified by fire, electrolytic, or chemical refining processes. These process treatments are grouped into their major categories in Table 1.1.

ROASTING

In most cases the metal-bearing concentrate is not in a chemical form from which the metal may be easily or economically removed by a simple direct reduction. It is necessary first to change it into some other compound that is more easily treated. Roasting is a preliminary chemical treatment that is commonly used, more specifically the oxidation roasting of sulfides to oxides, as many of the nonferrous metals are found as sulfide ores.

Roasting of sulfides is a process (gas–solid reaction) where air in large amounts, sometimes enriched with oxygen, is brought into contact with the sulfide mineral concentrates. This is done at an elevated temperature where oxygen will combine with sulfide sulfur to form gaseous SO_2 and with the metals to form metallic oxides. The solid product from roasting is called calcine.

This oxidation must be done without melting the charge, which would destroy the required maximum particle surface–oxidizing gas contact area. Stirring of the charge in

Table 1.1.   Nonreactive Metals, Pyrometallurgical Treatments

Roasting Methods

1.   Copper sulfide concentrate—hearth, fluid-bed, sintering
2.   Nickel sulfide concentrate—hearth, fluid-bed, sintering
3.   Lead sulfide concentrate—sintering

Smelting Methods

1a.   Copper sulfide roasted calcines—reverberatory, electric, blast furnace
1b.   Copper sulfides unroasted—reverberatory, oxygen reverberatory, electric, flash, continuous smelting
2a.   Nickel sulfide roasted calcines—reverberatory, oxygen reverberatory, electric, flash smelting
2b.   Nickel sulfides unroasted—electric, flash smelting, blast furnace
2c.   Nickel garnierite—low shaft blast furnace
2d.   Nickel oxide—blast furnace
3a.   Lead sulfide roasted calcines—blast furnace
3b.   Lead sulfide unroasted—electric
4a.   Cobalt oxide ores and concentrates—electric
4b.   Cobalt arsenical concentrate—blast furnace

Converting Methods

1a.   Copper furnace matte—air
1b.   Copper furnace matte and flotation concentrate—oxygen-enriched air
1c.   Copper flotation concentrate—continuous combination smelting–converting reactor
2.   Nickel furnace matte—air, oxygen-enriched air, oxygen rotating
3.   Lead metal-containing sulfur—air

Fire Refining Methods

1.   Copper converter blister—anode furnace
2.   Nickel converter matte—roasting and retorting, slow cooling—roasting and reduction
3.   Lead blast furnace metal—dross—soften—desilverize—dezinc—debismuthize

Electrolytic Refining Methods

1.   Anode copper—multiple system
2a.   Nickel metal anodes—multiple system—diaphragm cell—low voltage
2b.   Nickel matte anodes—multiple system—diaphragm cell—high voltage
3.   Softened lead anodes—multiple system

Chemical Refining Methods

1.   Nickel metal—carbonyl vapor extraction
2a.   Cobalt speiss—sulfatize—chemical precipitation
2b.   Cobalt white alloy—leach—chemical precipitation

some manner also ensures exposure of all particle surfaces to the oxidizing gas. The only exception to this general procedure is blast roasting (sintering), where the particle surfaces are partially melted and there is no stirring of the charge.

The degree of sulfur elimination is controlled by regulating the air supply to the roaster and by the degree of affinity the mineral elements have for sulfur and oxygen. Consequently minerals such as iron sulfide, which have a higher affinity for oxygen than for sulfur, may all be oxidized, while a copper mineral in the same roaster feed, with a greater affinity for sulfur than for oxygen, will emerge in the calcine still as a sulfide.

In the roasting process there is no clear-cut stage where all iron sulfide is oxidized, while minerals with lower oxygen affinities such as copper sulfide remain still as sulfides. The reactions then will overlap, and many will proceed simultaneously at a variety of process rates.

Roasting is essentially a surface reaction, with the oxide layer first formed remaining as a porous layer through which oxygen can pass into the still unreacted inner sulfide portion of the particle and through which the SO_2 gas then formed can pass out. This passage becomes more difficult as the porous oxide layer thickens, and there will be some reversing reactions in the particle interior as the concentrations of SO_2 build up:

$$MS + \tfrac{3}{2}O_2 = MO + SO_2$$

$$MO + SO_2 = MS + \tfrac{3}{2}O_2$$

This makes it difficult to remove the last interior amounts of sulfur. Particle size also is important, and large particles will take much longer to react to their centers.

The oxidizing roast of a sulfide concentrate is an exothermic reaction—it gives off heat. This heat of reaction helps to keep the roaster at the required roasting temperature so that the process can continue with little extra heat supplied by burning fuel. On occasion, autogenous roasting can be achieved when a high-sulfide roaster feed material has sufficient exothermic heat generated from its oxidation reaction to be self-propagating and to require no extraneous fuel. Autogenous roasting, for economic reasons and fuel economy, is practiced to whatever degree the roaster feed material will allow.

TYPES OF ROASTING

Oxidation of sulfide concentrates to obtain metal oxides and the agglomeration of fines into lumps which can be treated in shaft furnaces are the processes carried out during roasting.

The choice of the roasting process depends on the kind of smelting process to which the calcines are to be subjected after roasting. Roasting in multiple-hearth and fluid-bed furnaces requires fine feed material and furnishes fine calcines which are then treated in reverberatory, flash smelting, or electric furnaces. The multiple-hearth roaster is the oldest type, having been first developed in the late 1890s, and has found very wide acceptance. The fluid bed roaster is a more recent development, of the early 1950s, and is characterized by its exceptionally high capacity, a fluid bed roaster having an equivalent capacity of eight times that of a multiple-hearth type with the same hearth area. This innovation has removed the necessity of needing large numbers of multiple-hearth roasters in large tonnage operations because of the limited capacity of the individual roaster.

Sulfide concentrates that must be both desulphurized and agglomerated are usually roasted on blast roasters (sintering machines) rather than in multiple-hearth or fluid-bed roasters. This gives a coarse, porous, oxidized sinter cake as a product, and is then a smeltable shaft furnace feed material from the one single roasting operation.

Multiple-Hearth Roaster This unit consists of a number of superimposed, horizontal, circular, refractory hearths enclosed in a steel shell, with the feed material placed on the top hearth and worked downward to be discharged as roasted calcines from the bottom hearth. A central slowly rotating shaft turns air-cooled or water-cooled rabble arms on each hearth, which serve two purposes. The rotating rabble blades turn over the roaster charge to stir fresh material to the surface for the roasting, gas–solid, oxidizing reaction and also push the charge across the hearth to drop holes to be passed on down to the hearth below. The drop holes are located to be not directly below one another, but at the outer periphery of one hearth and at the center of the hearth below. The charge then follows a lengthy, zigzag path downward through the roaster, allowing time for the oxidizing reactions to take place.

As the feed material progresses downward in the roaster, it is heated by the rising hot gases from the exothermic roasting reaction taking place on the lower hearths, until finally this feed material is also heated to the reaction temperature, begins to burn, and oxidizes at a rapid rate. This reaction will continue until the roasted calcines are discharged from the bottom hearth of the roaster and cool in air to below the roasting reaction temperature.

Gas burners are provided on the lower hearths to ensure that reaction temperature is reached if the roasting is not autogenous. The flow of air into the roaster is regulated by opening doors on the lower hearths, and the natural draft provided sucks in air to supply oxygen for oxidizing.

Roaster capacity will average about 0.25 ton of pyrite (or equivalent sulfur content) per square foot (0.09 m^3) of hearth area per day and roasters range in sizes from four to 12 hearths of diameters 10 to 24 feet (3.05 to 7.32 m). The SO_2 concentration of the roaster gas will be 4.5 to 6.5%.

Fluid-Bed Roaster The furnace consists of a cylindrical brick-lined steel shell which is closed at the bottom by a grate. A wind box below the grate blows in air at a sufficient

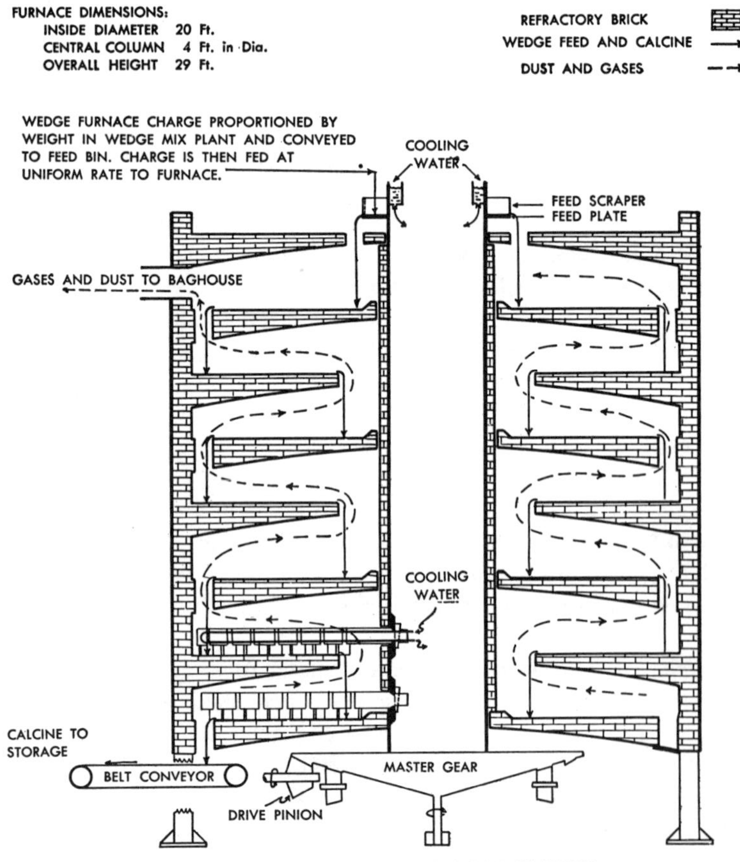

Figure 1.1. Multiple-hearth roaster. *Source:* United States Smelting, Mining and Refining Company.

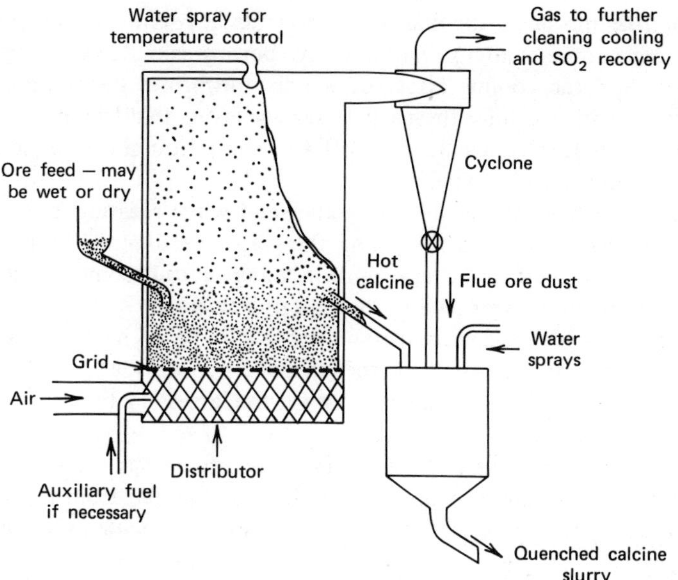

Figure 1.2. Fluid-bed roaster in operation, cutaway view. *Source:* J. D. Gilchrist, *Extraction Metallurgy*, Pergamon Press Ltd., Oxford, 1967.

volume and is distributed evenly by the grate to hold the solid feed particles in suspension and give excellent gas–solid contact on all surfaces.

A slurry, a suspension of solids in water, of material to be roasted, with the maximum particle size kept at about $\frac{1}{4}$ inch (6.3 mm), is continuously fed through a downpipe to the turbulent layer in the roaster. This turbulent layer with its solid particles in suspension has the flow characteristics of a fluid. If the feed material has mixed sizes and densities, the smallest and lightest particles will migrate to the top of the turbulent layer, while the larger and heavier particles collect at the bottom.

Part of the roasted calcines leave by a side discharge overflow pipe, and part is carried off in the effluent gases to be recovered as flue dust in a gas-cleaning system. Cooling coils remove the excess reaction heat from the turbulent layer, and in almost all cases this excess heat is used for steam production, with the cooling system of the roaster being connected to a waste heat boiler.

The oxidizing reaction is autogenous, and the high turbulence of the suspension and the resulting excellent gas–solid contact and heat exchange account for the very high reaction rate of the process and accompanying high capacity. This capacity will be on the order of 2.0 tons of pyrite feed material treated per day per square foot ($0.09\ \mathrm{m}^2$) of grate area. The SO_2 content of the roaster gas is 9 to 12%.

Blast Roaster (*Sintering*) A Sintering machine consists of a number of linked grate sections forming an endless belt which moves on rollers. A suction box is located under the linked grates, and the speed of the belt movement is adjustable.

A charge of fine feed material, generally $\frac{1}{2}$ inch (12.5 mm) in diameter or less, or preformed $\frac{1}{2}$ inch (12.5 mm) pellets, are moistened, mixed, and fed in a layer several inches deep onto the moving pallets ahead of the suction box. As the charged pallet comes over the suction wind box, the sulfides in the charge are ignited from above by a fuel-fired burner. The process requires no additional fuel, and the reaction temperature is maintained by the exothermic heat given off as the sulfide is oxidized by the air drawn through the charge.

The roasting zone travels downward through the pallet charge as the grates move forward over the sectionalized windbox, and the combustion zone gradually passes through the entire layer thickness from top to bottom before the roasted material is discharged from the sintering machine.

The high temperature of roasting heats the charge components high enough to make them sticky, adhering to one another and forming a strong porous cake. However the thinness of the charge layer and the cooling effect of the air drawn into the windbox prevent any extensive melting, and it is only the particle surface layers that become soft and sticky. Any molten material formed would shut off air penetration and terminate roasting, so excessive temperatures must be avoided.

At the end of the horizontal travel of the moving grates, and when roasting is completed to the bottom of the charge layer on the pallet, the grates are dumped inside a dust hood. The sinter cake is sized, with the coarse portion going on to become furnace or retort feed and the fines returned as revert feed to the sintering machine.

The capacity of sintering machines varies considerably from 1.7 ton of feed material per square foot (0.09 m^2) of grate area per day for the smaller 3$\frac{1}{2}$ foot wide by 45 foot long machines (1.06 × 13.5 m) to 0.3 ton per square foot for the largest 12 foot by 168 foot machines (3.6 × 50.4 m).

The sintering machine described above is a "downdraft" type, which has the suction windbox below the pallet grates and air is sucked down through the bed from top to bottom. There is a second type, the "updraft" type, which also has a wide industrial acceptance. With the "updraft" type the windbox is above the grate, drawing air up through the charge on the pallet. Ignition is made initially on a thin layer of feed placed on the grate. Then after ignition has started a thicker layer of feed material is added on top of the burning portion and burns upward from bottom to top as the pallet moves under the windbox toward the discharge end.

Grate speed for both types varies over quite a wide range, from 10 to 48 inches per minute (25 to 120 cm) and will depend on the degree of roasting and/or agglomeration desired, the bed depth, and the length of the machine.

Some metal industries have almost universally adopted one or the other of these two types of sintering machines, while other metal industries may use a mixture of the types.

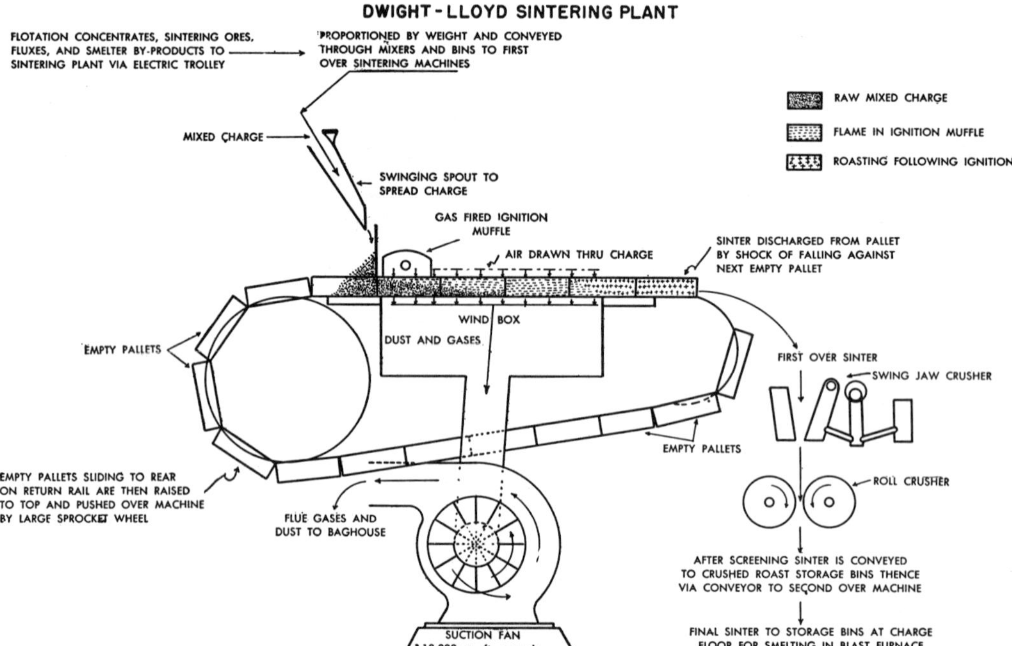

Figure 1.3. Dwight–Lloyd downdraft-type sintering plant. *Source:* United States Smelting, Mining and Refining Company.

The lead industry leans heavily toward the updraft machines, while the zinc industry uses the downdraft type to a great degree.

ROASTING PROCESSES

1. Copper Sulfide Concentrates are roasted in multiple-hearth roasters, fluid-bed roasters, and blast roasters (sintering machines), depending on the type of feed available to the roaster and the kind of smelting furnace process to follow. All three types of roasting will accomplish the same chemical changes, principally the oxidation of some combined sulfur and its removal as SO_2 gas.

The multiple-hearth and fluid-bed roasters are used for roasting fine beneficiated concentrates that are going to be smelted in a reverberatory or electric furnace, where the roasted calcines are still desired in a fine, separated state. On the other hand, blast roasting is only applied where a combination of both desulfurization and agglomeration is required, with the coarse sinter cake being fed to a blast furnace. Beneficiated concentrate or fine ore is the normal feed to the blast roasting process.

The stability of the copper sulfide in the roaster feed is somewhat greater than the stability of the contained iron sulfide, and the oxygen affinity for iron is rather larger than that for copper. Thus if a supply of oxygen is provided that is insufficient to oxidize all the sulfide constituents in the feed material, then the available oxygen will tend to combine preferentially with the iron before combining with the copper that is present. Consequently a considerable proportion (about half) of the iron is oxidized and removed as a component of the oxide slag in the following smelting operation, while substantially all of the copper remains in the roasted calcines still as a sulfide, Cu_2S:

$$3FeS_2 + 8O_2 = Fe_3O_4 + 6SO_2$$

Multiple-Hearth Roasters are arranged so that the calcines from approximately six roasters will feed one reverberatory furnace. Feed to the roasters is by conveyor belt from concentrate bins, and a common type of roaster would have two drying hearths and 10 interior roasting hearths. The steel shell of the roaster is $\frac{1}{2}$ inch (12.5 mm) steel plate, and

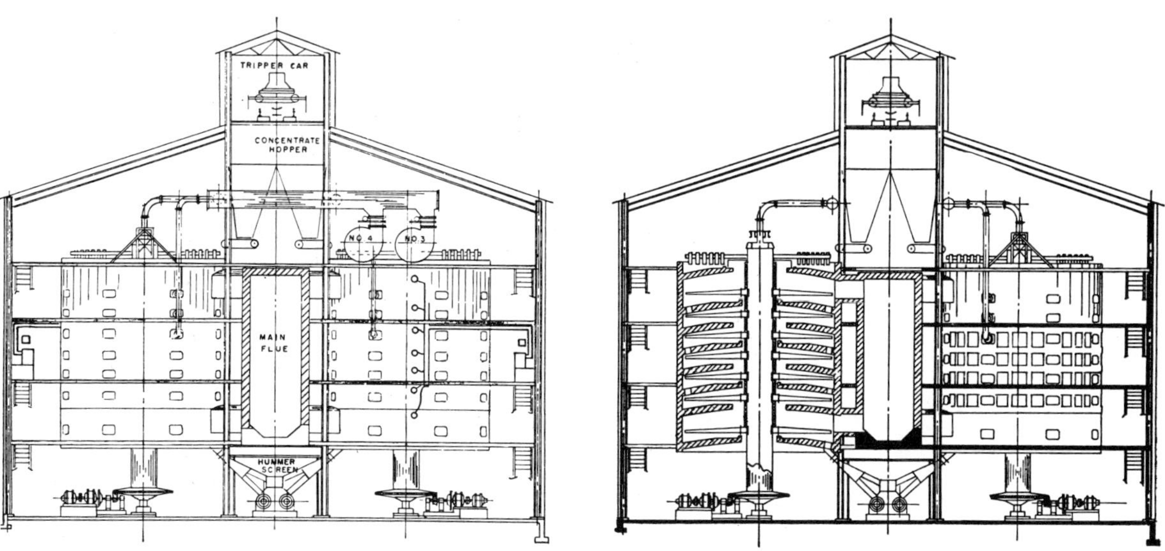

Figure 1.4. Multiple-hearth roaster plant. *Source:* B. Schneider, *Can. Inst. Min. Metal. Bull.*, Vol. 54, No. 592, 1961, pp. 587, 588.

the hearths and walls are of high-alumina fireclay brick. A 12-hearth roaster would be 22 feet in diameter and 40 feet high (6.7 × 12.2 m).

Speed of rotation of the rabble arms will be on the order of 1.5 rpm, and sufficient air will escape from air cooling in the rabble arms to provide a part of the oxygen needed for the combustion of the sulfides. The remaining air is sucked in through open doors on lower hearths.

An average roaster charge will be 250 tons of wet concentrate per day (0.5 to 1.0 metric ton of charge per m² of hearth area) containing about 22% sulfur. Fluxes for the following smelting processes are often added to the roaster feed to mix and preheat them with the roasted calcine, which is discharged from the bottom hearth at 1400°F (760°C).

Gas from the roaster goes to a dust precipitation plant, from which the recovered dust is sent to the reverberatory furnace for smelting. Because of the relatively large amount of air admitted during the roasting process, the SO_2 concentration in the final exhaust gas is low, 4.5 to 6.5%.

Fluid-Bed Roasters have replaced the multiple-hearth type to some degree because of their greater capacities, whereby one fluid bed roaster can replace eight multiple-hearth roasters.

One fairly typical fluid-bed roaster size has a cylindrical steel shell of $\frac{1}{2}$ inch (12.5 mm) steel plate, is 17 feet high with a conical top and wind box and 12 feet in diameter (5.18 × 3.66 m), and lined with high heat-duty firebrick. The grate has 164 tuyeres on 10 inch centers (25 cm), and air at 4 psi (27.6 kPa) is injected through the tuyeres.

At Copper Hill, Tennessee, the roaster is fed a slurry of 78% solids in water of concentrate filter cake mixed with fine silica, which is to be the flux during the following smelting operation. The slurry feed rate is varied as required to maintain the desired 1200°F (650°C) roasting temperature. Oxygen efficiency is close to 100%.

A considerable portion of the roaster charge (sometimes as much as 75 to 85%) leaves the roaster through the top opening along with the reaction gases. This gas travels in order

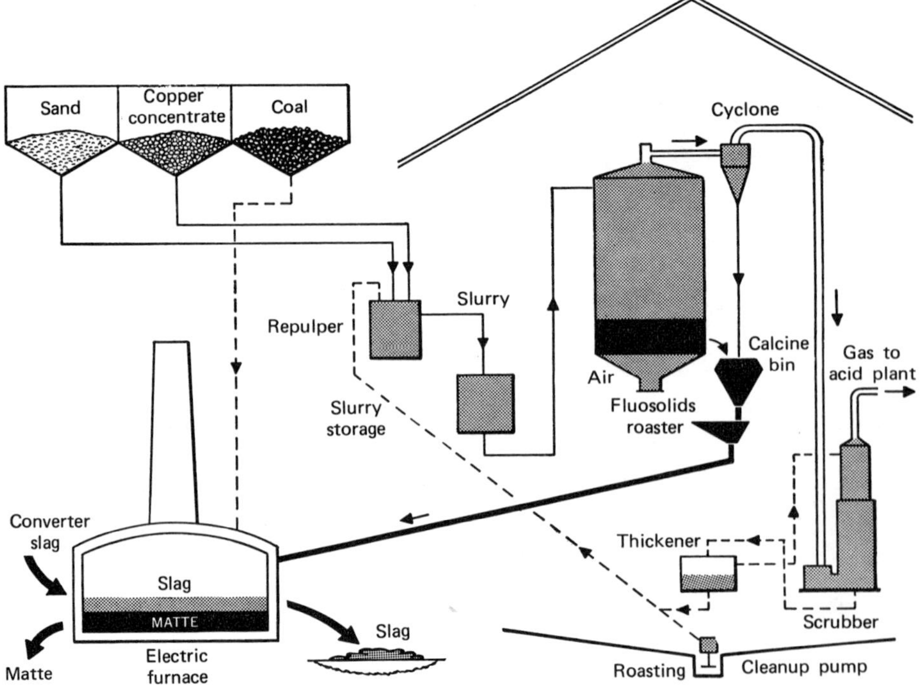

Figure 1.5. Copper Hill fluid-bed roasting and electric furnace smelting. *Source: Can. Inst. Min. Metall. Bull.*, Vol. 55, No. 608, 1962, p. 4., Courtesy of Dorr-Oliver Inc.

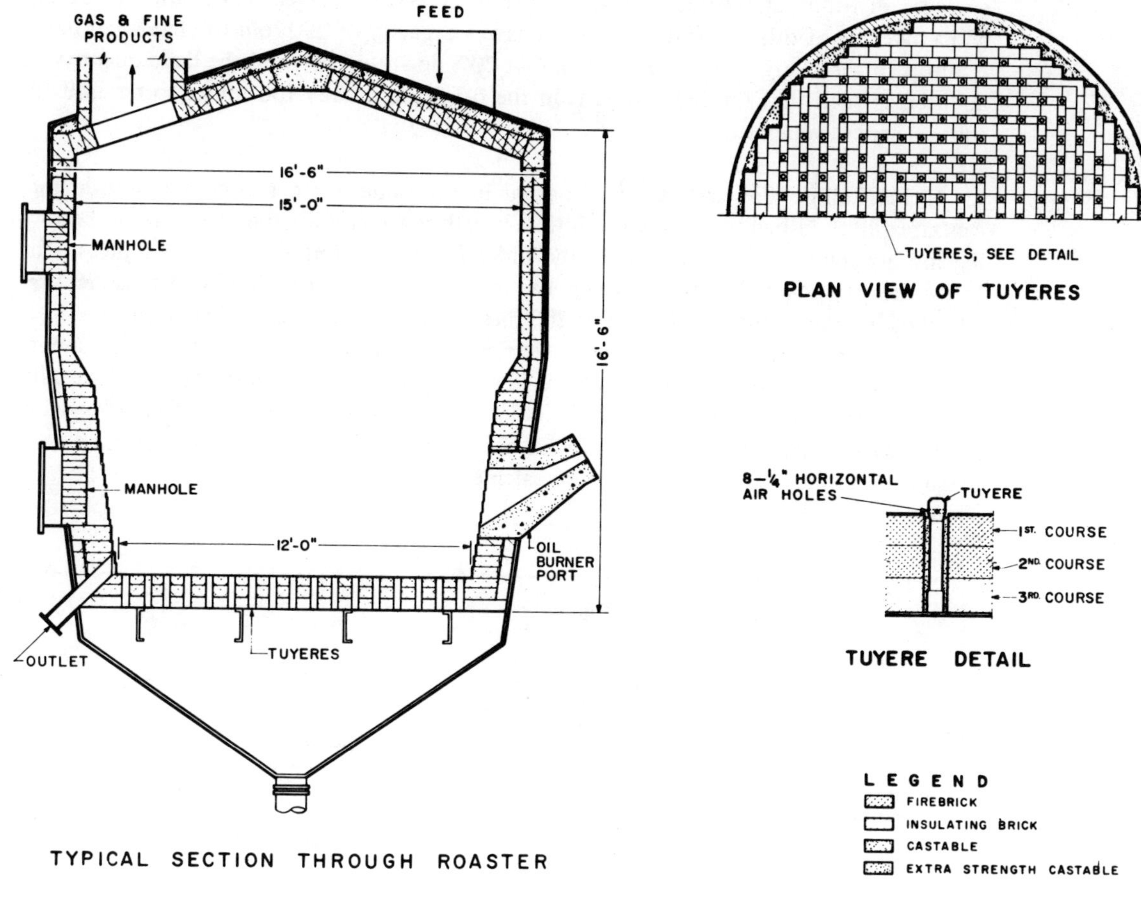

Figure 1.6. Sections of a fluid-bed roaster. *Source:* J. B. McConnell, I. P. Klassen, and S. A. Berkovich, Roasting and smelting practice at the Thompson plant of the International Nickel Company, Presented at the CIMM Annual Meeting, 1963.

through cyclone dust collectors, waste heat boilers, and sometimes electrostatic dust precipitators to recover the contained solids. Then the gas, of over 10% SO_2, passes on to the acid plant for sulfuric acid production. The recovered dust is added to the coarser calcines which overflows the roaster bed through a fluo-seal valve into a calcine bin.

By varying the air ratio, the amount of sulfur left in the calcine can be controlled to give almost any desired grade of matte.

*Blast Roasters (**Sintering Machines**)* are used in preparing agglomerated, roasted feed for blast furnace smelting; and as the blast furnaces have been decreasing greatly in number, so too have associated sintering operations. Only a few blast furnaces are still in current use, in Russia, Africa, and Japan.

Downdraft-type machines are conventionally used, and an average-size machine would have pallets 42 inches wide by 36 inches long (1.05 × 0.9 m) and travel a sintering distance of 30 to 50 feet (9.15 to 15.24 m).

A 6 inch (15 cm) layer of charge, made up of a mixture of fine ore, concentrate, flue dust, return sinter fines, and limestone flux, is usually carried on the grates, with a spreader distributing the feed evenly. Pallet speed is about 2 feet per minute (0.61 m), and a machine of this size and type would have a capacity of 250 tons of charge per day.

Sulfur in the charge is reduced from about 20% down to 14%, and the large quantity of air used is such that the SO_2 content in the off-gas is usually too dilute to be used for sulfuric acid production.

2. Nickel Sulfide Concentrates are roasted in the same manner as copper sulfide concentrates, in fluid-bed roasters and multiple-hearth roasters. Fluid-bed roasters are becoming more popular and are replacing the older hearth roasters in several instances. The International Nickel Company, which has recently converted to fluid-bed roasters, uses a feed of filter cake that is 64% minus 200 mesh in size and contains 10% moisture.

The reactions that occur with nickel sulfide, iron sulfide, and oxygen are also similar to that discussed with copper sulfides, in that oxygen again preferentially combines with iron, converting about half of the iron sulfide present into an oxide form to be slagged in the following smelting operation. As with the copper sulfide, almost all of the nickel still remains as sulfide, Ni_3S_2, in the roasted calcines.

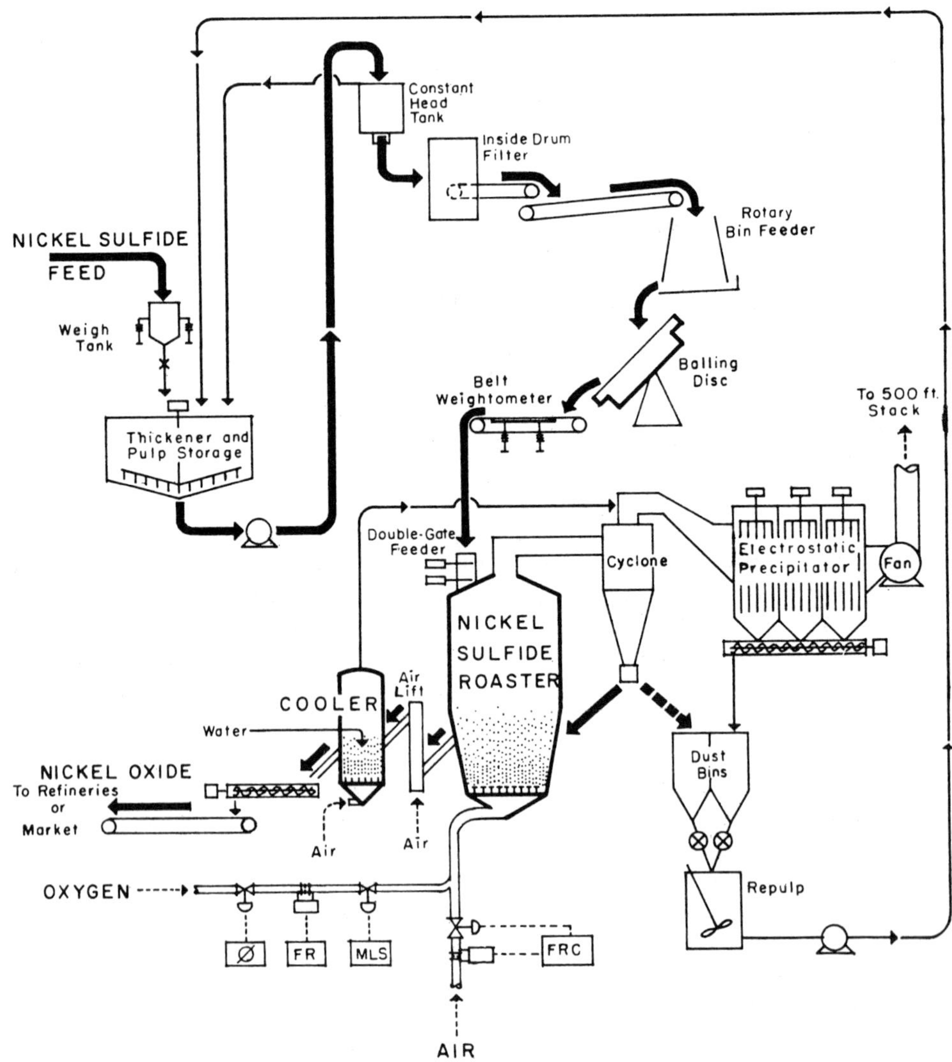

Figure 1.7. International Nickel Company fluid-bed roaster roasting nickel sulfide concentrate. *Source: J. Metall.*, Vol. 18, No. 4, 1966, p. 450.

3. Lead Sulfide Concentrates are roasted for the purpose of high desulfurization, and also agglomeration, to provide a suitable charge for the following blast furnace reduction smelting process.

Blast roasting, or sintering, is the roasting method used, and the type of sintering machines used are generally similar to those used for copper sulfide. These are mostly updraft machines, usually of the order 42 to 72 inches wide (1.05 to 1.8 m) and from 22 to 40 feet long (6.71 to 12.2 m), with a capacity in the larger machines of about

Table 1.2. Sinter Plant (Buick Lead Smelter), AMAX-Homestake: Operating Data

| | *Sinter Feed* | | |
Material	Weight (%)	Pb (%)	S (%)
Pb concentration	26	74.1	16.5
Iron flux	4	–	–
Lime flux	2	–	–
Silica flux	1	–	–
Gran. slag	15	3.5	1.0
Dust	2	50.0	8.0
Return sinter	50	50.0	2.8
	100	46.0	6.0

Analysis	Feed (%)	Final Sinter (%)
SiO_2	10.3	11.2
Fe	12.2	13.3
CaO	9.2	10.0
MgO	0.8	0.9
Zn	5.1	5.6
S	6.0	1.1
Pb	46.0	50.0
Cu	0.3	0.4
Approximate tons/h	100	45

Lurgi Updraft Sinter Machine		
Grate width	2.5	m
Grate length	24.0	m
Effective updraft area	60.0	m²
Grate speed ±	1	m/min
Bed depth (ign)	31	mm
Bed depth (total)	280	mm
Sinter production	1060	tons/day
Sulfur burning rate	1.37	tons/m²/D
Gas volume to acid plant	708	m³/min
Gas strength–% SO²	5.0	–5.5
Operating time	75%	

Source: C. H. Cotterill and J. M. Cigan, AIME World Symposium of Lead and Zinc, 1970, Vol. 2, pp. 754–756.

200 tons per day throughput. The bed is made up of lead sulfide concentrate, flue dust, returned sintered fines, and limestone flux for the following blast furnace operation. This latter is added to the sinter to be intimately mixed in and preheated before going to the furnace.

The advantage claimed for updraft sintering is that it prevents the grates from being clogged with molten lead formed during the roasting process by the reaction of lead sulfide with lead oxide and lead sulfate:

$$2PbO + PbS = 3Pb + SO_2$$

$$PbS + \tfrac{3}{2}O_2 = PbO + SO_2$$

$$PbS + PbSO_4 = 2Pb + 2SO_2$$

Also claimed are higher tonnage capacities due to improved charge permeability, lower wear on grate bars and pallets, and lower required fan pressures.

A problem in sintering lead sulfides is to reduce the sulfur content sufficiently, from approximately 12% sulfur down to 1% sulfur, to have the required oxidized blast furnace feed material and to accomplish this desulfurization without liberating sufficient exothermic heat to melt the sintering charge and have it drip down into the windbox. One remedy for this problem is double sintering. With this, the first sinter feed with 12% sulfur is passed over the sinter machine fairly quickly with light ignition, and no attempt is made to make a good sinter cake. Sulfur elimination from 12% down to 6% is accomplished. This first sinter is crushed to expose unoxidized sulfides; water is added, which on vaporization will leave a porous, cellular, sinter cake; and the mixture is resintered. Sulfur reduction on this second pass will be from 6% sulfur down to the approximate 1% sulfur desired in the final sintered calcine.

Generally enough sulfur is left (6%) after the first pass over the sintering machine to provide sufficient fuel to be ignited for the second pass. However in the event this first-pass sulfur is too low, a small amount of raw sulfide concentrates, coal or coke, can be mixed in to provide sufficient reaction heat during the second sintering pass.

As with other blast roasting operations, the quantity of air used so dilutes the SO_2 gas

Table 1.3. Sinter Plant Feed and Products, Glover Lead Smelter and Refinery, ASARCO

Composition of the Sinter Plant Feed

    27.5% concentrates
     1.4% fluxes
     7.0% purchased and generated secondaries
   16.3% granulated slag
   47.8% return sinter

Analysis

Ag%	Pb%	Cu%	SiO_2%	Fe%	CaO%	Zn%	S%
0.05	43.0	1.0	10.6	11.3	7.1	5.4	5.1

Composition of the Final Sinter

Pb%	Cu%	S%
46.3	1.08	1.7

Tons of feed to sinter plant per operating day—1125

Source: C. H. Cotterill and J. M. Cigan, Eds., AIME World Symposium of Lead and Zinc, 1970, Vol. 2, pp. 781 and 782.

Table 1.4. Dwight–Lloyd Sintering Machine Operations,
Hoboken Lead Smelter, Metallurgie Hoboken–Overpelt:
D.L. Machine Operating Details

Machine speed	0.5–2.5 m/min
Total windbox area	114 m^2
Pallet size	2 m by 0.68 m
Number of pallets per machine	116
Number of grates per pallet	50
Grate alloy	13% Cr steel
Rich gas windbox area	37 m^2 (boxes 2-3-4-5)
Recirculated gas windbox area	77 m^2 (other boxes)
Bed depth	10 to 16 cm
Sinter production rate	750 ton/day
Exhaust gas volume	
Rich gases	5–6 Nm^3/sec/machine
Poor gases recirculated	8–9 Nm^3/sec/machine
Ignition box area	1.8 m^2
Natural gas consumption	2.4 Nm^3/ton of sinter
Driving motor	29.5 hp

Source: C. H. Cotterill and J. M. Cigan, Eds., AIME World Symposium
of Lead and Zinc, Vol. 2, 1970, p. 836.

being produced that it is too low in concentration to be satisfactory for acid plant treatment and sulfuric acid production, unless the hearth area is sectioned to permit separate removal of the first roasting gases, which are high in SO_2 content and can be used. Overall the SO_2 content in the gases will run from 1.5 to 5.0%.

Another method of reducing the sulfur content to acceptable limits is by recycling and blending, as is carried out at the St. Joe Minerals Corporation Herculaneum smelter in Missouri. In this plant the incoming concentrate feed analyzing 9 to 11% sulfur is blended with sinter analyzing 1.2 to 1.7% sulfur, and the combined blend is then low enough in overall sulfur content that it can be sintered in one pass to an acceptable finished sinter averaging 1.4% sulfur.

SMELTING

Smelting is a concentration process where some of the impurities in the charge are gathered into a light waste product called a slag, which can be separated by gravitational flow from the heavier portion containing practically all of the desirable metal components.

The charge to a smelting furnace is made up principally of solids, though some molten material may also be charged in certain operations. The heat supplied to melt this solid charge can be fossil fuel, electricity, or, if a sulfide is charged, the exothermic heat from the oxidation of the charge itself. The furnace charge must be melted to the liquid state to permit gravitational separation of the slag and metal layers and also to facilitate the mobility and contact of the reacting compounds within the charge.

The slag components will be made up of the oxides in the charge, both those found naturally in the ore such as SiO_2 and those which have been converted to oxides during roasting such as Fe_3O_4. As these oxides have high melting points, higher than the metallic compounds also in the charge, it is necessary to add fluxes, most frequently SiO_2 or CaO, to combine with these oxides and form lower-melting slag-making compounds. It is a necessity to have the slag melted and quite liquid at the furnace reaction temperature so that the metallics can easily settle through it to collect in their lower, heavier layer, and also for ease in handling the slag and allowing it to be run out of the furnace as a liquid.

The other furnace products, in addition to the slag, are the heavy metallics layer which collects on the bottom of the furnace hearth and is also removed as a liquid and the gases of combustion, along with some volatiles. These latter, containing some dust from the fine material in the charge, pass on up through the furnace flue system into a dust collector where the solids are removed for recycling back through the furnace.

TYPES OF SMELTING

There are two main types of smelting—reduction smelting, which produces an impure molten metal and a molten slag from the reaction of a metallic oxide with a reducing agent; and matte smelting, which produces a molten mixture of metal sulfides and slag.

In reduction smelting the metallic values in the charge, as well as the slag-forming compounds, will be present as oxides. A reducing condition is arranged in the furnace whereby these metallic values, which have a greater reduction rate from oxides to metals than do the oxides in the gangue portion, will reduce to an impure metal and leave the gangue still as oxides to make up the slag.

Any type of furnace can be adapted to reduction smelting, but those most commonly used are the blast furnace and the electric furnace.

Matte smelting is rather different from reduction smelting, with matte being formed by the combination of the liquid sulfides of copper, nickel, iron, and cobalt into a homogeneous solution. The precious metals present and small amounts of other base metals are absorbed in the matte. The remaining portion of the charge, consisting of iron oxide, gangue, and siliceous flux, combines to form an oxide slag.

Matte smelting is done in a variety of types of furnaces, including the reverberatory furnace, blast furnace, electric furnace, flash smelting furnace, and the newest form, the continuous smelting process consisting of three furnaces in series, the first furnace for smelting, the second for slag cleaning, and the third furnace for converting.

Reverberatory Furnaces are used for the matte smelting of fine flotation concentrates, where a relatively quiet operation is desired that will not blow too much of the fine feed material out of the furnace with the exhaust gases.

The furnace is rectangular in shape, with a shallow hearth, arched roof, and length roughly four times the dimension of the width. These vary from 100 to 120 feet long (30.5 to 36.6 m) and 25 to 35 feet wide (7.62 to 10.67 m). Furnace refractory choice is important because of the high temperatures and corrosive nature of the molten charge. The roof is exposed to radiant heat from the hot liquid charge and the furnace burners as well as the eroding effect of hot gases carrying dust particles. It is made of silica brick if a sprung arch, or of suspended magnesite brick in a suspended arch. Suspended arches are generally preferred in the larger furnaces because of the excessive weight of the required refractories. Hearths, endwalls, and sidewalls are of silica brick, with courses of magnesite brick around the furnace at the slag line to minimize the effects of this very corrosive interface. In some of the more modern furnaces, magnesite is used for hearth and roof with chrome magnesite sides and endwalls. Vertical steel uprights down the sides of the furnace bolted to horizontal tie rods across the top of the furnace hold the whole furnace assembly rigid.

The reverberatory furnace is essentially a melting furnace. and this is accomplished by supplying heat from burners set in the endwall of the furnace. A variety of fuels can be

Table 1.5. Technical Data on Smelting Furnaces (English Units)

Smelting Furnace	Oxygen Flash Furnace	Hot-Charged Reverb. Furnace	Wet-Charged Reverb. Furnace	Water-Jacketed Blast Furnace	Top-Blown Rotary Converter
Furnace size, inside (ft)	20 × 78	24 × 110	24 × 110	4.2 × 20	9.2 diameter
Furnace fuel	Oxygen	Bunker C oil	Pulverized coal	Coke	Natural gas, oxygen, and air
Solid charge (stpd)					
Concentrate	1500	1116	691	405	240
Flux (80% SiO_2)	220	198	None	87	23
Pretreatment	Fluidized bed drying	Multiple-hearth roasting	None	Dwight–Lloyd sintering	None
% Sulfur removal	—	20	—	32	—
Specific fusion (stpd/ft^2)	1.10	0.50	0.26	—	—
Liquid converter slag (stpd)	None	574	270	138 (to settler)	None
Furnace products					
Matte (stpd)	920	1030	688	273	59
% Cu	43–45	33.0	30.2	44.2	97.3
% Ni	2.0	3.0	2.6	0.6	0.5
Slag (stpd)	620	784	240	327	156
% Cu	0.62	0.62	0.77	0.60	4.02
% Ni	0.13	0.12	0.22	0.20	0.73
% SiO_2	36.0	39.4	35.4	38.7	23.1

Source: J. C. Yannopoulos and J. C. Agarwal, Eds., *Extractive Metallurgy of Copper*, Vol. 1, The Metallurgical Society of AIME, 1976, p. 225.

Table 1.6. Technical Data on Smelting Furnaces (Metric Units)

Smelting Furnace	Oxygen Flash Furnace	Hot-Charged Reverb. Furnace	Wet-Charged Reverb. Furnace	Water-Jacketed Blast Furnace	Top-Blown Rotary Converter
Furnace size, inside (m)	6.1 X 23.8	7.3 X 33.5	7.3 X 33.5	1.3 X 6.1	2.8 diameter
Furnace fuel	Oxygen	Bunker C oil	Pulverized coal	Coke	Natural gas, oxygen, and air
Solid charge (mtpd)					
Concentrate	1360	1010	626	367	218
Flux (80% SiO_2)	200	180	None	79	21
Pretreatment	Fluidized-bed drying	Multiple-hearth roasting	None	Dwight–Lloyd sintering	None
% Sulfur removed	—	20	—	32	—
Specific fusion (mtpd/m^2)	10.8	4.9	2.6	—	—
Liquid converter slag (mtpd)	None	521	245	125 (to settler)	None
Furnace products					
Matte (mtpd)	834	934	624	248	54
% Cu	43.0–45.0	33.0	30.2	44.2	97.3
% Ni	2.0	3.0	2.6	0.6	0.5
Slag (mtpd)	562	711	218	297	141
% Cu	0.62	0.62	0.77	0.60	4.02
% Ni	0.13	0.12	0.22	0.20	0.73
% SiO_2	36.0	39.4	35.4	38.7	23.1

Source: J. C. Yannopoulos and J. C. Agarwal, Eds., *Extractive Metallurgy of Copper*, Vol. 1, The Metallurgical Society of AIME, 1976, p. 226.

used (pulverized coal, natural gas, oil) giving a long flame which extends half the length of the furnace. Part of the heat from this flame radiates directly on the charge lying on the furnace hearth below, and part radiates to the roof and sidewalls from where it is also reflected down on the charge. The temperature of the furnace will be about 2900°F (1600°C) at the firing end and 2200°F (1200°C) at the flue end. As the furnace exhaust gases carry such a large amount of sensible heat, the gases are passed over waste heat boilers or heat exchangers to recover some of this heat. Dust collection is also done before the gases are finally released.

Charging is done by running the furnace feed directly down through the roof through drop holes placed in a row along the edge of both sides. The charge drops down along either side of the furnace interior, and as it heats and melts flows down into the liquid pool on the hearth. The heavier matte will settle out into a bottom layer below the lighter layer of slag. The matte is tapped periodically from a side tap hole, while the slag is tapped off at the flue end of the furnace continuously. Both roasted calcines and unroasted concentrate can be used for feed, and the burner flames can be oxygen enriched to increase smelting capacity and decrease the fuel consumption per ton of charged material.

Flash Smelting Furnaces are one of the more recent types developed (after World War II) for large scale matte smelting, and are an autogeneous type that uses the oxidation of sulfides in the charge to supply exothermic heat to melt the furnace charge and reach reaction temperatures.

In these processes a fine, dry, unroasted, sulfide concentrate, and furnace flux material is blown into the hot reaction chamber of the furnace, to contact oxygen or pre-heated air which is also blown in. Part of the iron sulfide compound in the charge immediately flashes and oxidizes to FeO and SO_2 in a strong exothermic reaction

$$4CuFeS_2 + 5O_2 = 2Cu_2S \cdot FeS + 4SO_2 + 2FeO$$

The heat evolved is sufficient to melt the other fine particles of charge so that they all reach the hearth of the furnace as liquid droplets. This rain of molten particles separates on reaching the bath into the matte which settles to the bottom layer on the hearth and slag. The slag is formed from the reacted iron oxides, siliceous flux, and any gangue material in the furnace feed, and gathers in a layer above the matte.

The gases leaving the furnace are rich in SO_2 content (above 18%) and so are suitable for sulfuric acid plant treatment or production of liquid SO_2. They also carry a large amount of sensible heat which is recovered in waste heat boilers and reaction air pre-heaters. A considerable amount of dust is recovered in dust collectors. The matte and slag layers in the furnace are tapped into ladles.

There are two general types of flash smelting furnaces in use. One, developed by the International Nickel Company, is a reverberatory type unit with feed injection oxygen burners at both ends and the gas uptake flue relocated in the center of the furnace. All charge feeding is done through the oxygen injection burners, so the roof charging openings of a standard reverberatory design are not required. Matte is tapped from one side and slag, at an end.

The refractories used are the same as for a conventional reverberatory furnace, with magnesite brick roof and slag line and silica brick sides, ends, and hearth.

The second type of flash smelting furnace, the Outokumpu furnace, and the type first developed and more widely used, has a U-shaped design with a vertical reaction shaft at one end, a long, low, settling hearth in the middle, and a vertical gas uptake shaft at the other end.

Dry, unroasted concentrate and flux are injected along with preheated air into the top of the high, round reaction shaft. Flashing takes place immediately, and the liquid droplets fall down the shaft to the long, settling hearth where they settle out into layers of matte and slag. The exhaust gas leaves through a vertical gas uptake at the end of the settling hearth opposite to the location of the reaction shaft. These exhaust gases have to make a 90° angle to leave the settler and, in doing so, drop the molten particles which are still being carried along in the gas stream from the reaction shaft.

Matte is tapped from the reaction shaft end of the settling hearth and slag from the gas uptake end. Magnesite refractories are used throughout the furnace in the reaction shaft and settling hearth.

Electric Furnaces are used for both reduction and matte smelting. The common direct-arc, nonconduction hearth, three-electrode electric furnace is mostly used for reduction smelting and smaller matte smelting furnaces, while for large tonnage matte smelting the submerged arc-type resistance furnace with rectangular shape and six electrodes in line, three pairs individually connected, is most commonly used. A furnace similar to that for matte smelting is used for roast-reduction smelting by a combined electric furnace–flash smelting process.

The direct-arc furnace charges are heated principally by radiation from the arc, as current flows from the electrode to the charge, and especially where the arc strikes the charge. Some heat is also generated by the passage of current through the charge. The most common arc furnaces are three-phase types and use three electrodes, one connected to each phase. The charge then, in effect, completes the circuit for each pair of electrodes in turn.

The matte smelting furnace is not an arc furnace but is a resistance furnace, with the electrodes dipping into the slag layer. The slag resists the passage of current flowing between pairs of electrodes, heat is generated from this resistance, and smelting temperatures are produced. The less the electrodes dip into the slag, the greater the heat generated in the upper portions of the slag and the better the melting operation in the furnace. The transfer of heat from the slag to the newly added unmelted charge floating on its surface is partly by direct contact, but mostly by convection, as it is also to the matte below the slag layer.

Electric furnaces are used to smelt almost all combinations of roasted and unroasted concentrates, which can be charged as hot calcines, cold calcines, cold concentrate, or wet concentrate; and all these smelting operations produce matte of 40 to 50% metallic content.

The high cost of electrical power is a major consideration in electric furnace smelting, and as a consequence this type of smelting becomes more economically competitive when large sources of cheap electrical power are available or when costs for other fuels are very high. Electrical heat does however have inherent advantages over other types of fuels. Electrical energy can be converted into heat energy with high efficiency and gives easy, accurate temperature control. There are no fuel combustion gases, so the quantity of exit gas is less, making the recovery of sensible heat and dust from these flue gases much less complicated and also lowers air pollution.

The direct-arc reduction smelting furnaces are generally smaller in size, 6 feet in diameter (1.83 m), use graphite electrodes, are door charged, tilt to pour, and have magnesite linings with alumina roofs. They are generally similar to steel making electric furnaces.

The matte smelting electric furnace is however more similar in its operation to the reverberatory matte smelting furnace, with the chief difference that heat is supplied electrically rather than by burning a fuel.

The charge of fine concentrates and flux is fed through pipes in the roof down each side of the furnace near the walls and drops down on top of the slag. Here it melts and settles into the liquid bath as matte or slag. A large furnace will be 98 feet long, 23 feet wide, and 13 feet high (29.88 $\times$ 7.01 $\times$ 3.96 m); it has six self-baking Söderberg electrodes 4 feet (1.22 m) in diameter, 50 feet long (15.24 m), and 12 feet apart (3.66 m). Slag and matte are tapped intermittently from opposite ends of the furnace.

Refractories used in the electric furnace are quite different from the reverberatory furnace, because of the different location of the hottest zones in the two furnaces. In the electric furnace, as the hot zone is in the slag layer, the gas space above the charge is relatively cool, 1100°F (600°C), and a roof of cheap fireclay brick is adequate. On the other hand, the hot slag and matte from high electrical resistance heating require a high heat duty magnesite hearth. Silica sides and ends with a magnesite slag line area are adequate for these sections.

The electric furnace used in combination with flash smelting for a roast reduction smelting process is quite similar in design to the matte smelting electric furnace in that it is rectangular in shape, with its four electrodes placed in a row and connected in pairs. The electrodes dip into the slag layer, and here too heat is generated by the resistance to the passage of current flowing between the pairs of electrodes.

The furnace charge of fine dried concentrates and flux is charged through roof openings between the electrodes, and four air nozzles aim high-velocity streams of air tangentially against this incoming stream of feed material to swirl it into a vortex between the electrodes. This arrangement allows sufficient reaction time, while the feed particles are settling from the charging opening to the furnace slag layer, to burn off most of the sulfur and to oxidize and reduce a considerable part of the metallic elements in the charge. The liquid droplets from this flash smelting, and whatever portion of the charge that is still solid, collect on the hearth of the furnace, melt, and complete their oxidation–reduction reaction here.

A typical smelting furnace of this type is rectangular in shape, 44 feet long, 14 feet wide, 11 feet high (13.4 $\times$ 4.27 $\times$ 3.35 m), with a $5\frac{1}{2}$ foot height (1.68 m) from roof to slag layer. Magnesite brick is used in the hearth and chrome magnesite brick is used in the roof and walls. Four 40 inch diameter (100 cm) Söderberg electrodes are spaced equidistant down the length of the furnace and provide a power input of 8000 kVA. Both slag and metal are tapped intermittently, the slag from one end and the metal from the middle of a long side of the furnace.

A second type of electric matte flash smelting furnace with cyclone feeding has been developed in Russia, the KIVCET process. This furnace, of the same general shape as that just described, has fine concentrate feed mixed with pure oxygen fed into the furnace through a water-cooled cyclone positioned approximately one third along the furnace length from the off-gas flue end. Smelting takes place within the cyclone vortex, and the molten particles drop down to the furnace hearth, which is electrically resistance heated by a row of electrodes positioned in the end of the furnace opposite to the off-gas flue. A water-cooled partition divides the furnace into two segments, the separating chamber at one end with the off-gas flue and smelting cyclone, and the settling hearth at the other end containing the electrodes for heating.

Continuous Matte Smelting and Converting Furnaces consist of a series of three adjacent stationary hearth furnaces, with the first furnace in the set being the matte smelting furnace, the second a slag cleaning furnace, and the third a converting furnace to process the matte to semipure metal.

Concentrate and flux are injected along with air enriched to 25% oxygen through roof lances into the first and largest of the three furnaces, and matte smelting takes place.

Both matte and slag flow from the smelting furnace to an electric furnace settler for slag cleaning, with this being the second furnace in the series. Pyrite and coke are added to the molten charge to assist in the sedimentation of matte droplets settling out of the slag layer on top, down into the separated matte layer below. Slag containing 0.5% metal is continuously run off and discarded, while the matte containing about 50% metal flows on to the converting furnace.

The converting furnace, the third and last in the series, receives the matte from the slag cleaning furnace and through roof lances blows air and flux into the bath. This converts the matte to a semipure state containing 98 to 99% metal and produces a slag of 7 to 15% metal content which is returned to the matte smelting furnace.

Blast Furnaces are used both for copper and nickel matte smelting and for lead reduction smelting, with the general overall design and method of operation being quite similar for both cases. Both types of furnaces are shaft furnaces, rectangular in shape, and top charged and have refractory hearths with water-cooled hollow metal side and end jackets above the hearth. Rows of tuyeres down each side of the furnace are positioned through the water jackets above the hearth.

Blast furnaces require a coarse-sized feed that will not be blown out of the furnace by the strong air blast through the tuyeres. To accomplish this, prior roasting will be done by blast roasting to agglomerate fine flotation concentrates; revert fines will be pelletized, briquetted, or sintered to be agglomerated; or coarse, high-grade lump ore will be charged.

The Matte Smelting Blast Furnaces, of which there are now a considerably fewer number left in worldwide operation smelting nickel and copper ores and concentrates, are somewhat the larger in size of the two types and present designs smelting 1500 tons of charge per day have a rectangular crucible 15 inches deep (37.5 cm) and 5 feet wide (1.52 m), lined with chrome or magnesite brick. Other dimensions are 5 feet wide (1.52 m) and 20 feet long (6.1 m) at the tuyere line, which is 12 inches (30 cm) above the crucible, and 6 feet wide (1.83 m) and 20 feet long (6.1 m) at the top of the 14 foot high (4.27 m) sidewall water-cooled jackets.

Charge trains, dumping alternately from either side at the top of the furnace shaft, are loaded so that the first material charged to the furnace is coke, followed in order by sinter, briquettes, reverts, and high-grade lump furnace ore. The coke, which can be up to 10% of the weight of the solid charge, and to a lesser extent the exothermic heat from the oxidation of some iron sulfide to iron oxide, provides heat to smelt the charge in the furnace. The low-pressure air blast through the furnace tuyeres, 28 to 36 ounces (0.79 to 1.02 kg) pressure, supplies air to oxidize the carbon in the coke in a strong exothermic reaction to carbon dioxide:

$$C + O_2 = CO_2$$

and in a lesser exothermic reaction to carbon monoxide:

$$2C + O_2 = 2CO$$

As the cold charge descends in the furnace shaft, it is heated by the rising flow of hot gases from the exothermic burning of fuel above the tuyere line, until the temperature becomes high enough for the metal sulfides to melt and combine into liquid matte, while the oxides will melt along with fluxes and form a slag. FeO and SiO_2 are the main slag constituents; and if not enough SiO_2 is available in the gangue of the charge for proper combination with the FeO present, additional SiO_2 flux is added.

A liquid mixture of matte and slag runs down into the furnace crucible and out through a magnesite-lined trap spout at one end of the furnace into a refractory-lined settler. The settlers are either round, or rectangular with semicircular ends, and are made of sheet steel lined with chrome brick in typical dimensions of 42 feet long, 10 feet wide, and 5 feet deep ($12.8 \times 3.05 \times 1.52$ m). The heavier matte settles to the bottom of the settler from where it is tapped intermittently, while the lighter slag overflows in a steady stream from a spout near the top of the settler.

A partial deviation from the standard blast furnace design for matte smelting is the low-shaft matte smelting blast furnace, with an elliptical rather than rectangular shape, and the furnace shaft is brick lined and cooled by water flowing down the outside rather than having hollow steel water jackets for the shaft construction.

The Low-Shaft Blast Furnace for nickel matte smelting is tapped directly from the crucible with no external auxiliary settler and can also be sealed to the atmosphere and charged through a double bell and hopper system as is used with the iron blast furnace. If there is no sulfur in the ore, and this has to be added as a separate part of the charge in order to make matte, it is common to preheat the tuyere air in order to reduce the coke rate and to carry the additional sensible heat into the furnace. Hot furnace flue gases are used to preheat the entering tuyere air by burning the flue gases enriched with coke oven gas in preheating stoves.

Low-shaft furnace dimensions for a furnace smelting 800 tons of ore per day would be 24 feet long and 7 feet wide, with a shaft height of 16 feet from tuyeres to charge level ($7.32 \times 2.13 \times 4.88$ m).

Blast Furnaces Used for Reduction Smelting of lead are generally somewhat smaller with less throughput than those used for copper or nickel matte smelting. A typical 500 ton per day charge furnace would be rectangular in shape, 19 feet long (5.8 m) and 48 inches wide (1.22 m) at the tuyeres, and have a $2\frac{1}{2}$ foot deep hearth (0.69 m) and side wall water jackets 17 feet high (5.2 m), in either single or double tiers above the magnesite-lined crucible.

Charging is either from side dump cars through opened doors on either side of the furnace at the top of the shaft or from a bottom dump car directly down into the furnace, if it is an open-top type. Coke makes up from 10 to 14% of the furnace charge, along with oxide sinter, fluxes, and reverts; and it is the carbon monoxide formed from this coke reacting with the tuyere air that will supply the reducing conditions in the furnace and reduce the metal oxide sinter to an impure metal. The strong exothermic reactions of the coke and air producing CO and CO_2 raise the furnace charge to the smelting temperature and liquify the products formed.

With the deeper hearth, it is possible to begin separation of the different specific gravity layers inside the furnace before removing them. Consequently the heavy impure reduced metal collects on the bottom of the furnace hearth and is removed from here by a siphon tap. The lighter layers of slag, and matte and speiss, if also present, are run off from a tap hole at the top of the crucible into a settler. Separation takes place here between the lighter slag and the heavier matte, speiss, and reduced metal droplets which ran out of the furnace with the slag.

Because of the large volumes of air being blown through the various types of blast furnaces, there is considerable fine material carried off as flue dust in the furnace gases. This necessitates a dust collecting system to recover this flue dust before the furnace gases are allowed to escape.

SMELTING PROCESSES

1a. Copper Sulfide Roasted Calcines are smelted to matte and slag in reverberatory furnaces, electric furnaces, and blast furnaces. The smelting process in all these types of furnaces is essentially the same, so that when the charge of metal sulfides, iron oxide, gangue, and siliceous flux is melted together, the iron oxide, gangue, and silica combine together to form a lighter layer, the slag, which floats on top of the heavier layer of combined metal sulfides, the matte.

Table 1.7. The New Generation of North American Copper Smelters

Plant	Process	Designed Capacity (tons copper/year)	Startup	Comments
Noranda Mines Ltd., Noranda, Quebec	Noranda continuous smelting	50,000	1973	World's first continuous copper smelter
Inspiration Copper Co., Inspiration, Arizona	51 MVA electric furnace (dry charge) with Hoboken syphon converters	150,000	1974	World's largest electric copper smelting furnace
Anaconda Co. Butte, Montana	36 MVA electric furnace (calcine charge) with 38 ft diam. fluid-bed roaster	210,000	1975	World's largest calcine-charged electric copper smelting furnace
Phelps Dodge Corp. Hidalgo, New Mexico	Outokumpu flash furnace with reduction of SO_2 to elemental sulfur	100,000	1977	First North American Outokumpu flash smelter. First copper flash smelter to employ SO_2 reduction
Kennecott Copper Corp. Garfield, Utah	Noranda continuous smelting	298,000	1977	Second Noranda Process smelter, first in U.S.
Texasgulf Canada, Timmins, Ontario	Mitsubishi continuous smelting	130,000	1978	Second Mitsubishi Process smelter, first in North America
Afton Mines Ltd., Kamloops, British Columbia	Top-blown rotary converter	28,000	—	First commercial use of TBRC for copper smelting

Source: P. Tarassoff, Annual Review of Extractive Metallurgy–Pyrometallurgy, *J. Metall.* Vol. 28, No. 3, 1976, p. 12.

The matte is a homogeneous solution of copper, nickel, iron, cobalt, and sulfur, with smaller amounts of other base metals and the precious metals. Copper has by a slight margin the greatest affinity for sulfur of any of the metals normally found in these charges, and CuO has a lower stability than does Fe_3O_4, so all the copper present will combine with sulfur to form the stable sulfide Cu_2S. The quantity of iron in the matte can be controlled by adjusting the degree of oxidation in the smelting furnace charge by the preceding roasting operation. The iron oxidized during roasting will go into the furnace slag, while the iron remaining in the roasted calcines as sulfide will be part of the matte.

The more sulfur present in the furnace charge, in excess of the amount required to combine in the matte along with the copper, nickel, and cobalt, the greater will be the iron content and the lower will be the matte grade. Each smelting operation has its optimum matte grade, and as smelting is a concentrating as well as a combining process, smelting will be carried on until substantially all of the oxidized iron, predetermined during roasting, has been slagged off along with a considerable portion of the gangue and silica also present.

Only a small percentage of any copper oxide formed during roasting is lost in the furnace slag. This is due to the reaction between copper oxide and iron sulfide which returns the copper to sulfide, in which state it combines with the matte:

$$Cu_2O + FeS = Cu_2S + FeO$$

Even though smelting is a concentration process, the usually preferred product is a low-grade, high-iron matte of 35 to 45% copper rather than a very high-grade matte with only small amounts of residual iron. There are several reasons for preferentially producing a low-grade matte, most of which are associated with the high metal losses in a slag made along with a high-grade matte. A much larger volume of slag is made in the greater concentration effect of removing more iron to produce a high-grade matte. This larger slag volume flowing from the furnace carries with it more dissolved matte and more mechanically entrapped matte droplets, which will considerably decrease the overall copper recovery. The smaller volume of high-grade matte produced is not as efficient a collector of the precious metals, as there is a lesser amount of matte rain to combine with them; and there is an insufficient amount of iron sulfide left to provide the necessary exothermic heat through its oxidation for the converting operation to follow this furnace smelting process. The cost of roasting would also be increased, as a larger-capacity roasting plant would be needed to oxidize a larger percentage of the iron sulfides to iron oxides.

Reverberatory Furnaces are still the most commonly used smelting furnace in the copper industry generally, due to their adaptation to handle fine flotation concentrates, and are the type used by Centromin–Peru and Noranda, among others.

Pulverized coal, fuel oil, and natural gas are all used extensively as fuels, and the choice is determined by availability and cost. Natural gas is usually preferred for its ease of handling and cleanliness if the other factors are equal.

The usual furnace charge consists of hot roasted calcines, molten slag from the converters too high in copper to be discarded, fluxes, and such other copper-bearing materials as direct smelting ores, cement copper precipitates, flue dust, and other reverts. A typical 120 foot by 25 foot furnace (36.6×7.62 m) will have a daily charge of 900 tons of hot calcines, 60 tons of flux, and 400 tons of returned converter slag. From this charge will be produced 625 tons of matte analyzing 44% copper and about an equal amount of slag analyzing 0.3 to 0.4% copper.

The matte is tapped from the furnace periodically and conveyed in ladles by crane to the copper converters, while the slag is run off continuously into slag pots and hauled by truck or rail to the slag dump. Some of this slag is also now being treated in iron plants to produce high-grade reduced iron pellets for the steel industry.

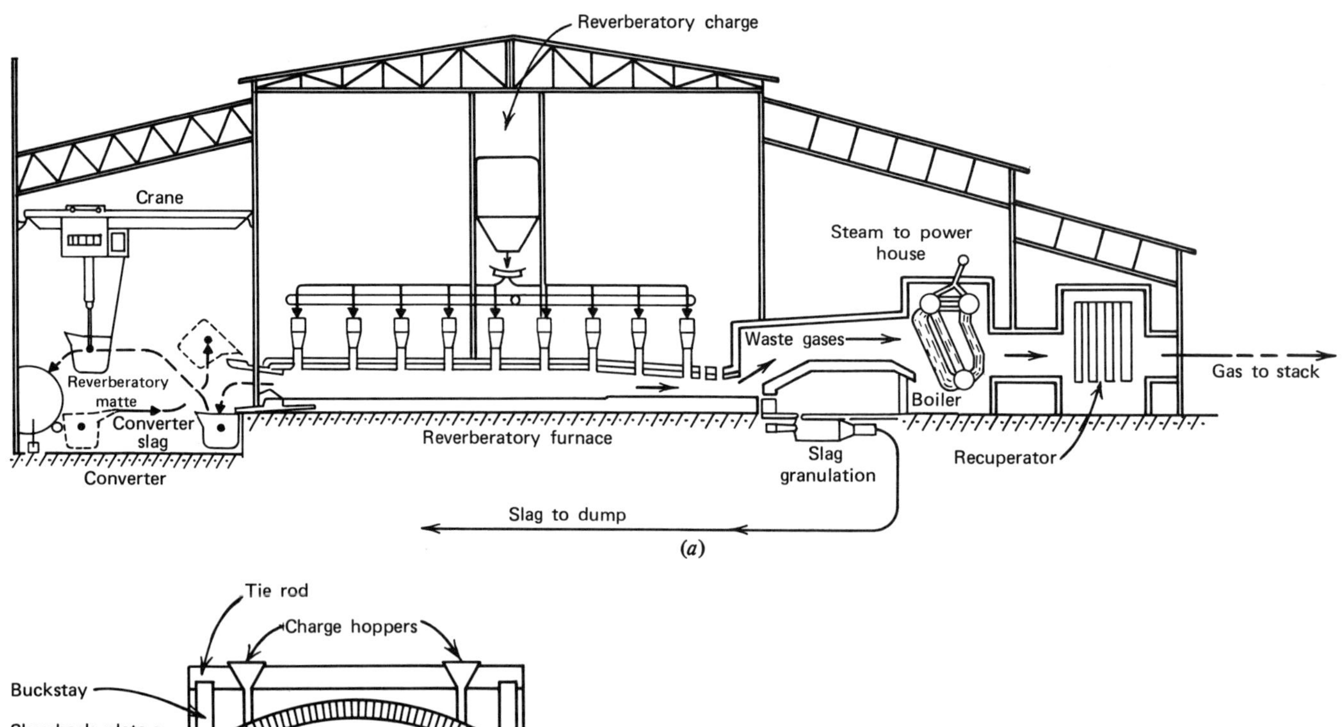

Figure 1.8. Copper reverberatory smelting furnace: (*a*) side view; (*b*) end view.

Approximate temperatures of the furnace charge and products will be $1120°F$ ($600°C$) for the hot calcines, $2250°F$ ($1230°C$) for the molten returned converter slag $2015°F$ ($1100°C$) for the reverberatory matte, and $2180°F$ ($1195°C$) for the reverberatory slag.

A major problem in reverberatory smelting is the formation and buildup of magnetite, Fe_3O_4, which can build up in a layer on the bottom of the furnace until it so restricts the volume of the furnace and obstructs the tapping that the furnace must eventually be shut down. The difficulty from magnetite is that copper smelting requires the oxidation of the iron and sulfur in the copper ore being treated and the combination of the oxidized iron with silica and other elements to form a fluid slag that will contain little copper and can be discarded. At the temperatures employed in copper smelting, Fe_3O_4 forms readily during the oxidation of FeS and is a stable oxide at these temperatures. Being chemically somewhat inert, Fe_3O_4 does not combine with SiO_2 to form part of the reverberatory slag and has only a limited solubility of 3 to 5% in the slag and 10 to 15% in the copper matte, making its removal from the furnace difficult.

If the Fe_3O_4 can be reduced to FeO, then this FeO formed will combine with SiO_2 in the normal way and go into the furnace slag. This reduction is dealt with in a number of ways, one by having sufficient FeS left after roasting, or adding unroasted concentrate high in FeS, to reduce the Fe_3O_4 to FeO:

$$FeS + 3Fe_3O_4 = 10FeO + SO_2$$

Ferrosilicon additions and adding green timbers through the furnace roof are also used to reduce the magnetite to ferrous oxide, while the recent trend to clean converter slag by flotation rather than recycling it back through the reverberatory reduces the considerable amount of Fe_3O_4 from this source.

Electric Furnaces have not been as widely used as reverberatory furnaces and until recently have been more or less restricted to countries with cheap hydroelectric power. However electric resistance heating in place of burning fossil fuels is now being looked on as a way to reduce appreciably gas volumes produced and with this to lessen greatly the problem of air pollution and gas cleanup associated with matte smelting. The Anaconda Company, for example, in its new Montana smelter designed to produce 210,000 tons of copper per year, has installed a 36 MVA electric matte smelting furnace to be fed roasted calcines produced in a 38 foot (11.59 m) diameter fluid-bed roaster.

There is much in common between the design and operation of the matte smelting reverberatory and electric furnaces. The charge is the same, hot-roasted calcines, molten returned converter slag, flux, and various other copper-bearing materials, and so is the general smelting process wherein the copper, nickel, iron, cobalt, and sulfur collect into a matte, while the iron oxide, gangue, and fluxes form a slag.

The electric furnace differs from the reverberatory furnace mainly in the manner in which heat is provided, by the resistance to electrical current flowing through the slag, and the varying resultant furnace conditions that this imparts on the charge.

The furnace electrodes normally used, and commonly placed with three to six electrodes in line, are the self-baking Söderberg type. These are made by feeding a carbonaceous paste into a sheet steel casing to form the electrode. This green paste is baked hard by heat from the electric current and heat rising from the hot zone of the furnace, so that when it is lowered into the furnace to contact and carry current to the slag layer it has been baked into a hard, monolithic column. As the electrode is eaten away in operation, new sections of casing are welded to the top projecting above the furnace and are filled with paste to be baked in turn. Electrode consumption is on the order of 6 pounds (2.7 kg) per ton of solid charge.

The charge fed through the roof at first spreads out as a layer of unsmelted calcines on top of the slag and is referred to as a "cold top." As it melts, the calcine gradually settles into the slag and separates into the matte and slag layers. To prevent calcine from floating out of the furnace with the liquid slag, any coarse, solid return material is fed into the last third of the furnace where it functions as a dam.

There is some turbulence in the slag flow due to overheated slag in the immediate areas of the electrodes, rising to the surface and flowing toward the cooler furnace walls. Part of the heat from this slag is transferred by convection to the calcines floating on its surface, and this slag movement, of some 5 to 10 cm per second, is advantageous for heat transfer from slag to calcine. However it does make settling out of matte particles more difficult and increases erosion of the furnace walls.

The electric furnaces that smelt copper calcines charge the hot calcines at 1100°F (600°C) and produce matte containing 40 to 50% copper. The slags run between 0.3 and 0.5% copper and are carried in slag pots to the dump. Both matte and slag are tapped periodically. Power consumption is 580 kW-hr per ton of solid charge and 400 kW-hr per ton of hot calcines. Smelting capacities are as high as 500 tons of charge per 24 hours. Off-gas temperatures run about 1300°F (700°C) and contain on the order of 5% SO_2.

Magnetite formation is also a problem in the electric furnace, though its behavior is much different than in the reverberatory furnace. In an electric furnace not only is the slag in turbulent motion but the matte is also subjected to electrodynamic forces, and the combination of these causes electric furnace matte to have a higher magnetite content than reverberatory matte. The magnetite at first forms an intermediate zone of high viscosity between the slag and matte layers and then settles to the furnace bottom, where it may remain in a fairly constant layer of up to 20 cm deep. The intermediate magnetite zone between the matte and the slag is especially objectionable to best furnace practice in that it prevents the matte droplets from settling out of the slag into the matte layer.

Magnetite in the electric furnace may be reduced or eliminated by reducing the Fe_3O_4 to FeO and fluxing the FeO produced with SiO_2 to go into the slag. This reduction of Fe_3O_4 is accomplished in several ways: by charging more coal and silica, by lowering the voltage, and by lowering the electrodes to nearer the matte surface.

Blast Furnaces were initially developed to smelt high-grade massive lump sulfide ores, which were smelted directly with only superficial benefication, such as hand picking. The advent of extensive beneficiation and resultant fine flotation concentrates brought about the development of the reverberatory and electric furnaces which could better handle this fine feed. However the blast furnace has not yet become entirely outdated, and with the innovation of sintering (blast roasting) to agglomerate the fine concentrate it is still an efficient smelting furnace, either on this sintered feed or briquettes. A few furnaces are still in use smelting copper ores in Africa and Japan.

The charge to the blast furnace consists of sintered calcines, with a minimum size of $1\frac{1}{2}$ inches (3.75 cm) coarse, lump ore, reverts, fluxes, and sized coke over $1\frac{1}{2}$ inches (3.75 cm). The coke ratio in the total charge will vary between 10 and 35%, depending on the percentage of sulfides in the smelting charge which will be oxidized and give off exothermic heat. With the higher coke consumptions the sulfur in the sintered calcines will be as low as 7 to 9%, and the copper matte grade produced will be very high (60%), due to large amounts of FeS being oxidized to FeO during the blast roasting and then this FeO being removed as slag by combining with SiO_2 during smelting. This then will leave a smaller amount of FeS to combine with Cu_2S in the matte and give a very high-grade copper matte.

A typical 20 foot long by 5 foot wide (6.1 × 1.52 m) water-jacketed furnace would have air supplied in volumes of 675,000 cubic feet (60,000 m^3) of air per hour at a pressure of 4.8 inches (120 mm) of mercury. The tuyeres for this furnace are distributed on the long sides of the furnace with one or two tuyeres per jacket section, 10 jackets per side. There is no tuyere in the jacket above the tap hole.

Products from the furnace are matte and slag, which in most furnaces are run off together into a refractory-lined settler to allow the two different specific gravity layers to settle and separate. The slag from the settler containing 0.3 to 0.5% copper flows off in a constant stream into slag pots and is carried to the dump. The 30 to 60% copper matte is tapped intermittently and conveyed in pots to the converters. Molten converter slag containing 4% copper is also returned to the settler, to recover its high copper content.

Some blast furnaces have been built with deeper than normal hearths [2 feet deep (0.61 m) compared to 12 inches (0.30 m)], have no exterior settlers, and effect a separation of matte and slag in the furnace hearth itself. The matte is tapped from the extreme bottom of the hearth, and the slag from a higher point below the tuyeres. Converter slag in this case is returned to the blast furnace as a solid part of the charge.

Dust collectors of various types are used to recover the flue dust swept off in the furnace gases. These are often two-stage units to recover first the coarser dust particles and then the fines. Combined, these dust particles may be up to 3% of the total solids charged. They are returned to the sintering plant for agglomeration.

Magnetite is not an excessive problem in the blast furnace, due to the strongly reducing conditions tending at least partly to reduce the Fe_3O_4 present to FeO, and this combines with SiO_2 as a slag:

$$2FeO + SiO_2 = 2FeO \cdot SiO_2$$

To combine with the reduced FeO, the SiO_2 content of blast furnace slags is generally high, from 25 to 30%, and it will be provided for in the furnace charge as part of the gangue or as added flux.

Typical furnace operating temperatures would be 2420°F (1325°C) for the combined matte and slag stream leaving the furnace, 2265°F (1240°C) for the matte tapped from the settler, and 2350°F (1290°C) for the slag flowing from the settler.

1b. Copper Sulfide Unroasted Concentrates are smelted to matte and slag in reverberatory furnaces, oxygen-enriched reverberatory furnaces, electric furnaces, flash smelting furnaces, and the new three furnaces in a series continuous melting process. The basic smelting reactions and combinations are the same as for the matte smelting of roasted sulfide calcines, in that the products of smelting will be a liquid matte containing the metallic metal sulfides of copper, nickel, cobalt, and iron and a slag made up of molten iron oxide, gangue, and siliceous flux.

Direct smelting of unroasted concentrates, both wet and dried, has a rather different approach depending on the type of smelting that is done. The reverberatory and electric furnaces using raw concentrates are primarily concerned with a saving on overall processing time, a saving on the cost of operating a roasting facility, and in reducing the amount of material handling. Flash smelting, on the other hand, utilizes the exothermic reaction of the high FeS content of the unroasted concentrate as the smelting fuel for an autogenous semipyritic reaction and needs the unroasted concentrate for this reason. The continuous smelting process combines both of these advantages and delivers a smelting rate, per unit volume of smelting furnace, six times greater than a reverberatory furnace and two to four times greater than a flash smelting furnace.

Reverberatory Furnaces in common smelting practice are similar in design and general operation for both roasted and unroasted feed material, though a comparison of the two furnace operations shows that a hot calcine-charged furnace can smelt nearly double the amount of feed material and smelt this with 43% less fuel than a furnace charged with cold, wet concentrate.

However the copper content of the slag from the wet-charge furnace will be lower than the slag from the hot-charge furnace, 0.23% copper as against 0.37%. Also, the magnetite content for both the matte and slag will be higher for the hot-charged furnace, 17.3% Fe_3O_4 in the matte and 11.0% in the slag for the hot-charge, compared to 11.2% Fe_3O_4 in the matte and 9.0% in the slag for the wet-charge furnace. These of course are due to the formation of magnetite and copper oxide during roasting and their subsequent pickup in both matte and slag.

It is difficult to make a direct overall comparison of the metallurgical and economic features of hot-charge and wet-charge furnaces because of these entirely different types of charges and furnace conditions. It is possible though to compare the costs of material treated in each of the overall processes, roasting plus hot-charge smelting against wet-charge smelting. This points out that the wet-charge smelting is somewhat more costly, due mainly to the increased fuel used, but this higher cost is partly offset by the lower slag losses in the wet-charge furnace.

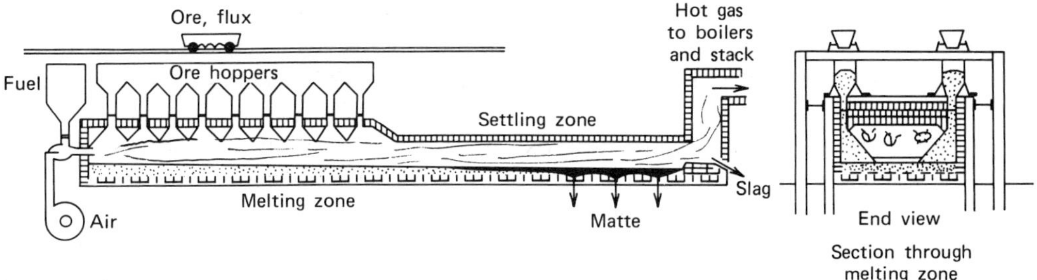

Figure 1.9. Schematic views of a copper matte smelting reverberatory furnace. *Source:* Extraction Metallurgy, by J. D. Gilchrist, Pergamon Press Ltd., Oxford, 1967.

Table 1.8.   Reverberatory Furnace Smelting of Mixed Unroasted Concentrate and Roasted Calcines, Hindustan Copper Ltd.: Reverberatory Furnace Performance[a]

Dry Charge Treated	Weight (tonnes)
Concentrate	2974
Calcine	413
Roaster flue dust	24
Recycled secondaries (matte and slag)	1348
Blister slag	95
Boiler flue dust	57
Converter flue dust	7
Converter chippings	20
Dust from stack	2
	4940
Liquid converter slag	1582
	6522
Estimated matte produced	3576
Estimated dump slag produced	2897
Estimated coal consumed	930
Estimated total heat value in terms of coal recovered as waste-heat steam	480
Tonnes of dry charge per tonne coal	5.31

[a]Conventional smelting: typical for 1 month.

Source: J. C. Yannopoulos and J. C. Agarwal, Eds., *Extractive Metallurgy of Copper*, Vol. 1, The Metallurgical Society of AIME, 1976, p. 202.

Dusting is less when wet-charging, and this eliminates the unpleasant working conditions from gas and dust on the furnace charge floor and also in the roaster building, which is now not needed. Flue dust from both wet- and hot-charge furnaces is considerable and must be collected by electrostatic precipitators or some other types of dust collectors. It was thought at first that a wet charge would greatly decrease the amount of flue dust, but this has not been the case. This is partly due to the burning of more fuel to maintain the smelting rate and the gas from the burners then escaping the furnace at a greater volume and higher velocity and carrying considerably more flue dust than first expected. The Kennecott Copper Corporation uses unroasted concentrate feed to reverberatory furnaces, while Noranda Mines, Ltd., uses roasted feed.

Oxygen-Enriched Reverberatory Furnaces are now being used by several major copper smelters on a production basis to increase smelting capacity, with two of the latest to add this innovation being the smelter at Onahama, Japan, and El Teniente's Caletones smelter in Chile.

Several techniques have been used for oxygen enrichment, and two of the most successful are by adding oxygen to endwall burners either by (a) premixing oxygen and combustion air, or (b) by undershot oxygen enrichment. Premixing increases flame temperature uniformly, while the undershot technique selectively enriches the underside of the conventional flame and by doing so radiates and concentrates extra heat downward toward the liquid bath and the cold concentrate charge. The mixing of oxygen and combustion

Table 1.9. Reverberatory Furnace Smelting Analysis of Charge and Products (Hindustan Copper Ltd.)

Material	%Cu	%Fe	%S	%Ni	%SiO$_2$	%Al$_2$O$_3$	%CaO	%MgO	%H$_2$O
Concentrate (dry basis)	26	28	30	0.42	6.5	1.8	0.5	0.7	—
Concentrate moisture	—	—	—	—	—	—	—	—	6.0
Calcine	29	30	4	0.45	—	—	—	—	—
Process secondaries	30.2	24.2	12.50	—	10.30	—	—	—	—
Dump slag	0.41	43.2	1.0	0.08	33.4	7.5	1.5	2.1	—
Matte	43	29	24	2.7	—	—	—	—	—
Converter slag	4	53	0.6	1.9	27	1.3	1.1	—	—
Blister slag	45.05	13.8	1.0	15	9	—	—	—	—
Converter flux	—	2.06	—	—	85.1	2.42	0.83	0.51	—
Blister copper	97.2	0.04	0.05	1.1	0.06 (insol.)	—	—	—	—
Boiler flue dust	4	10	3	—	46	—	3.5	—	—
Converter flue dust	45	7.5	13	4	6 (insol.)	0.3	—	—	—
Converter chippings	37	30	4	10	6	—	—	—	—

Source: J. C. Yannopoulos and J. C. Agarwal, Eds., *Extractive Metallurgy of Copper*, Vol. 1, The Metallurgical Society of AIME, 1976, p. 204.

air is not as complete with undershot oxygen, but this disadvantage may be counterbalanced by its more effective placement of the extra heat.

In some cases it also may be desirable to enrich selectively some of the endwall burners to higher levels than others, in order to particularly concentrate heat in colder or slower-moving zones of the furnace. This selectivity is easily accomplished with either type of enrichment technique, both of which are relatively simple to install on a conventional reverberatory furnace.

Roof lance burners using a combination of oxygen and fuel oil have also been successfully used.

In either case, with endwall or roof lance burners, there are different designs of burners that can be selected. One of these delivers a soft, wide-coverage flame from a burner located within a few feet of the cold-charge material. This flame impinges directly on the cold charge, smelts the concentrates quickly, and is a very effective use of oxygen and fuel. Where it is not practical to position the flame close to the cold charge, a long-reaching, narrow flame with higher flame velocity can be aimed at the cold material from a distance; and while not as efficient, it is still a reasonable alternative.

The Onahama, Japan, smelter increased its capacity by 25% using two burners per furnace with an oxygen equivalency of 22.5% of the combustion air, while the El Teniente, Chile, smelter increased its smelting rate by 71% using seven burners per furnace and oxygen equivalent to 43% of the combustion air.

Electric Matte Smelting Furnaces are used to smelt both wet and dry concentrates, and as with reverberatory wet-charge matte smelting the fuel consumption, in this case electricity, is much higher than for hot, dry charging. An electric furnace charged with cold, wet concentrate needs 6.6 kW-hr per ton of solid charge, while a furnace charged with hot, roasted calcines at 1100°F (600°C) needs only 4.4 kW-hr per ton of solids. A furnace with a mixed charge of wet concentrate and hot calcines needs energy midway between, of 5.5 kW-hr per ton of solids.

The largest electric matte smelting furnace using dry concentrate feed is at the new

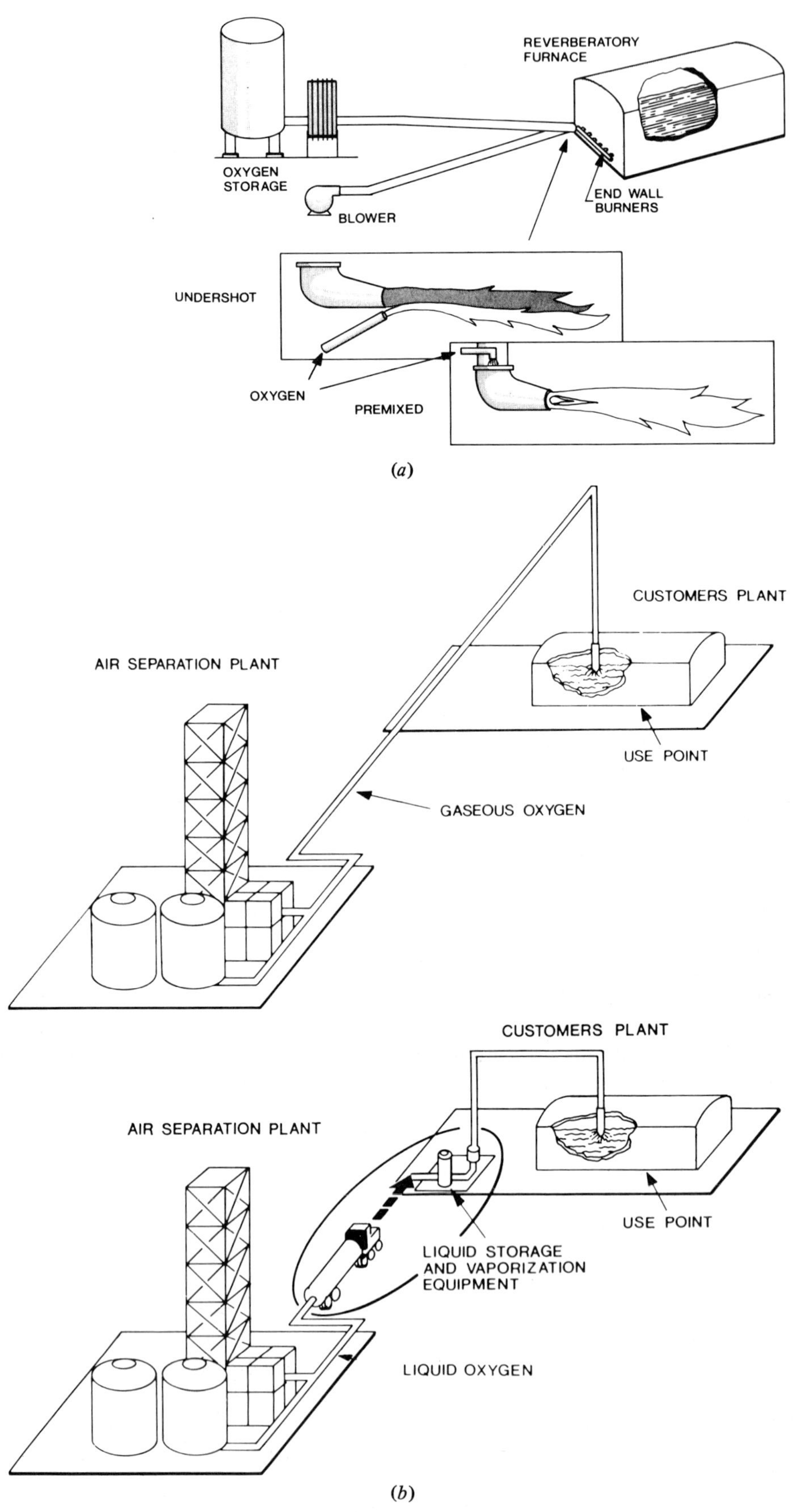

Figure 1.10. Reverberatory furnace oxygen enrichment techniques: (*a*) end wall burners; (*b*) roof lance burners. *Source:* Courtesy of Air Products and Chemicals, Inc., Allentown, Pennsylvania.

36

**Table 1.10.   Results of Reverberatory Furnace Smelting with Oxygen:
Reverberatory Furnace Oxygen Results**

Amount of production increase	25%
Oxygen consumption	1000 to 1200 SCF/ton charge
Method of oxygen utilization	
Conventional burner	22.5% to 24.0% oxygen
Enrichment or supplementary oxy-fuel burners	10% to 15% of conventional firing rate
Fuel savings	0.6 to 1.0 MM Btu/ton charge
Increase in waste gas SO_2	1.0%

Source: Oxygen in Copper Smelting, Air Products and Chemicals, Inc., Allentown, Pennsylvania, 1972, p. 6.

Inspiration Copper Company smelter in Arizona. This is a 51 MVA furnace in a smelter designed to produce 150,000 tons of copper per year. The Copper Hill property at Duck-town, Tennessee, and the Mulfulira smelter in Zambia have also installed new electric smelting furnaces. In all cases the gas volumes from matte smelting can be drastically reduced by using electrical resistance heating instead of burning fossil fuels, with its attendant decrease in volumes of gas to clean and reduced air pollution. Matte grades produced will run from 40 to 50% copper, and slag analysis will be in the range of 0.5% copper.

Electric matte smelting furnaces are also used for a type of roast-reduction flash smelt-ing, where fine dried concentrate is charged through roof openings and swirled into a vortex to give oxidation–reduction reaction smelting time, either by high-velocity streams of air directed tangentially against the incoming stream of feed or by adding the feed along with oxygen through a cyclone.

This process is the Russian-developed KIVCET cyclone smelting process for impure copper concentrates, the latest stage of which is now used in the Irtyph Polymetal Com-bine at Gluboke in Kazakhstan, Russia. This plant has a rated daily capacity of 350 tons of 24% copper sulfide concentrates containing some lead and zinc.

The concentrate feed for the smelting cyclone on this furnace must have a particle size below 0.04 inch (1 mm) and a moisture content of not over 2% and must be free flow-ing. It is metered axially into the smelting cyclone from above through closed conveyors and combines with oxygen gas which enters the cyclone tangentially. Reaction takes

Figure 1.11. Electric copper matte smelting fur-nace, operating on concentrate feed, 51 MVA—1600 tons per day, Inspiration Consolidated Copper Company. *Source:* Courtesy of Elkem-Sprigerverket a/s (Inc.)

place between the oxygen and concentrate, similar to that which occurs in conventional flash smelting, and the melted droplets of charge leave the outlet in the bottom of the cyclone in a rotating vortex which passes down into the furnace chamber where the liquid melt and reaction gases are separated.

The furnace is the conventional rectangular shape with alumina brick roof and upper walls and chrome magnesite hearth and is divided into two sections by a water-cooled copper partitioning wall projecting down into the bath from the roof. This wall divides the furnace into two sections, one slightly longer than the other, with section lengths of approximately 35 feet and 25.4 feet (11.5 m and 8.3 m) and a total hearth length of 60.4 feet (19.8 m). The shorter length is the smelting section known as the separating chamber, and into this projects the water-cooled, double-jacketed smelting cyclone, positioned midway in the furnace roof between the partitioning wall and the large gas flue, which also is located at this end of the furnace. The cyclone is 4.27 feet (1.4 m) at its largest internal diameter and has an internal height of 6.1 feet (2.02 m). The furnace width is 16.47 feet (5.4 m).

The dividing wall prevents the reaction gases from passing out of the smelting section of the furnace, and these pass out through the vertical off-gas flue positioned at the end of the furnace, at a temperature of 2282° to 2462°F (1250° to 1350°C). The gases pass through a water-cooled shaft designed as an indirect heat exchanger and enter an electro-

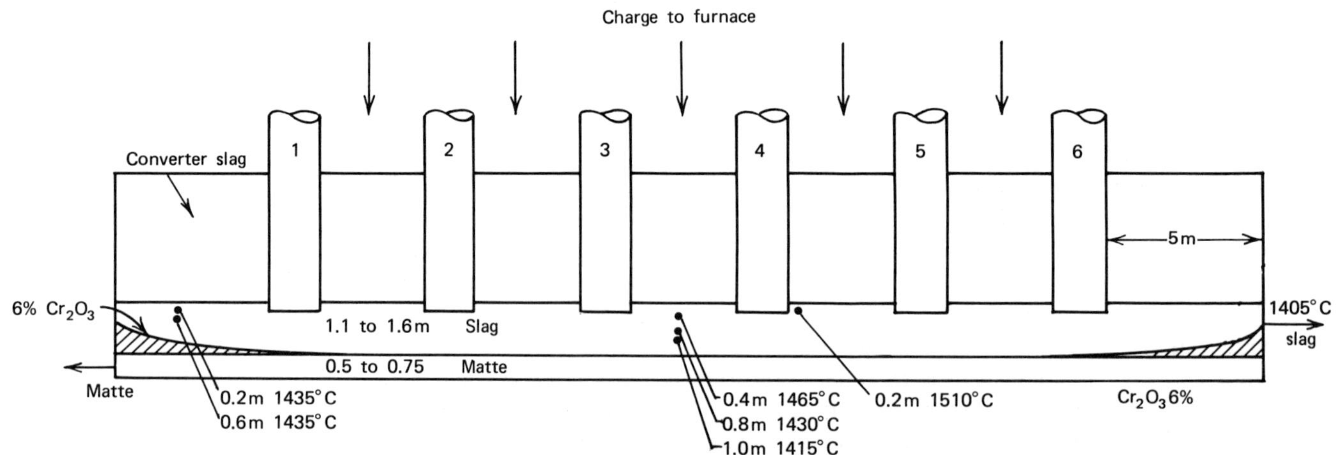

CHARGE AND PRODUCTS

Slag (%)		Matte (%)		Converter Slag (%)		Pelletized Concentrate (%)		Limestone (%)	
MgO	14.92	Cu	8.88	CaO	0.65	Fe	16.39	CaO	46.41
CaO	16.68	Ni	15.1	MgO	0.67	Cu	2.13	MgO	0.55
FeO	18.01	Fe	42.1	Al_2O_3	1.16	Ni	4.16	SiO_2	10.89
Fe_2O_3	0.3	S	28.59	FeO	60.1	S	10.34	Al_2O_3	0.65
SiO_2	42.79	Cr	0.091	SiO_2	35.2	CaO	3.29	FeO	0.41
Cr_2O_3	0.39			Cr_2O_3	0.28	MgO	15.96		
Cu	0.058					Al_2O_3	3.29		
Ni	0.079					SiO_2	35.75		
S	0.34					Cr_2O_3	0.31		
						H_2O	6		

Figure 1.12. Comparative positional temperatures of slag and matte in electric furnace smelting of unroasted copper–nickel concentrates, Anglo Am. Corp., So. Africa. *Source:* J. C. Yannopoulos and J. C. Agarwal, Eds., *Extractive Metallurgy of Copper*, Met. Soc. of AIME, Vol. 2, 1976, p. 276.

Table 1.11. Oxygen and Nonoxygen Reverberatory Furnace
Smelting of Unroasted Concentrates, Onahama Smelting
and Refining Co.: Reverberatory Furnace Data[a]

	Dec. 1970	Jan. 1972
Concentrate smelted (tons)	24,769	29,919
Silicious flux smelted (tons)	3,279	4,038
Limestone smelted (tons)	2,059	2,171
Reverts smelted (tons)	571	779
Total solid charge (tons)	30,678	36,907
Fuel oil consumed (kl)	4,723	4,870
Oxygen consumed (stp m^3)	–	667,058
Matte produced (tons)	20,354	25,180
Matte grade (%)	34.5	34.4
Slag produced (tons)	20,443	25,936
Copper in slag (%)	0.46	0.47

Source: J. C. Yannopoulos and J. C. Agarwal, Eds., *Extractive Metallurgy of Copper*, Vol. 1, The Metallurgical Society of AIME, 1976, p. 161.

[a]All weights are on dry basis.

static precipitator at 932° to 1112°F (500° to 600°C) to remove the solid particles for recycling to the furnace, after which the cleaned gas containing 80 to 85% SO$_2$ is suitable for the production of liquid sulfur dioxide or elemental sulfur.

The settling hearth of the furnace is on the side of the partitioning wall away from the smelting cyclone, and this is the longer 35 foot (11.5 m) section of the furnace. Three equispaced electrodes project through the roof into the bath and keep the bath in motion by convection induced by the direct resistance heating. The melt flows continuously into the settling hearth from the separating chamber, where it is considerably diluted by the fully reduced slag. A matte layer spreads over the bottom of the furnace and is tapped periodically at the end of the separating chamber of the furnace, while the slag layer that collects above the matte is run off at the opposite end of the furnace on the long side of the settling hearth.

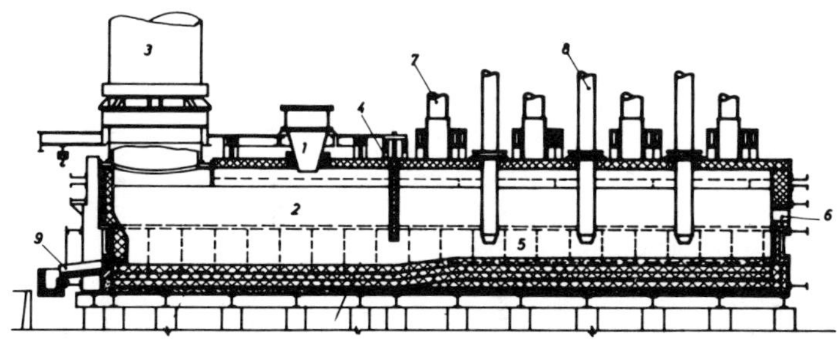

1. Smelting cyclone
2. Separating chamber
3. Cyclone waste gas
4. Partition wall
5. Settling reduction hearth
6. Waste gas from the hearth
7. Feed of reductant
8. Electrical resistance heating
9. Tap hole for the matte

Figure 1.13. KIVCET process electric copper matte smelting furnace. *Source: J. Metall.*, Vol. 28, No. 7, 1976, p. 7.

Table 1.12. Examples of Complex Copper Concentrates That Can Be Smelted by the KIVCET Process

Type Concentrate	1 Cu(Ni)	2 Ni(Cu)	3 (Ni)(Cu)	4 CuZn(Pb)	5 CuZn(Pb)	6 (Cu)(Zn)(Pb)	7 ZnCu(Pb)
Concentrate analysis (%)							
Cu	26.45	3.82	2.1	25.6	14.26	8.2	8–10
Ni	1.51	5.5	4.2	—	—	—	—
Co	—	0.14	0.13	—	—	—	—
Zn	—	—	—	10.00	8.75	7.0	20–23
Pb	—	—	—	1.7	2.46	1.6	32–43
Fe	28.8	32.6	23.9	24.0	27.6	30.6	19–22
S	30.5	21.4	15.2	33.0	34.0	42.5	32–36
Copper matte analysis (%)							
Cu	45–50	—	8.32	50	45	35	3.5
Ni	2.5–3.0	45	16.9	—	—	—	—
Co	—	—	0.4	—	—	—	—
Zn	—	—	—	2.5	2.3	3.5	—
Pb	—	—	—	2.0	3.0	5.0	—
Yield in the matte (%)							
Cu	98.5	97.5	96.2	99.1	98.2	96.3	94
Ni	97.0	98.8	97.8	—	—	—	—
Co	—	89.3	80.8	—	—	—	—
Zn	—	—	—	12.7	8.5	11.3	—
Pb	—	—	—	60.0	38.0	70.6	—
Metal contents of slag (%)							
Cu	0.35	—	0.07	0.35	0.35	0.35	—
Ni	0.02–0.04	—	0.09	—	—	—	—
Zn	—	—	—	3.5	3.0	3.5	—
Pb	—	—	—	0.2	0.2	0.2	—
Metal yield in the oxide/condensate (%)							
Zn	—	—	—	71.0	68.8	48.5	72–75
Pb	—	—	—	34.1	56.0	18.8	—

Source: G. Melcher, E. Müller and H. Weigel, The KIVCET Cyclone Smelting Process for Impure Copper Concentrates, *J. Metall.*, Vol. 28, No. 7, 1976, p. 8.

Table 1.13. Operating Results of a KIVCET Smelter with a Throughput of 300 to 350 Tonnes/Day of Cu–Zn Concentrates[a]

	Rich	Poor
Amount (t)	100	100
Analysis (%)		
Cu	25.6	8.2
Zn	10	7
Pb	1.7	1.6
Fe	24	30.6
S	33	42.5
SiO_2	1.4	1.5
CaO	0.15	
Quartz sand (t/100 t conc.)	14–16 (94–96% SiO_2)	24.4 (95% SiO_2)
Limestone (t/100 t conc.)	8–9 (50% CaO)	9.1 (60% CaO)
Copper matte (t/100 t conc.)	50.8	22.5
Analysis (%)		
Cu	50	35
Zn	2.5	3.5
Pb	2.0	5.0
Fe	20	31
S	22	18
Production (%)		
Cu	99.1	96.3
Zn	12.7	11.3
Pb	60.0	70.6
Zinc oxide (t/100 t conc.)	9.9	4.8
Analysis (%)		
Zn	71.7	70.8
Pb	5.8	6.3
Production (%)		
Zn	71.0	48.5
Pb	34.1	18.8
Slag (t/100 t conc.)	45	79
Analysis (%)		
Cu	0.35	0.35
Zn	3.5	3.5
Pb	0.2	0.2
Fe	30.5	30.5
SiO_2	38	31.5
CaO	9	7
Production (%)		
Cu	0.8	3.4
Zn	15.8	39.6
Pb	5.3	19.0
Oxygen (92–95% O_2) (m^3 at NTP/t conc.)	210	300
Electrode graphite (t/100 conc.)	0.4	0.4
Screened coke (t/100 t conc.)	3.5	3.2
Gas (80% SO_2) (m^3 at NTP/t conc.)	180	270
Electric power (kW-hr/t conc.)	300	300

Source: G. Melcher, E. Müller and H. Weigel, the KIVCET Cyclone Smelting Process for Impure Copper Concentrates, *J. Metall.*, Vol. 28, No. 7, p. 8.

[a]Cu–Zn flotation concentrate; t = tonnes; conc. = concentrate.

Autogenous smelting can be maintained if the sulfide sulfur in the chalcopyrite concentrate is 25%; if it is below this, fuel must be added. The furnace temperature is between 2462° and 2552°F (1350° to 1400°C), with matte 50% copper at 99.1% recovery being produced from rich 25.6% copper concentrate and matte 35% copper at 96.3% recovery from low-grade 8.2% concentrate. The copper content of the slag remains consistent at a quite low figure of 0.35%, regardless of whether the concentrate feed is rich or lean.

Flash Smelting is accomplished by feeding fine dried concentrates along with a stream of oxygen or preheated air into a furnace of special design, where the concentrate roasts and smelts itself in a single autogenous operation. The sulfide concentrate is flash smelted by burning some of its sulfur content while suspended in the oxidizing atmosphere of the oxygen, or heated air, injected with it. The liquid droplets of smelted charge then fall to the bottom of the furnace and separate out into the two layers of matte and slag.

There are two quite different designs of furnaces, one with a tall vertical reaction chamber attached to a long, low settler, while the other type is shaped like a rectangular reverberatory furnace, except for a central flue for escaping gases, as represented by the Finnish Outokumpu Oy and Canadian International Nickel Company furnaces.

The First Shaft Type of Furnace had its beginning at Outokumpu in Finland and is now operating or under construction in 10 countries at 18 different sites. The Japanese Taunano installation is one of the more modern and smelts 500 tons per day of dried concentrate, after screening out particles of over 0.2 inch (5 mm) in diameter, which are crushed and recycled. The hot, dried concentrate is injected into the top of a reaction shaft 10.3 feet in diameter and 36 feet high (3.14 × 11 m), along with preheated air. The air is heated by an oil burner and by hot exhaust gases from the furnace settler and enters the furnace at 780°F (416°C). In some installations the air is enriched to contain 35% oxygen, in which case the air preheating is reduced to 390°F (200°C).

Enough of the sulfur in the feed particles oxidizes to provide sufficient exothermic heat to melt the fine concentrate, which arrives at the bottom of the reaction shaft as a liquid

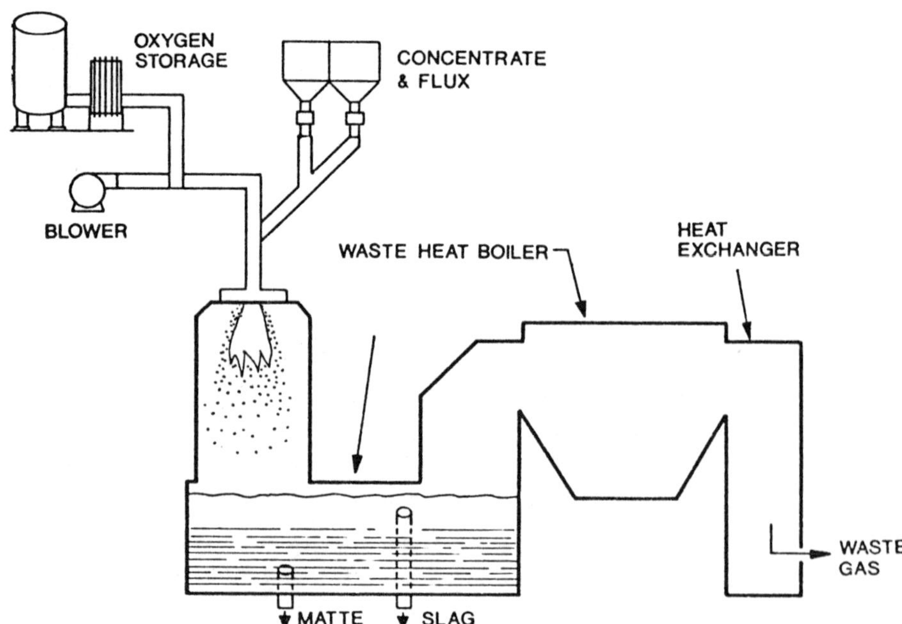

Figure 1.14. Outokumpu flash smelting furnace. *Source:* Courtesy of Air Products and Chemicals, Inc., Allentown, Pennsylvania.

droplet:

$$4CuFeS_2 + 5O_2 = 2Cu_2S \cdot FeS + 4SO_2 + 2FeO$$

The fine rain of molten drops gathers in the settler attached at right angles to the bottom of the reaction shaft and here separates into layers of matte and slag. The settler is 40.5 feet long, 12.5 feet wide, and 7 feet high ($12.35 \times 3.81 \times 2.13$ m).

Matte and slag are tapped from the settler. The matte, 48% copper, goes on to the converters, while the slag, too high in values to be discarded, goes to an electric furnace process to be fluxed, slagged, and cleaned. The final discard slag from the electric furnace is 0.47% copper, and the recovered matte is added to the matte from the flash smelting settler. Converter slag of 4% copper content is not returned to the flash smelting settler for cleaning but is treated in a separate electric furnace operation to recover the contained values.

Magnetite in the flash smelting furnace is a problem and has a bearing on the large amount of copper carried out in the settler slag. The flash furnace slag is 15% magnetite, and the retreated electric furnace slag is 13%.

Gases leaving the furnace through an uptake 5 feet wide by 5 feet high (1.52×1.52 m) at the opposite end of the settler to the smelting chamber are at a temperature of 2350°F (1290°C) and carry large amounts of sensible heat, which is first recovered in a waste heat boiler driving a steam turbine, then in a recuperator to preheat the air going to the furnace reaction shaft burner. Approximately 9% of the furnace charge is also carried off in the exhaust gas as flue dust, and a system of cyclone and electrostatic precipitators recovers this and returns it to the concentrate drier before the smelting chamber.

Comparative temperatures in the process are 2120°F (1160°C) for the settler matte and 2300°F (1260°C) for the settler slag.

The Reverberatory Type of Flash Smelting Furnace was developed by the International Nickel Company. Smelting 1500 tons per day, this furnace has sets of horizontal injection charging burners at both ends, and a stream of oxygen is used as the oxidizing gas. Fine, dried concentrate is injected with a stream of oxygen at one end of the furnace, it flash smelts in suspension, and drops to the hearth to separate into layers of matte and slag.

To clean the slag produced, a small quantity of pyrrhotite (practically barren iron sulfide) along with silica and oxygen is injected and flash smelted at the opposite end of the furnace. This keeps the temperature high at this end of the furnace, while the rain of molten iron sulfide droplets showers down on the slag and washes any trapped drops of matte through it into the matte layer below.

The iron sulfide also reacts with any copper oxide which may be present, forming Cu_2S which combines with matte and FeO that dissolves in the slag:

$$Cu_2O + FeS = Cu_2S + FeO$$

The furnace gases produced from both flash smelting processes are very rich in SO_2 and are excellent sources for sulfuric acid plant treatment; or where extremely high grade (80% SO_2), they can be condensed directly and marketed as liquid SO_2.

Continuous Matte Smelting and Converting Furnaces is a new development of the Mitsubishi Metal Corporation and is known as the Mitsubishi Process. This is the largest development in continuous copper smelting, and two of these three-furnace smelters are in existence, the first built in Japan and producing 18,000 tons of copper per year, and the second at Texasgulf, Inc., in Canada, producing 130,000 tons per year.

The process delivers a very high smelting rate on unroasted concentrate feed and combines the advantages of a reverberatory furnace using unroasted feed and a flash furnace

Table 1.14. Analysis of Charge and Products—Copper Flash Smelting Furnace (Outokumpu Type), Hindustan Copper Ltd.

Materials	%Cu	%Fe	%S	%Ni	$\%SiO_2$	$\%Al_2O_3$	%CaO	%MgO	$\%H_2O$
Concentrate (dry basis)	22.20	29.00	27.87	0.49	14.72	0.53	0.25	0.87	—
Wet concentrate moisture	—	—	—	—	—	—	—	—	8.36
Dry concentrate moisture	—	—	—	—	—	—	—	—	0.146
Flash furnace matte	50.5	22.5	22.30	1.1	—	—	—	—	—
Flash furnace slag	1.65	44.02	—	0.35	27.1	6.8	0.51	1.23	—
Flash furnace dust	16.32	20.50	9.40	0.50	23.0	8.9	0.67	1.54	16.6
Flash furnace ground secondaries	34.2	24.2	9.57	—	12.30	—	—	—	—
Slag-cleaning furnace matte	53.7	17.5	22.38	1.65	—	—	—	—	—
Dump slag	0.78	44.6	1.41	0.15	29.7	7.82	0.61	0.52	—
Converter slag	4.28	45.5	—	—	28.65	—	—	—	—
Blister slag	45.5	—	—	3.48	—	—	—	—	—
Blister copper	98.26	—	—	0.50	—	—	—	—	—
Flash furnace gas analysis at I.D. fan delivery	$\%SO_2$	$\%CO_2$	%CO	$\%SO_3$	$\%O_2$	$\%H_2O$	$\%N_2$		
	5.2	4.2	0.10	0.01	8.9	5.1	Balance		

Source: J. C. Yannopoulos and J. C. Agarwal, Eds., *Extractive Metallurgy of Copper*, Vol. 1, The Metallurgical Society of AIME, 1976, p. 211.

**Table 1.15.   Flash Furnace Performance
(Typical for 1 Month), Hindustan Copper Ltd.**

Concentrate treated	4972 tonnes
Milled secondaries treated	475 tonnes
Matte produced	2463 tonnes
Matte temperature	1170°C
Slag produced	2188 tonnes
Slag temperature	1260°C

Source: J. C. Yannopoulos and J. C. Agarwal, Eds., *Extractive Metallurgy of Copper*, Vol. 1, The Metallurgical Society of AIME, 1976, p. 210.

**Table 1.16.   Flash Smelting in Outokumpu Type Furnace
(Typical Assays in Harjavalta Copper Smelter)**

	%Cu	%Ni	%Fe	%S	%SiO$_2$
Concentrate feed	21.9	0.11	30.3	32.0	8.6
Typical flash furnace matte	64.1	0.85	10.6	21.5	—
Flash furnace high-grade matte	78.5	0.75	0.9	19.0	—
Typical flash furnace slag	1.5	0.05	44.4	1.6	26.6
Concentrate from slag flotation	20.8	0.54	29.9	10.8	15.6
Tailing from slag flotation	0.3	0.10	44.0	0.4	—

Source: J. C. Yannopoulos and J. C. Agarwal, Eds., *Extractive Metallurgy of Copper*, Vol. 1, The Metallurgical Society of AIME, 1976, p. 493.

using oxygen-enriched air. This gives a smelting rate about six times greater than that of a reverberatory furnace and two to four times greater than a flash smelting furnace, per unit volume of smelting furnace.

Comparative Smelting Rates

	(tonnes/day/m^3 of furnace volume)
Mitsubishi process	4.7
Reverberatory, unroasted calcines	0.8
Reverberatory, roasted calcines	1.0
Flash smelting	2.0

Fuel consumption is also decreased, with the Mitsubishi process using 1.5 to 2.0 million Btus per ton of concentrate compared to reverberatory fuel consumption ranging between 4.5 and 6.5 million Btus.

There are three separate metallurgical stages, each of which takes place in adjacent stationary hearth furnaces. In the first stage, matte smelting is carried out in the largest of the three furnaces. Unroasted copper concentrate, fluxes, and air enriched to 25% oxygen and revert converter slag are charged into the smelting furnace through lances installed vertically down through the furnace roof and extending to just above the bath surface. This feeding through lances encourages a high smelting rate, simplifies furnace maintenance and design, and allows flexibility in charging.

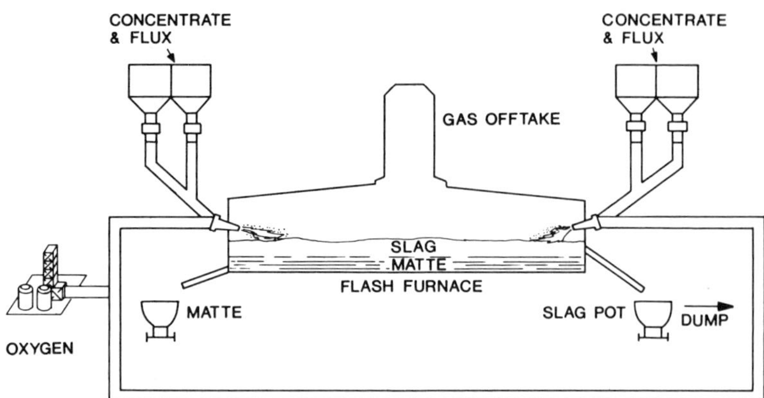

Figure 1.15. International Nickel Company flash smelting furnace. *Source:* Courtesy of Air Products and Chemicals, Inc., Allentown, Pennsylvania.

The matte and slag produced in the smelting furnace flows on to a slag cleaning furnace which is electrically heated and acts as a settler both to clean the slag and separate the matte and slag into layers. Pyrite and coke are added to facilitate matte droplets settling from the slag layer into the matte. Slag analyzing 0.4 to 0.5% copper is run off continuously to be granulated in water and then discarded, while matte of about 60% copper content flows on to the converting furnace. Operation of the smelting furnace with a very thin slag layer is claimed to be a major factor in achieving a low copper value in the discarded slag.

The converting furnace is the last in the series, and it too is equipped with roof lances to introduce air and CaO flux to the bath. The air is oxygen enriched to 22 to 25% to increase the conversion rate of matte to blister copper, which analyzes 98 to 99% copper and is continuously tapped from the furnace. The slag contains 7 to 15% copper and must be recycled back to the matte smelting furnace. Before being charged, the converter slag is cooled and crushed. This slag is a break from the traditional SiO_2–FeO–Fe_2O_3 copper smelting slag system in that CaO is added as flux instead of the usual SiO_2, and a CaO–FeO–Fe_2O_3 slag is produced instead.

The Mitsubishi process is very highly automated with the process basically under computer control, and an on-line feed-forward computer controls the inputs of air and flux

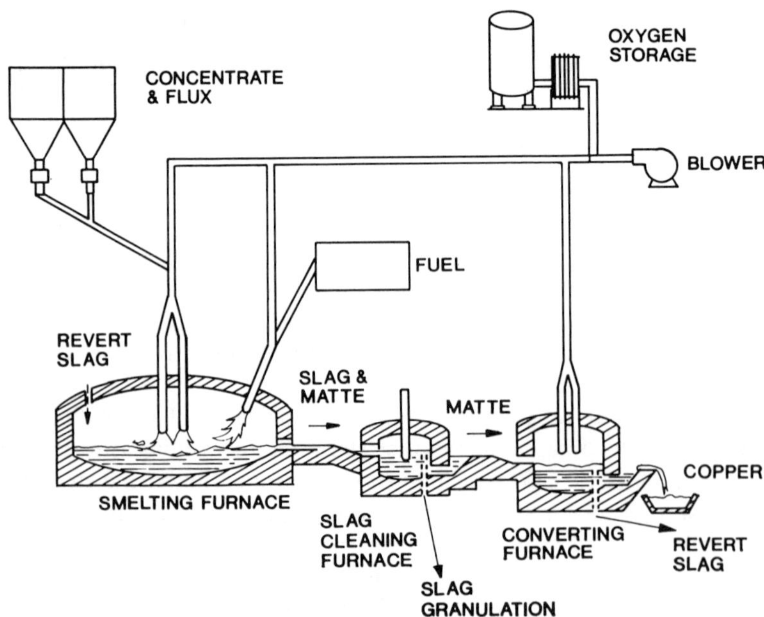

Figure 1.16. Mitsubishi process for continuous copper matte smelting and converting. *Source:* Courtesy of Air Products and Chemicals, Inc., Allentown, Pennsylvania.

to the converting furnace, and the feed-forward control, now based on accurate information concerning input amounts and compositions, simplifies the difficulties of measuring the flow volumes, temperature, and composition of the matte and slag.

The feed-back control, based on an analysis of the products formed (matte, discard slag, and converter slag) and their measured temperatures, is used once each hour by the computer to correct inaccuracies in the feed-forward control.

2a. Nickel Sulfide Roasted Calcines are smelted to matte in reverberatory furnaces, oxygen-enriched reverberatory furnaces, flash smelting furnaces, and electric furnaces, in a smelting process very similar in all respects to the matte smelting of copper sulfide calcines. Nickel and iron sulfides, Ni_3S_2 and FeS, combine to form a matte, and any precious metals present also combine with the matte.

The iron oxide, which is a product of the prior roasting operation, will combine with the silica, in the gangue and also added as flux, to form a slag in the smelting furnace. This iron silicate slag is practically immiscible in the matte, and as it has a specific gravity of 3 compared to 5 for the matte, the matte and slag readily separate into two layers, the slag floating above the matte.

Reverberatory Furnaces are similar in size and design to those used in copper matte smelting and are used by the International Nickel Company. A furnace 110 feet long by 30 feet wide (33.5 × 9.15 m) will smelt 1500 tons of solid charge per day, in addition to 700 tons of molten slag returned from the converters for cleaning and poured in at the firing end of the furnace.

The calcine charge from the roasters is fed through the furnace roof and drops along both sides of the furnace interior, where it is melted by the heat from the furnace burners and flows down into the center to separate into layers of matte and slag.

Combustion at the burner end of the furnace generates temperatures in the 3000°F (1650°C) range, and this drops to 2400°F (1320°C) by the time the hot gases have traveled the length of the furnace and are about to escape up the flue. The flue gases carry large amounts of sensible heat and considerable flue dust, both of which are recovered. The sensible heat is recovered in waste heat boilers and preheaters for the furnace combustion air, while dust collectors salvage the flue dust and return it to the furnace.

The settled matte layer is tapped periodically from the side of the furnace, and the long settling length of over 100 feet (30.5 m) lets the matte have time to settle through the slag before this slag is run off at the flue end of the furnace. The slag is collected in slag pots and carried either to the slag dump or to a plant to recover the iron for high-grade pellets used in steel making.

Very little oxidized nickel is gathered up in the slag and lost, due to its reduction back to Ni_3S_2 by FeS in the charge, and the nickel sulfide formed then being gathered into the matte:

$$9NiO + 7FeS = 3Ni_3S_2 + 7FeO + SO_2$$

while the resulting FeO makes up part of the slag.

A relatively low-grade matte is produced, 13% combined nickel with copper when copper is also present, with 0.3% combined nickel and copper contained in the furnace slag. These slags also contain magnetite, which is undesirable as there is a tendency for the magnetite to build up in a layer on the bottom of the furnace and to decrease appreciably the operating volume as well as interfere with tapping. Steam poling, by running steam pipes down each side of the furnace into the both, helps to agitate the bath and promote the reactions between oxides and sulfides to lower the magnetite content:

$$3Fe_3O_4 + FeS = 10FeO + SO_2$$

Table 1.17.   Furnace Dimensions—Mitsubishi Continuous Copper Smelting, Mitsubishi Metal Corporation

Smelting Furnace

Furnace shape	Oval
Size (inside brickwork)	
Short diameter (m)	7.0
Long diameter (m)	10.25
Number of lances	10
Number of burners	4
Depth of bath (mm)	800

Slag Cleaning Furnace

Furnace shape	Oval
Size (inside brickwork)	
Short diameter (m)	4.0
Long diameter (m)	7.0
Hearth area (m^3)	25.0
Transformer capacity (kVA)	1200
Number of electrodes (prebaked graphite)	3
Diameter of electrode (mm)	350
Thickness of slag layer (mm)	600
Power requirement (kW-hr/mt slag)	31

Converting Furnace

Furnace shape	Oval
Size (inside brickwork)	
Short diameter (m)	6.0
Long diameter (m)	9.0
Number of lances	8
Number of burners	1
Depth of bath (mm)	800

Source: J. C. Yannopoulos and J. C. Agarwal, Eds., *Extractive Metallurgy of Copper*, Vol. 1, The Metallurgical Society of AIME, 1976, p. 447, 448, 449.

The addition of scrap iron, ferrosilicon, or green poles is also used to reduce this Fe_3O_4 to FeO, which can then be combined in the slag with SiO_2 and removed from the furnace without excessive metal losses.

Oxygen Reverberatory Furnaces are similar in design to the conventional reverberatory furnace, with the exception that part of the combustion air is replaced by oxygen. This permits the combustion of a larger amount of fuel, at an increased fuel efficiency, producing more heat to smelt more solid charge and increasing the smelting throughput rate of tonnage treated. Twenty-five percent oxygen enrichment of combustion air has increased furnace capacities up to 30%.

About 45% of the gross fuel value of an air-fuel fired reverberatory furnace leaves the furnace as sensible heat in the unreacting nitrogen portion of the air. The amount of heat carried out of the furnace by a ton of nitrogen is roughly the same amount of heat required to smelt a ton of solid charge. Consequently enriching the combustion air with oxygen and replacing part of the nitrogen makes this much more heat available for smelting the solid charge.

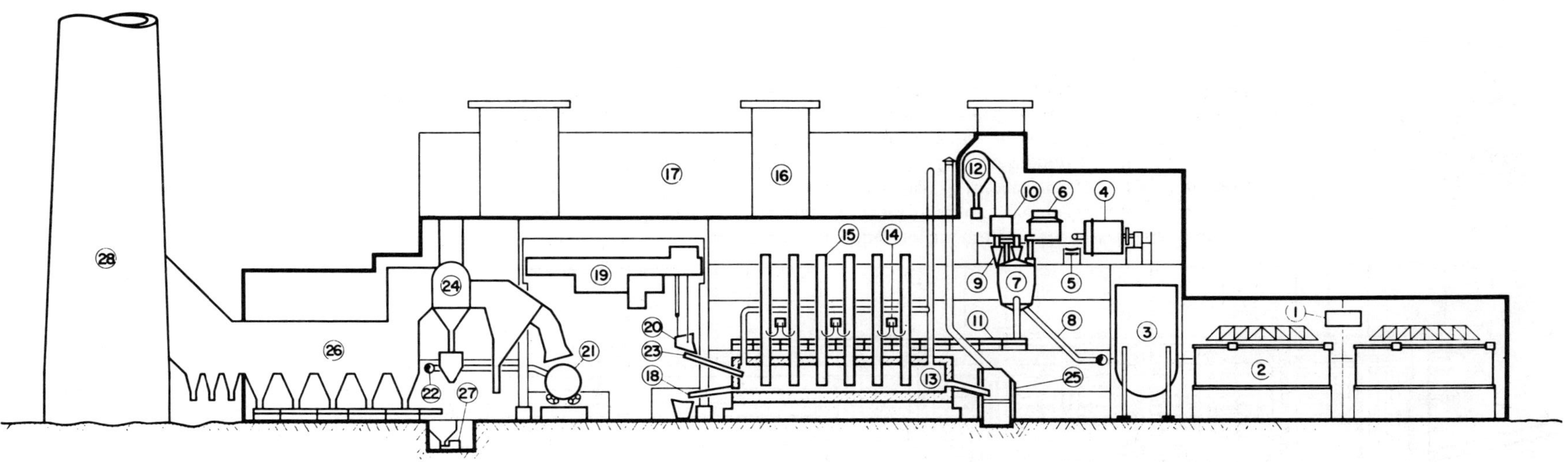

Figure 1.17. Section through electric furnace nickel matte smelter: **1,** distributor; **2,** tray thickener 60 feet diam.; **3,** pulp storage 27 feet × 49 feet; **4,** drum filter 14 feet × 16 feet; **5,** conveyor and weightometer; **6,** 50 ton rotary bin 14 feet × 17 feet; **7,** fluid-bed roaster nominal 14 feet × 16 feet; **8,** 4 P.S.I. air; **9,** cyclones 4 feet diam.; **10,** waste heat boilers; **11,** drag conveyor 36 feet × 20 feet × 135 feet; **12,** roaster flue; **13,** electric furnace 90 feet × 22 feet; **14,** power supply; **15,** electrode 4 feet diam. × 50 feet; **16,** ventilator; **17,** cross flue; **18,** matte tapping launder; **19,** 60 ton crane; **20,** ladel; **21,** Bessemer converter 13 feet 6 inches × 35 feet; **22,** 15 P.S.I. air; **23,** slag return launder; **24,** converter flue; **25,** slag granulator; **26,** Cottrell; **27,** dust pump; **28,** 500 feet stack. *Source:* J. B. McConnell, I. P. Klassen, and S. A. Berkovich, Roasting and smelting practice at the Thompson plant of the International Nickel Company, Presented at the CIMM Annual Meeting, 1963.

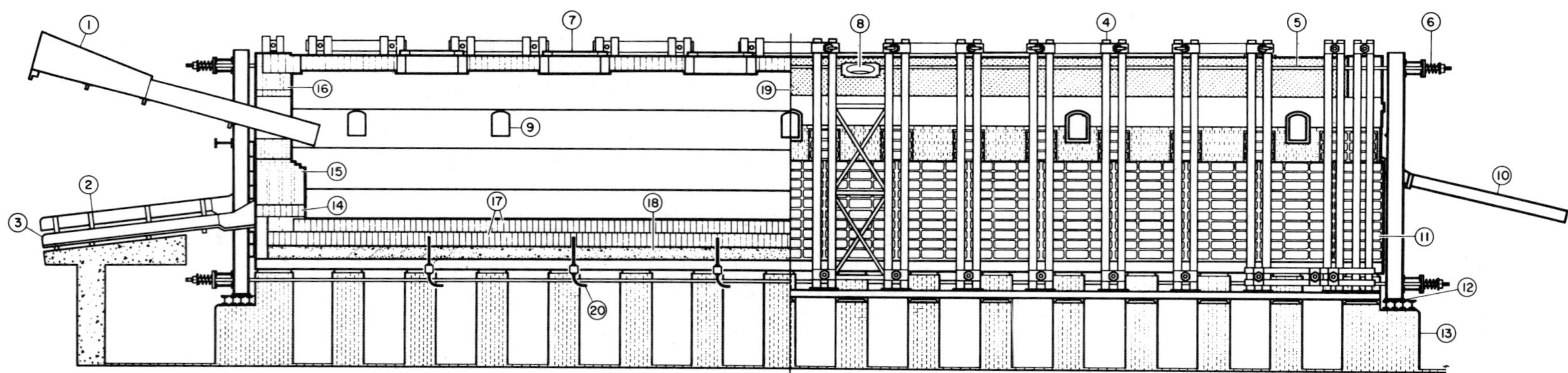

Figure 1.18. Longitudinal section of nickel matte smelting electric furnace: **1**, converter slag launder; **2**, matte launder; **3**, drain out launder; **4**, buckstay, 8 inches I-beam 21 feet high; **5**, tie rod 2 inches diam.; **6**, tie rod spring $15\frac{1}{4}$ inches long; **7**, electrode opening; **8**, furnace vent; **9**, inspection door; **10**, slag launder; **11**, cast iron support plate; **12**, slip joint; **13**, furnace support pier 9 feet 5 inches high; **14**, tapping blocks; **15**, magnesite refractories; **16**, firebrick; **17**, magnesite bottom arches; **18**, magnesite ramming mass; **19**, roof firebrick; **20**, thermocouple. *Source:* J. B. McConnell, I. P. Klassen, and S. A. Berkovich, Roasting and smelting practice at the Thompson plant of the International Nickel Company, Presented at the CIMM Annual Meeting, 1963.

Electric Furnaces, such as those installed by the International Nickel Company at its Thompson Smelter, are quite similar to reverberatory furnaces for matte smelting of both copper and nickel sulfide calcines. The furnaces are classed as resistance furnaces of the submerged-arc type. As such, the slag acts as the heating element with the electrodes projecting down into the slag layer, and heat is generated by the resistance to current flowing between electrodes. Heat is transferred from the slag to the floating calcines and silica flux charged through the furnace roof by direct contact and by convection; and once melted, the matte compounds, being heavier, will sink to the bottom of the crucible, leaving the slag in the upper section. Both matte and slag layers will each be approximately 25 inches deep (0.625 m), in a furnace 98 feet long, 28 feet wide, and 13 feet high (29.88 $\times$ 8.54 $\times$ 3.96 m).

The furnace design is to have a complete magnesite crucible, with a firebrick sprung-arch roof. Six 4 foot diameter (1.22 m) Söderberg electrodes, 50 feet long (15.24 m) and each weighing 20 tons, are paired to constitute one phase of a three-phase circuit. The electrodes are spaced down the furnace length at 12 foot centers (3.66 m) between phase pairs and at $12\frac{1}{3}$ foot centers (3.76 m) between phases. The electrodes are held, slipped, and regulated by hydraulically operated spring-mounted clamps and jacks.

Three 18 MVA single-phase transformers, one for each pair of electrodes, provides the power for smelting, and the matte is tapped at a temperature of 2040°F (1115°C), while the slag is tapped at 2350°F (1290°C). Daily smelting capacity of a furnace will be approximately 850 tons of calcines, plus 450 tons of returned converter slag for cleaning. The grade of matte produced will be 16% nickel and 1% copper, and the slag losses 0.25% combined nickel, copper, and cobalt. Matte and slag are tapped at opposite ends of the furnace, with the matte transferred to the converters and the slag going to the dump.

The main smelting reaction is the reduction of magnetite produced during roasting to ferrous oxide by the unroasted iron sulfide:

$$3Fe_3O_4 + FeS = 10FeO + SO_2$$

and the reaction of the FeO produced with silica in the gangue and flux to form an iron silicate slag. However all the magnetite is not reduced to FeO, and some Fe_3O_4 is dissolved by both matte and slag. Magnetite also has a tendency to form a highly viscous intermediate layer between the matte and slag, which is a deterrent to the droplets of matte trying to settle out of the slag layer into the matte layer.

A circulating flow of liquid charge upward at the hot zone beside the electrodes, laterally close to the surface over to the furnace sidewalls, then down and back toward the electrodes through the lower part of the bath, is of some aid in dispersing this magnetite layer. Magnetite can also be decreased by reducing it to FeO to be slagged, and this can be accomplished by charging coal and silica, lowering the voltage, and lowering the electrodes down close to the matte layer.

As there are no combustion gases from the electric furnace, the amount of exhaust gases is relatively small, and this simplifies the recovery of sensible heat and flue dust from them.

Blast Furnace Matte Smelting had remained a popular smelting method for nickel sulfide ore and nickel–copper sulfide concentrates until the mid-1970s, when the International Nickel Company closed its Coniston smelter which used blast furnaces, and Falconbridge Nickel Mines, Ltd., the world's second largest nickel producer, rebuilt its smelter to replace the sintering plant and blast furnaces with a fluid-bed roasting facility and two electric matte smelting furnaces. This revised smelting program was put into partial operation in 1978 and was scheduled to reach full production by 1980.

The changes in design in the Falconbridge smelter will give a significant reduction in emissions of SO_2 and improve working conditions in the plant, with the reduction of sulfur emissions expected to be on the order of one fifth. This will be from a present level of 500 tons per day to about 400 tons per day, with both electric furnaces in production.

These nickel blast furnaces were very similar in design and operation to those used for copper matte smelting and produced a settler matte analyzing 13.5% combined nickel and copper and a settler slag going to the slag dump analyzing 0.26% combined nickel and copper.

2b. Nickel Sulfide Unroasted Concentrates are smelted to matte and slag in flash smelting furnaces, electric furnaces, and blast furnaces. As with the comparative smelting of

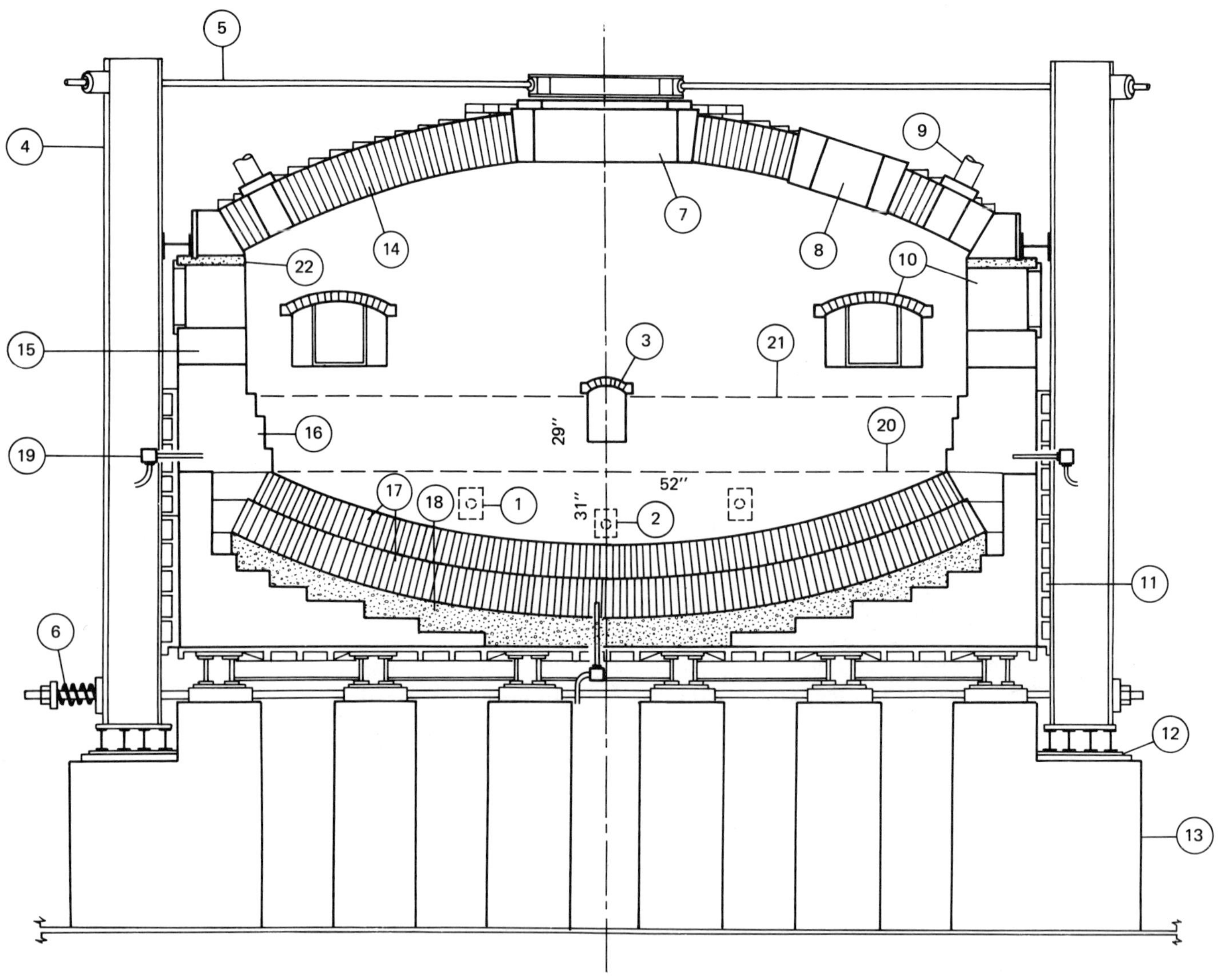

Figure 1.19. Cross section of nickel matte smelting electric furnace: **1**, matte tapping hole; **2**, drain out hole; **3**, slag skimming hole; **4**, buckstay; **5**, tie rod 2 inches diam.; **6**, tie rod spring $15\frac{1}{4}$ inches long; **7**, electrode opening; **8**, furnace vent; **9**, fettling pipe; **10**, inspection door; **11**, cast iron support plates; **12**, slip joint; **13**, furnace support pier 9 feet 5 inches high; **14**, roof firebrick (span 23 feet $8\frac{3}{4}$ inches, rise 3 feet 4 inches); **15**, wall firebrick; **16**, wall magnesite refractories; **17**, magnesite bottom arches, depth 2 feet 5 inches; **18**, magnesite ramming mass; **19**, thermocouple; **20**, average matte level; **21**, average slag level; **22**, roof expansion joint. *Source:* J. B. McConnell, I. P. Klassen, and S. A. Berkovich, Roasting and smelting practice at the Thompson plant of the International Nickel Company, Presented at the CIMM Annual Meeting, 1963.

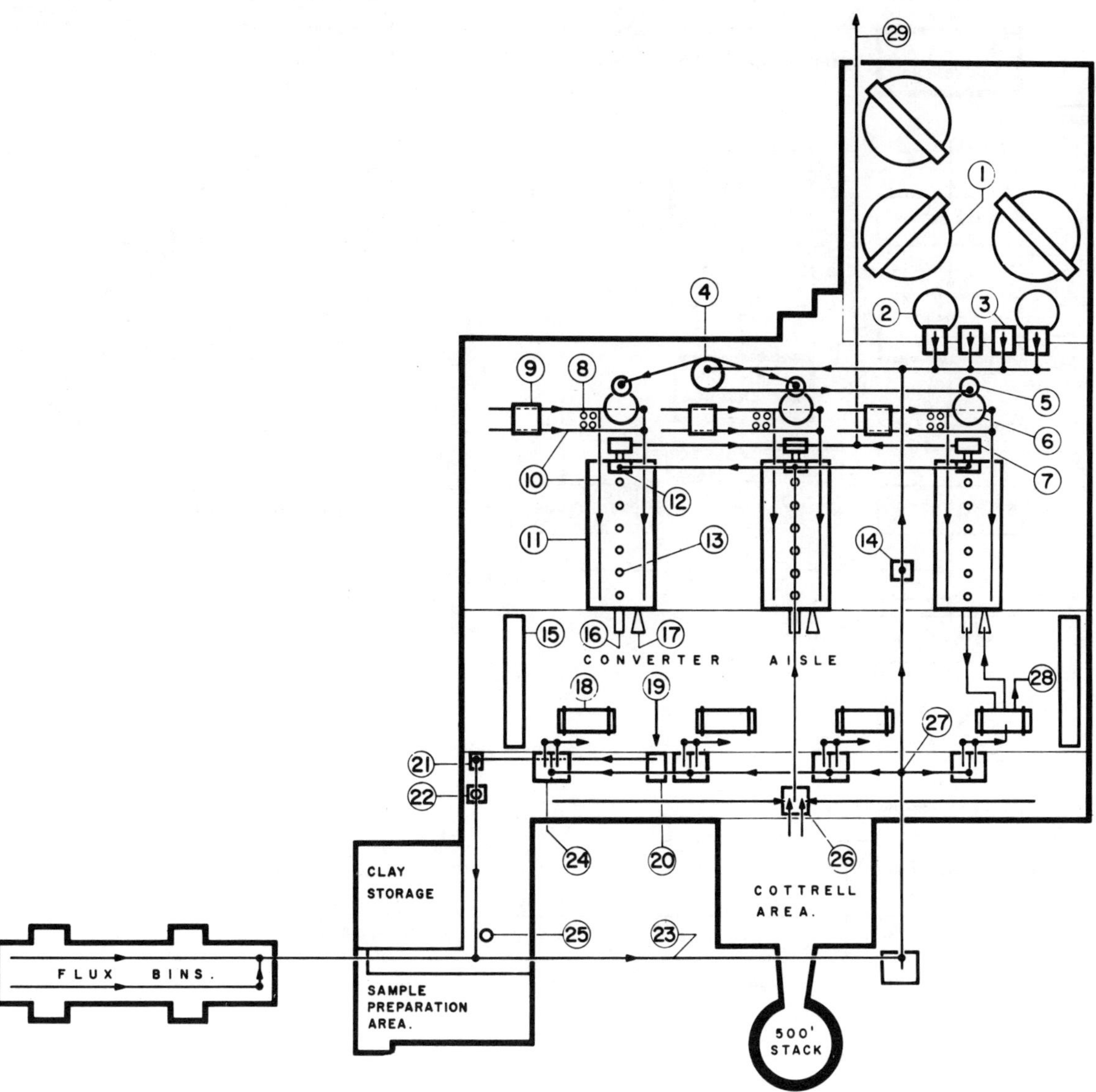

Figure 1.20. Nickel matte smelter with electric furnaces, plan view: **1,** tray thickener 60 feet diam.; **2,** pulp storage tank 27 feet × 49 feet; **3,** drum filter 14 feet × 16 feet; **4,** 100 ton rotary bin 18 feet × 21 feet; **5,** 50 ton rotary bin 14 feet × 17 feet; **6,** fluid-bed roaster nominal 14 feet × 16 feet; **7,** slag granulator; **8,** cyclone 4 feet diam.; **9,** waste heat boiler; **10,** dray conveyor 36 feet × 20 feet × 135 feet; **11,** electric furnace 90 feet × 22 feet; **12,** dust hopper; **13,** electrode 4 feet diam. × 50 feet; **14,** flux surge bin; **15,** 60 ton overhead crane; **16,** matte tapping launder; **17,** converter slag return launder; **18,** converter 13 feet 6 inches × 35 feet; **19,** smelter scraps; **20,** Traylor crusher; **21,** 100 ton storage bin; **22,** Symons cone crusher; **23,** flux and scrap conveyor; **24,** flux and scrap bin; **25,** clay mill; **26,** dust bin; **27,** 3-way chute; **28,** Bessemer matte to anode casting; **29,** granulated slag to disposal area. *Source:* J. B. McConnell, I. P. Klassen, and S. A. Berkovich, Roasting and smelting practice at the Thompson plant of the International Nickel Company, Presented at the CIMM Annual Meeting, 1963.

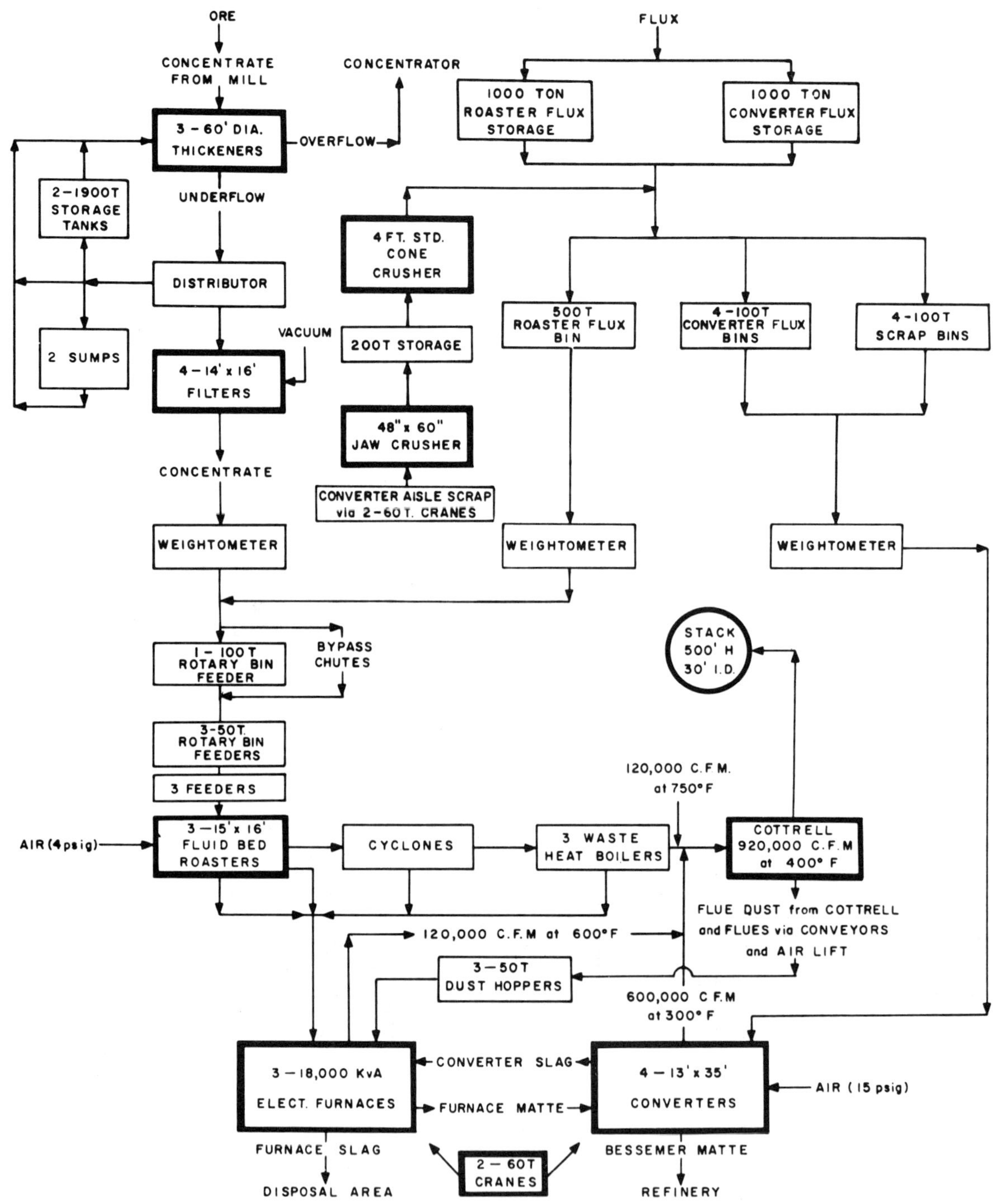

Figure 1.21. Nickel matte smelter flow sheet. *Source:* J. B. McConnell, I. P. Klassen, and S. A. Berkovich, Roasting and smelting practice at the Thompson plant of the International Nickel Company, Presented at the CIMM Annual Meeting, 1963.

roasted and unroasted copper sulfide concentrates, the smelting reactions and combinations are the same when roasted and unroasted nickel sulfides are smelted. In all cases the metallic metal sulfides and precious metals combine into a matte layer, while the waste rock gangue, iron oxides, and fluxes will form a slag.

Similarly also as in the treatment for raw copper concentrates the electric furnace is used for smelting raw nickel concentrates when the primary concern is a saving in eliminating roasting and other preliminary operations and reducing processing time and material handling. Flash smelting, on the other hand, takes advantage of a high sulfur content in the feed material which can be used as fuel for the process.

Electric Furnaces for direct smelting of raw nickel sulfide ores are used successfully in a situation where the ore is rather high grade for present times, mixed pentlandite and chalcopyrite 4 to 5% nickel and 2.5 to 2.8% copper, and very finely disseminated. Direct smelting without extensive beneficiation eliminates the metal losses from this beneficiation step, and the ore is sufficiently high grade to compare with the concentrate product made from a lower-grade ore.

Submerged arc-type furnaces with six $3\frac{1}{2}$ foot diameter (1.07 m) Söderberg electrodes in a row are also used for this type of smelting. Furnace construction is conventional with a magnesite hearth and lower side walls, while the upper side walls and roof are fireclay brick. Dimensions are 71 feet long, 18 feet wide, and $10\frac{1}{4}$ feet deep (21.6 × 5.49 × 3.13 m), with a smelting capacity of 500 tons per day and electrode consumption of 8 pounds (3.6 kg) per ton of charge at 20,000 kVA total power input.

The ore is crushed to approximately $\frac{1}{2}$ inch in size (12.5 mm) before charging into the furnace, where it smelts to matte containing 12 to 18% nickel and 6 to 8% copper and slag containing 0.2% nickel and 0.1% copper. The slag and matte are tapped periodically, the slag being run off at a temperature of 2480°F (1360°C) and discarded at the slag dump. The matte is taken and converted.

Magnetite presents the same problem, and is taken care of in the same manner, as in the other electric furnace matte smelting processes.

Flash Smelting of unroasted nickel concentrates has been done by mining companies which have experience in smelting raw copper concentrates, such as Outokumpu at the

Figure 1.22. Interior of electric matte smelting furnace prior to start-up. *Source:* J. A. Persson and D. G. Treihard, *J. Metall.*, Vol. 25, No. 1, 1973, p. 37.

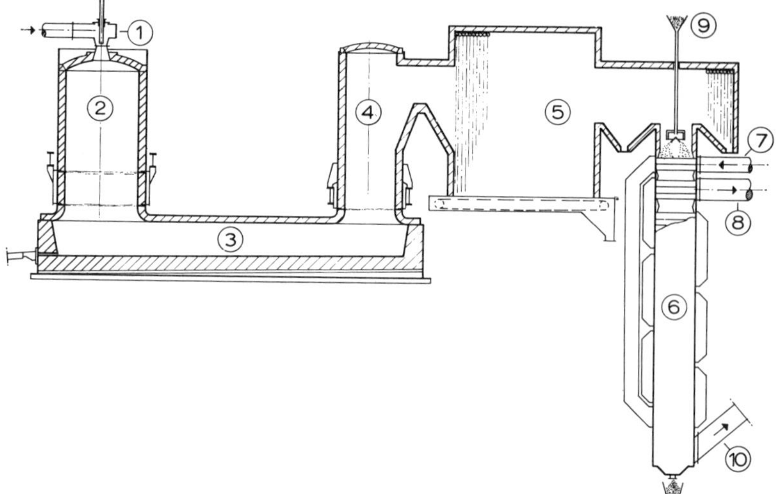

Figure 1.23. Flash smelting furnace, Outokumpu type. (1) Concentrate Burner; (2) Reaction Shaft; (3) Settler; (4) Uptake; (5) Waste Heat Radiation Boiler; (6) Heat Exchanger; (7) Cold Air Inlet; (8) Hot Air to Concentrate Burner; (9) Shot Cleaning; (10) Flue to Cottrell. *Source:* T. Toivanen and P. O. Grönqvist, *Can. Inst. Min. Metall. Bull.*, Vol. 57, No. 626, 1964, p. 653.

Harjavalta smelter, and these have copied the same general procedure used for their copper sulfides.

Fine, dried flotation concentrates of pentlandite–pyrrhotite containing about 6% nickel and 0.5% copper ore are mixed with quartz sand flux and injected with 900°F (480°C) preheated air into the top of the reaction shaft of a flash smelting furnace. The sulfides flash in the shaft, and a rain of molten particles which form matte and slag settle down on the hearth below. The matte, of approximately 35% nickel, is tapped off and sent to the converters. The slag is taken to an electric slag-cleaning furnace to recover contained matte and then goes to the slag dump. The converter slag is also cleaned in this electric furnace and is not returned to the flash furnace settler as is frequently the practice with other types of smelting furnaces.

Furnace design and construction are approximately the same as those used in flash smelting raw copper concentrates, with the reaction shaft dimensions being 12 feet in diameter and 26 feet high (3.66 × 7.93 m). Fuel oil is burned in the reaction shaft to make up any furnace heat losses and to ensure that the reaction temperature is reached to have autogenous smelting of the small feed particles in the reaction shaft.

Blast Furnace smelting is carried out at the Niihama plant of the Sumitomo Metals Mining Company, Japan's largest nonferrous metal producer. New Caledonian oxide ore is

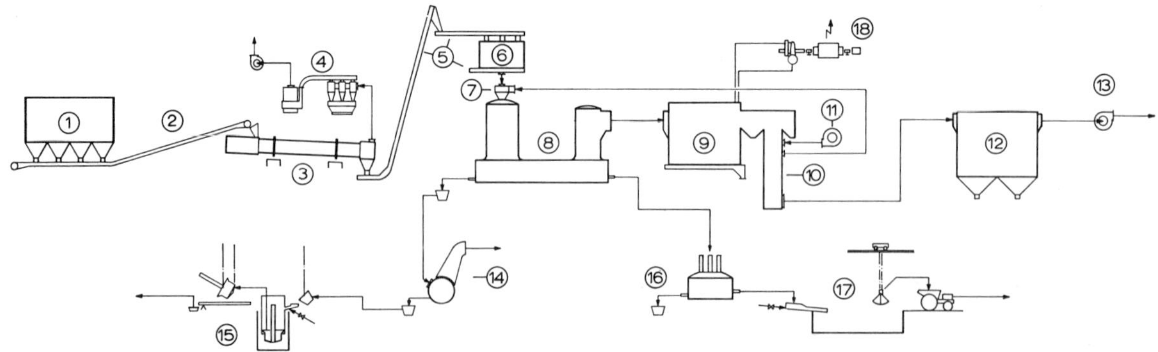

Figure 1.24. Nickel smelter using Outokumpu-type flash smelting furnace. (1) Concentrate Storage; (2) Belt Conveyor; (3) Dryer; (4) Dryer Dust Collectors; (5) Redler Conveyors; (6) Feed Hopper; (7) Concentrate Burner; (8) Flash Smelting Furnace; (9) Waste Heat Radiation Boiler; (10) Heat Exchanger; (11) Primary Air Fan; (12) Cottrell; (13) Exhausting Air Fans; (14) Converter; (15) Granulating of Ni Matte; (16) Slag Cleaning Furnace; (17) Granulating of Slag; (18) Turbogenerator. *Source:* T. Toivanen and P. O. Grönqvist, *Can. Inst. Min. Metall. Bull.*, Vol. 57, No. 626, 1964, p. 653.

mixed with Canadian sulfide concentrate in a 70 : 30 ratio, and the mixture is briquetted and charged to a blast furnace with coke and flux. Smelting takes place, and a matte is produced which contains about 27% nickel.

2c. Nickel Garnierite is a hydrated silicate of iron and magnesia in which nickel oxide replaces part of the iron oxide and the magnesia. In many cases the treatment of non-sulfide ores is more difficult than for sulfide ores, and when matte smelting of nickel sulfide ores was a process practiced from early times and had considerable processing experience accumulated, it was natural to use this experience to smelt the nonsulfide garnierite to matte by adding a sulfidizing agent gypsum.

Gypsum, hydrated calcium sulfate, $CaSO_4 \cdot 2H_2O$, was chosen as a source of sulfur because it was iron free, and in addition the calcium oxide content could serve as a useful fluxing material in the smelting operation. The type of furnace used is the low-shaft hot-blast blast furnace; and as the charge of sintered ore, coke, gypsum, and limestone moves down the furnace shaft, it is heated, reduced, and melted to form matte and slag. The gypsum (calcium sulfate) is reduced to calcium sulfide:

$$CaSO_4 + 4CO = CaS + 4CO_2$$

and this reacts with the nickel, iron, and cobalt oxides (NiO, FeO, and CoO) to form a matte of mixed nickel, iron, and cobalt sulfides and calcium oxide which goes into the slag. The slag composition is made up of a calcium–iron–magnesium–silicate mixture.

Reduction in the furnace is achieved by the burning of coke at the tuyeres which provides the CO reducing atmosphere necessary to transform the calcium sulfate to calcium sulfide. The more coke and gypsum added to the charge, the greater will be the amount of CaS formed to react with the matte-making metallic oxides and the lower the grade of matte produced. Nickel and cobalt have greater affinities for sulfur than does iron, which has the greatest affinity of the three for oxygen. Consequently iron oxide will be the last to react with CaS to form a matte constituent, and if a large amount of $CaSO_4$ is charged to the furnace, with its subsequent large reduction to CaS, this excess CaS combines with the FeO available to produce larger quantities of FeS which are absorbed in the matte and lower its grade. The proper balance of matte grade and makeup has to be adjusted for in the furnace charge, taking into consideration that FeS is needed for exo-thermic fuel in the following converting operation but that any great excess of FeS over these heating requirements increases the cost of slagging these larger amounts of iron waste material. Also to be considered in the balance of matte grade is that the higher the grade of matte, the more nickel is lost in the furnace slag by occlusion and solution.

Low-Shaft Blast Furnaces with hot-blast, in use since 1963, have partly replaced the standard water-jacketed blast furnace common until this time. Le Nickel's Doniambo Smelter has this type of furnace.

Nickel ore analyzing 3% combined nickel and cobalt in the form of NiO and CoO is dried, dehydrated, and agglomerated on sintering machines and charged to the low-shaft furnace as hot sinter at 650°F (340°C). The furnace charge also includes coke between $1\frac{1}{4}$ and 4 inches in size (3.125 to 10 cm); limestone, which is a flux for the high-magnesia slag produced; and gypsum, which is the sulfidizing agent needed to produce nickel matte. The matte grade produced will depend on the Fe–Ni ratio in the ore and on the amounts of coke and gypsum added to the charge to produce, after the reduction of $CaSO_4$ to CaS, greater or lesser amounts of FeS in the matte. Normally the matte will analyze in the range of 27% combined Ni + Co, 63% Fe, and 10% S, while the slag will be in the range of 0.3% combined Ni + Co.

The low-shaft furnace, having a smelting capacity of 800 tons of dry ore per day, is elliptical in cross section, 24 feet long, and 7 feet wide (7.32 × 2.13 m); and, as with other shaft furnace designs, the side walls taper out slightly from hearth to charge floor.

The elliptical shape was chosen rather than the more common rectangular shape so as to get the best strength and life performance from the magnesia brick which lines the hearth and part of the shaft while still maintaining a relatively small distance between the opposite rows of tuyeres. There are eight tuyeres to a side, projecting 4 inches (10 cm) into the furnace, while matte is tapped periodically from the side of the hearth. The slag is taken to the dump and the matte goes to the next stage of converting.

The shaft height is 16 feet (4.88 m) from tuyere level to charge level, and the shaft, above the lower zones where magnesia brick is needed to withstand the corrosion of the liquid high-magnesia slag, is lined with silica–alumina brick. Water running down the outside steel shell of the furnace shaft provides external cooling.

Charging is done through two double-bell feed hoppers, which effectively seal the furnace and close it to the atmosphere. The furnace flue gases leaving at 600°F (315°C) contain 20% CO; and after two-stage cleaning to remove the flue dust, this furnace gas is enriched with coke oven gas and burned in blast stoves to preheat the tuyere air blown into the furnace to 1600°F (870°C).

As there is no combined sulfur in this ore as is found in sulfide ores, there is no furnace heat provided by the exothermic oxidation of excess sulfur as would be expected to some degree in the blast furnace smelting of a sulfide ore. Consequently the coke in the charge must supply all the heat for the 2500°F (1370°C) reaction temperature required, and the addition of sensible heat to the furnace by preheating the tuyere air gives a saving in operating costs due to a large (one third) decrease in the coke used. High blast temperatures also superheat the slag, decreasing its viscosity and allowing improved matte recovery by better matte and slag phase separation in the furnace crucible.

2d. Nickel Oxide ore is smelted to matte in blast furnace operations at three smelters in the Russian Urals, at Ufaley, Rezh, and Orsk. As with the nickel garnierite, sulfur in some form must be added to the oxide ore to form a matte during smelting, and gypsum or pyrite are the sulfur-containing materials used.

The preparation of the blast furnace charge at each of the smelters is different and depends on the physical condition of the ore. At Ufaley the ore is crushed and dried to 10% moisture in rotary driers, after which it is sized in rotary trommels. The larger sizes from the trommels are smelted directly, while the smaller sizes are briquetted first and then

Figure 1.25. Charge floor, copper–nickel blast furnace. *Source:* Reproduced with the permission of Falconbridge Nickel Mines Limited.

charged to the furnaces. At Orsk the fines lack the proper plasticity for good briquetting and are agglomerated on traveling grate sintering machines. The furnace charges consist of coarse ore and briquettes or sinter, coke, limestone flux, and gypsum or pyrite.

The blast furnaces are the conventional water-jacketed type, equipped with settlers, and are similar in major respects to those used for copper matte smelting. A low-grade matte is produced to minimize furnace slag losses, and the settler slag, which is granulated before discarding, contains up to 0.2% nickel.

The matte grade produced at Rezh is lower than at the other two smelters, and it is sent to Ufaley for further treatment.

	Matte Analysis			
	%Ni	%Co	%Fe	%S
Ufaley and Orsk	17	0.7	56	18
Rezh	10	0.4	65	20

3a. Lead Sulfide Roasted Calcines are almost exclusively smelted to slag and unrefined lead metal in water-jacketed blast furnaces using coke for fuel. The prior roasting-sintering process both agglomerates and removes most of the sulfur from the mechanically mixed furnace charge of sulfide concentrates, undersized screened sinter, recycled flue dust, and smelter reverts, along with the required furnace fluxes. Air blown through the furnace tuyeres burns the coke to CO and CO_2 at a temperature of 2550°F (1440°C). This hot CO formed, along with hot solid carbon from the coke, reduces the sintered lead oxide to lead metal, which melts at this temperature and is tapped from the bottom of the furnace crucible:

$$PbO + CO = Pb + CO_2$$

$$CO_2 + C = 2CO$$

Most lead ores are found with silica or limestone, and this combines with either a CaO or SiO_2 flux and with FeO from roasting to form a liquid slag layer above the metal. Any copper, iron, cobalt, or nickel present in the furnace charge combines with sulfur or arsenic to form matte or speiss. Matte and speiss have intermediate specific gravities

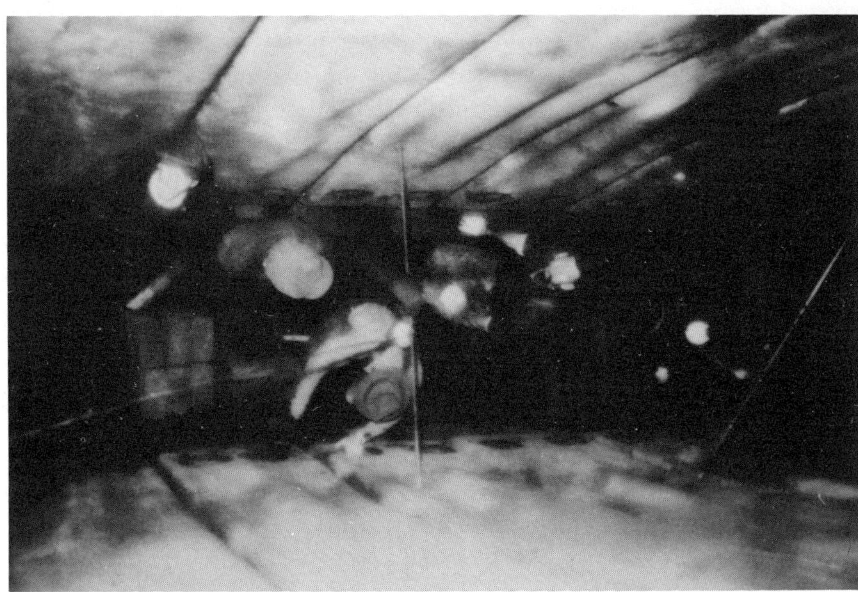

Figure 1.26. Repairing refractory hearth water-jacketed copper–nickel blast furnace. *Source:* Reproduced with the permission of Falconbridge Nickel Mines Limited.

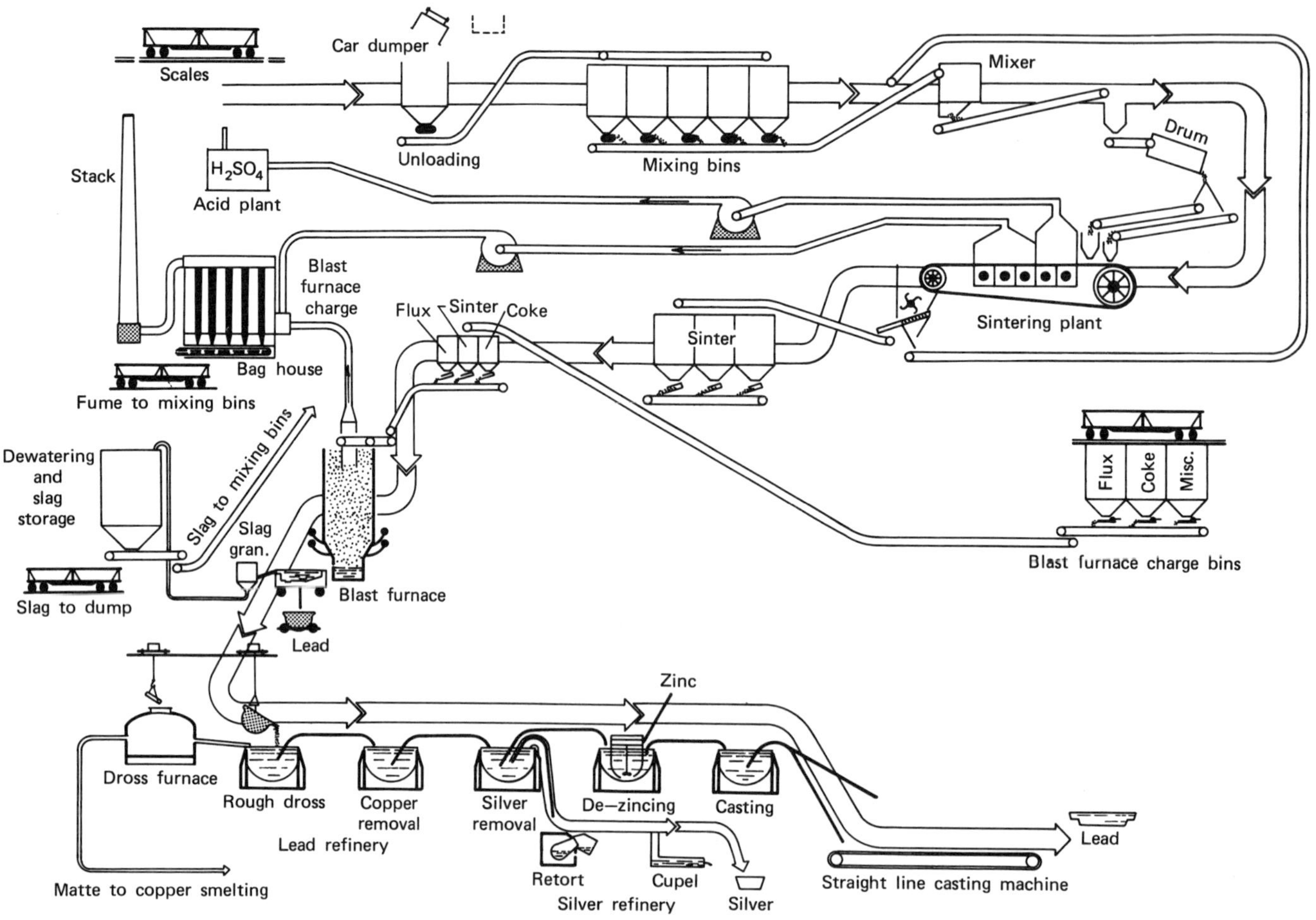

Figure 1.27. Flow sheet of Herculaneum lead smelter. *Source:* Courtesy of St. Joe Minerals Corporation.

between those for the lead and slag, (lead, 11; speiss, 6; matte, 5.2; slag, 3.6) and collect in liquid layers between the slag and lead. These are tapped from the furnace with the slag and separated from it in one or two external settlers. If there is only a small amount of matte, it is all soluble in the lead and there will be no separate matte layer formed. Any heavy precious metals are also soluble in the liquid lead layer, while zinc in the charge accumulates in the slag as ZnO.

The lead blast furnace operation shows lead recoveries on the order of 97 to 99% of the lead contained in the charge and leaves a relatively small margin for process improvement. Consequently, most recent modernization projects have mostly dealt with charging and charge preparation, roasting methods, flue dust collection, and material handling and have been very effective in increasing the tonnage smelted per unit furnace.

Blast Furnaces, because of their high percent of lead recovery (97 to 99%), have remained the standard reduction smelting method for processing roasted lead sulfide concentrates, although there are two separate designs of furnaces that are both widely used. These are the closed-top type with side doors on the charging floor through which the furnace charge is dumped, with the space above the charge floor enclosed and connected to a flue; the other is the open-top type with the furnace flue take-off below the charging floor and the whole top of the furnace able to be slid open to accommodate a large drop-bottom

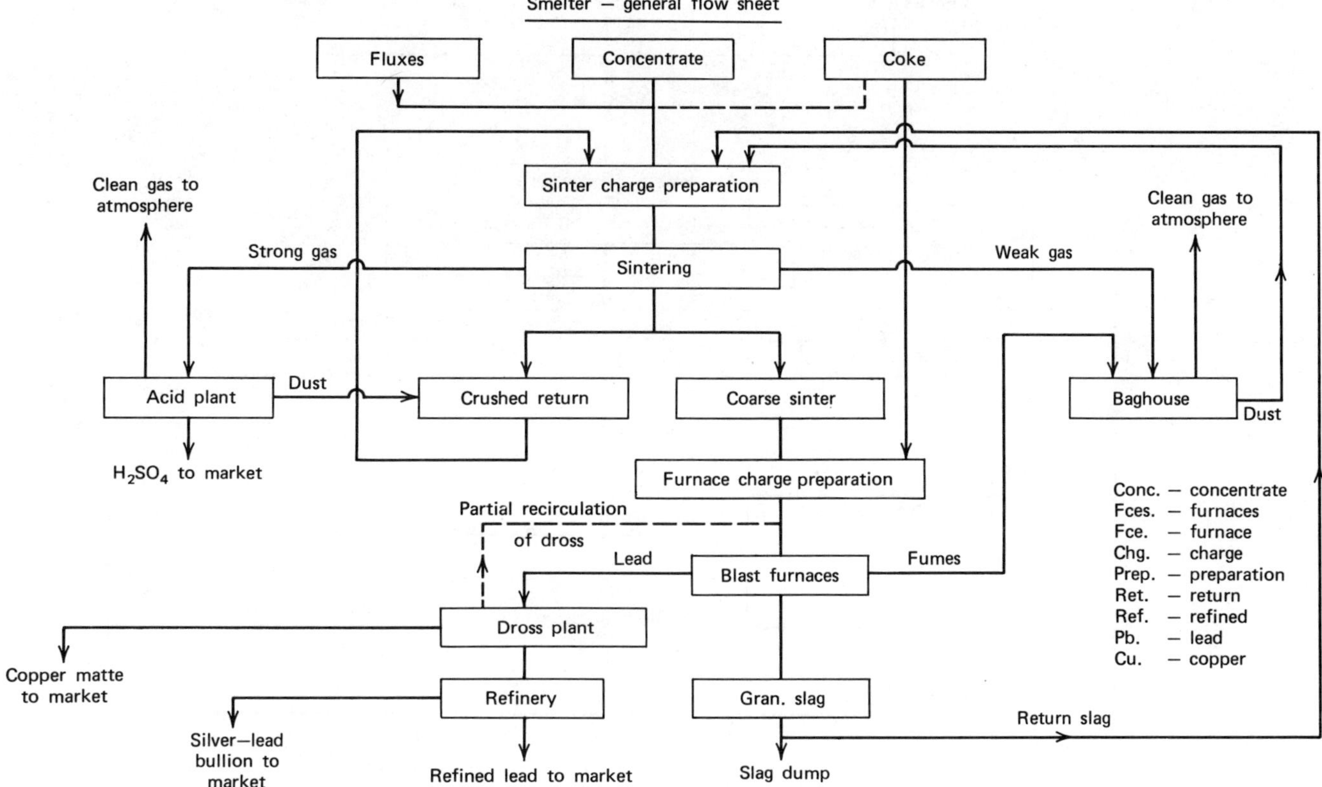

Figure 1.28. Flow sheet Buick lead smelter, AMAX Homestake. *Source:* C. H. Cotterill and J. M. Cigan, Eds., AIME World Symposium of Lead and Zinc, Vol. 2, 1970, p. 740.

charge car. The flue draft for the open-top furnace must be such that the furnace fumes are carried off even when the top is open for charging. This type of open-top charging minimizes charge segregation in the furnace shaft which has a tendency to cause crusting and channeling. The furnace interior is also more exposed to be able to bar down any accretions that build up on the interior furnace walls and cause operating irregularities.

Design improvements to increase throughput that have been tried on some furnaces but have not yet been universally adopted are continuous tapping of lead and slag and inverting the bosh of the furnace so that the bosh zone flares out from top to bottom, rather than the reverse, which is customary.

Continuous tapping reduces the amount of liquid slag and lead retained in the furnace crucible and consequently decreases the coke requirement in the charge to keep this larger amount of product molten. There is also an increase in the flexibility of furnace operations by freeing space for other charge materials which formerly was taken up by the additional coke. The reverse bosh-type furnace has vertical walls below which is an outward sloping bosh to a wide hearth. This shape tends to keep the smelting zone low in the furnace which lessens the buildup of sidewall accretions and lowers the coke consumption to give generally improved furnace conditions and higher tonnages smelted.

Sinter of 40 to 50% lead, containing roasted concentrates, recycled sinter fines, flue dust, and smelter reverts, along with flux for slag making, is screened at $1\frac{1}{2}$ inches (3.75 cm) and the oversize charged to the furnace along with coke screened at plus $\frac{3}{8}$ inch size (9.4 mm). The coke added will vary between 8 and 14% of the charge. Often a small amount of scrap iron (1%) is also charged which provides iron to decompose any remain-

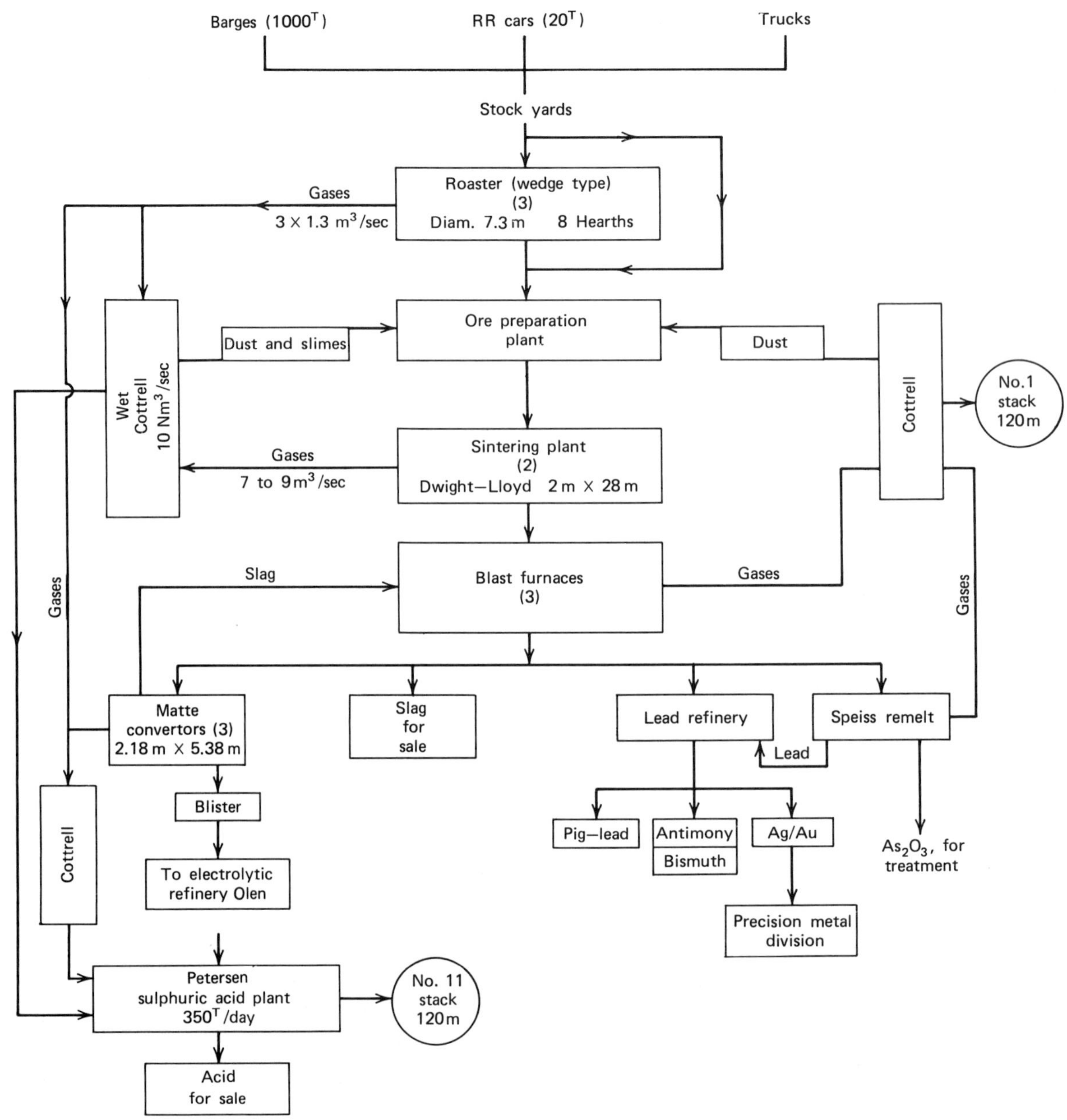

Figure 1.29. Lead smelter operations, N. V. Metallurgie Hoboken S. A. *Source:* C. H. Cotterill and J. M. Cigan, Eds., AIME World Symposium of Lead and Zinc, Vol. 2, 1970, p. 830.

ing lead sulfide as well as iron for slag making and also will combine with any crusts high in arsenic to form lower melting point speiss which can be run out of the furnace.

The reduced liquid lead that collects in the bottom of the furnace crucible and dissolves up to 5% copper in the form of matte as well as any precious metals is tapped off at 1700°F (930°C) through a side siphon tap into a lead well beside the furnace. The siphon tap serves to keep the level of lead in the crucible constant. Matte and speiss, if present, and slag are tapped from the front of the furnace through a slag tap, which is placed a little above the level of the lead in the crucible, and run into a small refractory lined settler, 9 by 4 by 2 feet (2.74 × 1.22 × 0.61 m).

Figure 1.30. Water-jacketed lead blast furnace (Hoboken). *Source:* Courtesy of Metallurgie Hoboken–Overpelt.

The speiss, matte, and slag are relatively insoluble in one another and separate into liquid layers. The speiss is an arsenide of iron, cobalt, and nickel with a specific gravity of 6, the matte a sulfide of copper, iron, cobalt, and nickel with a specific gravity of 5.2, and the slag a silicate of iron, lime, magnesia, and alumina with a specific gravity of 3.6. Any entrained lead and most of the matte and speiss are removed through tap holes in the

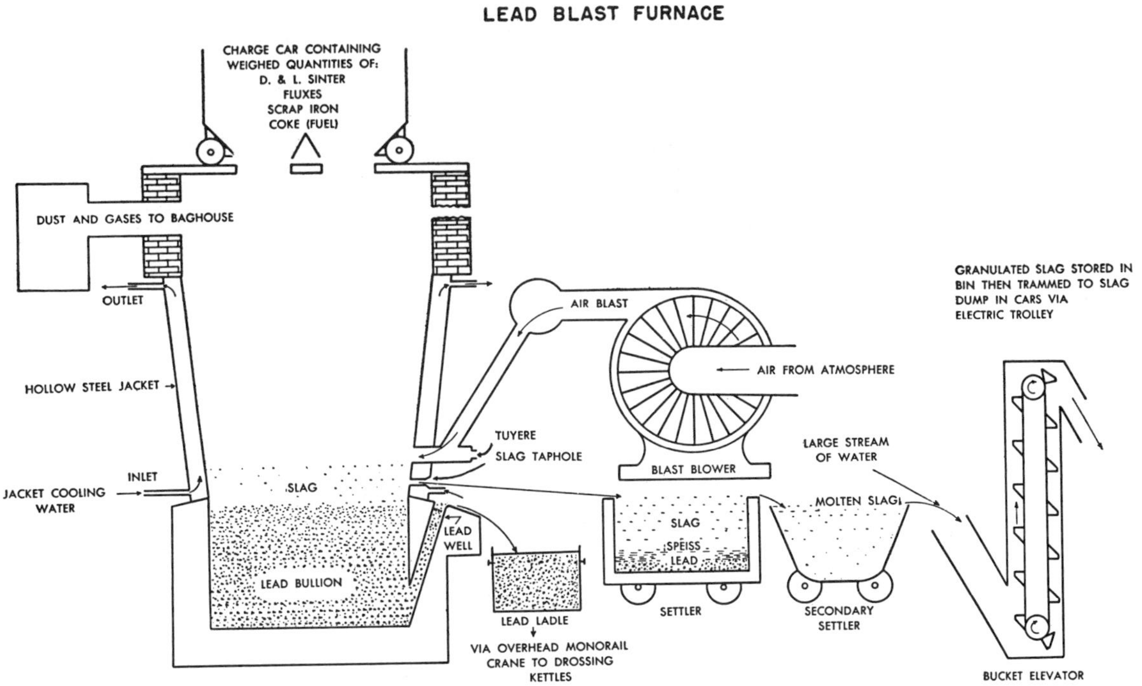

Figure 1.31. Lead blast furnace, open top type. *Source:* United States Smelting Mining and Refining Company.

**Table 1.18.   Lead Blast Furnace Operating Data, Buick Lead Smelter,
AMAX-Homestake**

1. *Furnace Dimensions*

Length (m)	6.40
Width (lower tuyeres) (m)	1.52
Width (upper tuyeres) (m)	3.04
Tuyere area (lower) (m^2)	9.75
Height of column (m)	5.9
Number of tuyeres (lower)	28
Number of tuyeres (upper)	38
Diam. of tuyeres (lower) (mm)	102
Diam. of tuyeres (upper) (mm)	76
Jacket water volume (liters/sec)	44
Jacket water temperature in (°C)	38
Jacket water temperature out (°C)	55

2. *Average Performance—30 Day Period*

Tons per day	755
Tons per m^2 per day	77.5
Vol. air blown (total) (Nm^3/min)	315
Pressure (lower tuyeres) (cm/H$_2$O)	156–200
Tons bullion per day	371
Percent coke on furnace charge	9.7
Percent fixed carbon	8.7
Gas offtake to baghouse (m^3/min)	1130
Top temperature (°C)	121
Draft (cm/H$_2$O)	−6

3. *Blowers* (2)

Make: American Standard Centrifugal	
Rated volume (m^3/min)	425
Rated pressure (cm/H$_2$O)	335
Drive	Direct
Horsepower	350
rpm	3600

4. *Slag Analysis*

%SiO$_2$	%FeO	%CaO	%MgO	%Zn	%S	%Al$_2$O$_3$	%Pb	%Cu
22.9	31.0	18.0	3.5	11.5	1.0	5.8	3.5	0.1

Source: C. H. Cotterill and J. M. Cigan, Eds., AIME World Symposium
of Lead and Zinc, Vol. 2, 1970, p. 766.

side of the settler, whereupon the slag is run into a second smaller settler with any still
entrained matte recovered here before the slag is removed.

The slag may not yet be discarded, and if it still contains some 16% Zn and up to 2%
Pb, it is treated at several smelters to recover some of these values before finally going on
to the slag dump. A slag fuming furnace is used for this cleaning operation, which is
essentially a water-jacketed blast furnace without a crucible. The charge is molten slag

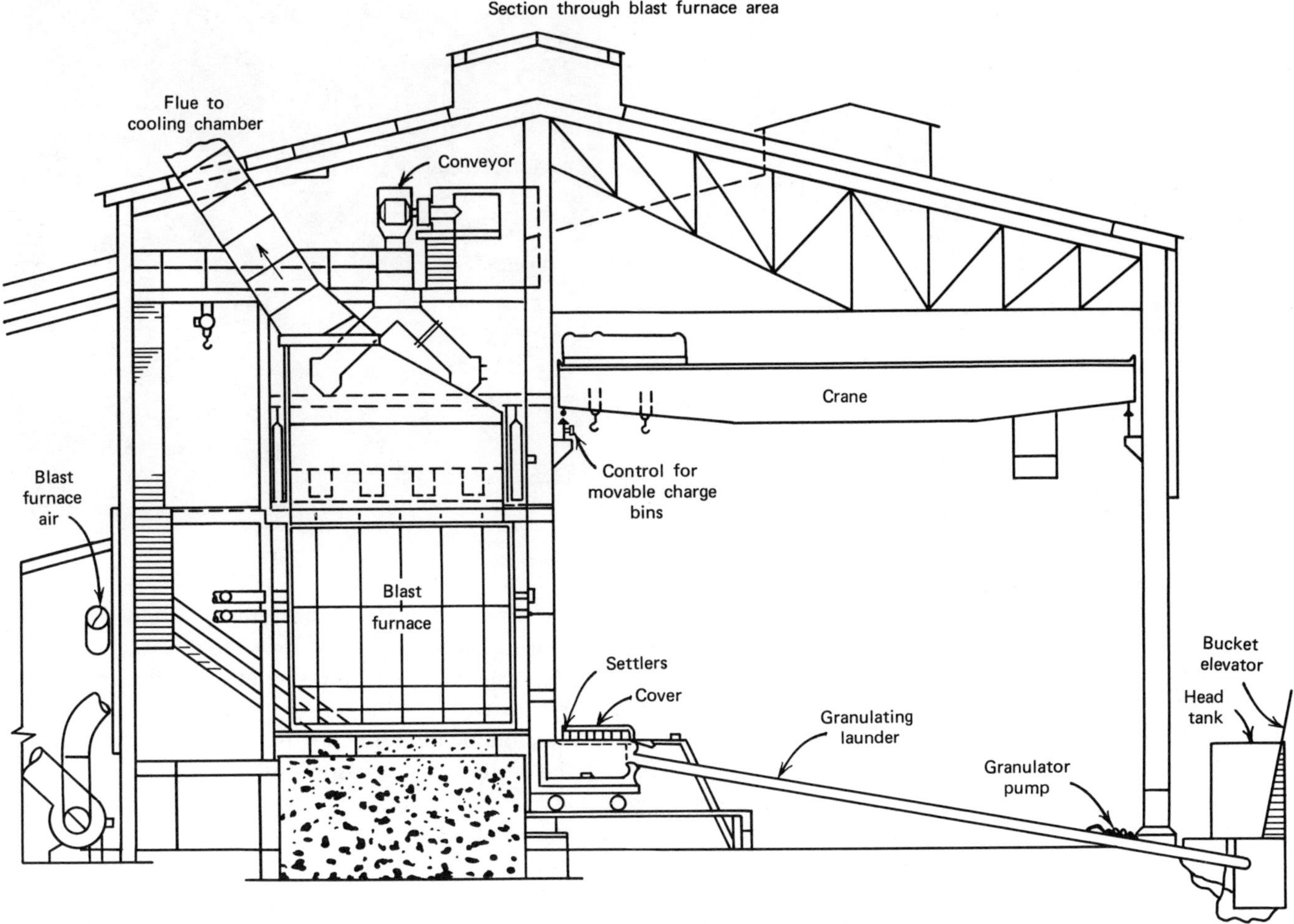

Figure 1.32. Section through Buick lead smelter blast furnace, AMAX Homestake. *Source:* C. H. Cotterill and J. M. Cigan, Eds., AIME World Symposium of Lead and Zinc, Vol. 2, 1970, p. 764.

from the second settler, sometimes mixed with solid slag from the dump to be retreated, while air and pulverized coal are blown in through the tuyeres, where they form CO gas. This strongly reducing atmosphere largely reduces the lead and zinc oxides in the slag to metallics, which vaporize at the temperature of the fuming furnace. These metallic vapors, or fume, are carried out the furnace flues, through long cooling pipes, and as they condense and reoxidize on cooling, are caught and collected in dust collectors. The treated slag, with practically all of the lead and two thirds of the zinc recovered, is run off at the bottom of the fuming furnace and taken to the slag dump.

Because of the lower temperature requirements for smelting lead as compared to copper or nickel, the refractory requirements in the lead blast furnace are not as stringent, and a high heat duty fireclay brick is usually adequate for crucible construction. Magnesite brick in the upper five or six courses of the crucible walls effectively counteracts any tendency to more severe corrosion in this part of the crucible where the various product layers meet.

Furnace temperature is kept down to prevent loss of lead by volatilization, and the width between the tuyeres is also somewhat restricted as the excessive blast pressure that

Figure 1.33. Tapping lead from water-jacketed blast furnace (Hoboken). *Source:* Courtesy of Metallurgie Hoboken–Overpelt.

would be needed to penetrate to the center of a very wide furnace would also cause excessive volatilization of lead. Modern furnaces are 19 feet long and 4 feet wide, with a 17 foot shaft height (5.8 × 1.22 × 5.2 m), and a blast pressure at the tuyeres of 2.2 psi (15.1 kPa), to smelt 500 tons of charge per day.

With the restricted operating temperature of the lead blast furnace, it is especially imperative that the slag composition be such that it is easily fusible and fluid at 2100°F (1150°C) to permit good separation of lead, matte, and speiss from it. Suitable fluxes are added to achieve the required low melting slag.

Dust and fume collecting systems are also quite necessary to collect both the normal flue dust and the low melting point metals here evolved as vapors, which chill to solid form as they pass through the long cooling furnace flues.

A few blast furnaces have been operated on 2.5 to 5% oxygen-enriched air to the tuyeres and have reported 15 to 25% increased smelting rates with a coke savings of up to 10%.

Table 1.19.   Lead Blast Furnace Slag and Dross, Glover Lead Smelter and Refinery, ASARCO

Blast Furnace Slag Produced

%Pb	%Cu	%SiO$_2$	%Fe	%CaO	%Zn	%Al$_2$O$_3$	%MgO
3.8	0.26	23.5	24.6	12.1	10.8	4.4	5.0

Rough Dross Produced

%Pb	%Cu
62.3	20.48

Smelted Blast Furnace Per Operating Day	665 tons
Lead Produced From Blast Furnace Per Operating Day	310 tons

Source: C. H. Cotterill and J. M. Cigan, Eds., AIME World Symposium of Lead and Zinc, Vol. 2, 1976, p. 786.

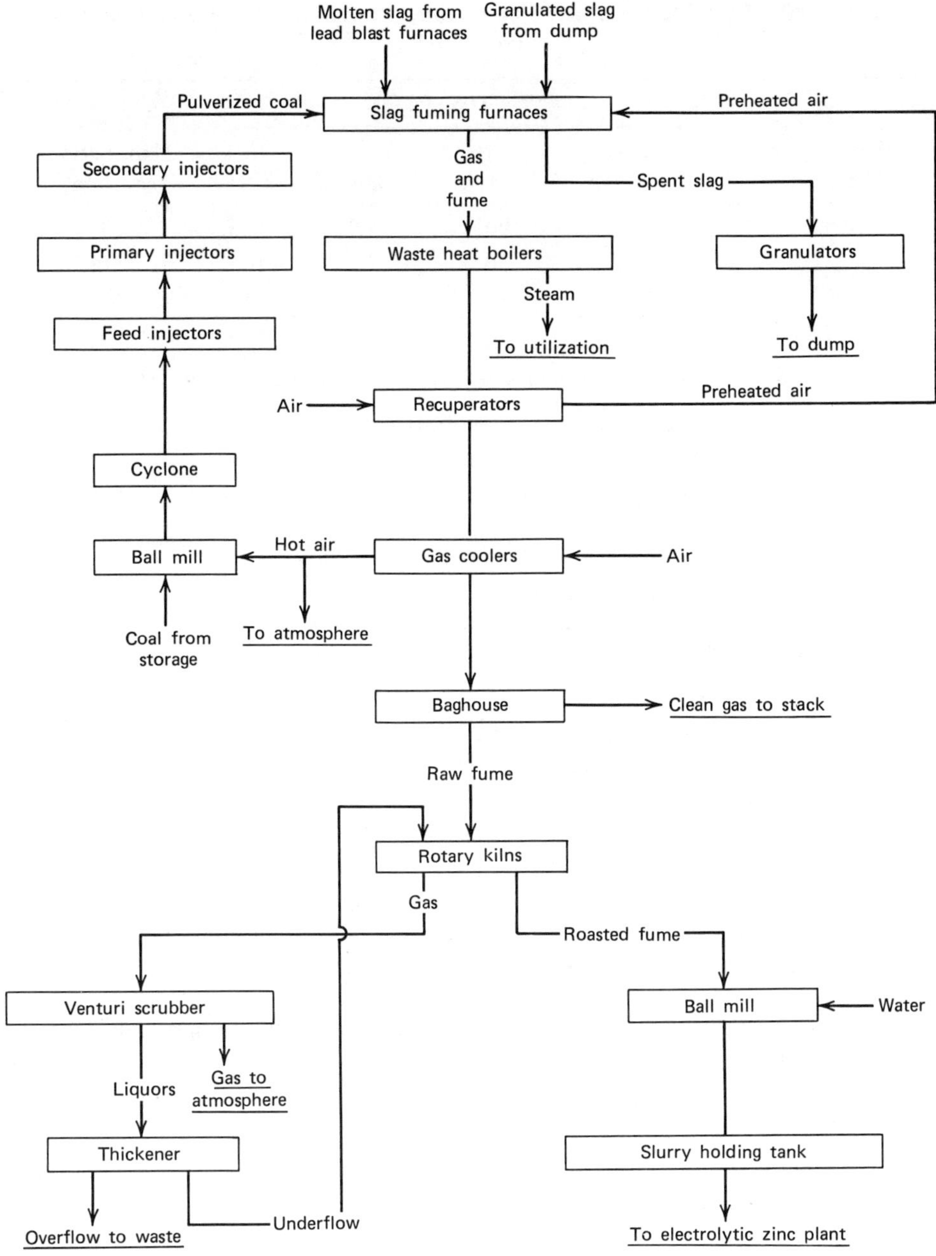

Figure 1.34. Slag fuming furnace treatment of lead blast furnace slag, Broken Hill Associated Smelters, Port Pirie. *Source:* C. H. Cotterill and J. M. Cigan, Eds., AIME World Symposium of Lead and Zinc, Vol. 2, 1970, p. 811.

3b. Lead Sulfide Unroasted Concentrate

Electric Furnaces in combination with flash smelting are used in the newest process for smelting lead sulfides and employ an electric furnace of the immersed electrode slag-resistance type very similar to those in use for the matte smelting of copper and nickel sulfides. This process was developed at Boliden's Röunskär works in Sweden and was the first process of its type commercially developed.

The furnace has magnesite brick in the semirounded bottom and chrome magnesite brick in the walls and roof. Pressure is applied by springs on the walls around the furnace to keep the side and bottom refractories in place and not be floated out by the heavy

Table 1.20.   Lead Blast Furnace Slag Fuming, Broken Hill Associated Smelters, Port Pirie[a]

| Hot Slag on Charge (%) | Cycle Time (min) | | | | Zinc Elimination (kg/min) |
	Charging and Heating	Fuming	Tapping	Total	
100	20	90	15	125	58
75	40	90	15	145	50
50	60	90	15	165	44
25	80	90	15	185	39

Typical Operating Results for a 7 Day Week for a Single Furnace

Molten slag treated (incl. skull) (tonnes)	2276
Granulated slag treated (tonnes)	598
Total slag treated (tonnes)	2874
Number of charges	62
Average weight per charge (tonnes)	46.3
Average cycle time (min)	147
Operating time efficiency (%)	90.2
Coal consumption (tonnes)	457
Coal rate (kg/min)	50.3
Average zinc in input slag (%)	18.3
Average zinc in spent slag (%)	3.3
Zinc elimination (%)	85.7
Average lead in input slag (%)	2.6
Average lead in spent slag (%)	0.03
Est. weight of fume produced (tonnes)	670
Zinc elimination rate (kg/min)	50.0
Av. coal per kg. of zinc eliminated (kg)	1.02

Source: C. H. Cotterill and J. M. Cigan, Eds., AIME World Symposium of Lead and Zinc, Vol. 2, 1970, p. 812.

[a]Under these operating conditions, the zinc in spent slag is reduced to 3.3% representing approximately 86% zinc elimination from the input slag.

metal bath. The furnace is 44 feet long and 14 feet wide, and there are 11 feet from roof to crucible (13.3 × 4.3 × 3.35 m) and $5\frac{1}{2}$ feet from roof to slag line (1.65 m). Four Söderberg electrodes are connected in pairs and are 40 inches (1 m) in diameter with the electrode paste filling steel shells. The power input is 8000 kVA and heats the slag layer to 2460°F (1350°C).

Blended, unroasted, 77% minus 325 mesh lead sulfide concentrate dried to 2% moisture, along with recovered flue dust and limestone, are charged through roof openings between the two sets of electrodes at the end of the furnace opposite to the gas take-off flue. Air nozzles are inserted through these roof-charging openings and aim horizontal jets of air against the incoming streams of feed material, deflecting them into two vortices in the two spaces between the three electrodes. Four air streams directed tangentially shape a whirling action between the electrodes without having to depend on any confining support from the furnace walls.

This arrangement gives sufficient reaction time, even in the short $5\frac{1}{2}$ foot distance (1.65 m) between the roof and the slag layer to oxidize and burn most of the sulfur to

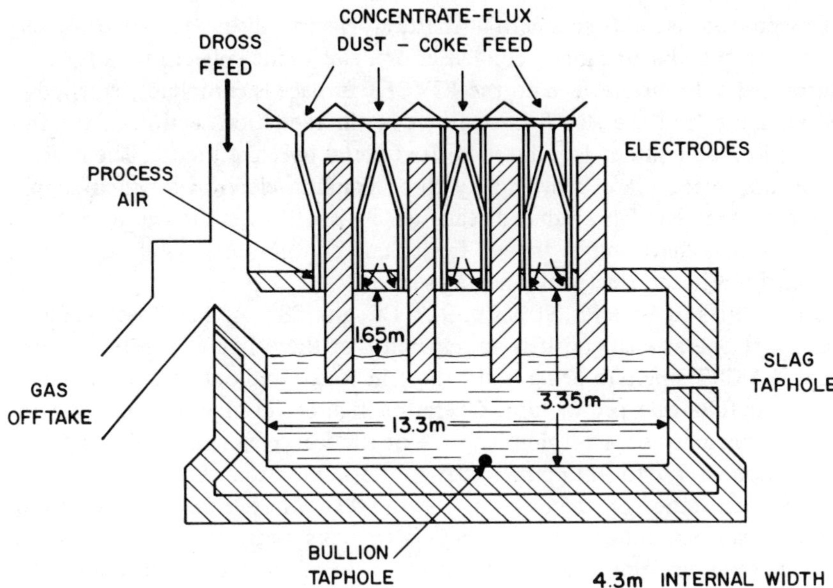

Figure 1.35. Electric furnace for lead smelting. *Source:* A. G. Matyas and P. J. Mackey, *J. Metall.,* Vol. 28, No. 11, 1976, p. 12.

SO_2 and oxidize the lead, iron, and zinc in the charge. The temperature in the flash combustion zone is 2100°F (1150°C). After this flash smelting, the liquid droplets, and whatever portion is still solid, collect in the slag bath where the roast-reduction reactions are completed to oxidize the remaining PbS with PbO in the returned flue dust and slag and reduce the PbO to Pb:

$$2PbS + 3O_2 = 2PbO + 2SO_2$$

$$PbS + 2PbO = 3Pb + 3O_2$$

Thereupon the slag and lead separate into two layers, the oxide slag containing iron, calcium, and zinc on top, and the liquid lead below. Coke breeze is charged periodically on top of the slag in order to reduce the lead oxide content of the slag, and cause this reduced lead oxide to settle down into the lead metal layer.

The lead sulfide concentrate charged is 75% lead; and from this, 98% is recovered as lead containing about 3% sulfur, and is tapped intermittently from the tap hole in the middle of the furnace side in 70 ton batches. The slag is also taken off periodically from tap holes at the end of the furnace opposite to the gas take-off flue. This slag will contain 4.5% PbO, despite the addition of coke breeze to keep unreduced PbO to a minimum, and 20% ZnO, and is therefore sent to a slag fuming plant to recover these values before discarding at the slag dump.

Because of the large volumes of high-velocity air blown in through the swirling jets for flash smelting [4500 cubic feet (127.3 m³) per minute at a velocity of 200 feet (61 m) per second at 125°F (50°C)] and because of the fineness of the dry furnace charge, the departing furnace gases carry off up to 100 tons of flue dust per day, which is recovered in dust collectors and recycled back to the furnace feed bins. The flue gas which leaves the furnace at 1925°F (1050°C) first has part of its sensible heat recovered in waste heat boilers. Then, after dust removal, the gas is transferred to a sulfuric acid plant to utilize its 8% SO_2 value as feed material in acid making.

The plant's initial capacity was 47,000 mtpy, and this has been gradually increased in several stages to 80,000 mtpy.

The KIVCET Process developed in Russia, a process similar to the process described for electric furnace copper smelting, combines sintering, blast furnacing, and slag fuming all in one unit. The process has a flash smelting shaft, an electrically heated zone for slag deleading and dezincing, and a zinc metal condenser or a zinc oxide collection unit.

The dried unroasted concentrate feed to the KIVCET furnace is completely smelted in the flash smelting shaft, and the slag formed flows to the reduction section of the furnace, where its lead oxide content is reduced to lead metal by coke breeze. The heating requirements for this section are provided by three graphite electrodes which supply resistance heating to the slag. The reduced lead, which is of low sulfur content, flows back under the smelting shaft and is tapped from there continuously, while the slag is tapped intermittently from the electrically heated section.

Overall metal recoveries of 98% Pb, 90% Zn, 92% Cu, and 88% Ag have been made in semicommercial plant operations, and it is anticipated that these figures can be retained when the several KIVCET smelters being built, three in Russia and one each in Germany and Bolivia, get into full scale production. Operation figures for the commercial pilot plant include coke consumption averaging 3 to 4% of the lead concentrate feed weight, electrode consumption of 7.7 to 9.9 pounds (3.5 to 4.5 kg) per ton of lead concentrate, and electrical power consumed on the order of 450 to 500 kW-hr per ton of concentrate.

The hot furnace gases pass into waste heat boilers to make steam for plant use or are used to dry concentrate smelting feed and then go on to particulate removal and SO_2 recovery. The dust and fume, averaging 25% of the charge, are returned to the smelting chamber. With 17 to 22% of the charge being sulfur, smelting is autogenous, and if the sulfur content and the associated smelting rate increase, this return of cold feed can be adjusted to stabilize the smelting rate and keep the temperature practically constant.

The percent of SO_2 in the gases is high and ideal for SO_2 recovery. If oxygen is used instead of air for smelting, 97% of the sulfur in the charge is recovered as SO_2 at a concentration of 50 to 60%.

The QSL lead recovery process is the latest direct lead smelting attempt and is a development from the original Queneau–Schuhmann process with participation by St. Joe Minerals and Lurgi. Pelletized lead concentrates are fed into a long, narrow rotary converter, which has double-tube tuyeres. The center tube in the tuyere injects oxygen, while the outer tube injects a protective gas, which can consist of purified waste gases from the operation. Lead is tapped continuously from the feed end and slag continuously from the opposite end, after passing tuyeres along the converter length with lower and lower oxygen content (activities) from mixing the oxygen with coal dust or other reducing gases.

4a.   Cobalt Oxide Ores and Calcines, as at Katanga in Zaire, are smelted in single-phase, three triangular-spaced electrode steelmaking-type electric furnaces. Lump oxide ore, coarse concentrates, sintered fines, lump lime as a flux, and coke as a conductor and reducer are charged into the furnace. The products of the reduction are a slag and two alloys. The slag is tapped off and discarded, while the alloys are poured off together into ladles where they separate into two molten alloys of different densities. The top layer, called the white alloy and containing 42% cobalt and 15% copper, is further processed to recover the cobalt, while the bottom red alloy of 89% copper and 4.5% cobalt is furnace refined to recover its copper.

The cobalt oxides and concentrates for this electric furnace smelting are combined from three main sources. One of these sources is the returned high-cobalt slag (15% Co) from the furnace refining of the red alloy; another is the concentrates (6 to 14% Co) separated during the beneficiation of copper oxide ores with low cobalt content; and the third is from copper oxide ores that have locally high cobalt concentrations which permits selective mining of these areas to give an ore of from 4 to 12% cobalt content.

Electric Furnaces 6 feet in diameter (1.83 m) operating on 600 kW single-phase current with three graphite electrodes are used for the reduction smelting of the combined oxide feed materials. A composite feed is made up from the three cobalt oxide sources—ore, concentrate, and slag—with the fines below $\frac{3}{8}$ inch (9.4 mm) being screened out and agglomerated by sintering before charging. The furnaces do not operate satisfactorily if there is an appreciable quantity of fines in the charge, especially if there is a substantial fire loss on heating.

Lump ore, coarse concentrate, coarse screened sinter, lump lime flux, and coke as a conductor and reducer are charged into the furnace. Melting the charge in a reducing atmosphere produces cobalt, copper, and iron in the metallic state, and these combine as two alloys which contain 93% of the cobalt and copper available in the charge, along with part of the iron and some silicon. The two alloys are tapped together into a ladle where the white alloy (42% Co, 15% Cu, 34% Fe, 2% Si) has the lower density and forms the upper layer, while the heavier red alloy (89% Cu and 4.5% Co) forms the bottom layer. After solidification the ladles are dumped and the two layers split apart, or they can be separated while still molten by tapping the red alloy from the bottom of the ladle and pouring off the white alloy from over the ladle lip.

The silicon in the white alloy is kept above 1.5% to ensure it being lighter than the red alloy, and this is accomplished by charging sufficient coke of a small enough size to provide the required reduction needed of SiO_2 to Si. The Si produced is absorbed in the white alloy.

The slag from the gangue material in the charge and the lime flux added will analyse 0.3% Cu and 0.4% Co. This is run out through a water-cooled tap hole and taken to the dump.

4b. Cobalt Arsenical Ores and Concentrates are smelted in small blast furnaces of the conventional rectangular-shaped, side tuyere type used for copper sulfide smelting. Crushed raw ore, briquetted flue dust and roasted ore, returned slag, fluxes, and coke are charged to the blast furnace and produce silver bullion, speiss, matte, and slag in proportion to the charge makeup. The flue gases from the top of the blast furnace contain flue dust which is settled out of the gas stream and briquetted to return as blast furnace charge, and arsenous oxide fume which is blown on by a fan to collect in a bag house.

Blast Furnaces have been found to be satisfactory smelting units for arsenical cobalt ores and concentrates because they can both directly smelt the cobalt, nickel, and copper to matte and speiss and remove considerable gangue material as slag, while also oxidizing a considerable amount of the contained arsenic to arsenous oxide fume which is carried off in the blast furnace flue system and collected when cooled in a bag house. Some contained sulfur in the feed material may also be oxidized, minimizing the amount of matte formed, which is not wanted in large quantities.

Cobalt ores and concentrates containing up to 11% cobalt, with lesser amounts of nickel and copper, arsenic as high as 30%, and sulfur as high as 10%, are charged to a small 6 foot long by 3 foot wide (1.83 × 0.92 m) blast furnace. Coke, briquetted flue dust, return slag, as well as limestone and silica for fluxes as required, are also charged.

The charge must contain sufficient arsenic to combine with the cobalt and nickel and form speiss, otherwise they will be lost in the slag. Copper, cobalt, and nickel have the greatest affinities for sulfur, and these will form matte with generally a higher ratio of cobalt to nickel than is found in the speiss. The high affinity for oxygen has iron being oxidized first before the other metallics, cobalt, nickel, and copper, and this assists in the charged iron largely oxidizing and going into the slag instead of into the speiss or the matte.

The furnace crucible is tapped periodically, and a mixture of matte, speiss, and bullion is run out into cast iron pots where it is allowed to solidify. The pots are then dumped, and the layers of matte on top, speiss in the middle, and bullion on the bottom which have separated are easily broken apart. The speiss will contain most of the cobalt and nickel and be the largest product, though when speiss and matte are forming at the same time it is impossible to prevent cobalt, nickel, copper, and iron from dividing between the two. A typical speiss will assay 20% Co, 12% Ni, 2% Cu, 18% Fe, and 23% As, and a typical matte will assay 9% Co, 4% Ni, 12% Cu, 27% Fe, and 23% S.

Slag, which is below 1% combined Co + Ni, is run off continually and is either discarded or partially returned to the furnace as a charge conditioner or for further cleaning, as charging space permits.

Gases, flume, and dust from the blast furnace pass through a downcomer from the furnace to brick flues where most of the coarse flue dust settles out. The remaining arsenic fume and fine flue dust are blown by a fan into a bag house where cooled arsenic is recovered as As_2O_3.

CONVERTING

Converting is the second and final stage in smelting sulfide ores or concentrates and is also a concentration operation, as was smelting. The liquid metallic sulfide phase of matte which was produced by furnace smelting has had most of the waste rock and part of the iron content removed as slag during the smelting step, which left the matte as a complex but homogeneous solution of copper, nickel, cobalt, iron, and sulfur, with small amounts of precious metals and other base metals.

The matte must now be processed to remove most of the still unwanted constituents, principally the iron and sulfur, and this is done in a batch, sometimes two-stage process known as converting. By blowing air, oxygen, or oxygen-enriched air into the charge of liquid matte in the converter, a process of selective oxidation takes place whereby those constituents of the matte with the greatest affinities for oxygen are oxidized first and can be removed.

The first, or "white metal," stage of converting is the rapid oxidation of the iron sulfide in the matte to iron oxide and sulfur dioxide gas, as iron has the greatest affinity for oxygen of all the matte components. Enough silica is added to combine with this iron oxide produced and form an iron silicate slag which is poured off. A large amount of exothermic heat is produced from this oxidation, and it is quite sufficient to keep the converter charge liquid and up to the reaction temperatures required.

The second converting stage, which can be used only with metals that do not oxidize too readily, is to oxidize the sulfur from the remaining metallic sulfides left in the matte after most of the iron sulfide has been oxidized and the slag made from it poured off. Blowing is now continued until the sulfur in the "white metal" has been oxidized to SO_2, leaving relatively pure metal, which is poured out of the converter and taken to be refined.

This second converting stage cannot be used if the white metal has a high affinity for oxygen and will form an oxide instead of remaining in the metallic state after its sulfur content has been oxidized to sulfur dioxide. In this case the white metal is the final converter product and is poured off and transferred to the next step of refining.

TYPES OF CONVERTING

The purpose of converting is to eliminate by oxidation and slagging, and by oxidation and removal as a volatile gas, as much of the iron and sulfur, respectively, as possible from the matte made in the prior smelting of ores and concentrates. As much as 80% of the matte charge in a converter may be removed in this way, and as the oxidized impurities are removed in increments of several steps of first blowing for a period to oxidize and then removing the amount of oxide slag that has been produced, the overall operation of converting matte to metals, or to white metal, is one that takes several hours.

The slag produced, due to the increasing richness of the percent metallics in the converter charge as the process proceeds, contains an appreciable amount of metal values and must be retreated to recover these. Also, as the ores and concentrates have already had one stage of concentration in their furnace smelting, the volume of matte now being treated is considerably less than the amount of charge which was put in the smelting furnace. Consequently, even with the high percentage of impurities removed during converting, the volume of slag from the converter is much less than the volume of slag from the smelting furnace. Converter slag is almost always taken back and poured still liquid into the smelting furnace for recovery of its metal values, though some is also treated by slow cooling, crushing, grinding, and selective flotation to recover the contained values.

It is the usual practice during the many-step batch operation of converting to add fresh liquid matte from the smelting furnace to replace the converter slag which has just been poured off. This fresh matte addition provides fuel to keep the converter charge liquid and at reaction temperature by the exothermic heat from its contained sulfur oxidation. White metal, or metal, gradually accumulates in the converter as the iron and sulfur in the matte are removed until the converter charge is completely in this state, and it is then removed for the final refining operations to give pure metal.

Converting is a relatively standardized operation with air blown through tuyeres into a stationary horizontal drum-shaped vessel being the conventional process. However evolution is proceeding in converter operation and designs with oxygen and oxygen-enriched air supplementing air alone, and vertical rotating converters with top-blow oxygen lances replacing the standard form are recent innovations.

The high exothermic heat produced in converting can be utilized to some degree for the direct smelting of high-grade ores and concentrates, as well as scrap reverts and skulls put back into the process to recover their metal values.

The most recent processes combine smelting and converting in a continuous operation so that both are carried out simultaneously in a single reactor. The several advantages of this combined treatment are that the two separate operations, smelting and converting, can be consolidated into one overall process carried out in only one reactor vessel, and also that excess exothermic heat from the converting reactions can now be fully recovered by being absorbed and utilized in the smelting part of the process. In addition, the escaping flue gas containing SO_2 from the unroasted sulfide concentrate feed is only minimally diluted by gases from auxiliary burners, so that it leaves the reactor with a high SO_2 content which is economically best suited for sulfuric acid manufacture.

Two different styles of combined smelting–converting reactors have been developed. One of these is a cylindrical reactor somewhat related to the conventional horizontal converter, while the other is a fixed-position furnace type more of a smelting furnace design.

Horizontal Converters of the Peirce-Smith Horizontal Drum Type are the conventional type most frequently used for converting matte. These converters are steel cylinders, magnesite brick lined, lying in a horizontal plane on rollers and can be rotated on the long axis through an arc of 120°. An opening at the center of the top is used to add charge and pour off slag and product, at which time the converter is rotated so that the opening is rolled forward from the upright position. The operating position is with the open mouth upright and under a fume hood which will carry off the gas and dust from the process.

Liquid matte from the smelting furnace is poured into the converter, when it is "turned down" with the open mouth in its forward position, filling the converter about half full. Compressed air is then turned on through a row of tuyeres blowing into the converter along the back and positioned so that when the converter is now rotated upward to its upright operating position these tuyeres are beneath the surface of the liquid matte and are blowing into it.

Oxidation of iron sulfides in the matte commences, and siliceous material is added at intervals as needed for flux to combine with the iron oxide being formed and make an iron–silicate slag. When sufficient slag has been made the converter is "turned down" to pour off this slag, and the compressed air is shut off when the tuyeres rise out of the molten bath. The slag containing 2 to 5% of metallic values is taken back to the smelting furnace for cleaning or, in the most recent trend, is cooled, ground, and treated by flotation to recover the metal values. Fresh matte is added to the converter to replace the volume of slag removed, the compressed air is turned on, the converter rotated to its upright operating position, and the processing cycle is repeated. When after many cycles the converter is filled with white metal or metallic metal, whichever is to be the final product, the converter is "turned down" and the finished charge poured out into ladles to be taken on to the next step of refining.

The main exothermic reaction in the converter,

$$2FeS + 3O_2 = 2FeO + 2SO_2$$

provides sufficient heat to keep the charge liquid and at the operating temperature without the addition of any other fuel. The slag making reaction is

$$2FeO + SiO_2 = 2FeO \cdot SiO_2$$

Converter dimensions are on the order of the length being approximately two to two and a half times the diameter, so that converters are common in such sizes as 13 feet in diameter by 30 feet long (3.96 × 9.15 m), 12 feet in diameter by 28 feet long (3.66 × 8.55 m), and 13 feet in diameter by 35 feet long (3.96 × 10.67 m).

Flue gases collected by the hood above the open converter mouth during blowing are taken to dust collectors from where the recovered flue dust is recycled back into the process. The gases from the beginning of the blowing period are richer in SO_2 than those at the end of the blow, which are quite lean. The rich SO_2 gases can be taken off separately and used for sulfuric acid plant source material.

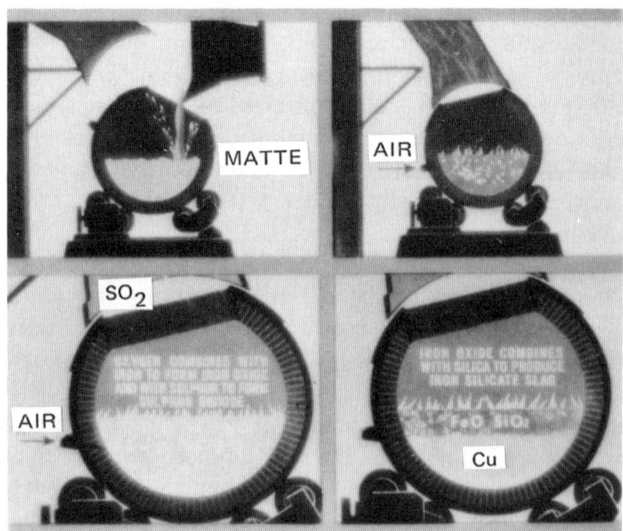

Figure 1.36. Peirce–Smith converting operation. *Source:* Courtesy of Inco Limited.

Hoboken (or Siphon) Converters are a recent innovation, which have been installed in a number of smelters, to further improve the collection of SO_2 gases and minimize air pollution, beyond the effectiveness of the hoods that are standard equipment on Peirce-Smith converters.

The Hoboken converter draws off the product gases high in SO_2 through a flue connected axially to one end of the converter, with a "goose neck" or "siphon" letting the gases flow freely from the converter at all phases of the operation, without any liquid droplets spilling into the gas-collecting flue system.

This allows the converter to be sealed during the blows by closing off its mouth, which then both prevents SO_2 escaping into the atmosphere and any dilution of SO_2 gas by infiltrating air.

As there are no hoods or flues obstructing the converter's mouth, some charging can be carried out through a retractable spout during the blow, and this somewhat reduces the down time. However the pouring in of matte and the removal of molten slag and metal still require that this be done through the mouth and necessitate stoppage in converter blowing.

Top-Blown Rotary Converters (TBRC) are a new type quite similar in design to the Kaldo furnace used in oxygen steelmaking. The converter is cylindrical in shape, closed at one end, and with an open cone at the other. The vessel is a steel shell, magnesite lined, which is rotated at speeds of from 5 to 40 rpm, and is inclined in its operating position at 17° to the horizontal. The converter rests on two sets of roller bearings and can be tilted through 360° to bring the mouth to various positions for slagging off, flux additions, and pouring of the final finished charge.

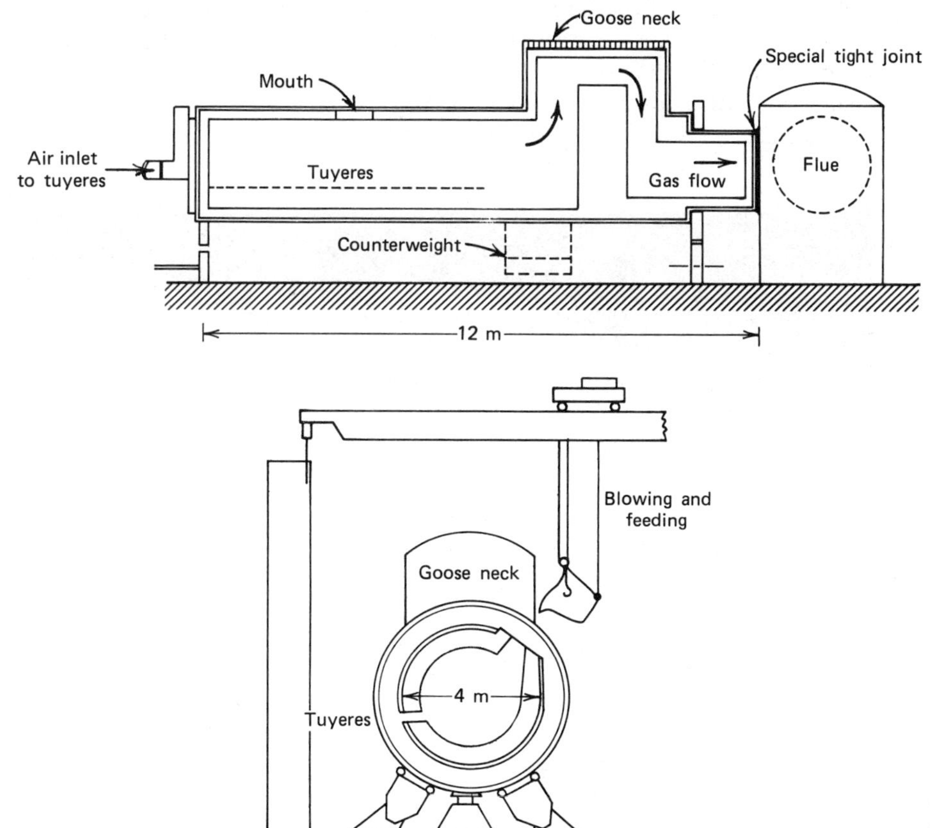

Figure 1.37. Hoboken "siphon" converter, side and end views. *Source:* A. K. Biswas and W. G. Davenport, *Extractive Metallurgy of Copper*, Pergamon Press, Oxford, 1976, p. 198.

Figure 1.38. Assembling Hoboken-type converter. *Source:* Courtesy of Metallurgie Hoboken–Overpelt.

After the vessel is charged with matte through the open end, it is inclined at its 17° operating position, rotation is begun, and a water cooled lance is inserted through the mouth to direct a jet of oxygen on the surface of the rotating bath. Oxidation of the iron sulfide quickly takes place, and silica flux is added to form an iron silicate slag, which is poured off and returned to the smelting furnace. Fresh furnace matte is now added to replace the volume of slag removed, and the blowing, oxidation, and slag-making cycle is repeated until the converter is filled with metallic metal. This final product is poured off and taken to be refined.

Rotating converters used for nickel are 18 feet long with an outside diameter of $11\frac{1}{2}$ feet (5.49 × 3.51 m) and normally operate below their maximum rotation speeds of 40

Figure 1.39. Hoboken-type copper converter in operation (Hoboken). *Source:* Metallurgie Hoboken–Overpelt.

rpm until just before the end of the blow when a short period of maximum rotation will be given. The combination of surface blowing from the lance and mechanically induced turbulence of the bath from rotation are thought to improve the performance of the essential functions of the converter, while at the same time making the two actions of blowing and agitation independent of each other and increasing the process flexibility.

Smelting–Converting Reactors are of two different designs, one of which, now in commercial operation, is the *Noranda process* consisting of a horizontal cylindrical furnace with a depression in the center into which the metal product is collected and then tapped, and a raised hearth at one end from which the slag is run off. The opposite end of the reactor from the raised slag hearth is where the smelting–converting reactions take place, and a row of air tuyeres is positioned here along the furnace for approximately two fifths of its length.

The reactor is a steel shell, chrome magnesite brick lined, and can be rotated 54° to bring the tuyeres out of the bath. A hood over a central top opening in the shell collects the escaping gases, which are water-spray cooled and blown on to dust collectors. The cleaned, high SO_2-content gas passes to a sulfuric acid plant, while the recovered dust is pelletized and recycled back to the reactor. Burners at either end of the shell provide auxiliary process heat if required, and a feed port with a belt slinger at the tuyere end of the reactor distributes pelletized concentrate and flux over the surface of the bath in the smelting–converting zone.

The bath in the vessel consists of a top layer of slag, an intermediate layer of high-grade matte, and a bottom layer of metallic metal which settles into the depression in the middle of the unit. Air blown through the tuyeres gives an intense mixing action in the bath to provide the high transfer of exothermic heat needed for the matte smelting phase, as well as providing oxygen for the converting phase.

The reactions which take place are similar to those which would occur in separate matte smelting and converting operations, with the pelletized concentrate charge first smelting to matte and then the sulfides in this matte being oxidized by converting. Iron sulfide oxidizes first and forms an iron-silicate slag with the silica flux added as part of the charge, while the remaining metal sulfide is then oxidized to metal. The slag is too high in metal values to be discarded and must be processed first to recover these values, while the metallic metal product is sent on to be refined.

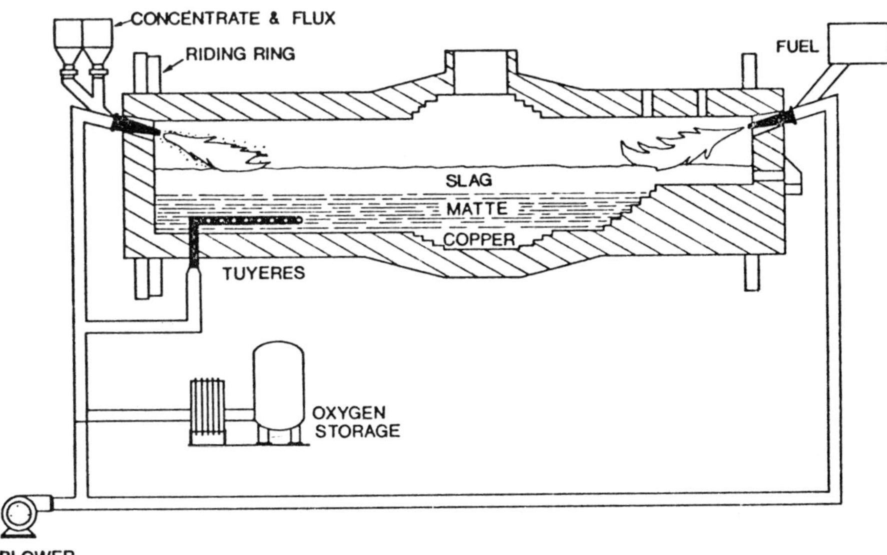

Figure 1.40. Cylindrical rotating smelting-converting reactor, Noranda process. *Source:* Courtesy of Air Products and Chemicals, Inc., Allentown, Pennsylvania.

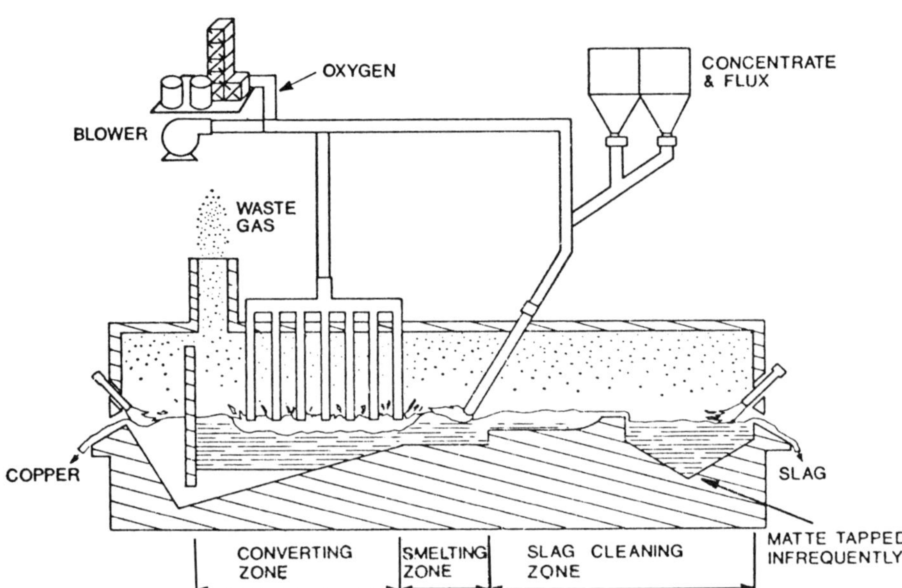

Figure 1.41. Stationary furnace smelting-converting reactor, Worcra process. *Source:* Courtesy of Air Products and Chemicals, Inc., Allentown, Pennsylvania.

A cylindrical reactor of a size capable of processing 100 tons of concentrate per day has an inside length of $32\frac{1}{2}$ feet (9.90 m) with a 7 foot (2.13 m) internal diameter at the feed end. There are 13 tuyeres 2 inches (5 cm) in diameter on 6 inch (15 cm) centers in the smelting and converting section.

The second design type of smelting–converting reactor, *the Worcra*, which has not yet been commercially adopted, is a stationary furnace type, which in a size treating 72 tons of concentrate per day is constructed in a U-shape with the concentrate and flux being fed in at the bottom of the U. One leg of the U-shape is fitted with lance-type tuyeres, and it is here that the converting reactions take place, with the same three layers of slag, matte, and metallic metal forming. The metal is removed from the top end of this leg through a wall extending down to the bottom of the furnace.

The opposite leg serves for slag cleaning, and entrained matte and metal settle out by gravity in this relatively quiet zone. The slag overflows a weir at the top of the leg and enters a slag well, from where it is tapped continuously, and is then discarded.

The slag cleaning leg is 8 feet wide by 31 feet long (2.44 × 9.45 m), while the converting leg is 6 feet wide and 28 feet long (1.83 × 8.54 m), with a 10 foot diameter (3.05 m) smelting–charging area adjacent to it at the bottom corner of the U-shaped furnace.

The gas offtake flue is positioned over the converting zone, and the gas high in SO_2 content is first passed through a dust-cleaning system and then on to the sulfuric acid plant.

Straight-line furnaces, rather than U-shaped ones, are also being considered. With these the charging and smelting will take place in the center of the furnace, with the converting zone and tuyere lances at one end and the slag cleaning taking place at the opposite end. The flue gases will be removed from the converting end.

CONVERTING PROCESSES

1a. Copper Sulfide Furnace Matte is converted to blister copper, 98% copper or higher, in Peirce–Smith converters most commonly found in the 30 by 13 foot (9.15 × 3.96 m) size. The converters are steel shells lined with magnesite or chrome magnesite brick and have drive gears and riding rings at the ends for rotation from the vertical operating position to the horizontal position used for charging and pouring.

The tuyeres, some 48 to 52, are placed in a row along the back of the converter so that

Figure 1.42. Smelter converter aisle with Peirce–Smith converters. *Source:* Courtesy of Inco Limited.

they are submerged and are blowing into the bath when the converter is in its upright operating position. Compressed air is blown into the converter bath at 13 to 14 psi (89.6 to 96.5 kPa); higher pressures are undesirable, in that air would be forced through the bath too rapidly for most effective oxidation to occur and the bath agitation would be so great that some of the molten charge would slop and splash out. The tuyere openings into the converter have a tendency to become plugged with solidified molten charge which is cooled by the tuyere air blast. To keep the tuyeres open and operating efficiently, they must be frequently cleaned by ramming a steel poker through them. This operation is called punching and is now largely done by mechanical devices operating on compressed air which can punch the whole row of tuyeres in about 30 seconds. The blowing rate to a converter will be on the order of 22,000 cubic feet (623 m^3) per minute, and the tuyere openings are $1\frac{1}{2}$ to 2 inches in diameter (3.75 to 5 cm). Frequently the last few tuyeres at each end of the converter are not used in order to minimize the blast from these tuyeres cutting into and rapidly eroding the refractory lining on the converter ends.

It is most convenient to have the converters positioned close to the smelting furnace so that the transfer ladles transporting furnace matte to the converters have only a short distance to travel, reducing the transfer time and also reducing chilling of the matte in the ladle. Cooled matte will solidify in layers against the ladle shell and remain in the ladle as a so-called "skull" when the remaining still liquid matte is poured out. The buildup of these skulls in the ladle reduces the effective ladle capacity and also eventually necessitates a separate operation of breaking these skulls out of the ladles and then returning them as cold copper bearing material either to the converter or to the smelting furnace.

Converting of copper matte is a two-stage, batch operation. It is two-stage in that the first or white metal stage of converting concerns the rapid oxidation of the iron sulfide in the matte to iron oxide and sulfur dioxide. The iron oxide is slagged off after combining with SiO_2 flux, leaving molten copper sulfide, known at this stage as white metal. The second stage is to continue the blowing and to oxidize the copper sulfide to metallic copper. The operation is a batch process in that a converter charge of matte analyzing from 25 to 55% copper, depending on whether a high- or low-grade furnace matte was produced, will be blown and slagged to remove the iron and sulfur until the matte has been purified to blister copper analyzing over 98% copper.

Table 1.21. Production Data for Copper Converters, International Nickel Co.

	Copper Converter Productivity	
	Per Charge (English)	Per Charge (metric)
Flash furnace matte	184 tons	167 tonnes
% Cu + Ni	45–47	45–47
Matte separation	80 tons	73 tonnes
Copper concentrate		
%Cu	73	73
%Ni	5	5
Scrap	15 tons	14 tonnes
Flux (70% SiO_2)	35 tons	32 tonnes
Metal blown (Cu + Ni)	147 tons	133 tonnes
Blister copper	106 tons	96 tonnes
Cycle hours (cast to cast)	24 hours	24 hours
Blast rate	18000 Scfm	29000 Nm^3/hr
Blast pressure	15 psi	100 kPa
Avg. oxygen content	30 wt-%	30 wt-%
Percent blowing time	40%	40%
Charges per refractory campaign	110 days	110 days
Metal blown per campaign (Cu + Ni)	16170 tons	14630 tonnes
Blister copper produced per campaign	11660 tons	10560 tonnes

Source: J. C. Yannopoulos and J. C. Agarwal, Eds., *Extractive Metallurgy of Copper*, Vol. 2, The Metallurgical Society of AIME, 1976, p. 231.

Matte is essentially FeS and Cu_2S, and it is these two compounds which are involved in the two stages of converting. The principal reaction in the first stage is the oxidation of the iron sulfide,

$$2FeS + 3O_2 = 2FeO + 2SO_2$$

followed immediately by the slag making reaction

$$FeO + SiO_2 = FeO \cdot SiO_2$$

Any Cu_2S oxidized to Cu_2O in this stage will be reduced back to Cu_2S by the reaction

$$Cu_2O + FeS = Cu_2S + FeO$$

After the slag is removed, the second stage of converting takes place in which Cu_2S is oxidized to metallic copper:

$$Cu_2S + O_2 = 2Cu + SO_2$$

There may be some oxidation of copper to Cu_2O during this period; but if there is, the Cu_2O is reduced back to Cu by the Cu_2S still present:

$$2Cu_2O + Cu_2S = 6Cu + SO_2$$

The total weight of matte treated during a complete converter batch cycle from matte to blister copper will be on the order of 160 tons, and combined with this for fluxing will be 50 tons of siliceous flux; and 40 to 50 tons of cold, solid copper-bearing material such as copper scrap, matte, slag, skulls, and other reverts which are also returned for salvaging metal values. The products from this complete charge will be 65 tons of blister copper, with the remainder passing off as slag or volatile gases.

Table 1.22. Converter Operations, Hindustan Copper, Ltd.:
Converter Performance (Typical for 1 Month)

Matte treated	2924 tonnes
Quartz flux consumed	521 tonnes
Cold secondaries treated	158 tonnes
Copper scrap treated	86 tonnes
No. of converter charges made	313
Blister copper produced	860 tonnes
Converter slag produced	1582 tonnes
Blowing time	1149 hr 55 min
Slagging time	109 hr 45 min
	1259 hr 40 min
Normal converter campaign	225 blows
Blister produced per campaign	550 tonnes
Refractories consumed	2.5–3 kg/tonne blister

Source: J. C. Yannopoulos and J. C. Agarwal, Eds., *Extractive Metallurgy of Copper*, Vol. 2, The Metallurgical Society of AIME, 1976, p. 203.

The sequence of converter operations is first to charge a few tons of revert scrap into the empty converter in order to protect the refractories, and on top of this protective layer to pour in 60 to 75 tons of liquid furnace matte at 1900°F (1040°C). Blowing is started and the converter turned to its upright operating position. Then after 10 minutes, 12 to 14 tons of SiO_2 flux is added to form the first slag. The operating temperature will be in the 2200°F (1200°C) range, and punching is done 60% of the time at the start in order to keep the tuyeres open.

After approximately 1 hour of blowing the converter is turned down, the air is shut off, and 40 to 50 tons of slag are skimmed off and carried back to the smelting furnace. Another 35 tons of molten furnace matte is now added to bring up the converter volume, along with a few tons of scrap, and a second hour-long blowing cycle is repeated. Punching to keep the tuyeres open is not required as often as with the first so-called "green charge," and tuyere blowing is freer. A somewhat smaller amount of flux, 8 tons, is added during the second blow, as there is now a lesser amount of iron to be oxidized and slagged.

This complete cycle is repeated through a total of half a dozen hour-long blows, with reducing amounts of revert scrap and silica flux added until the last blow will have only some 2 tons of flux added to it. The final product of blister copper is now poured from the converter and taken on to the next operation of refining. The complete batch conversion from matte to blister copper takes on the order of 12 hours.

Converter slag contains appreciable amounts of copper, from 2 to 5%, and is too high in values to be discarded without recovering these. Prills of matte and copper are carried out with the slag, which is skimmed off. Matte entanglement with magnetite, also skimmed off with the slag, and the fact that the slag is at times rather sticky to hold up metallic values sinking through it both contribute to high slag losses. The usual method of slag cleaning is to return the converter slag, still liquid, to the smelting furnace.

Considerable magnetite (Fe_3O_4) is produced during converting, as Fe_3O_4 forms readily from the further oxidation of FeO during the intensive initial oxidation of FeS to FeO, and Fe_3O_4 is a stable oxide at the 2200°F (1200°C) operating temperature of the converter. Being chemically rather inert, the Fe_3O_4 will not readily combine with the SiO_2 flux to form a slag, but it does have a solubility in the converter slag of some 15 to 30%,

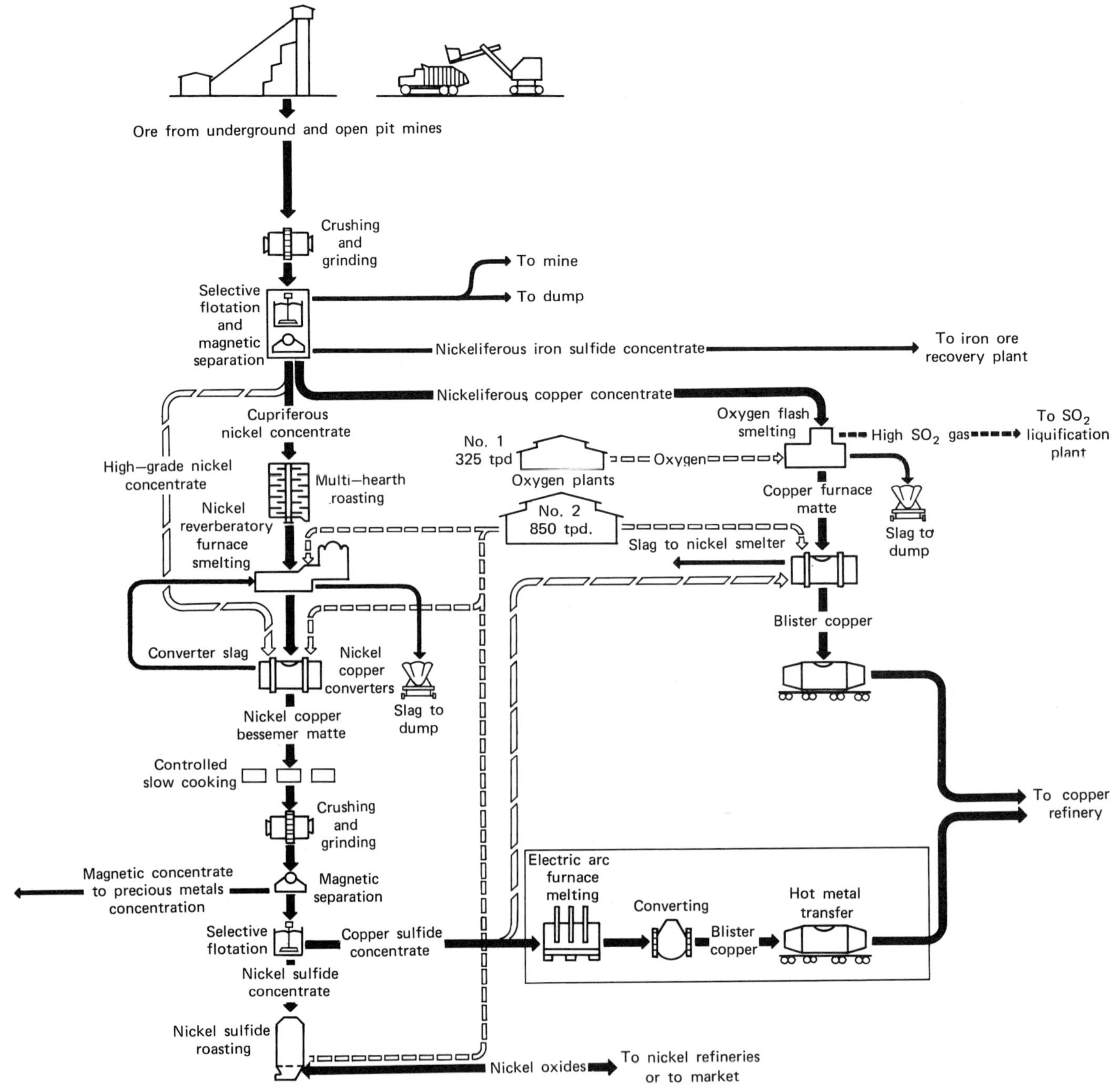

Figure 1.43. Nickel–copper smelting, the International Nickel Company. *Source: J. Metall.*, Vol. 18, No. 4, 1966, p. 442.

depending on the slag temperature. This Fe_3O_4 is carried back to the smelting furnace with the converter slag, where it has a tendency to settle out in the bottom of the furnace; and if the buildup is not checked, it can be severe enough to cause the shutdown of the furnace due to limiting the space necessary in the furnace for proper fuel combustion and room for the slag and matte layers to form and separate.

A fairly high percentage of magnetite is a necessity to protect the converter linings and prevent rapid erosion. However this has to be balanced against too high a percentage of magnetite carried out in solution in the converter slags, with its associated problems to the smelting furnace. It has been found that to control the amounts of Fe_3O_4 in the converter certain definite amounts of free excess SiO_2 are necessary to combine with the FeO, produced on oxidizing FeS, in order to make a slag of $FeO \cdot SiO_2$ before the FeO can be further oxidized to Fe_3O_4. Also, if sufficient FeS is also available, it will reduce the Fe_3O_4 and permit it to form a slag with SiO_2:

$$3Fe_3O_4 + FeS + 5SiO_2 = 5Fe_2SiO_4 + SO_2$$

Another solution to this carrying back of magnetite to the smelting furnace and its subsequent buildup is not to recycle the converter slag, but instead to cool, crush, and grind it and then clean it of copper content by froth flotation. This new trend has the advantage of better control of hearth accretions and increased capacity in the smelting furnace, as well as lower overall copper loss from the smelting–converting operation.

With the large volumes of air being blown into the converter and the type of reaction proceeding, large amounts of flue gas containing considerable flue dust as well as SO_2 pass off into the converter flue system. The dust is removed in dust collectors and recycled back into the process, and the gas, if high enough in SO_2 content, is used for sulfuric acid production.

1b. Copper Sulfide Combined Matte Converting and Concentrate Direct Smelting are converting operations using oxygen enrichment in the air blast which speeds up the oxidation process changing FeS and Cu_2S to FeO and metallic Cu, and also provides additional heat of reaction which is used to smelt copper concentrate directly in the converter and treat it without having had prior furnace smelting to matte.

The Hitachi smelter in Japan used this process (in conjunction with a blast furnace) from the late 1950s until the early 1970s, and Kennecott Copper (Utah) also used it for several years. Currently, the International Nickel Company (Sudbury) is making use of this method to smelt Cu_2S from their nickel and copper matte separation operation.

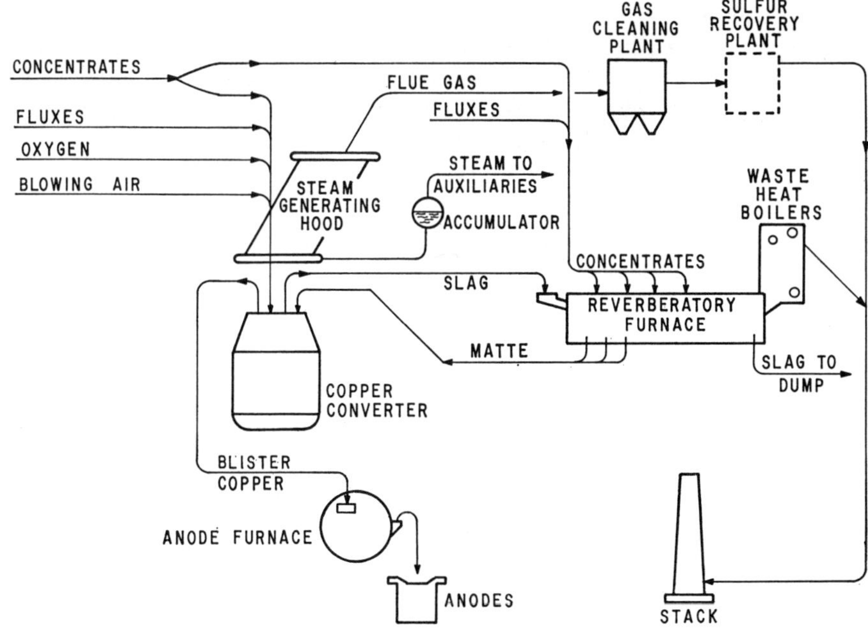

Figure 1.44. Converter smelting of copper concentrates in oxygen enriched converter. *Source:* R. P. Ehrlich, Ed., *Copper Metallurgy*, Extractive Met. Div. Symposium on Copper Metallurgy, AIME, 1970, p. 93.

Table 1.23.   Oxygen Converter Results

Oxygen enrichment levels	25–30% oxygen
Blowing time reductions	15–30%
Increase in waste gas SO_2	At least 2%

Cold Material Melted	Melting Ratio (tons material/ton enrichment oxygen)
Cold dope	3–8
Precipitate (cement) copper	4–8
Copper scrap	6–12

Source: E. J. Harbison and J. A. Davidson, *Oxygen in Copper Smelting*, Air Products and Chemicals, Inc., Allentown, Pennsylvania, 1972, p. 5.

The chief disadvantage in this type of treatment is in the high rate of refractory wear at the tuyere region from the overheating due to the oxygen-enriched blast, which affects the overall economics of the converting step.

Standard 13 by 30 foot Peirce–Smith converters (3.96×9.15 m), with chrome magnesite refractory linings and modified to add oxygen enrichment to the tuyere air blast, are used for this process. A pressure gun is also installed to feed moist concentrate and cement copper precipitate, if also available, varying from 6.5 to 12.5% moisture. Greatest feeding ease is with mixtures of less than 10% moisture.

Oxygen enrichment has been established at 25 to 30%, and oxygen enrichment through the air blast is begun with the period of concentrate feeding and continued throughout this period. Blowing is continued for some time following the concentrate charging, sometimes with oxygen enrichment, until all the solid feed material has been reasonably digested in the existing liquid pool of matte.

Concentrate feed rates vary from 1 to 2 tons per minute through the low-pressure air feeding gun, and a wide distribution of particles is achieved both in the gas above the surface of the liquid bath and at the bath surface. This provides a suitable kinetic environment for both rapid smelting and the oxidation of the iron and sulfur. The water in the moist feed is quickly vaporized and causes no problem with entrapped steam in the bath. A smelting rate of $6\frac{1}{2}$ to $10\frac{1}{2}$ tons of concentrate on a dry weight basis per ton of oxygen enrichment has been established as the ratio of oxygen consumption.

Because of the difference in the type of charge from the ordinary converting operation, with up to two fifths of the metallic charge being fine concentrate, precipitate, and scrap, the dust carry-off in the flue gases will be 40% higher than for normal practice where the metallic charge is made up principally only of molten furnace matte. The flue dust is recovered in dust collectors and recycled back into the smelting furnace.

Gas analysis show that during the periods of oxygen enrichment the SO_2 concentration increases as the oxidation rate in the converter is higher. Similarly, when concentrate is fed during a period of oxygen enrichment, the SO_2 concentration in the flue gases increases still further, due to the fact that the amount of sulfur in the bath has been increased with this type of charge and the availability of sulfur to oxidize to SO_2 is greater. As the damp feed material does not smelt immediately, the oxidizing reaction and period of high SO_2 concentration in the flue gases will extend beyond the period when feeding is being carried out. Normal SO_2 gas content will be in the range of 4.5% SO_2, while this increases to 6.5 to 9.5% SO_2 during periods of concentrate charging and oxygen enrichment, with an accompanying rise in gas temperature of up to 150°F (65°C).

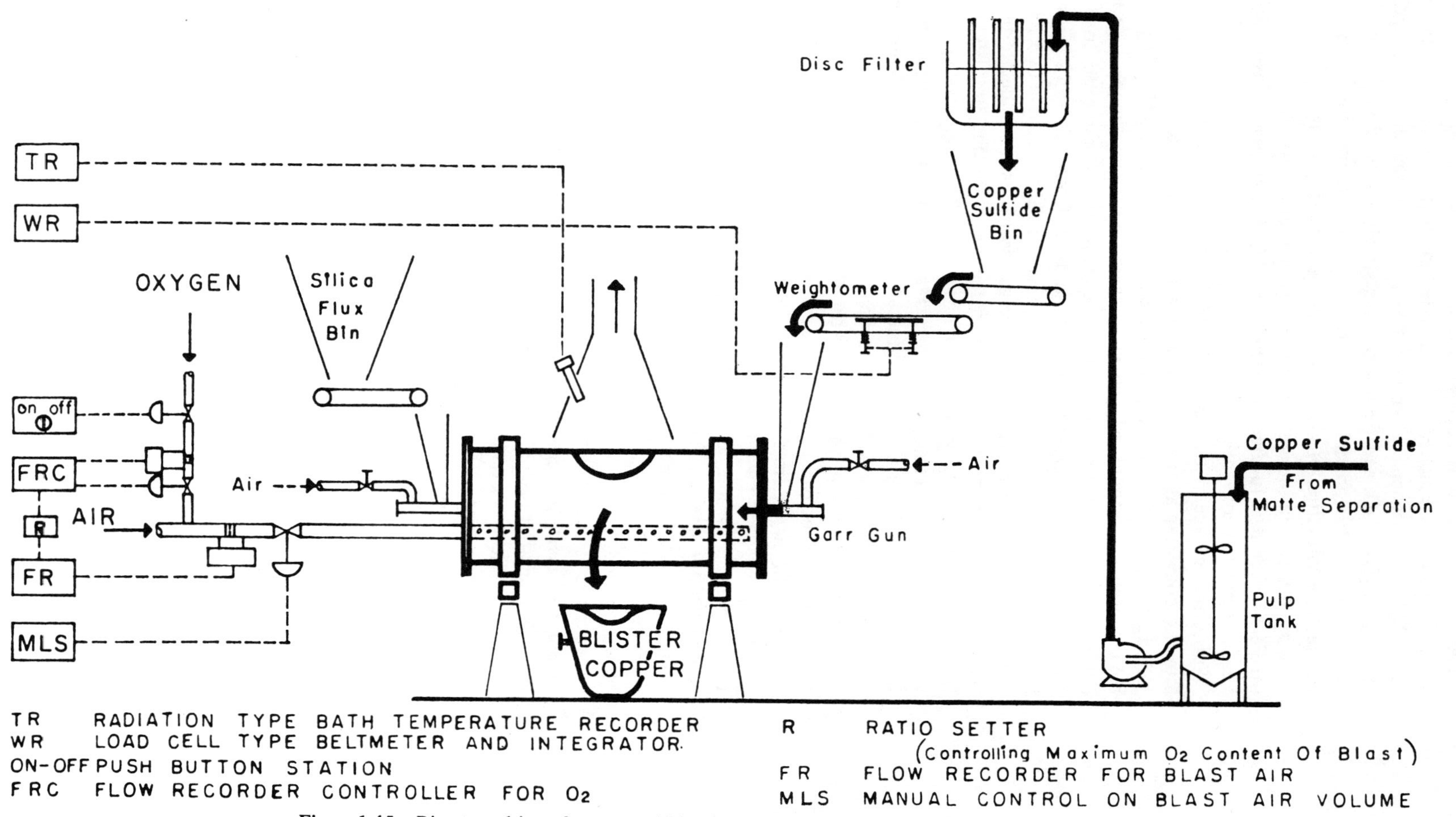

Figure 1.45. Direct smelting of copper sulfide with oxygen-enriched air in copper converters, International Nickel Company. *Source: J. Metall.*, Vol. 18, No. 4, 1966, p. 451.

The initial charge to the converter is matte, and concentrate is added to this after blowing has begun, to be melted and assimilated in the liquid bath. The amount of slag will be somewhat greater when concentrate is part of the charge, due to the additional impurities of SiO_2, Al_2O_3, CaO, and FeO, which are largely removed as slag during the smelting to matte but are still present in the green concentrate making up part of this converter charge. The formation of magnetite is not excessive, probably partly due to the amounts of SiO_2 available in the concentrate charge in addition to the SiO_2 flux added. These slag off FeO before it is further oxidized to Fe_3O_4, and compares favorably with amounts of Fe_3O_4 formed in conventional converter operations.

Sufficient furnace matte, some 60 to 70 tons of 35 to 40% copper content, is poured into the converter at the beginning of each charge to cover the tuyeres and provide a sufficient bath to absorb and smelt the concentrate to be added later. The matte is then blown, with or without oxygen enrichment, until the temperature is up to 2050°F (1120°C), at which time raw moist concentrate, of some 30 to 35% copper, or a concentrate–cement copper precipitate mixture, from 6.5 to 12.5% moisture, is blown into the converter through a low-pressure air gun at a feed rate of 1 to 2 tons per minute until 60 to 70 tons has been added. Oxygen enrichment is added to the tuyere air during the concentrate feeding period and for some time following, until all the material charged has been reasonably digested in the molten bath. The temperature rise from exothermic oxidation reactions during blowing increases sufficiently, especially during periods of high oxygen enrichment, so that reductions in enrichment levels must be made to maintain reasonable thermal conditions in the converter.

Slag formation begins shortly after the start of the concentrate feeding as 10% of the concentrate weight is silica. This, and an additional 10 tons of flux averaging 70% SiO_2, combine with the oxidized FeO and make a slag.

Blowing is carried on for $2\frac{1}{2}$ hours, with 28 to 30% oxygen enrichment for 75 to 85% of the blowing period, after which time slag analyzing 3.75% copper is skimmed and returned to the smelting furnace for cleaning.

The converter is now recharged for the second blow with 35 tons matte, 30 tons concentrate, and 10 tons flux and blown for $1\frac{1}{2}$ hours with O_2 enrichment for 75% of the blowing period.

This cycle is repeated for a total of four or five blows, with gradual decreases in the amounts of matte and concentrate charged as the converter fills with white metal and the slag skimmed after each blow decreases to leave less room for fresh charge.

The final blow converts the white metal now filling the converter to metallic blister copper, and no matte or concentrate is charged before this blow. If any charge is added here it will be a few tons of material already high in copper metallics, such as anode rejects or cement copper precipitates. This last blow is of 3 hour duration, and oxygen enrichment is either not used at all or, if used, is added for no more than one third of the blowing period. Blister copper of approximately 99% copper content is the final converter product and is taken to the next operation of refining.

1c. Copper Sulfide Concentrate Continuous Combination Smelting-Converting combines both smelting and converting into one operational step, so that a pelletized or nodulized unroasted concentrate charge is smelted and converted simultaneously to give products of metallic blister copper and slag.

Both the smelting and converting operations follow the same reactions as when the two are conventionally carried out as separate individual steps in the overall processing of beneficiated concentrate to blister copper, with the smelting stage first melting the charge and then in the liquid state having the Cu_2S and FeS combine as matte. The heat required for smelting is mostly supplied by excess exothermic heat from the second operational stage, which is converting.

In converting the first reaction to occur is the highly exothermic reaction where iron sulfide is oxidized to iron oxide and sulfur dioxide by the oxygen in the air blown in through the tuyeres,

$$2FeS + 3O_2 = 2FeO + 2SO_2$$

and it is largely heat from this reaction which is utilized in the smelting. The iron oxide formed then combines with the silica flux to form a slag, while the SO_2 goes off as flue gas.

When the $FeO \cdot SiO_2$ slag has formed, the second stage of converting takes place, with the Cu_2S in its turn being oxidized to metallic blister copper:

$$Cu_2S + O_2 = 2Cu + SO_2$$

Both of these two later reactions of slag formation and Cu_2S oxidation are also exothermic and add to the total exothermic heat available in the reactor.

While the overall reactions in the two different types of process reactors are similar, the products differ to some degree in that, while both produce blister copper of high grade as the metallic product, the slag from the rotating horizontal cylindrical furnace is high in metal values and must be processed to be cleaned, the slag from the stationary furnace is low in values and can be discarded directly.

Cylindrical Rotating Reactors, developed by Noranda Mines, Ltd., and also used by the Kennecott Copper Company, are similar in some general aspects to the conventional drum-shaped Peirce–Smith converter in that the smelting–converting reactor is also

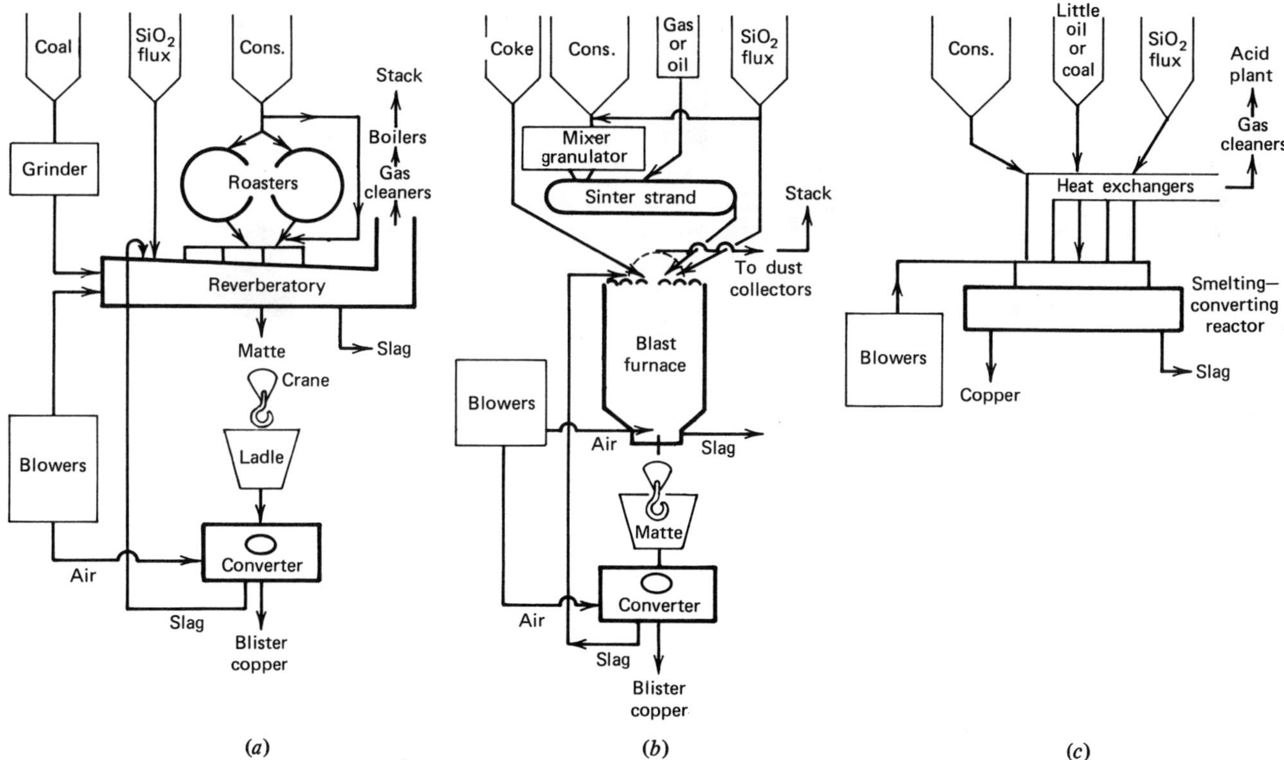

Figure 1.46. Diagrammatic comparison of (*a*) roasting–reverberatory–converter smelting; (*b*) sintering–blast furnace–converter smelting; (*c*) continuous smelting–converting. *Source:* R. P. Ehrlich, Ed., Copper Metallurgy, Extractive Met. Div. Symposium on Copper Metallurgy, AIME, 1970, p. 210.

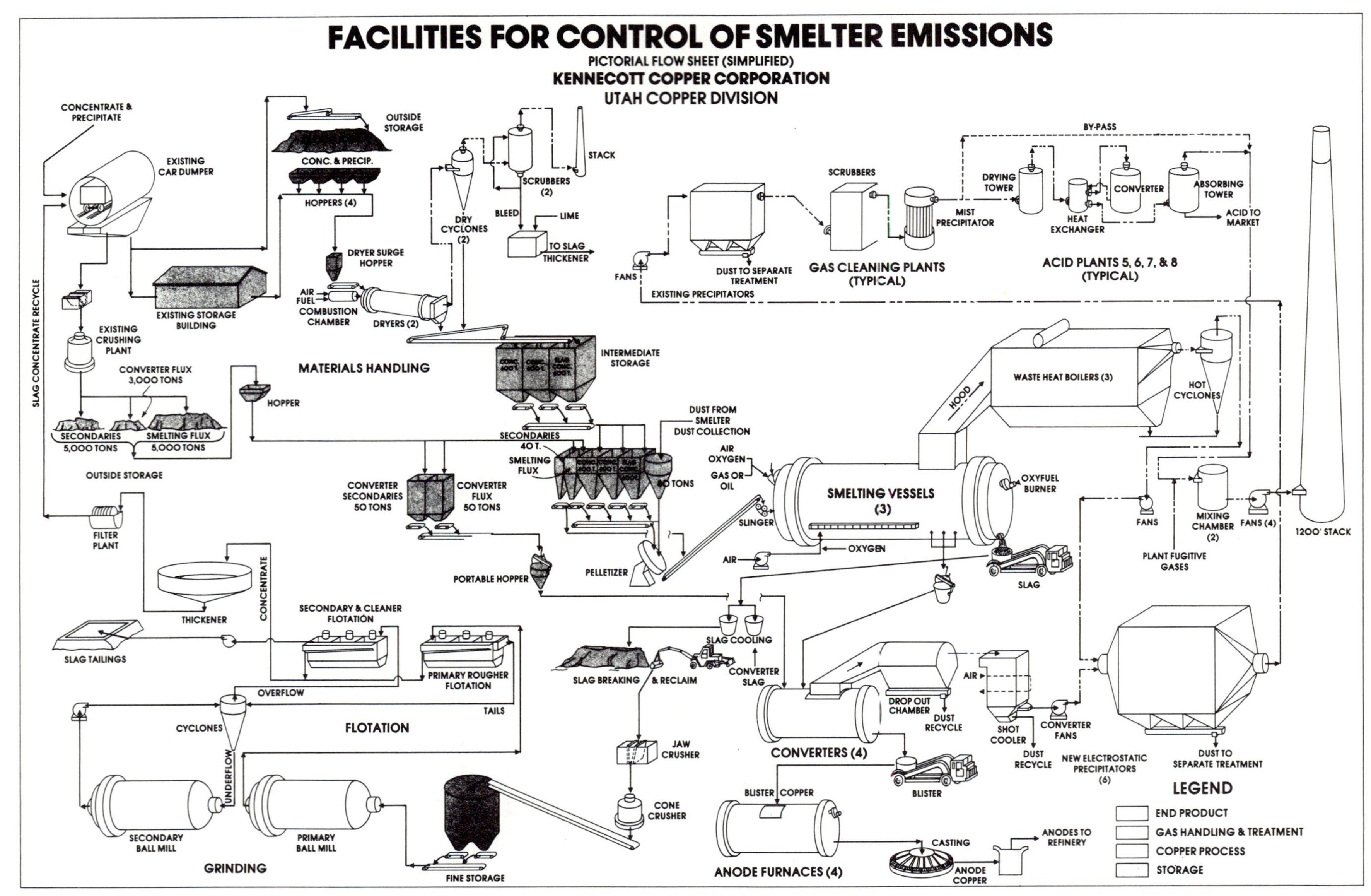

Figure 1.47. Flow sheet of Utah Copper Division smelter using Noranda-type smelting–converting reactors. *Source:* Courtesy of Kennecott Copper Corporation.

cylindrical, can be rotated to raise the tuyeres out of the bath, and has tuyeres in a row along part of the long side of the vessel.

The construction of the reactor is a steel shell with a chrome magnesite brick lining and in a size to process 100 tons of pelletized concentrate feed per day has an inside length of $32\frac{1}{2}$ feet (9.90 m) with a 7 foot (2.13 m) inside diameter at the feed end. [The most recent design has been scaled upward to produce 200 tons copper from 800 tons concentrate per day and is 67 feet (20.4 m) long.] A sump with two 1 inch (2.5 cm) tap holes at its bottom is located in the middle of the reactor, and it is here that the blister copper, which is the metal product, collects and is tapped to be removed and eventually cast into anodes for transportation to the electrolytic refining operation which follows.

Beyond the sump depression the bottom slopes up to a height at the endwall of 4 feet (1.22 m) above the bottom of the sump. The slag tap hole is in this endwall, at the end opposite to where charging takes place, and both molten copper and slag can be tapped intermittently or continuously to maintain the bath at the desired operating level.

Gas burners are positioned in both endwalls of the vessel to provide additional heat to the bath and supplement the heat produced by the exothermic reactions. The burner at the feed end, where the cold pelletized charge is added, has three times the heating capacity of the burner located at the end where the hot slag is run off, 15×10^6 Btu per hour as compared to 5×10^6 Btu per hour (14.15×10^6 kJ to 4.72×10^6 kJ).

The cold charge consists of concentrate pellets $\frac{1}{4}$ to $\frac{1}{2}$ inch in size (6.25 to 12.5 mm), mixed with silica flux $\frac{1}{2}$ to 1 inch in size (12.5 to 25 mm). Flue dust is recycled into the process by adding it to the concentrate before pelletizing. A belt slinger throws the mixture of pellets and flux in a continuous stream into the end of the reactor through an open feed port. This wide distribution of the feed charge over the molten bath contributes to the fast melting and high smelting rate accomplished in the reactor.

Thirteen 2 inch (5 cm) tuyeres on 6 inch centers (15 cm) are positioned in a row for approximately two fifths of the vessel's length from the feed end. Air at 160°F (71°C) is blown through as many tuyeres as are required to add oxygen in stoichiometric proportions so that all the sulfur and iron in the charge are oxidized to produce blister copper and slag. By having the tuyeres at a sufficient depth below the bath surface (3 feet, 0.91 m), 95 to 100% of the oxygen available reacts with the matte, and this consistent high oxygen utilization makes it possible to predict accurately the amount of oxygen required for each ton of concentrate of a particular composition that is charged. The reactor can be rotated through 54° from the vertical to raise the tuyeres out of the bath.

Two additional tuyeres are located just past the copper sump toward the slag end of the

Table 1.24. Operational Heat Balance, Noranda Continuous Reactor, Noranda Mines Ltd.

Heat Input[a]	MM[b] Btu	Heat Output	MM Btu
Heat from oxidation of feed	3.46	Heat content	
Heat from fuel input		Copper at 2190°F	0.18
Gas	5.09	Slag at 2250°F	1.18
Oil	0.05	Dust at 2400°F	0.07
Coke	0.31	Off-gases at 2400°F	6.90
		Heat loss from shell and mouth	0.58
Total	8.91		8.91

Source: J. C. Yannopoulos and J. C. Agarwal, Eds., *Extractive Metallurgy of Copper,* Vol. 1, The Metallurgical Society of AIME, 1976, p. 475.

[a]Heat inputs and outputs expressed as MM Btu per dry ton of copper concentrate.
[b]MM = 1000×1000.

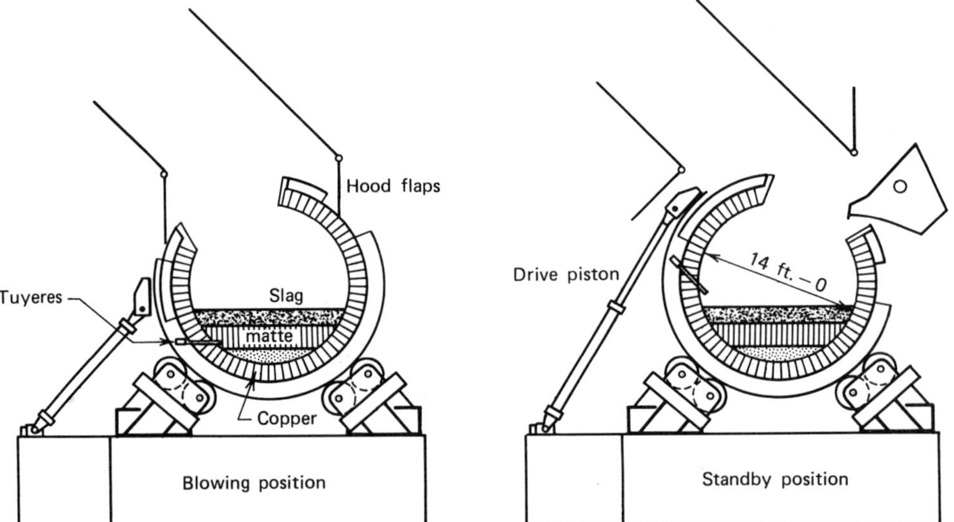

Figure 1.48. Section through Noranda smelting–converting reactor. *Source:* J. C. Yannopoulos and J. C. Agarwal, Eds., *Extractive Metallurgy of Copper*, Met. Soc. of AIME, Vol. 1, 1976, p. 464.

reactor, and these can be used if needed to blow either additional air for oxidation or natural gas for reduction of excess magnetite that may form.

Flue gas escapes from the vessel through the reactor mouth which is positioned above the copper sump. The gas passes into a water-cooled hood where it is cooled by water sprays and is then cleaned of solid particles by a settling chamber and an electrostatic precipitator in series. The collected dust is returned to the concentrate pelletizing operation for recycling through the process, and it is about 5% of the total charge input. The cleaned gas is an economically desirable source material for sulfuric acid manufacture.

The bath of the reactor is intensely active, with two processes occurring simultaneously

Table 1.25a. Production and Operational Data of the Noranda Continuous Reactor, Noranda Mines, Ltd.: Mass Balance—Matte Production Using Air

	Dry Tons	%H$_2$O	%Cu	%Fe	%SiO$_2$	%S	%Pb	%Zn
Input								
Copper concentrate	45,880	9.1	23.8	29.2	3.0	34.5	2.1	3.6
Slag concentrate	4,770	11.0	44.4	17.9	10.5	15.0	2.1	1.4
Flux	11,300	3.5	0.1	4.5	68.9	—	—	—
Evaporative cooler dust	400	—	47.1	11.5	6.4	16.5	4.5	2.2
Reverts	300	—	40.0	13.0	8.0	14.0	4.0	2.0
Precipitator dust	2,300	—	8.9	4.5	2.0	11.5	25.0	12.3
Output								
Matte	14.520	—	75.8	1.8	—	19.8	1.1	0.4
Slag	37,760	—	5.4	38.4	25.6	1.3	2.4	4.4
Evaporative cooler dust	400	—	47.1	11.5	6.4	16.5	4.5	2.2
Reverts	300	—	40.0	13.0	8.0	14.0	4.0	2.0
Precipitator dust	2,300	—	8.9	4.5	2.0	11.5	25.0	12.3

Source: J. C. Yannopoulos and J. C. Agarwal, Eds., *Extractive Metallurgy of Copper*, Vol. 1, The Metallurgical Society of AIME, 1976, p. 474.

Table 1.25b. Production and Operational Data of the Noranda Continuous Reactor, Noranda Mines, Ltd.: Mass Balance—Copper Production Using Air

	Dry Tons	%H_2O	Dry Basis					
			%Cu	%Fe	%SiO_2	%S	%Pb	%Zn
Input								
Copper concentrate	5210	9.9	27.9	26.4	5.5	29.6	1.9	4.5
Slag concentrates	1796	9.1	35.2	20.7	11.6	8.2	1.6	–
Flux	800	3.5	0.3	4.7	68.4	–	–	–
Precipitator dust	260	–	9.9	3.9	2.0	14.6	12.0	18.5
Output								
Copper	1525	–	97.0	0.2	–	1.9	0.4	–
Slag	5217	–	10.6	34.0	20.0	2.4	2.3	5.7
Evaporative cooler dust	100	–	54.2	7.9	5.6	13.1	1.2	3.2
Precipitator dust	260	–	9.9	3.9	2.0	14.6	12.0	18.5

Production Data

Instantaneous smelting rate—copper concentrate	803 tons/day
Average blowing rate	29,500 scfm
Average excess air to feed-end burner	1,300 scfm
Utilization efficiency of oxygen in converting air	95.5%
Overall fuel ratio (per ton copper concentrate)	5.45 MM Btu
Average feed-end burner rate	146 MM Btu/hr
Coke fines fed (instantaneous)	14 tons/day
Total off-gas volume (calculated)	64,000 scfm
Off-gas analysis (calculated, % wet basis)	SO_2, 6.7; O_2, 0.4; N_2, 73.5; CO_2, 5.2; H_2O, 14.2

Source: J. C. Yannopoulos and J. C. Agarwal, Eds., *Extractive Metallurgy of Copper,* Vol. 1, The Metallurgical Society of AIME, 1976, p. 479.

on a continuous basis in the same general zone of the furnace, namely, the concentrate being smelted to matte and the matte being converted to copper. The mixing action of the air from the tuyeres keeps the bath well stirred, and this assists in the high heat transfer of the exothermic heat generated in the converting process over to the smelting phase where it is absorbed, as well as bringing together the reacting components to form matte, slag, and blister copper.

As the bath reactions progress, three layers are formed according to their specific gravities, with a 12 to 15 inch (30 to 37.5 cm) thick slag layer at the top, this over a 29 to 34 inch (72.5 to 85 cm) thick matte layer, and a copper pool in the sump at the bottom of the reactor which is allowed to rise to 10 to 12 inches deep (25 to 30 cm) before it is tapped down to 4 to 5 inches (10 to 12.5 cm). The reactor temperature at the slag surface is 2250°F (1232°C), while the copper is tapped at 2200°F (1204°C) and the slag is tapped at 2230°F (1221°C). Furnace off-gas is at a temperature of 2400°F (1315°C) and contains 4% SO_2. When the burner air is enriched to 50% O_2 and the tuyere air to 35% O_2, the SO_2 content in the off-gas increases to 13%.

The copper from the reactor contains about 2% sulfur, which is considerably higher than the 0.02 to 0.1% sulfur in conventional converter copper. So while it is possible to oxidize this higher sulfur in an anode furnace, it is a slow operation which takes several hours, and it is the usual practice instead to pour the reactor copper into any standard converter during its copper finishing blow.

The beneficiated concentrate charged to the reactor averages 23 to 24% copper, and the matte produced is of very high grade, on the order of 75 to 80% copper. The products are blister copper containing 97.0% copper and 1.9% sulfur and slag analyzing 8 to 12% copper, most of which is present in small metal globules, with 20 to 30% Fe_3O_4.

The runoff slag is much too high in metal values to be discarded, despite the quiet cleaning portion of the hearth from which it flows, and can be treated by processing in either of two ways. One method is to charge the slag to a conventional Peirce–Smith converting operation where the converter is processing matte produced by an ordinary smelting furnace. The other method is to granulate the slag in water and then fine grind this to 90% minus 325 mesh. The ground slag is put through flotation cells which give a concentrate of about 47% copper and a discardable tailing of 0.5% copper. The tailings loss of copper is on the order of 1.65% of the total copper in the pelletized concentrate that is initially charged to the reactor.

Stationary Furnace Type Reactors, the Worcra Process, are similar in processing principles to the cylindrical rotating reactor but are very much different in design, being more of a conventional smelting type of furnace. The earliest operational furnace design, of a size treating 72 tons of dry concentrate feed per day, was a U-shaped furnace, although later, larger designs have been proposed with a long, narrow, rectangular, straight-line furnace shape. Development of this continuous smelting process has been by Howard Worner of Conzinc Rio Tinto in Australia, and while long test campaigns have been successfully run and the process has many desirable features, it has not been commercially adopted yet.

The U-shaped furnace has converting taking place in one leg of the U and slag cleaning taking place in the other leg. The converting zone is 6 feet wide by 28 feet long (1.83 × 8.54 m), and is 13 feet high (3.96 m), with a row of 10 sidewall air lances positioned down its outside wall. These lances angle down into the bath and are positioned to blow close to the bottom of the furnace and at about the center line of the converting leg. This introduction of air at a position remote from the side wall is to reduce refractory wear

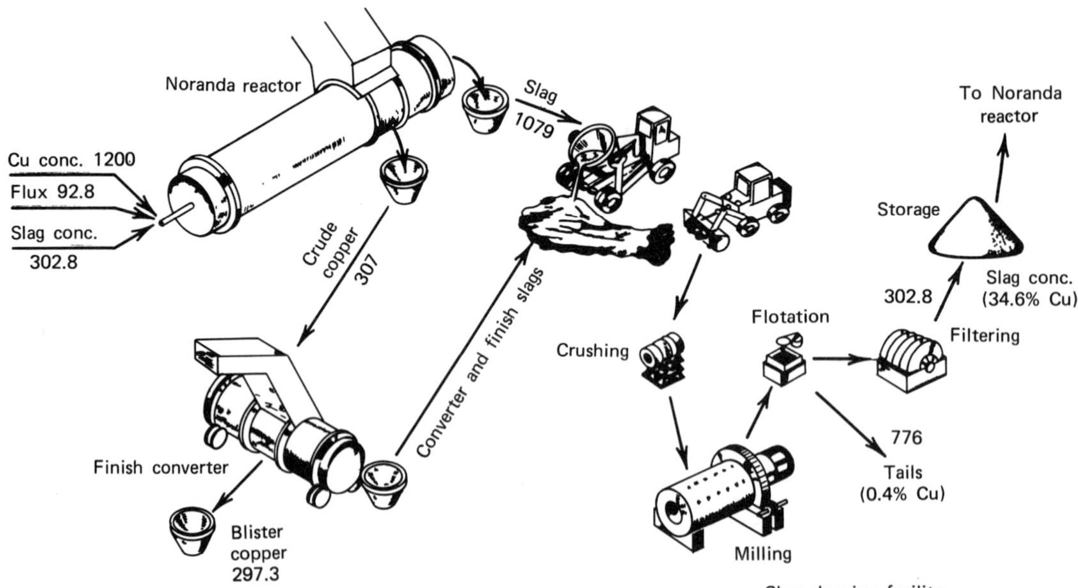

Figure 1.49. Slag cleaning process for the Noranda smelting–converting reactor (Noranda Mines Limited). *Source:* J. C. Yannopoulos and J. C. Agarwal, Eds., *Extractive Metallurgy of Copper*, Met. Soc. of AIME, Vol. 1, 1976, p. 357. (All flowrates in tons per day.)

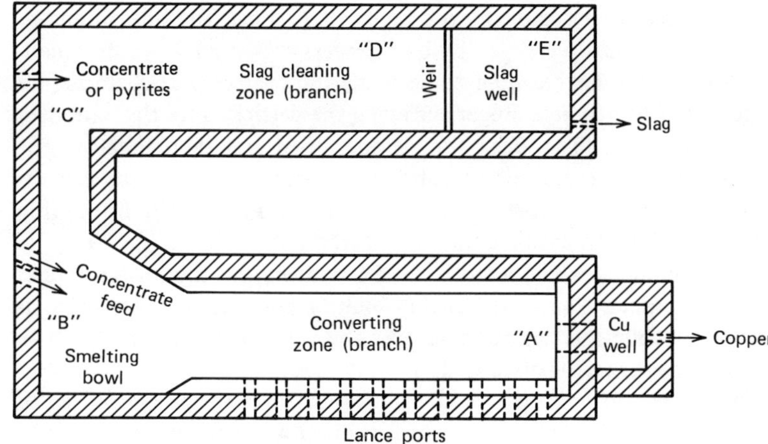

Figure 1.50. Plan view of U-shaped Worcra-type smelting–converting furnace. *Source:* R. P. Ehrlich, Ed., *Copper Metallurgy*, Extractive Met. Div. Symposium on Copper Metallurgy, AIME, 1970, p. 215.

and lengthen the furnace life, particularly if future trends are toward gaseous oxygen being blown in as the converter gas. The higher heat evolved in this case badly cuts and erodes the refractories if the gaseous oxygen is blown through a normal tuyere-type opening. The low positioning of the lances in the bath also gives a high gas–liquid reaction efficiency, so that oxygen efficiency is 100% relating to the oxygen in the air blown through the lances.

Conventional converting reactions occur in this zone, with matte being oxidized to blister copper and the iron oxide formed combining with the silica flux to make a slag. The three compound layers, slag, matte, and blister copper, then settle out according to their specific gravities. Sufficient exothermic heat is produced in the converting reactions to take care of both smelting and converting without more than minor need of auxiliary burners or supplemental oxygen.

The blister copper which is produced settles as a layer on the bottom of the converter leg of the furnace and flows continuously through an underpass in the endwall to a copper well. From this well the copper is surface tapped periodically and cast into 3500 pound blocks (1591 kg) to be shipped for refining.

The opposite leg of the furnace, which is 8 feet wide by 31 feet long (2.44 × 9.45 m) and $8\frac{1}{2}$ feet high (2.6 m), is used for slag cleaning and is a quiet zone where the entrained globules of matte and metal have an opportunity to settle out of the slag into their respective heavier layers below. After this settling, the slag overflows a weir at the top end of the furnace leg and runs into a slag well. From here it is tapped off continuously and, as it is low in metal values, can be discarded without any further metal cleaning treatments.

The furnace bottom slopes upward gradually from the converter zone endwall where the bath is 4 feet deep (1.22 m) to the weir at the end of the slag cleaning zone where there is just enough bath depth for the slag to overflow the weir and run into the slag well. This sloping bottom is an aid to having the matte and metal settling out of the slag slide down to the smelting and converting areas of the furnace and enter the zones of process reactions.

Smelting is carried out in an area 10 feet in diameter (3.05 m) located at the bottom corner of the converter leg on the section connecting the converting and slag-settling legs of the furnace.

The concentrate charge to the smelting-converting furnace is thoroughly dried in a rotary drier, and recycled flue dust along with a small amount of binder to give some nodulizing and pelletization are added at this time. The dried concentrate is screened, with one portion being smaller than $\frac{3}{16}$ inch (4.7 mm) and the other portion being larger than

$\frac{3}{16}$ inch (4.7 mm), with a maximum size of up to $\frac{3}{4}$ inch (18.75 mm). The fine, minus $\frac{3}{16}$ inch (4.7 mm) material is continuously charged by belt slingers at two points in the smelting zone and one position in the slag cleaning zone, while the coarser plus $\frac{3}{16}$ inch (4.7 mm) material is charged by a fourth belt slinger throwing the particles into the bath over the length of the converting zone. The silica flux is also dried and then added to the coarser concentrate being fed to the converting zone. This ensures that the flux is delivered to the area of the furnace where most of the iron is oxidized and positions the flux where it can most readily combine to produce the $FeO \cdot SiO_2$ slag.

The concentrate fed to the reactor will analyze 23.5% copper, and the blister copper produced will run 98.75% copper. The matte in the smelting zone will be very high, as much as 70 to 75% copper, while the slag running over the weir to the slag well will be on the average only 0.50% copper and is discarded.

The furnace gases, which pass off through a flue positioned over the converter leg of the furnace, carry all the sulfur in the charge except for minor residual amounts retained in the slag and blister copper, and these gases will contain on the order of 9 to 14% SO_2. The gases also carry off up to 4% of the total weight of the charge to the furnace, and this is removed in dust collectors and recycled back to the pelletizing driers which prepare the concentrate feed going to the reactor. The clean, high SO_2-content gas is ideal feed material for sulfuric acid manufacture and can be processed to produce this by-product.

Newer, larger furnace designs have been proposed to treat four to six times the yearly production of the furnace previously described, and the design of these seemingly will tend to be a long, narrow, rectangular straight-line furnace.

In this furnace design, the combined charging and smelting zone, with feeder lances, will be in the center portion, with the converting zone at one end and the slag settling zone at the other end. Air or air–oxygen lances will be inserted through one sidewall of the converting zone, and the bottom layer of blister copper formed will flow out into a copper well for removal.

The reactor bottom will slope upwards over the whole length of the furnace, from the deepest bath depth at the converting zone end wall to the shallowest bath depth at the opposite end where the cleaned slag overflows a weir into the slag well.

The roof may also slope, with a greater height over the smelting-charging and the converting zones than over the slag cleaning zone; while the flue gases will be removed through a roof port at the converting end of the furnace.

In many respects the Worcra Process looks to be superior to either the Noranda or Mitsubishi as it appears to do the whole continuous smelting operation in one furnace, and does not require that the slag be handled for recycling outside the furnace.

1d. Top-Blown Rotary Converters (TBRC Process) were initially developed to convert nickel matte to a crude nickel metal, with the top blowing from the oxygen lance generating the high temperature necessary to minimize nickel oxide formation, and this operation is discussed in detail in the section on nickel matte converting processes.

However some interest has been generated in copper converting by this process, although it is unlikely that top-blown oxygen converters will be used extensively for the normal converting of copper matte, as the conventional copper converting is autogenous with air blowing and the usual converting temperatures can be reached without difficulty.

The Afton Mines Smelter (British Columbia, Canada) is installing a TBRC process which will clean the slags resulting from a copper concentrate containing chalcocite and native copper being smelted and then converted to metallic copper.

2. Nickel Sulfide Furnace Matte is converted to nickel sulfide in Peirce–Smith-type converters, similar to those used for copper matte converting, by blowing the converter with

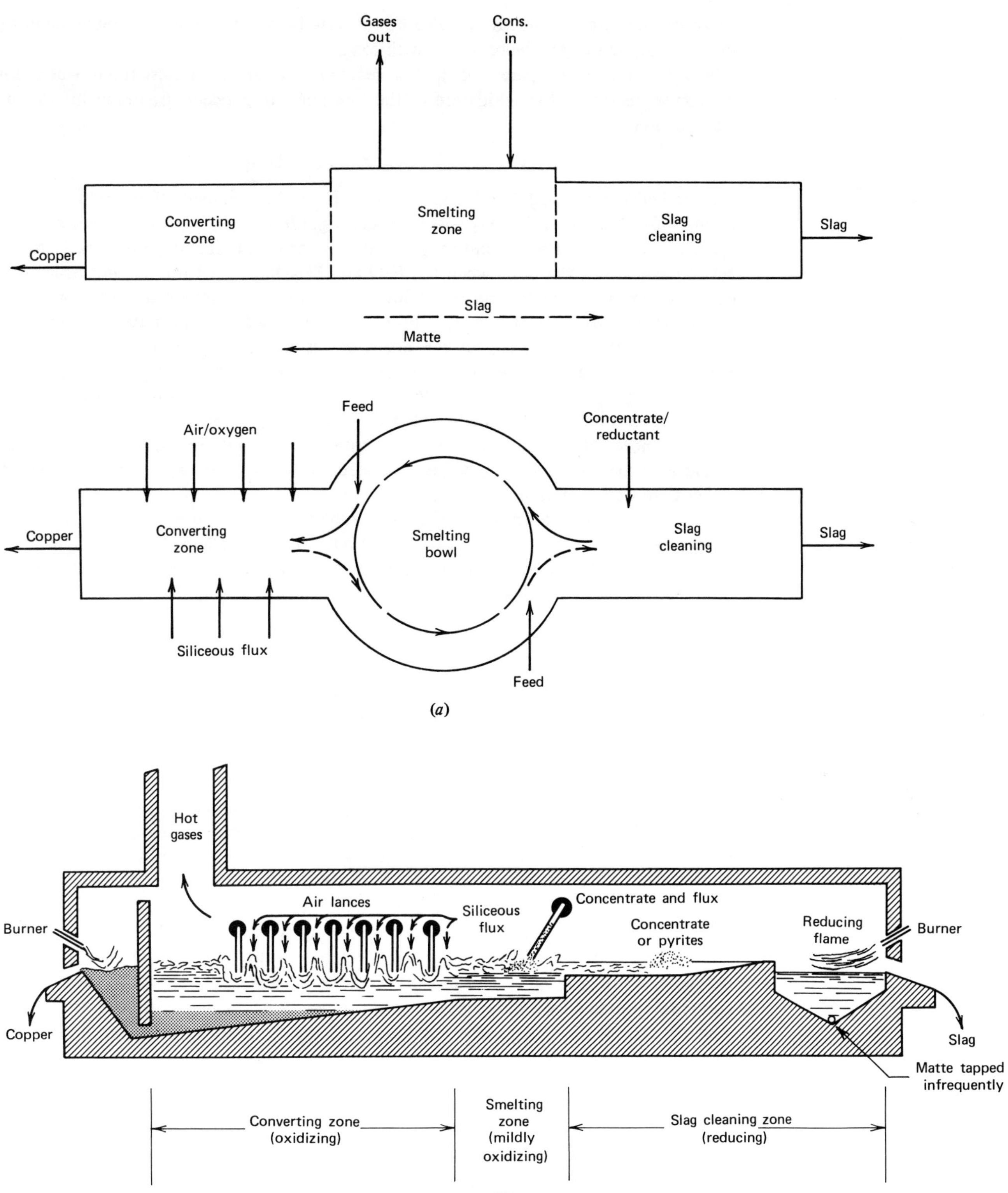

Figure 1.51. Worcra furnace: (*a*) straight-line form of smelting–converting furnace; (*b*) vertical section through furnace. *Source:* R. P. Ehrlich, Ed., *Copper Metallurgy*, Extractive Met. Div. Symposium on Copper Metallurgy, AIME, 1970, pp. 211, 213.

air or oxygen-enriched air. It can also be converted directly to liquid nickel metal in top-blown rotary converters being blown with oxygen.

In all the converting operations the first reactions are similar to those found when blowing copper matte, in that oxidation of the iron and sulfur occurs first in highly exothermic reactions,

$$2FeS + 3O_2 = 2FeO + 2SO_2$$

and the FeO forms a slag with the added SiO_2 flux to be skimmed off and returned to the smelting furnace, while the SO_2 passes off as a volatile substance in the flue gases.

When blowing in Peirce–Smith-type converters with air or oxygen-enriched air, it is customary to stop the blow when the FeO and SO_2 have been formed and to leave the converter product as molten nickel sulfide or a solution of mixed nickel sulfide with some copper sulfide, to be further treated by some other process. This product corresponds to the white metal phase in copper converting and is called converter matte, or sometimes Bessemer matte. The reason for suspending the blow at this stage is that nickel has a higher melting point than copper, 2650°F (1455°C) as compared to 1980°F (1084°C), and oxidizes much more readily. Therefore the temperature to keep metallic nickel liquid is more difficult to reach and maintain, and the rate of oxidation being greater would take metallic nickel, as it was produced by Ni_3S_2 oxidizing, on to the NiO stage rather than leaving it as metallic nickel.

In the top-blown rotary converter using oxygen with the turbulent bath promoting mixing and the gas stream directed into the bath giving excellent thermal efficiency, it is possible to obtain the higher temperatures needed and to go on to the direct oxidation of liquid nickel sulfide to the nickel metal stage,

$$Ni_3S_2 + 2O_2 = 3Ni + 2SO_2$$

in one process.

Horizontal Converters of the Peirce–Smith Type used for nickel matte converting are similar in size and construction to those used for copper converting. The charging, blowing, and slagging cycle is also similar, with 75 tons of a smelting furnace matte 13 to 18% combined nickel and copper being poured at 2000°F (1090°C) into a 13 by 35 (3.96 × 10.67 m) or 13 by 30 foot converter (3.96 × 9.15 m), and blown to produce FeO from the FeS in the matte. Six to 8 tons of siliceous flux plus some coarse ore is added at intervals as required to combine as a $FeO \cdot SiO_2$ slag, and this is skimmed off periodically to be returned to the smelting furnace for cleaning, as it contains some 2% combined nickel and copper values. The converter is then brought back up to its operating level with fresh molten furnace matte, and the blowing and slagging cycle is repeated.

Each blow requires about 35 minutes, and a temperature of 2250°F (1230°C) is reached before the blow is stopped and one or two ladles of slag are skimmed. The desired sulfur content is obtained by blowing until the matte reaches a predetermined temperature, which is achieved by the exothermic heat from the oxidation of sufficient sulfur to heat the converter charge to this temperature.

After a cycle of some 40 blows over a 24 hour period, the final converter matte will be 100 tons and will analyze 75 to 78% combined nickel and copper plus any precious metals. High-grade nickel ores will give a converter matte with as much as 75% nickel and only 3% copper, while lower-grade nickel and higher copper ores will give matte more of the proportion of 50% nickel and 25% copper.

Flue gases from the converter are cleaned of flue dust which is returned to the smelting furnace, while the cleaned gases are then discharged to the atmosphere through high stacks. As converter matte is the final product, still containing 22% sulfur, the SO_2 content of the gases is not sufficiently high for a sulfuric acid manufacturing source material.

Magnetite production is also similar to that found in copper converting.

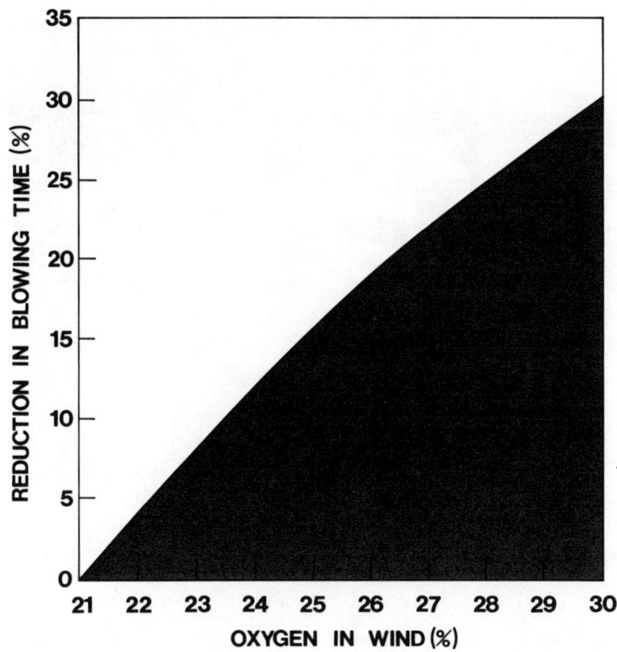

Figure 1.52. Effect of oxygen enrichment on converter blowing time. *Source:* Courtesy of Air Products and Chemicals, Inc., Allentown, Pennsylvania.

Oxygen-Enriched Horizontal Converters decrease the proportion of nitrogen present in the blowing air and make heat available to the converter that is normally carried out in the nitrogen of the flue gases. Consequently oxygen enrichment permits smelting in the converter of some cold charge, such as lump ore, sulfide concentrates, or revert materials, to utilize this excess heat and keep the temperature within reasonable operational limits not excessively detrimental to the refractory lining. Up to 3 tons of cold charge can be added for each ton of oxygen used.

The conversion reaction rate of FeS to FeO and SO_2 is also speeded up with oxygen enrichment, as the output of a converter is nearly directly proportional to the amount of oxygen blown through the charge. The oxygen content of the blast averages 28%, and the converter temperature reached is approximately 2300°F (1260°C). The metal content of the slags produced is similar to that from air blowing alone.

Top-Blowing Rotary Converters (TBRC), initially installed by the International Nickel Company, are the newest in converter design and are quite similar to the Kaldo rotating furnace developed for oxygen steelmaking in the 1950s. While first developed to convert nickel sulfide, they are now also being used or projected to convert copper matte and smelt secondary copper materials, as at Afton Mines' new copper smelter in British Columbia, Canada, and Peko-Wallsend's Tennant Creek smelter in Australia.

The converter has a cylindrical steel shell with a magnesite or chrome magnesite lining, is closed at one end, and has an open-ended cone shape at the other end. Two sets of roller-bearing running rings support the converter and rotate it on its long axis at any speed up to a maximum of 40 rpm. The whole unit is supported on trunnions and may be tilted through at least 180° for charging, flux additions, and slag skimming, and in its normal operating position will be inclined at approximately 17° (between 16° and 20°) to the horizontal, with the open end of the vessel pointing up.

Oxygen is supplied by a water-cooled lance which is inserted through the open mouth of the converter, and it is blown against the bath surface at a shallow angle which can be varied (22° to 30° to the horizontal). The lance normally has a single orifice 3 to 5 inches in diameter (7.5 to 12.5 cm), and oxygen is supplied at low velocity with an operating pressure of less than 40 psi (275.6 kPa). The gases from the converter are carried off in a

Figure 1.53. Installation of 226 ton top-blown rotary nickel converter. *Source:* Courtesy of Inco Limited.

fume hood connected to a flue system which fits down over the open mouth and substantially closes the vessel when blowing is in progress.

With the converter inclined at an angle of $17°$ to the horizontal, the liquid charge covers at least half of the back wall so that this part of the lining as well as the cylindrical part is all covered and submerged by the bath during each rotation. This counteracts the tendency for overheating of the refractory lining during the blowing cycle, as some of the radiant heat absorbed by the upper exposed portions of the refractory lining will be transferred to the melt during part of each rotation when the previously heated lining is now submerged.

The converter dimensions vary somewhat, with the diameter being approximately two thirds of the length. Units of 30 ton capacity are 18 feet long and $11\frac{1}{2}$ feet in diameter (5.49×3.51 m), and units of 100 ton capacity are 27 feet long and 18 feet in diameter (8.23×5.49 m).

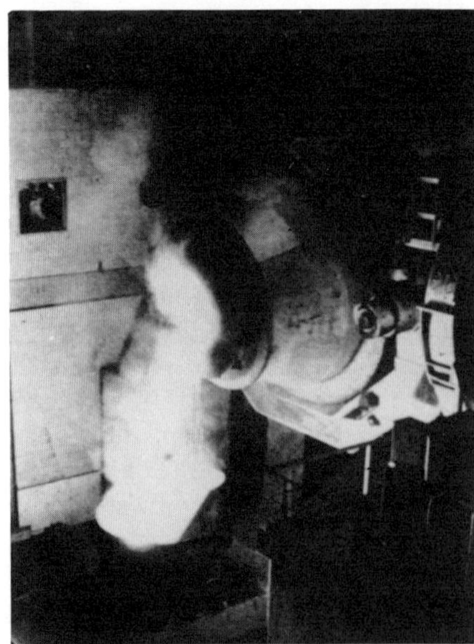

Figure 1.54. Top-blown rotary nickel converter pouring nickel product. *Source: J. Metall.*, Vol. 25, No. 1, 1973, p. 10.

The horizontal Peirce–Smith converter does not operate at a high enough temperature to allow nickel metal to form as nickel sulfide is oxidized. The converting temperature must be above 2516°F (1380°C) when the sulfur content of the bath is above 20% and increased to at least 2912°F (1600°C) when the sulfur has decreased to below 5%. Otherwise NiO will accumulate as the Ni metal that has formed reoxidizes:

$$Ni + O = NiO$$

In addition to the necessity of keeping the system for converting nickel at much higher temperatures than needed to convert to blister copper, it is also important that the bath be well mixed. This mixing is much less important in the conversion of white metal to blister copper, for as the copper-rich blister copper phase is produced, it sinks to the bottom of the converter away from the blast and minimizes any reoxidation of the copper metal. A phase of roughly constant sulfur–copper ratio is left and is presented to the blast throughout most of the process, as the two products—S off as SO_2 gas and Cu settling out as blister copper—are removed from it. On the other hand, when nickel sulfide is converted to metal the bath is a single phase of decreasing sulfur content. If it is not thoroughly mixed, the region of the bath exposed directly to the blast may be temporarily depleted of sulfur and saturated with oxygen, permitting liquid Ni metal to be oxidized to NiO crystals. The precipitation of fine crystals of NiO then increases the viscosity of the bath, which still further reduces the mixing rates and aggravates the situation even more.

Consequently the top-blowing rotary converter is superior for the converting of nickel matte to nickel metal because of the intimate contact and mixing between the gas, liquid nickel sulfide, and solid nickel oxide crystals supplied by the converter rotating and the greater heat evolved from exothermic oxidizing reactions by the use of gaseous oxygen for blowing rather than air.

The cycle of operation is the same for the top-blown rotary converter as for the side-blown Peirce–Smith type, in that liquid furnace matte is charged into the converter, rotation is started, the oxygen lance is lowered, blowing begins, and FeS in the matte is oxidized to FeO. Siliceous flux is added to make a slag, and when sufficient slag is produced it is skimmed off and returned to the smelting furnace for cleaning. This cycle is repeated until after several blowing periods the converter is full of nickel metal, which is poured off and taken to be refined.

Figure 1.55. Top-blown rotary nickel converter. *Source:* Courtesy of Inco Limited.

Some of the difficulties encountered with high-purity nickel matte are alleviated when the matte also contains appreciable quantities of copper. The metallic nickel–copper alloy produced in this case has a melting point lowered in roughly direct proportion to the amount of copper present, and the copper also serves to lower the activity of nickel in the bath and reduce its tendency to oxidize to NiO.

Because of the greater sulfur removal in this converter operation, with nickel being converted to metal rather than left as nickel sulfide, the converter gases are higher in SO_2 content, and after removing the flue dust, which is briquetted and returned to the converter, the gas can be used for sulfuric acid manufacture.

3. Lead Metal Containing Sulfur, as processed at Boliden's Rönnskär Works, is blown in a converter to oxidize the sulfur and to separate the dross and matte from the deoxidized lead.

Fine lead sulfide concentrate, 77% minus 325 mesh, is first flash smelted in an electric furnace above the liquid bath to react most of the PbS with air to give metallic Pb and SO_2. The remaining unoxidized PbS reaches the slag bath, here to undergo the roast-

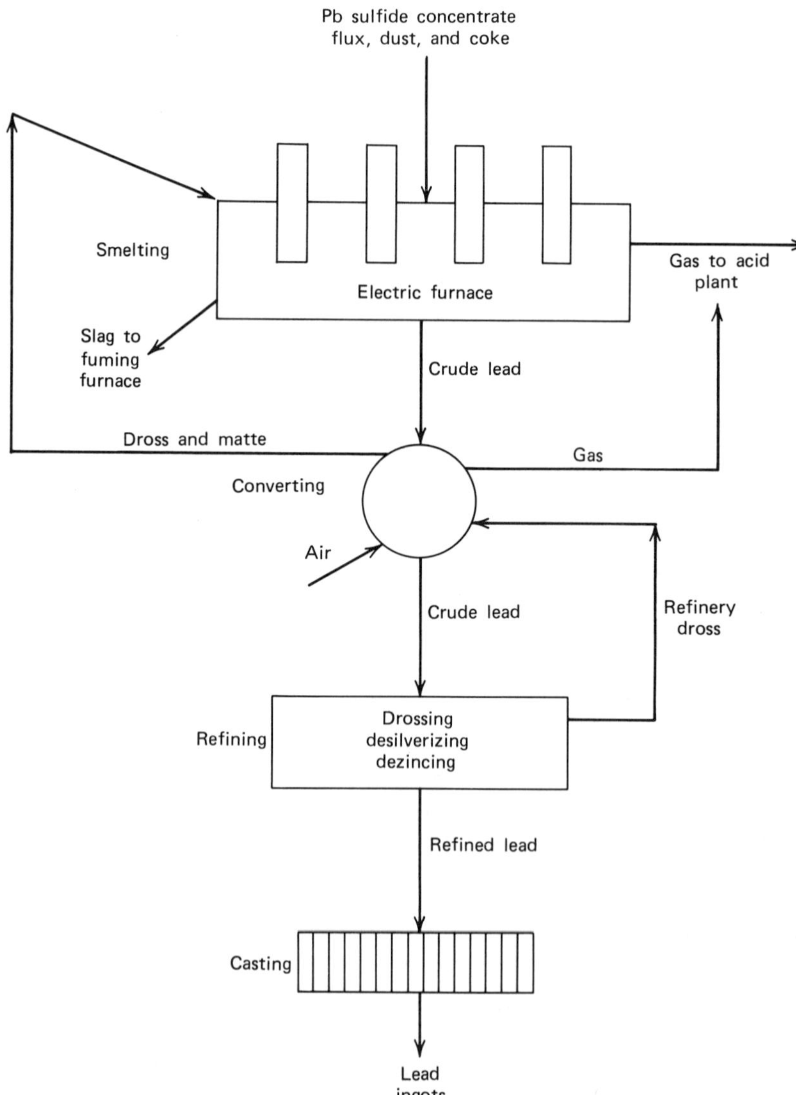

Figure 1.56. Lead smelter with electric furnace smelting and converting (Boliden).

reduction reaction with PbO in the returned flue dust and be reduced to lead bullion still containing 3% sulfur.

The lead is drawn off periodically, 70 tons at a time, and transferred along with refinery dross to $7\frac{1}{2}$ by 14 foot (2.28 × 4.27 m) alumina brick-lined Peirce–Smith converters, where it is air blown for 40 minutes at the rate of 3000 cubic feet (85 m^3) per minute. The lead is then allowed to cool in the converter for 7 to 9 hours, during which time matte and dross separate from the bottom layer of liquid lead. The dross and matte are raked out and recycled back to the smelting furnace. The crude lead, which lost its sulfur as SO$_2$ during the blowing period, is taken on to be further refined.

Converter gases are cleaned in a dust collecting system and the flue dust is recycled back into the process at the smelting furnace.

REFINING

Refining is the final operation to remove, and usually recover, the last amounts of im-purities still remaining after the major extractive processes have concentrated the valuable metal constituent and mostly separated it from the large amount of associated gangue material. The refining operation is generally considered separately from the extractive process, and in many cases the two plants are long distances apart with the crude metal from the smelter being shipped to the refinery for final treatment. This is often practical because of the great reduction in bulk during the smelting operations, removing most of the gangue impurities as slag and leaving only a relatively small tonnage of semipure metal to be shipped to the refinery.

The type of refining that is used to purify the impure metal coming from the smelter is balanced against the different impurities and their amounts to be removed, whether or not the impurities are to be recovered as by-products, and against the desired purity of the final refined metal product. To accomplish this, three separate types of refining are used—fire, electrolytic, and chemical—and in certain cases a combination of preliminary fire refining along with electrolytic refining.

FIRE REFINING

Fire refining is the last stage of purification for many metals in which the crude metal product from smelting, or from smelting followed by converting, is refined to produce an acceptable commercial grade of relatively pure metal. Or it may be an intermediate refin-ing step between smelting and pure metal, where a second stage of another type of refin-ing will follow fire refining. In either case it is always a batch operation.

The impurities to be removed by fire refining are generally in relatively small amounts, fractions of a percent. These were present in the original ore and remained with the metal during any prior roasting and smelting treatments. These impurities, although often only in minute amounts, impart properties to metals much different than the pure metal is expected to have and will considerably change such properties as ductility, electrical conductivity, brittleness, mechanical working ability, and corrosion. These types of im-purities are removed by selective oxidation, where air or oxidizing agents are added to the crude liquid metal and the impurities are then removed as an oxide slag or volatile oxide gas, as the impurities will oxidize before the metal.

Precious metals cannot be removed in this fashion as they have a slow oxidation rate and will oxidize more slowly than the metal being refined.

When fire refining is used for an intermediate refining step, it can be applied in two ways. These are, first, to remove some impurities which oxidize easily, leaving those impurities which do not oxidize easily to be recovered by another type of refining operation. Secondly, selective crystallization during slow cooling from the molten condition will concentrate certain values into definite crystalline combinations. Also, some compounds which are highly mutually soluble and form homogeneous solutions at high temperatures become less soluble as the temperature is lowered and will reject certain compounds as a solid dross floating on the remaining liquid metal.

TYPES OF FIRE REFINING

The purpose of fire refining is to eliminate impurities which are either detrimental to the physical properties of the metal, or the impurities themselves have a definite commercial value and are being removed from the crude metal in order to be collected as a valuable by-product. The amounts of impurities being removed are comparatively small as the crude metal has been smelted, or smelted and converted, and the great bulk of the more easily removable impurities has already been taken off in these prior operations. The tonnage of crude metal to be refined is also small for the same reason, and a ton of flotation concentrate analyzing 22% copper will shrink to some 420 pounds (191 kg) of 99% blister copper to be fire refined.

Refining operations removing the impurities as an oxide slag and volatile oxide gases are carried out in small reverberatory furnaces of sturdy construction. These are smaller than reverberatory smelting furnaces because of the smaller amount of material being treated and are stoutly constructed due to the increased specific gravity of the metal as it has been concentrated through the various stages of smelting and the lighter gangue materials removed. *Oxidation* in the liquid bath is carried out by adding some form of gaseous oxygen or an oxidizing chemical reagent. The slag produced is small in volume but high in values and is treated to recover these.

Drossing, where cooling precipitates a no longer soluble compound as the liquid bath cools, is done in small reverberatory furnaces or in large metal open-top kettles that can be heated from below. The dross is also treated to recover its values.

Slow Cooling for selective crystallization is done in insulated trays where the temperature can be controlled during cooling.

Metals that are still sulfides after smelting are fire refined in two stages, the first being to *dead roast the sulfide to an oxide*, next to *reduce this oxide to metal* in roasters with a reducing gas such as hydrogen or water gas, or with charcoal in retorts. Only the volatile impurities are removed in this treatment. The charge is not liquified, and so no slag is made.

Refining Furnaces, such as are used for copper, are conventional small reverberatory types usually with magnesite or silica refractory linings. The charge will be on the order of 300 tons of metal to be refined, and steel pipes are inserted into the bath and air or steam blown in to agitate the bath and oxidize the impurities, which either go off as volatile oxides or form a slag which is skimmed. Because of the values in these two products, they are collected to be reprocessed.

It is most convenient if the refinery is adjacent to the smelter so that hot liquid metal can be transferred quickly and directly from one operation to the other. The time required for furnace refining with a liquid metal charge is about 14 hours, including time for charging and casting. If the refinery is distant from the smelter and the crude metal has to be shipped as solid slabs and remelted in the refining furnace, then several more hours will be added to the furnace time for melting the solid charge.

All refining operations are batch processes, and one complete furnace charge of metal will be treated at a time from crude to refined metal.

Drossing Kettles, as are used in lead processing, are open-topped containers usually semispherical in shape, constructed of cast iron or welded steel plate, and heated from below. Capacities in the kettles run from 50 up to 300 tons, and the charge will be added to the kettle in a molten condition. Only moderate temperatures are required in drossing kettles, and it is infrequent that temperatures much above 1100°F (593°C) are needed for the low melting materials that lend themselves to drossing.

When the kettle is full, it is agitated by mechanical impellers, or air, which give a stirring motion to the bath and speed up the drossing operating. The kettle temperature is lowered and many impurities which are less soluble at the lower temperature rise to the surface of the bath to form a lumpy dross, which carries with it large quantities of entrained molten metal. The dross is removed by dipping it out in a basin with a perforated bottom and then is usually returned to the primary smelting operation. The perforated bottomed basin allows molten metal picked up with the dross to drain back into the kettle.

Drossing is a batch operation, and one kettle full of liquid metal is treated at a time.

Cooling Trays, developed for nickel–copper matte, are used for the controlled slow cooling of liquid metal mixtures which are homogeneous at a high temperature but will allow preferential crystallization and phase separation of the several different compounds in the melt as it is slowly cooled and solidified. The growth of larger grain sizes of these different phases is promoted as is their grain separation when cold, with a marked tendency to fracture along the grain boundaries rather than through the grains.

The nickel sulfide phase, after physical separation, can be used as-is for electrolytic anodes, or it can be roasted to oxide in fluid-bed roasters.

The oxide calcine can be treated in two ways, one of which is to reduce it with hydrogen in a fluid-bed reactor to give a semipure (90%) Ni metal which can be used directly for alloying. The other treatment is to reduce the oxide to impure metal in a hearth roaster by means of a reducing gas and then further refine the metal by a chemical method.

Retorts, as also used for nickel, are used to reduce oxides which have been roasted from sulfides and which contain only minor amounts of impurities to final refined metal. The metal oxide is briquetted into small, round, flat slugs, and these are charged with charcoal for the reduction of the oxide to metal into the top of tall retorts. The retorts have a small sectional area to ensure good uniform heat transfer into the center of the charge from external burners placed outside the retort silicon carbide walls and heating through it.

Retorting is a continuous operation, with feed being charged at the top and product being removed from the bottom in a steady progression. As the metal oxide, in close proximity with charcoal, reaches the reaction temperature of 2400°F (1316°C), it is reduced to metal by the carbon. The reaction is in the solid state, and there is no melting of the charge or slag formation. The solid refined metal slugs produced are discharged from the bottom of the retort and cooled in gas-tight coolers to prevent any surface oxidation while still at an elevated temperature. Then, when cool, they are separated from the excess charcoal still remaining after the reduction reaction was completed.

FIRE REFINING PROCESSES

1. **Converted Blister Copper** along with certain grades of copper scrap, cement copper, and cathode copper is fire refined to remove many of the contained impurities. The fire refining process, which employs oxidation, fluxing, and reduction, is based on the weak affinity of copper for oxygen as compared with the strong affinity of the impurities for

oxygen. The precious metals cannot be removed by fire refining as they have even a weaker affinity for oxygen than does copper.

A small 300 ton capacity reverberatory furnace, oil or gas fired, magnesite lined, with a shorter and deeper hearth than the reverberatory smelting furnace, is one type used; or chrome magnesite or magnesite-lined rotary furnaces, 30 feet long by 13 feet in diameter (9.15 × 3.96 m) holding 240 tons, heated by coal, oil, or gas burners, can also be used.

The operation is standard in either type of furnace and consists of charging either ladles of molten blister copper directly from the adjoining smelter or charging slabs of solid blister copper shipped from a distant smelter. Charging some 10 ladles of liquid blister copper will take 1 hour, while it will take from 2 to 4 hours to load the furnace with solid slabs.

Melting time of the charge in the furnace is also related to the type of charge. If liquid converter blister copper is added there is no time needed for furnace melting, while with solid slabs these will take 8 to 12 hours to melt.

As soon as the charge is liquid, oxidation of the impurities and slag removal begins, which takes from 2 to 4 hours. Compressed air at 8 to 10 pounds pressure (55 to 69 kPa) is introduced under the molten metal surface through $1\frac{1}{2}$ inch diameter (3.75 cm) steel pipes that are often protected by a coating of asbestos and fire clay cement. The compressed air blown into the bath causes a bubbling and agitation action which exposes as much molten blister copper as possible to oxidation, and cuprous oxide is formed. This blowing period is referred to as "flapping." The copper oxide formed dissolves in the metal bath, saturates it with oxygen, and oxidizes many of the contained impurities. Copper oxide reacts with the copper sulfide present to form SO_2 which escapes as a gas and almost completely eliminates any sulfur present. Zinc, tin, and iron are also nearly completely eliminated as oxides, as are lead, arsenic, and antimony. Silica flux is added to form a slag with the oxides formed that are not volatile. Nickel, selenium, tellurium, and bismuth, in addition to the precious metals, do not lend themselves to removal by fire refining and are generally removed by electrolytic refining.

This sequence is continued as long as slag is made to be skimmed off and until the bath is saturated with copper oxide. At this stage the copper is referred to as "set copper," and a cooled sample is easily spotted visually from the depressed "set" of the surface on cooling and the fact that it fractures easily and has a brick-red, coarse crystalline surface on being broken. This "set copper" contains from 6 to 10% Cu_2O, giving up to 0.9% oxygen in solution. The bath skimmed clean of slag is covered with charcoal or fine coke to prevent further oxidation, and the next process of reduction is begun.

The stage of reduction of the copper oxide in the bath to copper metal takes from 2 to 4 hours and reduces the set copper oxygen content from 0.9% to the required tough-pitch copper requirement of about 0.1 to 0.03%. The conventional reducing material is green wood poles, preferably some 24 feet long and 18 inches in diameter (7.32 × 0.45 m), which are inserted into the bath below the surface and allowed to burn away. Newer methods of reduction have used reformed natural gas and ammonia in place of the green poles, and there have also been proposals to substitute propane and butane as well. All of these reducing agents accomplish the same purpose, to convert the copper oxide in the bath to copper. The release of steam in the metal bath from the green poles, or the bubbling of reducing gas from a lance inserted into the bath, stirs the molten liquid and speeds up the reducing reaction.

Small test samples are cast at intervals during the reduction stage, and the surface and fracture of these samples indicates when reduction is sufficient and the "tough-pitch" copper stage has been reached. At this point the sample will have a slight "crown," and the fracture will show crystals of copper with small amounts of Cu–Cu_2O eutectic at the grain boundaries.

Once the correct pitch has been reached for the copper, casting takes place and will occupy 4 to 6 hours to discharge the furnace load of refined copper. In some refineries it is not customary to cover the furnace bath with charcoal during this casting period, and oxygen will be absorbed to give the final tough pitch copper as cast a content of 0.13 to 0.15% oxygen. Casting is done into molds in a straight-line endless chain type, or into molds in a large 20 to 40 foot diameter (6.1 to 12.2 m) wheel with 15 to 30 molds revolving in a horizontal plane. The copper flows from the furnace into a tilting ladle which fills the molds as they are brought into position below the ladle lip. The molds, which are made of copper, are dressed with an alumina slurry to prevent the cast liquid metal from sticking to them; and a variety of shapes of finished castings are made, including billets and various ingots, as well as anodes to be further electrolytically refined. The billets, wire bars, and ingots are removed by tilting the molds and dumping the solidified copper shapes into a water-filled cooling tank, while the anodes, which are 40 inches square (1.03 m^2) and 2 inches thick (5 cm) and weigh 600 pounds (272.7 kg), are raised in the casting wheel molds by a cast iron pushup pin with a steel head and are removed to a cooling tank by a crane. The precious metals and other trace metal impurities such as nickel, selenium, and tellurium that could not be removed by fire refining will be removed in electrolytic refining; and the fire-refined anodes are now a tough, homogeneous copper that will not fracture during the handling and treatment they are subjected to during electrolytic refining.

2. Nickel Sulfide Converter Matte consists of approximately 48% nickel, 27% copper, 22% sulfur, and lesser amounts of cobalt, iron, selenium, arsenic, and lead. The matte is not blown past the white metal stage in the air-blown Peirce–Smith side-blown converter because of the probability that nickel metal will be oxidized at the converting temperature of 2190°F (1200°C) and produce high slag losses.

Retort Reduction produces refined metal rondelles, 99.25% nickel, as is carried out by the LeHavre plant of Le Nickel. The first process of roasting the converter matte is carried out in two stages, with the matte being crushed and ground and then roasted in a fluid-bed reactor for the first stage:

$$2Ni_3S_2 + 7O_2 = 6NiO + 4SO_2$$

This reaction is exothermic, with the oxidation reaction providing sufficient heat, and the 22% sulfur content of the matte fed into the roaster being reduced to 0.5% in the discharged calcines.

In the second stage of roasting the calcines are put through a rotary kiln, countercurrent to a flow of hot oxidizing gas, and the sulfur content is further reduced to 0.002%. The final nickel oxide product is taken to a rotary cooler, then screened, with the oversize being ground in a ball mill and re-added to the undersize portion.

The nickel oxide powder is briquetted into rondelles $1\frac{1}{4}$ inches in diameter and $\frac{3}{4}$ inch high (3.125×1.875 cm); these are dried, then mixed with charcoal, and charged into the top of a reduction retort. The retort is a vertical, gas-tight unit 7 feet long, 1 foot wide, and 25 feet high ($2.3 \times 0.3 \times 7.62$ m), of silicon carbide brick for good heat conductivity, and is heated externally by producer gas and preheated air burners at several levels down the 25 foot height (7.62 m) of the retort. The narrow 1 foot width (0.3 m) ensures good heat transfer throughout the whole charge.

A continuous flow of preheated feed is added to the retort at the top; this descends slowly through the hot zone of the unit; and at 2400°F (1316°C), nickel oxide is reduced by the charcoal to nickel metal. Refined metal rondelles and excess charcoal are continuously discharged from the bottom of the retort by a roll extractor into a gas-tight cooler, from which after cooling the rondelles are separated from the nonmagnetic charcoal by passing over a magnetic pulley, then are cleaned and polished for shipment to market.

Controlled Slow Cooling, developed by the International Nickel Company, is another method used to treat nickel–copper converter matte. With this method, finished matte is poured from the air-blown Peirce–Smith converters at about 1800°F (980°C) and transported in ladles to cooling trays. These trays are fire clay molds 12 feet long, 8 feet wide, and 2 feet deep (3.66 × 2.44 × 0.61 m), with 45° angled sides to facilitate easy removal of the solidified cake which will weigh 25 tons. As soon as the mold is filled with matte, it is covered with an insulated steel cover and left for three days, in which time the matte cools to slightly below 900°F (480°C). After this time the insulating cover is removed and the ingot is cooled for a fourth day in air, bringing its temperature down to about 400°F (204°C); then it is removed from the mold.

Because of the long, four-day residence time of converter matte cooling in the trays, a large number of trays must be available for a continuing process on a large tonnage basis.

Slow cooling of matte permits phase separation and promotes larger grain size growth. Down to 1700°F (927°C), nickel, copper, and sulfur in the molten matte are completely miscible, but as the liquid cools to 1690°F (921°C), solid copper sulfide (Cu_2S) crystals begin to form. As the temperature drops still further, more Cu_2S crystallizes, leaving a liquid phase richer in nickel as the copper is removed from it. As more copper is precipitated from the liquid melt, the tendency is for it to combine with and enlarge the existing Cu_2S crystals rather than to form new nuclei, and so the existing grains of copper sulfide are enlarged. The slow cooling rate enhances this tendency toward grain growth.

At a temperature of about 1290°F (700°C), a nickel–copper alloy begins to precipitate; and at 1067°F (630°C), nickel sulfide (Ni_3S_2) also begins to solidify. The temperature of the matte remains at this 1067°F (630°C) temperature, the freezing point, until solidification is complete and the liquid matte has all transformed into crystals of copper sulfide, nickel–copper alloy, or nickel sulfide. At the solidification temperature, the solubility of nickel in the solid copper sulfide is less than 0.5%, and the solubility of copper in the solid nickel sulfide (designated as βNi_3S_2) is 6%.

As slow cooling progresses further, the next major change occurs at 968°F (520°C), where βNi_3S_2 undergoes a structural transformation to αNi_3S_2, and the temperature remains at 968°F (520°C) until this transformation is complete. Copper sulfide and the nickel–copper alloy are much less soluble in αNi_3S_2 than in βNi_3S_2 and will be continuously rejected from the αNi_3S_2 as cooling proceeds down to 700°F (370°C). At this

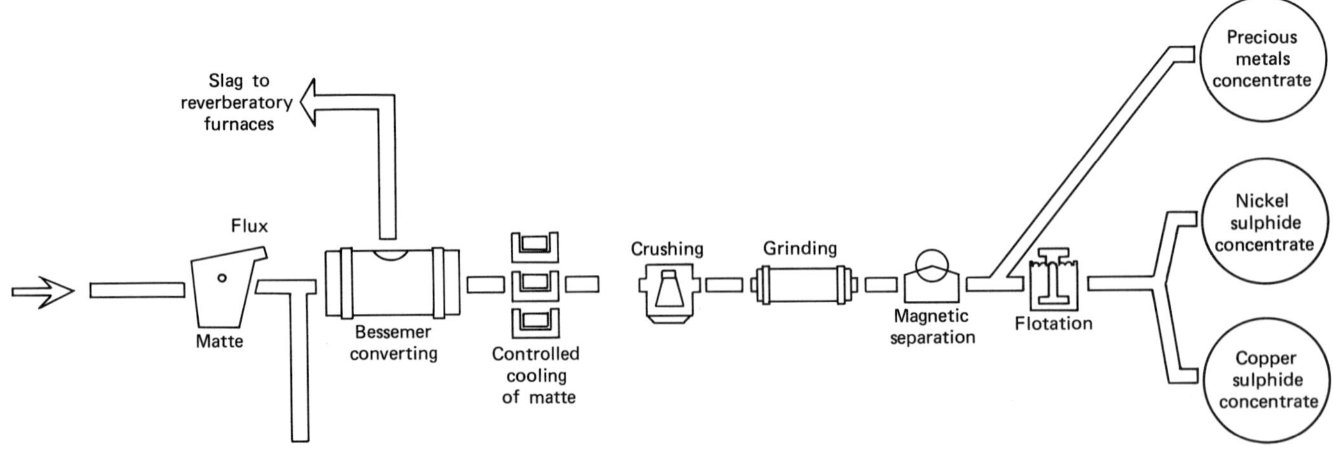

Figure 1.57. Refining by controlled slow cooling. *Source:* Courtesy of Inco Limited.

temperature, less than 0.5% copper sulfide still remains in the nickel sulfide phase, and this is the last of any significant segregation between the two phases.

The metallic nickel–copper alloy formed amounts to about 10% by weight of the solidified ingot and contains the gold and platinum metals as well as approximately 20% of the available copper in the matte. Silver, selenium, and tellurium, because of their greater affinity for sulfur, tend to concentrate in the crystals of copper sulfide rather than in the metallic alloy phase.

After another day of cooling to 400°F (204°C), when it can be easily handled, the solidified 25 ton ingot is broken into chunks by a pile driver, then crushed and ground in a series of jaw and gyratory crushers and rod and ball mills to a final product of 99% minus 200 mesh. All three phases tend to fracture along their grain boundaries and so are liberated from each other by the crushing and grinding. Separation of the three phases is accomplished by removing the magnetic nickel–copper alloy in magnetic separators and separating the remaining two Cu_2S and Ni_3S_2 phases from each other by froth flotation.

The copper sulfide concentrate phase analyzes 73% copper and 5% nickel and is taken to be smelted in converters treating copper smelting furnace matte.

The nickel sulfide concentrate phase analyzes 73% nickel and 0.6% copper and is either (1) cast into anodes for direct electrolytic refining or (2) further fire refined by an oxidizing roast to convert the nickel sulfide to oxide, followed by a reduction to convert the oxide to metal. The reduction is done either by hydrogen in a fluid bed reactor or by a reducing gas mixture of hydrogen and carbon monoxide in a hearth roaster.

Hydrogen Reduction. The first stage, to dead roast the nickel sulfide concentrate to oxide, is carried out in fluid-bed roasters 15 feet in diameter and 35 feet high (4.57 × 10.67 m), operating at a gas velocity of 7 feet per second (2.3 m) and having a capacity of 300 tons of sulfide feed per day. Roasting has to be carried out above 1800°F (982°C) in order to produce a low sulfur calcine, and the usual operating practice is to roast at close to 2000°F (1093°C):

$$2Ni_3S_2 + 7O_2 = 6NiO + 4SO_2$$

This high roasting temperature, combined with the fineness of the feed (99% minus 200 mesh), presents several operating problems. One of these is that the very fine feed will produce excessive flue dust carryoff unless the fluidizing gas velocity is kept very low, which in turn will give a low tonnage throughput and limit the roaster capacity. Another problem is the high roasting temperature of 2000°F (1093°C), which is much above the 1200°F (650°C) temperature where nickel sulfide begins to soften and become sticky, although nickel oxide does not melt until well above this roasting temperature. The large exposed surface area of the very fine feed will still further accentuate the problem, as speed of roasting and the excessive exothermic heat generated results in the fine particles fusing into a sticky, semiliquid mass, which will then preclude any further roasting.

The operational problems are overcome by agglomerating the fine charge into $\frac{1}{4}$ inch pellets (6.25 mm), containing 8% moisture, and mixed from 90% Ni_3S_2 flotation concentrate and 10% returned recovered roaster flue dust, by weight. The larger, heavier pellets have sufficient mass not to be blown out of the roaster as the finer flotation concentrate would be, and the roasting reaction rate with the larger pellet particles is slowed sufficiently that excessive fusion does not take place. The major part of the pellet oxidizes rapidly, but some sulfur remains in the core and is only removed, down to the 0.2% sulfur permitted in the calcines, after a long retention time of up to 8 hours in the roaster. This long roaster residence time is obtained by maintaining a deep fluidizing bed, up to 15 feet deep (4.57 m), which will contain as much as 100 tons of material.

This calcine can be given another stage of fluid-bed roasting to lower the sulfur content to 0.01%, by using oxygen-enriched air for fluidizing. Then the resulting NiO is reduced with hydrogen in another fluid-bed reactor to give a product of 90% Ni metal and less than 0.005% S, which is used directly as an alloying element in steelmaking.

Hydrogen–Carbon Monoxide Reduction. A second more common method of treating the roasted nickel oxide is to reduce it to metal with hydrogen-rich carbon monoxide gas in a type of multiple-hearth roaster, 40 feet high and 6 feet in diameter, with 21 hearths (12.2 × 1.86 m):

$$H_2 + NiO = Ni + H_2O$$

The reducing gas, analysing 51% hydrogen and 40% carbon monoxide, is made by passing steam over hot coke and coal and is then fed into the top of gas-tight roasters along with the nickel oxide to be reduced. A reduction temperature of 800°F (427°C) is maintained in the 21 hearth roaster by burning producer gas inside the hollow hearths. Rotating rabble arms plow the solid material across the hearths and through drop holes to the hearth below. The roasted solids which finally discharge from the bottom of the roaster are passed through a second, similar unit to complete the reduction to nickel metal granules.

These impure metal granules are further refined by a chemical method.

3. Blast Furnace Lead, which is the valuable product from the lead blast furnace, along with some matte and speiss, still contains small amounts of base metal impurities, arsenic, antimony, tin, bismuth, and copper, that must be removed before the lead is in its final, soft, marketable form. Many lead ores also contain silver, a worthwhile byproduct to be extracted. All of these are removed in a series of fire refining operations done on a relatively small scale in drossing kettles or small reverberatory furnaces.

Drossing is the first stage of refining and is carried out in open-top iron kettles or small reverberatory furnaces, either of which has capacities up to 200 tons of liquid metal. Drossing consists of holding the lead molten at a low temperature to allow impurities which are dissolved in the lead at a high temperature in the blast furnace to separate and rise to the surface as their solubility in lead at lower temperature decreases.

Figure 1.58. Skimming a lead kettle and removing dross. *Source:* Courtesy of ASARCO, Inc.

Blast furnace lead at 1700°F (927°C) is transferred to the drossing plant and placed in kettles or a small reverberatory furnace. The temperature is lowered to 660°F (350°C), just above the melting point of lead, and some of the impurities, including a considerable portion of the copper, rise to the surface and are skimmed off, while the temperature is held steady at this point. Mechanical or air agitation assists the drossing action; and if there is still copper to be removed, this agitation, along with the addition of elemental sulfur, causes the remaining copper to rise to the surface as a black copper sulfide powder, which is skimmed off.

The dross which has been skimmed is charged into a small oil-fired reverberatory furnace 22 feet long by 9 feet wide (6.71 × 2.74 m), smelting 120 tons per day. Sodium carbonate and coke breeze are added to the furnace as fluxes to form matte and speiss, which separate into two separate layers, and some slag which floats on top. Liquid lead collects on the bottom of the furnace and is tapped periodically to return it to the drossing kettle. The matte and speiss are also periodically removed and go to the copper smelter. The slag is returned to the lead blast furnace. This reverberatory furnace operates at 2375°F (1300°C) and the dross cleaning operation takes some 5 to 6 hours. The flue gases go through dust collectors, and the recovered flue dust is returned to the blast furnace sinter plant.

Softening is the second fire refining stage following drossing and is undertaken to remove the remaining 1.5% of arsenic, antimony, and tin still present in the lead after drossing. These impurities make lead hard and brittle rather than soft and malleable, as expected in its marketable form, and also interfere with the operation of desilverization which follows softening.

A process of selective oxidation can be used to remove these three impurities, as they

Table 1.26. Dross Plant Operating Data, Buick Lead Smelter (AMAX-Homestake): Operating Data

1. *Production*	(Tons/Day)
Dross	50
Matte	14
Lead	35
Pb–Ni–Cu Alloy	1

2. *Analysis*[a]

	%Pb	%Cu	%Fe	%S	%Ni	%Co
Dross	65.0	16.0	0.7	5.0	0.8	0.3
Matte	15.0	53.0	1.5	16.0	1.5	0.6
Pb–Ni–Cu Alloy	50.0	14.0	0.1	8.0	18.0	1.2

3. *Dross Reverberatory Furnace*

Size (L × W)	2.1 m × 6.4 m
Bath depth, total	101.6 cm
Normal bath matte	25.4 cm
Normal bath Pb	76.2 cm
Fuel consumption	0.88 million cal/ton

Source: C. H. Cotterill and J. M. Cigan, Eds., AIME World Symposium of Lead and Zinc, Vol. 2, 1972, p. 769.

[a]Based on recycle of dross through blast furnace.

Figure 1.59. Removing dross from a lead kettle. *Source:* Courtesy of ASARCO, Inc.

all oxidize more readily than lead. Bismuth is also frequently present, but as it has little oxygen affinity, even less than lead, it cannot be removed at this stage.

After drossing the temperature of the lead is raised and the bath is air agitated to induce oxidation. The tin, arsenic, and antimony are oxidized, and the oxides which are insoluble rise to the surface to be skimmed off.

Softening is done in small reverberatory furnaces of up to 300 ton capacity and measuring 28 feet long by $13\frac{1}{2}$ feet wide (8.54 × 4.12 m). Water-cooled jackets are often extensively used, with high-alumina firebrick laid on the inside of these jackets, with this combination preventing the quite active corrosion at the slag line which would otherwise occur. The furnace is sturdily built to contain the considerable weight of the lead metal charge, and the bath depth is kept at some $2\frac{1}{2}$ feet (0.76 m).

Softening can be done as either a batch or a continuous operation, with the same general reactions of the oxidation of tin, arsenic, and antimony taking place in both cases.

With the batch operation, which can take up to 24 hours to treat and discharge a furnace full of drossed lead, lead from the drossing kettles is pumped to the reverberatory furnace where the temperature is raised to 930°F (500°C) and compressed air lances blow into the bath to oxidize the impurities. Arsenic, tin, and antimony, along with some lead, are oxidized and rise to the surface as skims, to be scraped off. One, two, or three skims are made, depending on the amount of impurities and their order of oxidation. The first skimming contains most of the tin as an oxide, SnO, or as a tin antimonate, $Sn_3(SbO_4)_2$, and may also be mixed with some lead stannate, $2PbO \cdot SnO_2$, or lead antimonate, $3PbO \cdot Sb_2O_5$. The second skim consists of lead antimonate, and the third skim of lead arsenate, $3PbO \cdot As_2O_5$, with some excess PbO. The furnace temperature is sometimes lowered shortly before skimming in order to have the skimmings more readily collectable.

In continuous lead softening an equilibrium is maintained between liquid lead and liquid lead oxide slag in a reverberatory furnace at about 1470°F (800°C). Impure lead from the drossing kettles is added at one end of the furnace, and air is blown into the bath to oxidize some lead to lead oxide for slag and also to stir the bath and improve reaction rates. Antimony, arsenic, and tin are rapidly oxidized by the lead oxide slag and are held in solution in it, to run from the furnace at the end opposite to that used for charging:

$$5PbO + 2Sb = 5Pb + Sb_2O_5$$

Softened lead that collects in a layer under the slag is also run off in a continuous stream at this same discharge end of the furnace.

A second method of softening, the *Harris process*, is used by the St. Joe Minerals Corporation at their Herculaneum Smelter and by AMAX-Homestake at their Buick Smelter. Both of these concentrates have low arsenic-antimony-tin contents, and both plants do their softening following desilverizing and just before the lead is cast into pigs or ingots for market.

The Harris process is a chemical method for the removal of the arsenic-antimony-tin impurities and consists of heating the lead to a temperature just above its melting point and then bringing it into contact with a molten mixture of sodium hydroxide (NaOH) and an oxidizing agent, sodium nitrate ($NaNO_3$).

The process is carried out above the kettle in a portable unit incorporating a pump and a chemical treatment tank which can be lifted by a crane and moved from kettle to kettle. The tank is immersed in the lead in the kettle, and the pump lifts the lead and showers it through the molten salt until it is softened, by means of the impurities first being oxidized and then fixed as soda salts which remain dissolved in the salt layer:

$$10NaNO_3 + 6As = 5Na_2O + 3As_2O_5 + 10NO$$

$$4NaNO_3 + 3Sn = 2Na_2O + 3SnO_2 + 4NO$$

Figure 1.60. Harris process for lead softening (Hoboken). *Source:* Courtesy of Metallurgie Hoboken–Overpelt.

Then

$$Na_2O + Sb_2O_5 = 2NaSbO_3$$

$$Na_2O + As_2O_5 = 2NaAsO_3$$

$$Na_2O + SnO_2 = Na_2SnO_3$$

The caustic dross can be returned to the blast furnace, as is done at these two smelters, or can be treated chemically to recover the salts it contains in a marketable form.

Desilverizing Softened Lead, known as the *Parkes process*, also removes any gold present, though this is found in much smaller quantities than silver associated with lead ores. While the precious metals are not objectionable impurities, they are valuable by-products and any appreciable amounts are economically recoverable by combining as an alloy with zinc metal which is stirred into the bath. The precious metals have a stronger affinity for zinc than for lead, and lead and zinc are only slightly soluble in each other. So when the zinc–precious metal alloy has a higher melting point and solidifies first to float as a crust on the still liquid lead, a practical method of separation can be carried out.

Batch desilverizing is carried out in open-top steel kettles, varying in size from 60 to 135 tons capacity, with an arrangement for heating the kettle from below.

Softened lead is pumped to the desilverizing kettle at 1400°F (760°C), and unsaturated crusts from a previous kettle which contain unused zinc are stirred into the lead, saturating it with zinc. The kettle temperature is lowered to 932°F (500°C) and held there for several hours, during which time the excess zinc, no longer soluble in the lead at this lower temperature, combines with the gold and silver to form a saturated crust containing all the gold and silver that the zinc can hold. This crust, containing the major value of Ag_2Zn_3, is skimmed, pressed to squeeze out the entrained liquid lead, and sent on to retorts where the zinc will be distilled off, leaving the precious metals.

Some silver will still be left in the lead, and this is removed by stirring in new slab zinc in considerable excess of that required to alloy with the previous metals left and, on cooling close to the melting point of lead 621°F (327°C), produces an unsaturated crust containing a great deal of unused zinc. This crust is skimmed, but not squeezed to remove entrained lead, and is returned as the initial source of zinc in a fresh kettle of softened lead to be treated.

Figure 1.61. Removing dross from a lead kettle with mechanical scraper (Hoboken). *Source:* Courtesy of Metallurgie Hoboken-Overpelt.

Table 1.27. Typical Analysis of Lead, Herculaneum Lead Smelter, St. Joe Minerals Corporation

	Doe Run (%)	St. Joe Chemical (%)
Silver	0.0005	0.005
Copper	0.0000	0.05
Bismuth	0.0001	0.0001
Arsenic	0.0000	0.0000
Tin	0.0000	0.0000
Iron	0.0000	0.0000
Zinc	0.0000	0.0000
Cadmium	0.0003	0.0003
Nickel	0.0000	0.0003
Lead	99.99+	99.94

Source: A Metal for the Future, St. Joe Lead, St. Joe Minerals Corporation.

The desilverized lead is now treated to remove the excess zinc still in solution in the liquid lead.

Continuous desilverizing is done in a tall, closed cast iron kettle 23 feet high and 10 feet in diameter (7.01 × 3.05 m), divided into four separately heated sections so that a temperature gradient can be maintained of 1112°F (600°C) at the top, down to 604°F (318°C) at the bottom, which is the melting point of the lead–zinc eutectic 99.5% Pb and 0.5% Zn.

The upper 3 feet (0.91 m) of the vessel are filled with liquid zinc and the remainder with molten lead. Soften lead at 1200°F (650°C) is fed into the top of the vessel and passes down through the zinc layer to join the lead layer below. An equal amount of silver-free lead to that entering is drawn off from the bottom of the vessel by a siphon tap, so that the vessel is always full.

Silver in the freshly added lead forms a compound with the zinc as it passes through the top zinc layer. This compound is dissolved in the lead, but as the lead passes on downward into the cooler zones of the vessel, the compound becomes less soluble and comes

Table 1.28. Analysis of Refined Lead, Naoshima Lead Smelter (Mitsubishi-Cominco): Typical Assay of Refined Lead

%Ag	Trace
%Cu	0.0008
%Zn	0.0002
%Fe	0.0002
%As	Trace
%Sn	Trace
%Sb	Trace
%Bi	0.0002
%Pb	99.998 up

Source: C. H. Cotterill and J. M. Cigan, Eds., AIME World Symposium of Lead and Zinc, Vol. 2, 1970, p. 862.

out of solution in the lead. It then rises to the surface where it dissolves in the layer of molten zinc. When this zinc layer builds up to 6000 ounces (170.5 kg) of silver per ton, it begins to show signs of freezing, so is ladled out and sent to the silver refinery, while a fresh top layer of zinc is added to the continuous desilverizing kettle.

Dezincing of the 0.5 to 0.7% zinc dissolved in the desilverized lead must be done before the lead is in marketable condition. This zinc removal is accomplished successfully in

Table 1.29.   Fire Refining of Lead to Recover Silver at Buick Lead Smelter, AMAX-Homestake: Operating Data

	%Cu	%Ag	%Zn
1. *Analysis*			
a. Decopperizing			
Lead bullion to refinery	0.080	0.009	—
After decopperization	0.004	0.010	0.30
Dry dross removed	5.70	0.034	21.20
Zinc head to kettle	0.56	2.040	12.50
b. 1st Desilverization			
Lead to kettle	0.0040	0.010	0.30
"Cold" sample	0.0003	0.0007	0.60
"Pumping" sample	0.0003	0.0023	0.70
Zinc heads removed	0.50	2.04	12.50
Zinc heads concentration	0.40	2.50	10.00
c. 2nd Desilverization to dezincing	0.0003	0.0007	0.57
d. Dezincing to final refining	0.0003	0.0007	0.06
e. Refined lead to market			

%Ag	%Cu	%Zn	%As %Sb %Sn %Bi	%Pb
0.0007	0.0003	0.0003	Nil	+99.99

f. Silver concentration	%Cu	%Ag	%Zn
Zn concentration	0.40	2.50	10.0
Conc. heads	0.34	5.5	8.0
Pressed dross	1.00	9.0	10.0
Retort bullion	1.5	12.0	1.2

2. *Reagent Use*	Refined Pb (kg/ton)
New zinc	4.9
Reclaimed zinc	2.0
Caustic	.5
Niter	.1

3. *Kettles*	No.	Capacity (tons)
Decopperizing	1	200
Desilverizing	3	250
Dezincing	2	200
Refining and casting	2	200

4. *Vacuum Pumps*

2 Model KT 300
1 Model KT 500

Source: C. H. Cotterill and J. M. Cigan, Eds., AIME World Symposium of Lead and Zinc, Vol. 2, 1970, p. 774.

several ways. It can be removed by vacuum distillation and condensing the metallic zinc as it distills out, by blowing chlorine gas through the melt to form zinc chloride and by oxidizing with air to form zinc oxide.

The latter two processes are quite similar in overall processing. In the one instance chlorine gas combines with zinc to form a greyish zinc chloride scum which rises to the surface of the lead and is skimmed off. In the oxidizing process air or steam blown into the bath oxidizes the zinc, which rises to the surface and is also skimmed off. A 100 ton kettle of lead can be chlorine dezinced in about 4 hours from 0.6 to 0.005% zinc content.

Vacuum dezincing is the most recent of these processes and is in quite general use. It is accomplished by fitting a water-cooled vacuum head down on the standard kettle filled with desilverized lead and heating the kettle to 1100°F (593°C). The bath is stirred, and at this temperature and under vacuum, the zinc in the lead over a 5 or 6 hour period will distill off and condense on the undersurface of the water-cooled lid. This zinc is in a pure form and can be reused immediately in the desilverizing process. Zinc removal is from 0.6% down to 0.02% in the lead.

Debismuthizing is the final stage of fire refining lead and is known as the *Kroll–Betterton process*. It is the process for removing bismuth from desilverized, dezinced lead and makes use of the affinity of bismuth for calcium and magnesium, forming alloys with them when they are dissolved in the lead bath. The principle of bismuth removal is quite similar to that of the Parkes process for desilverizing, for the bismuth–calcium and bismuth–magnesium alloys that form are insoluble in a lead bath saturated with calcium and magnesium and rise to the surface as a crust which is skimmed off.

A kettle of lead at 780°F (416°C) has a calcium–lead alloy (3 to 4% Ca) and pig magnesium added to it, and is stirred well to mix in these added alloys as they melt. The temperature is lowered to below 750°F (400°C), and a dross of Ca_3Bi_2 and Mg_3Bi_2 forms on the surface and is skimmed off. The operation is repeated with a lower skimming

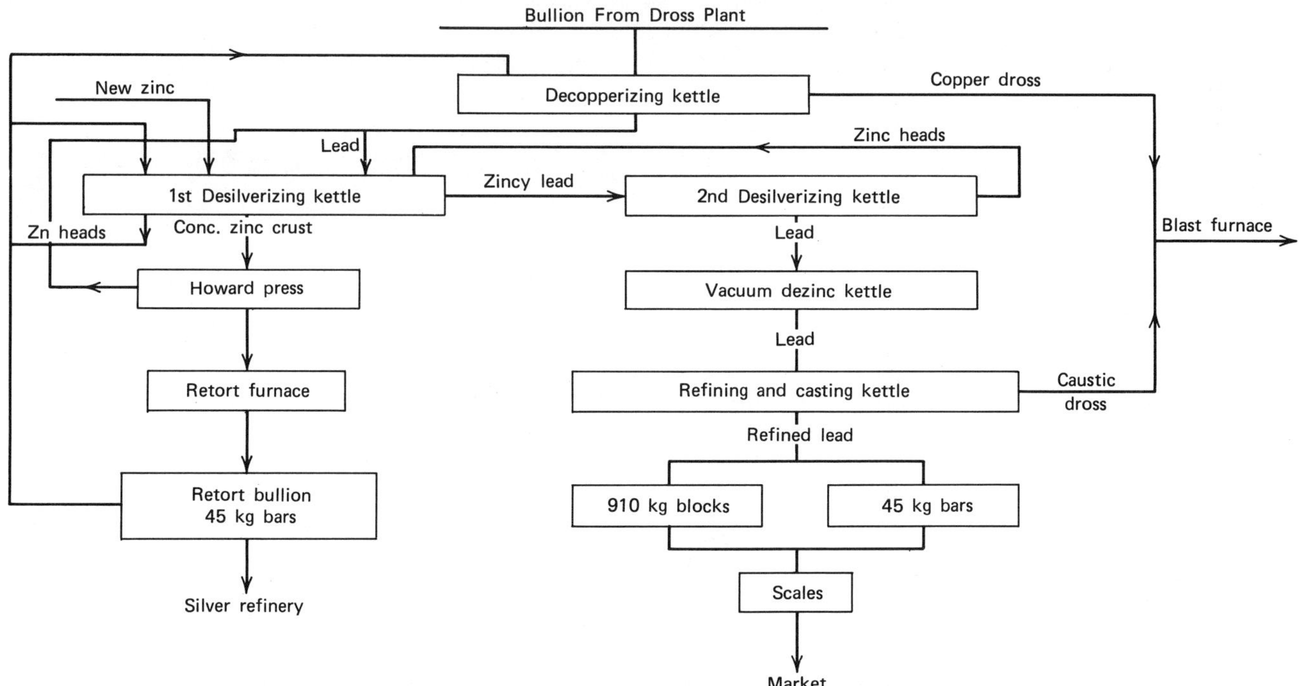

Figure 1.62. Lead refinery flowsheet, Buick lead smelter, AMAX Homestake. *Source:* C. H. Cotterill and J. M. Cigan, Eds., AIME World Symposium of Lead and Zinc, Vol. 2, 1970, p. 771.

temperature, down closer to the melting point of lead, and a second dross is removed. If the drosses contain over 20% bismuth they are removed and treated to recover their bismuth content. However if they are below 20% bismuth, they are reused to debismuthize successive batches of lead until the bismuth content is built up to above 20%.

The final debismuthized lead will contain below 0.05% bismuth; and the residual calcium and magnesium it contains can be removed by blowing steam or chlorine into the bath to form oxides or chlorides which rise as a dross to be skimmed off.

ELECTROLYTIC REFINING

This is a system of refining which gives the ultimate in purity of metal product and also permits the recovery of valuable impurities such as the precious metals, selenides, and tellurides, which are not recoverable in fire refining.

The metal to be refined is cast into a slab which is made the anode of the electrical system, while a thin sheet of the same metal in pure form is made the cathode. If these two pieces of metal are placed in an appropriate solution, one which readily carries an electrical current, and a direct current is impressed to flow from an electrical power source to the anode, from the anode through the electrolyte solution to the cathode, and from the cathode back to the electrical source to complete the circuit, metal atoms will then be transferred from anode to cathode. This metal transfer is accomplished by the metal at the anode, along with some of its associated impurities dissolving in the electrolyte as current flows through the cell. An electrical field exists between the two metal electrodes, with a flow of positively charged ions migrating toward the cathode and negatively charged ions migrating toward the anode. The metal dissolved from the anode is in the form of positively charged metallic ions in the electrolyte, which will migrate to the cathode where they give up their charge, become neutralized, and deposit as metal atoms.

The behavior of the impurities in the anode metal depends on whether they are higher or lower in the electromotive force series than the metal being electrolytically refined. Those metallic impurities that are higher in the series generally also go into solution as charged metallic ions, while those lower in the series do not dissolve but drop to the bottom of the tank to form anode mud. The dissolved impurities, which will also deposite

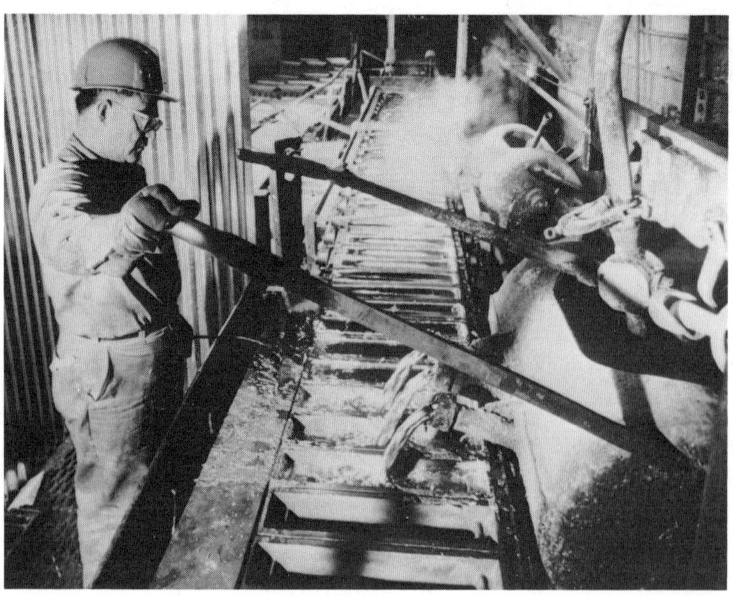

Figure 1.63. Casting bars of refined lead. *Source:* Courtesy of ASARCO, Inc.

at the cathode, if present in any appreciable quantity will upset the efficiency of the refining process.

Electromigration alone would be too slow and inefficient in transporting metal ions from anode to cathode, and this transportation is aided by a variety of physical methods to keep the electrolyte well mixed and in motion, and also by the pumping of electrolyte in and out of the cell as it is taken off for purification and then returned.

TYPES OF ELECTROLYTIC REFINING

Electrolytic refining both eliminates small amounts of impurities which alter the physical properties of a pure metal and permits most of these impurities to be recovered as by-products, which in many instances have a considerable value.

In most cases the metal to be electrolytically refined has had some degree of fire refining, partially to refine it and also to produce a homogeneous product which is cast into anodes. These impure metal anodes are hung alternately at carefully spaced intervals with pure metal cathodes in a concrete cell containing an electrolyte solution. When direct current is applied, the metal dissolves from the anode into the electrolyte and at the same time an equivalent amount of metal deposits out of the solution at the cathode.

Many of the usual impurities such as gold, silver, the platinum group metals, selenides, and tellurides have a lower position in the electromotive force series than metals such as copper or lead that are commonly processed by electrolytic refining. These impurities do not go into solution in the electrolyte but fall to the bottom of the cell as insoluble anode mud, to be dug out periodically and treated to recover the values it contains.

Other common impurities such as antimony, arsenic, and bismuth have a higher position in the series, higher than or very close to that of the metal being refined, and as a result these impurities will often go into solution along with the ions of this metal. If these and other impurities also following this same procedure contaminate the electrolyte or give an objectionably impure product depositing at the cathode, it necessitates removal of the electrolyte, purification of these contaminants, and then returning it to the electrolytic refining cell.

Electrolytes with difficult impurities are treated in compartmented diaphragm cells

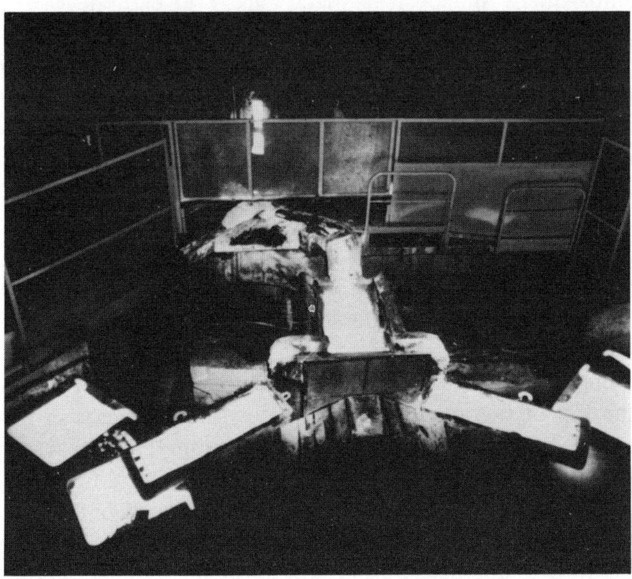

Figure 1.64. Copper anode casting on two casting wheels (Olen).
Source: Courtesy of Metallurgie Hoboken–Overpelt.

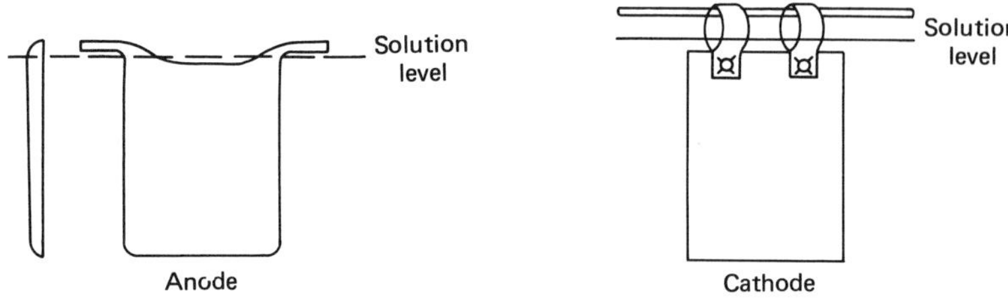

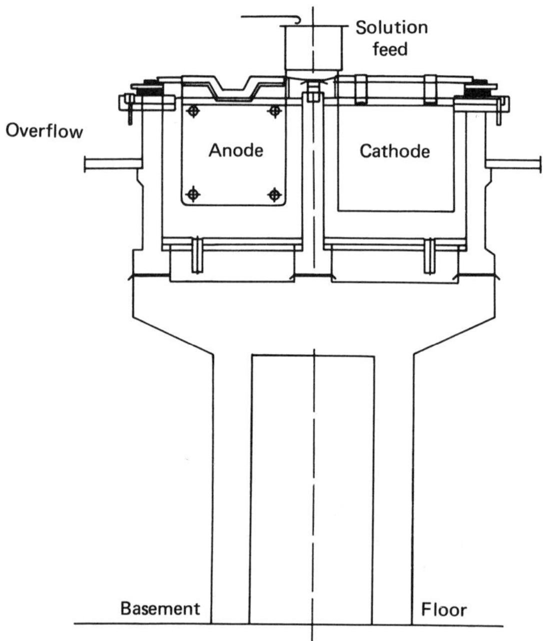

Figure 1.65. Position of anode and cathode in multiple-circuit cell. *Source:* P. Queneau, Ed., *Extractive Metallurgy of Copper, Nickel and Cobalt*, Interscience, New York, 1961, p. 337.

where contaminated electrolyte is taken out of the anode compartment, purified, and then returned to the cathode compartment to deposit its metal ions at the cathode.

Standard Cells have open-topped rectangular tanks with flat bottoms, constructed of a variety of materials, concrete or wood, all lined with lead sheet, pitch, asphalt, or in some cases plastic. The corrosion-resistant lining protects the tanks from attack by the high-conductivity electrolyte solution which generally contains an acid or some other corrosive compound. Sulfuric acid is one of the more common electrolyte components, as is also hydrofluosilic acid.

The impure metal anodes, which are cast as slabs following some considerable degree of fire refining, are relatively large and heavy and weigh from 300 to 600 pounds (135 to 270 kg), depending on the specific gravity of the metal being treated. The anodes are cast with surfaces as smooth as possible to avoid projections reaching out, making contact with other electrodes, and causing short circuits when in position in the electrolytic tanks. Side lugs are cast in at the top of the anode both for suspension purposes in the cell and also to carry current, as one lug rests directly on the current supplying busbar.

Cathodes are of pure metal, slightly larger in area than the anodes with 1 or 2 inches of overlap (2.5 to 5 cm) on sides and bottom, and are made only thick enough (a fraction of an inch, a few mm) to be stiff enough to be straightened flat and not warp, causing electrode touching and short circuiting in the cell. Suspension loops are attached on one

Figure 1.66. Copper electrolytic refining tankhouse under construction at Kilwezi, Zaire. *Source:* Courtesy of Metallurgie Hoboken–Overpelt.

end of the cathode, and a metal rod through these loops supports the metal cathode sheet and also makes contact with a busbar for current flow.

The cathodes and anodes are prespaced on a rack, with one more cathode than anode in the load. The anode spacing will vary from $3\frac{1}{2}$ inches to $5\frac{1}{2}$ inches (8.75 to 13.75 cm), with a cathode placed between each anode and a cathode at each end of the rack. This arrangement provides that each anode has a cathode positioned on each side, and transfer of metal ions from the anode to the cathode is most efficiently accomplished.

The Walker multiple system is the electrical arrangement most commonly used, and in this system the current flows to a copper busbar on one side of the tank on which rests one support lug of each of the anodes in the cell. The current then travels through the

Table 1.30. Standard Electrode Potentials for the Common Nonferrous Metals

Element	Anodic Reaction	Reversible Electrode Potential (V)
Sodium	$Na = Na^+ + e$	-2.71
Zinc	$Zn = Zn^{2+} + 2e$	-0.76
Iron	$Fe = Fe^{2+} + 2e$	-0.44
Cobalt	$Co = Co^{2+} + 2e$	-0.28
Nickel	$Ni = Ni^{2+} + 2e$	-0.25
Lead	$Pb = Pb^{2+}$	-0.13
Hydrogen	$H_2 = 2H^+ + 2e$	0.00
Copper	$Cu = Cu^{2+} + 2e$	$+0.34$
	$Cu = Cu^+ + e$	$+0.52$
Silver	$Ag = Ag^+ + e$	$+0.80$
Palladium	$Pd = Pd^{2+} + 2e$	$+0.99$
Platinum	$Pt = Pt^{2+} + 2e$	$+1.2$
Oxygen	$2H_2O = O_2 + 4H^+ + 4e$	$+1.23$
Chlorine	$2Cl^- = Cl_2 + 2e$	$+1.36$
Gold	$Au = Au^{3+} + 3e$	$+1.50$

anodes down into the cell and across the electrolyte solution to the cathodes, and out through the cathode support rods to where one end of each support rod rests on a second copper busbar on the opposite side of the cell from where the anode lug–busbar connection was made. This second busbar feeds current to the anodes of the second tank, and the connections and current flow are repeated here.

The electrolyte which carries the current between the anodes and cathodes in the cell and provides a medium for transportation of pure metal ions from the anode to deposit at the cathode is usually an acid solution. Portions of electrolyte have to be withdrawn from the cells periodically for purification of dissolved impurities, which are also soluble from the anode in addition to the major metal being refined. If not removed, the concentration of these impurities can build up in the electrolyte to a level where they cause cell operation difficulties, or they can deposit at the cathode and lower the purity of the refined depositing metal. Purification is commonly accomplished by selective cementation or chemical precipitation of the impurity, after which the purified electrolyte is returned to the cell.

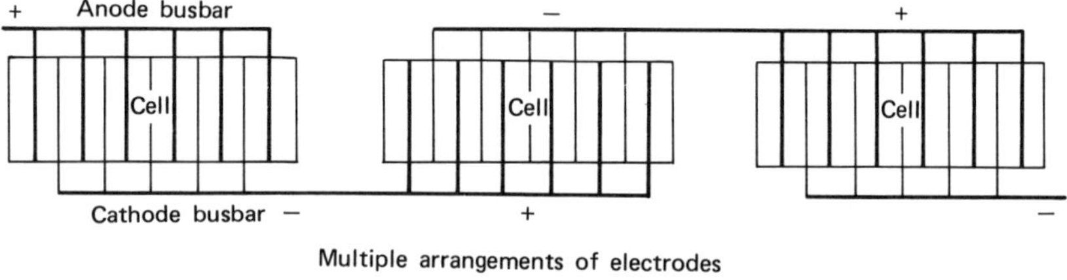

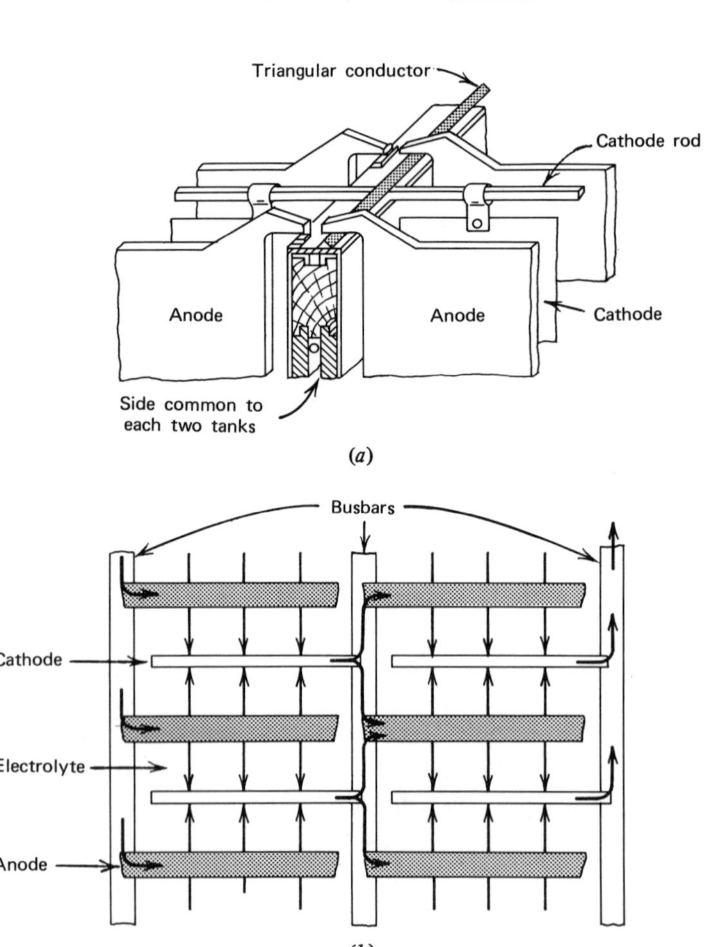

Figure 1.67. Multiple circuit cells. (*a*) Arrangement of anodes and cathodes in multiple-circuit cells. *Source:* W. H. Dennis, *Metallurgy of the Non-ferrous Metals*, Pitman, London, 1961, p. 132. (*b*) Flow of electrical current in multiple-circuit cells. *Source:* J. R. Boldt, Jr., and P. Queneau, *The Winning of Nickel*, Metheun and Company Ltd., London, 1967. Copyright Inco Limited, 1967, p. 347.

Current density is one of the more important factors in electrolytic refining; in general, the lower the current density used, the higher will be the grade of refined metal produced and the cheaper will be the power costs for electrolysis. However this condition would require a very large plant with much capital tied up in both equipment and metals because of the low production rate. A compromise must then be decided on where maximum production, consistent with an acceptable purity of refined metal product, is accomplished. A copper refinery would use 3 to 5 amperes per square foot (0.09 m^2) of cathode surface for highest purity, and 40 amperes for greatest production, with most copper refineries compromising on a current density of 15 to 18 amperes per square foot (0.09 m^2) of cathode surface.

Cathodes are replaced about every 14 to 15 days, with anodes being replaced every 28 to 30 days, so that one set of anodes in a cell will transfer metallic ions to two successive sets of cathodes in the same cell. The anodes are not completely dissolved, and some 8 to 15% of the original weight is removed at the end of the anode cycle to be washed and returned to the anode furnace as scrap, which is remelted and cast into fresh anodes. The cathodes can be sheared and sold directly to the market for alloy manufacturing or are melted and cast into ingots, bars, or slabs and sold in this form. The insoluble portion of the anode which dropped to the bottom of the cell as anode slimes is dug out and treated to recover the by-products it contains.

Diaphragm Cells are used in special cases where impurities are very close in the electromotive force series to the major metal being refined, so that there would be a strong tendency for these impurities going into solution to plate out at the cathode and contaminate the refined metal. To prevent this, diaphragms of canvas-covered boxes on wooden frames are placed in the cell and so positioned that anode and cathode chambers are formed. The electrolyte from the anode chamber, which will contain both the metal ions and impurity ions, is run off and treated to remove the impurities it contains, after which the purified electrolyte is returned to the cathode compartment to deposit at the cathode the pure metal atoms still remaining.

Because of the space taken up with the compartment boxes, the diaphragm cell tanks are somewhat more crowded than the same-sized standard cell tank. Therefore the diaphragm cell will have a load of only 30 anodes and 31 cathodes as compared to 38 anodes and 39 cathodes for a standard cell tank of the same dimensions. The Walker multiple-system circuit is used for both types of cells.

The general operation of standard and diaphragm cells is very similar. The anodes and cathodes are prepared and treated in the same way, as are the products of refined cathode-deposited metal. anode scrap, and anode mud.

ELECTROLYTIC REFINING PROCESSES

1. Anode Copper is electrolytically refined to produce the high grade of pure copper required by the electrical industry and to recover the precious metals which it contains as well as the other metal by-products, namely, nickel, bismuth, selenium, and tellurium. Close to 75% of the world silver production is a by-product of copper and lead refining, and many precious metals are recovered from ores far too lean in value to be treated for their precious metal value alone.

The raw material for electrolytic refining is blister copper which has been fire refined to remove most of the oxygen and sulfur and smaller quantities of base metals. From the fire-refining anode furnace, the copper is cast into anodes varying from 28 by 18 by 3 inches ($70 \times 45 \times 7.5$ cm) at 350 pounds (159.1 kg) to 36 by 36 by $1\frac{1}{2}$ inches ($90 \times 90 \times 3.75$ cm) at 600 pounds (270 kg), depending on the variations of refinery practices. The anodes contain about 99.4% copper and have 0.6% total impurities.

Cathode starting sheets that form the basis for cathode growth in the cell are thin sheets, $\frac{1}{16}$ inch thick (1.56 mm) weighing some 8 to 13 pounds (3.64 to 5.91 kg) of electrodeposited pure copper formed on special polished, oiled blanks in stripper cells. The cathodes are 1 or 2 inches (2.5 to 5 cm) larger in length and wider than the anodes.

The tanks are most frequently made of concrete lined with lead, asphalt, or plastic to withstand the corrosive electrolyte solution. Into this is placed a rack of anodes and cathodes, with 38 anodes and 39 cathodes in a tank of dimensions of $11\frac{1}{4}$ by $3\frac{1}{2}$ by $3\frac{3}{4}$ feet ($3.39 \times 1.07 \times 1.14$ m), The anodes are charged in a unit of 38 pieces, and so are the cathodes in a unit of 39. Both anodes and cathodes are prespaced, with the anode spacing varying from $3\frac{1}{2}$ to $5\frac{1}{2}$ inches (8.75 to 13.75 cm), center to center, and the cathodes fitting centrally between each anode and also positioned at either end of the anode rack. Two crops of cathodes are produced from each load of anodes, and a cathode will have grown to $\frac{1}{2}$ inch thickness (1.25 cm) and 250 pounds weight (113.6 kg) when it is removed after 14 days. The anode after 28 days is down to 8 to 15% of its original weight, when it is pulled, washed, and returned to the anode furnace for remelting and casting into fresh anodes.

Anode slimes from the tanks amount to 9 to 10 pounds (4.1 to 4.5 kg) per ton anodes dissolved. These are removed periodically and sent to the precious metals refinery to recover the contained values.

Theoretically the electrolyte of 40 grams per litre of $CuSO_4$ and 200 grams per litre of H_2SO_4 acts only as a medium in which the reaction takes place, and there should be almost no change in the concentration of copper in the solution. However in operation the copper concentration tends to increase, and it is common practice to use a few cells (liberators) in the circuit with insoluble lead anodes and copper cathodes to plate out the excess copper and maintain the electrolyte concentration at the level desired. Circulation of the electrolyte is also necessary, and this is usually done by admitting electrolyte at the bottom of the cell and removing it at the top. This circulation minimizes the tendency of the solution to stratify due to the fact that at the anode where copper is passing into solution the electrolyte is a saturated solution of high specific gravity copper sulfate which will sink. Meanwhile at the cathode, copper is being removed from the solution and leaving a dilute sulfuric acid solution of low specific gravity which will rise. As a result, unless there is continuous mixing, two separate solution layers will form in the cell and interfere considerably with its successful operation.

Sufficient electrolyte is also drawn off for purification at regular intervals so that the chief impurities of nickel and arsenic do not build up beyond tolerance limits, as the degree of soluble impurities in the electrolyte has a direct bearing on the purity of the cathode product. Three cells with lead anodes and copper cathodes connected in series are used for the purification process. The first cell produces good cathode copper; the second produces poorer quality copper, which goes to the anode furnace for melting; and the third produces a lower-grade sludge high in arsenic, which is returned to the smelter. The decopperized solution, now consisting essentially of nickel sulfate and sulfuric acid, is concentrated in lead evaporators for 6 to 7 hours until the acid concentration is up to 1000 grams per litre. Upon cooling, nickel is precipitated as crude anhydrous nickel sulfate, which is removed by filtering. The reclaimed, purified acid solution is returned to the tank house electrolyte system.

The Walker multiple system is the cell arrangement commonly used for copper refining. Tank voltages are kept to a minimum with close spacing of the anodes and will average around 0.25 volt. The current density is on the order of 16 to 20 amperes per square foot (0.09 m^2) of cathode surface, at a current efficiency of over 90%. The amperage follows Faraday's law, so that at 100% current efficiency 1 ampere-hour will deposit 1.186 grams copper, with copper acting as a divalent metal. The circulating electrolyte is heated in steam-heated hot wells to 140°F (60°C), then pumped to head tanks from which it is

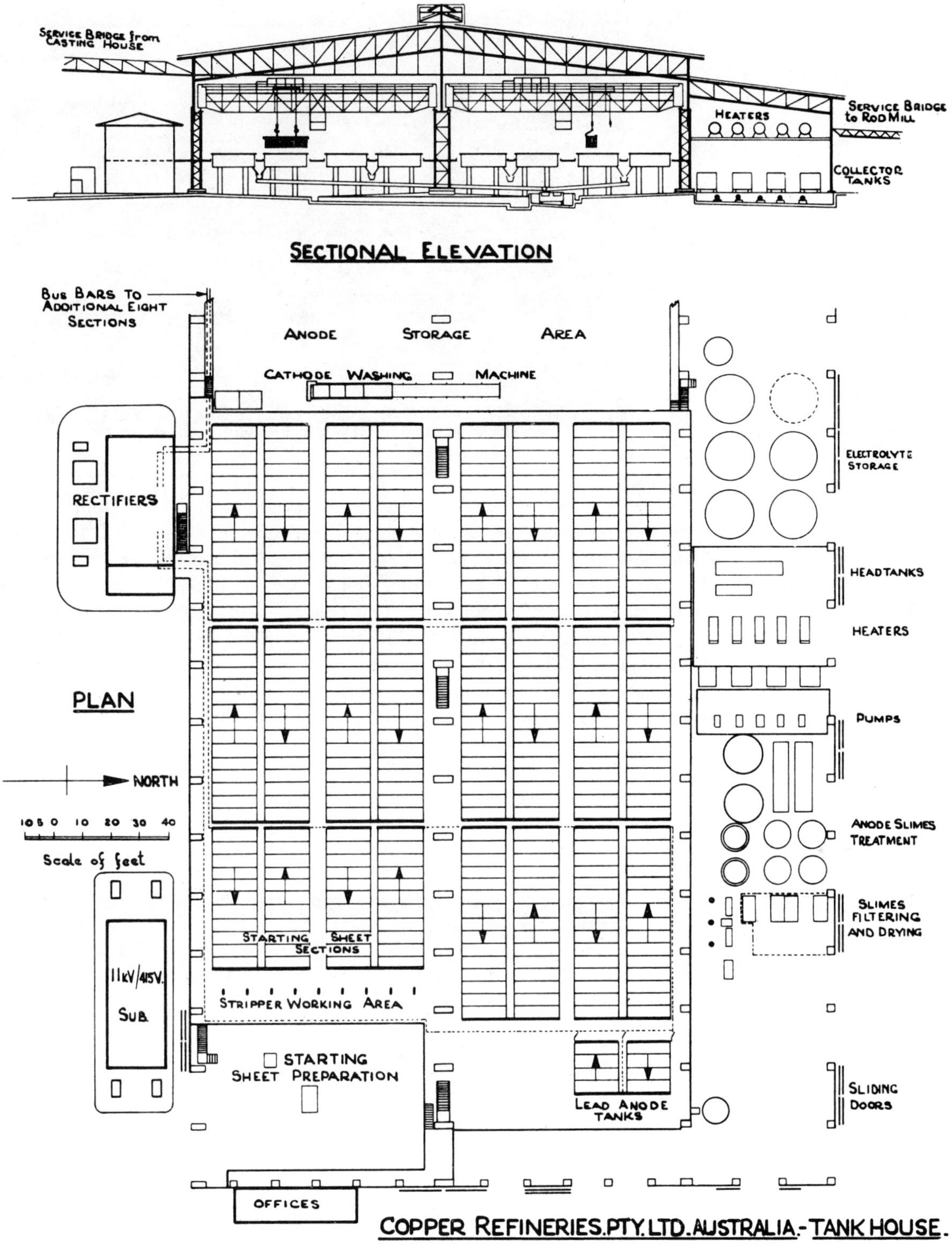

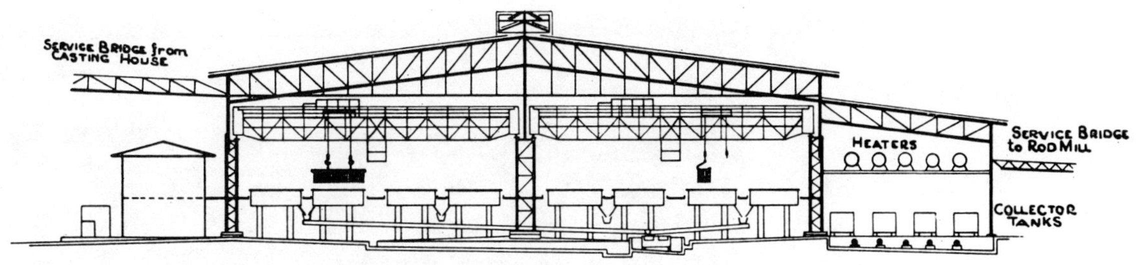

Figure 1.68. Electrolytic copper refining tankhouse. *Source:* J. C. Saint Smith, and J. C. Jenkins, The design, construction, and operation of the Townsville Australia copper refinery, Presented at the AIME Annual Meeting, 1962.

Figure 1.69. Partial view of copper electrolytic refining tankhouse (Olen). *Source: Courtesy of Metallurgie Hoboken–Overpelt.*

circulated by gravity to each depositing tank at a rate of 2 to 4 gallons per minute (9 to 18 litres). The electrolyte being circulated is withdrawn at the top of the tank and added at the bottom, in the opposite direction of flow to that of the settling anode slimes. However this upward solution flow is not enough to affect the normal downward travel of the insoluble anode components that collect on the tank bottom. In addition to heating the electrolyte to improve cell efficiency, glue and salt are also added in small quantities. The glue, in amounts of 0.1 pound (45.5 g) per ton of copper deposited, improves the character of the cathode deposit, while salt, sufficient to maintain a concentration of 0.012 to 0.03 grams of chloride per litre, will precipitate any silver, bismuth, or antimony which may be present in the solution.

Cathode copper can be marketed directly as sheared cathodes to be used in alloying or can be melted in direct arc-type electric furnaces or reverberatory furnaces and cast into other desired marketable shapes of wire bars, ingots, billets, or slabs. Electric furnaces are preferred because there is no fuel contamination such as occurs with reverberatory

Figure 1.70. Casting refined copper wire bars. *Source:* Courtesy of ASARCO, Inc.

furnace melting, and blowing, poling, and slag making to remove this contamination are unnecessary. A charcoal cover prevents oxidation in a tilting pour hearth beside the melting furnace, and the refined copper analyzing 99.98% is cast into molds on rotating, horizontal casting wheels. The oxygen content of tough pitch copper, 0.03% oxygen, can also be obtained, if desired, during this melting and casting stage by blowing air into the bath to an 0.9% oxygen level, then reducing the bath by poling back to 0.03%.

2a. Nickel Metal Anodes, as processed by most of the nickel refineries, are made by roasting nickel sulfide to nickel oxide, then furnace reducing the nickel oxide to metal, with coke used as the reducing agent. The impure metal is tapped from the furnace and cast into anode molds, as is done for copper anodes. The anodes are 38 by 24 by 2 inches (95 $\times$ 60 $\times$ 5 cm), weigh 580 pounds (264 kg), and are cast with two lugs for support in the cells. The metal analysis of the anodes is rather below that of copper anodes and will only average 95% combined nickel, copper, and cobalt.

The sulfur content of the anode determines to a large degree the amount of insoluble anode mud, and consequently is kept to a low degree, around 1%. This is due to the metallic phase of the anode dissolving selectively during cell operation, leaving the sulfides undissolved. If the anodes contain as high as 10% sulfur, the undissolved sulfides

Table 1.31. Standard Copper Cell Operations, Tamano Smelter, Hibi Kyodo Smelting Co., Japan

Item	P.R.C. Cell	Stripper Cell
Electrolytic Current		
Current density (A/m^2)	340	220
Process	P.R.C.	Normal
Current efficiency (%)	95	95
Cathode		
Size (mm)	1000 $\times$ 1000	1000 $\times$ 1000 (starter sheet)
Weight (kg)	151	6.5 (starter sheet)
Cycle (days)	9	1
Spacing (center to center) (mm)	100	100
Number (sheets per cell)	45	45
Anode		
Size (mm)	980 $\times$ 960 $\times$ 41	1040 $\times$ 1040 $\times$ 38
Weight (kg)	350	370
Cycle (days)	18	27
Number (sheets per cell)	46	46
Electrolyte		
Copper concentration (gpl)	40–45	40–45
Sulfuric acid concentration (gpl)	185–195	185–195
Temperature ($^\circ$C)	64	55
Circulating volume (l/per min)	40	25

Source: J. C. Yannopoulos and J. C. Agarwal, Eds., *Extractive Metallurgy of Copper*, Vol. 1, The Metallurgical Society of AIME, 1976, p. 530.

Table 1.32.   Operational Data, Electrolytic Copper Refining at Tamano Smelter, Hibi Kyodo Smelting Co., Japan

Operational Results and Labor Requirement

Anode charged	9185 mtpm
Electrolytic copper produced	7785 mtpm
Decoppered slime	16 mtpm
Power consumption	377 kW-hr/ton electrolytic copper
Steam consumption	0.183 mt/ton electrolytic copper
Labor requirement	0.848 man-hr/ton electrolytic copper
Current efficiency	95.8%
Anode scrap ratio	14.4%
Anode slimes fall	0.4% of dissolved anode
Current density	326 A/m^2
Unit cell voltage	0.339 V

Assay of Electrolytic Copper

%Cu	>99.99
%Ag	0.0008
%Pb	<0.0001
%As	<0.0001
%Sb	<0.0001
%Bi	<0.0001
%Fe	<0.0001
%Ni	<0.0001
%S	0.0009

Source: J. C. Yannopoulos and J. C. Agarwal, Eds., *Extractive Metallurgy of Copper*, Vol. 1, The Metallurgical Society of AIME, 1976, p. 531, 538.

would be as much as 30% of the original anode weight, and this along with some 10% anode scrap also to be reprocessed would give a very large amount of anode to be re-treated and drastically cut down on the cell throughput.

The position of nickel in the electromotive series makes it necessary to remove various other metals (copper, iron, cobalt, and arsenic) which are all close to nickel in the series. These will also go into solution in the anolyte, and unless the electrolyte is purified they will deposit along with the nickel at the cathode and lower its purity. This is prevented by having the cell divided by the pure nickel cathodes being suspended inside boxes of porous cotton duck on wood frames. The impure anolyte solution is removed from the tanks and put through a purification system to remove the offending impurities, after which it is returned to the cathode compartment to deposit its still remaining nickel ions at the cathode. Catholyte purity is maintained by having a 1 to 2 inch (2.5 to 5 cm) greater hydrostatic head of solution in the cathode compartment than in the anode compartment, so that when the purified electrolyte has deposited its nickel ions at the cathode the solution will flow through the cotton diaphragm into the anode compart-ment to replenish its content of metallic ions at the anode and at the same time wash back those metallic ions that are attempting to migrate directly from the anode to the cathode.

In the anolyte purification process, copper is removed first by cementation with nickel powder. The copper in solution is precipitated as copper powder, and an equivalent

Table 1.33. Typical Electrolytic Copper Cell Analyses, Onahama Smelter and Refinery, Mitsubishi Metal Corp., Osaka, Japan

	Anode (%)	Cathode (ppm)			Electrolyte (g/l)	
		Nos. 1 and 2	No. 3		Nos. 1 and 2	No. 3
Cu	99.57	99.99 (%)	99.99 (%)	H_2SO_4	200	200
Pb	0.076	1.5	1.3	Fe	1.04	0.92
Zn	0.001	0.3	0.3	Ni	17	16
Fe	0.025	2.5	2.4	Sn	0.001	0.001
Ni	0.070	0.7	0.7	As	3.55	3.58
Se	0.030	0.4	0.4	Sb	0.35	0.34
Te	0.006	0.3	0.3	Bi	0.10	0.10
As	0.044	0.4	0.4	Cu	43	45
Sb	0.011	0.7	0.8	Zn	0.54	0.43
Bi	0.006	0.3	0.3			
Sn	0.002	0.5	0.4			
S	0.007	4	4			
Insol.	0.020	–	–			
Au	0.000026	0.1	0.1			
Ag	0.000399	5.0	5.0			

Source: J. C. Yannopoulos and J. C. Agarwal, Eds., *Extractive Metallurgy of Copper,* Vol. 1, The Metallurgical Society of AIME, 1976, p. 601.

quantity of nickel goes into solution to replace it. The cement copper is filtered off and sent to the copper smelter. Iron is removed next by vigorously aerating the solution to oxidize the iron and have it precipitate as ferric hydroxide, $Fe(OH)_3$. Some nickel and cobalt are hydrolyzed and precipitate with the iron, and much of the arsenic is carried along as well. The filtered precipitates are returned to the smelting furnace to recover the nickel and cobalt values. The final purification step is the treatment of the filtered solution with chlorine gas; and the final precipitate from this, cobaltic hydroxide $[Co(OH)_3]$, along with arsenic and lead, is removed by filtering and sent to the cobalt refinery. Purification is now complete, and only the pH of the solution must be adjusted with nickel carbonate to 5.2, neutralizing the hydrogen ion that results from the purification hydrolysis, before the solution is pumped back to the refining cell cathode compartment.

The pure electrolyte now used for deposition of nickel at the cathode contains about 60 grams nickel per litre in a solution of predominately chloride but also containing boric acid and sodium sulfate. One pass through the cathode compartment deposits 12 to 14 grams per litre from this solution, and the cathodes are removed after eight to ten days weighing some 125 pounds (57 kg). Most of the cathodes analyzing 99.93% are sheared into pieces 2 by 2 inches (5 × 5 cm) and marketed in this form.

The anodes are left in the tank during the deposition of nickel metal on three successive cathodes, after which time the 10 to 20% of the anode remaining is removed, washed, and returned for remelting to the anode furnace. All the constituents of the anode dissolve in the cell, except the sulfides and precious metals, which collect on the tank bottom as anode slimes. These are removed periodically and treated for precious metals recovery.

The circuit connections are the Walker multiple system, and tank construction is conventional mastic or asphalt-lined concrete. Precautions are taken to keep the electrolyte out of contact with metals which might dissolve and be further contaminants in it, so the

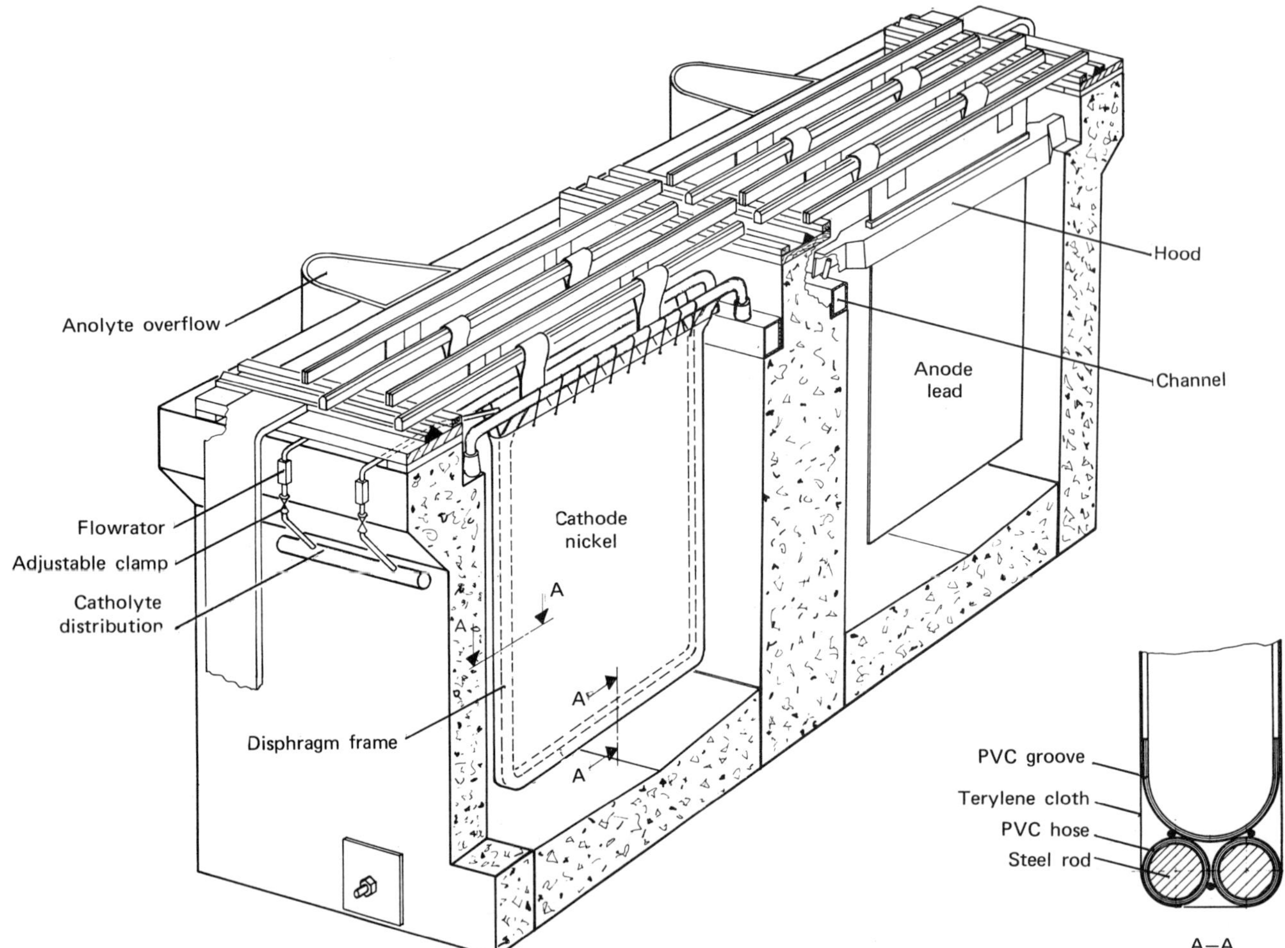

Figure 1.71. Diaphragm cell—electrolytic nickel refining. *Source:* T. Toivanen and P. O. Grönqvist, *Can. Inst. Min. Metall. Bull.*, Vol. 57, No. 626, 1964, p. 657.

solution is transported in lines constructed of hard rubber, plastic, or wood. Current densities are on the order of 16 amperes per square foot (0.09 m^2) of cathode surface, and it is possible to use these higher current densities in a chloride electrolyte and still obtain good, uniform cathode deposition. Current efficiency will be 95%, operating voltage is 2.5 volts per cell, and electrolyte temperature is held at around 135°F (57°C).

2b. Nickel Matte Anodes, treated in a process developed by the International Nickel Company, are made by casting converter matte analyzing 75% nickel and 20% sulfur directly from the converter into anode molds. These anodes are approximately the same size as nickel metal anodes but are more brittle and require quite careful cooling and handling. Slow cooling with the anodes held at 950°F (510°C) for 12 hours completes the phase transformation from βNi$_3$S$_2$ to αNi$_3$S$_2$. After this the anodes can be cooled, without danger of cracking, down to 400°F (204°C) over another 24 hours and then left in air to cool down to ambient temperature.

The tanks and electrical circuits are the same for matte and metal anodes, except that the cell temperature is a bit higher at 145°F (63°C) due to an increased cell voltage to as high as 4 volts, which increases the heat input to the electrolyte and requires a higher melting synthetic resin tank lining in place of asphalt. The same sulfate–chloride electro-

Figure 1.72. Electrolytic nickel refining tanks showing the position of the anodes, canvas boxes, and cathodes. The tanks are built in pairs with a common center wall. To make an electrical connection, all the cathodes in one tank rest on a common busbar with all the anodes in the next tank. *Source:* Courtesy of Inco Limited.

Figure 1.73. Removing slimes containing precious metals from electrolytic refining tanks. *Source:* Courtesy of Inco Limited.

Figure 1.74. Nickel refinery tankhouse, Thompson Plant. The tankhouse contains 684 tanks, each of which produces approximately 350 pounds nickel (772 kg) per day. *Source:* Courtesy of Inco Limited.

lyte is also used for both. The cathodes are pure nickel sheet in conventional cotton duck cathode compartments, and the anodes are enclosed in woven dynel bags treated with neoprene after weaving.

Sulfide anode corrosion involves the oxidation of sulfide sulfur to elemental sulfur at an anode potenial of 1.2 volts, as compared to the anode potential of 0.2 volt required for the anodic corrosion of metal anodes, with an overall cell reaction of

$$Ni_3S_2 = 3Ni + 2S$$

The elemental sulfur formed during anode corrosion remains on the anode surface as a soft coating containing more than 95% sulfur. This elemental sulfur, along with the precious metals and selenium, collects in the dynel bag surrounding the anode. Cobalt, copper, iron, and arsenic dissolve in the anolyte along with nickel, and the sulfide anodes corrode away quite smoothly.

The maintenance of high-purity cathode deposition from matte anodes again necessitates electrolyte purification to eliminate copper, iron, arsenic, lead, and cobalt in essentially the same manner as was employed with metal anodes. The major difference in the two processes is that the rate of nickel ions going into solution is lower with the sulfide

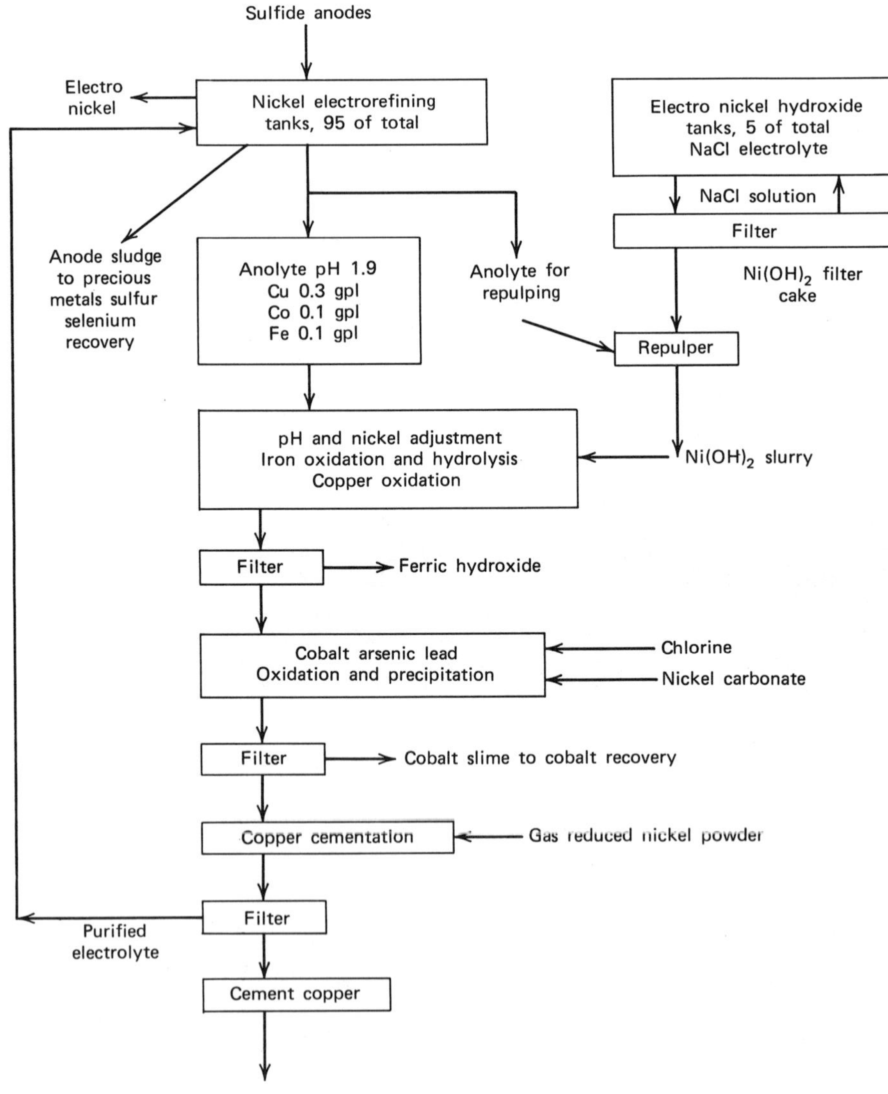

Figure 1.75. Flow sheet showing the production of electrolytic nickel from sulfide (matte) anodes. *Source:* Courtesy of Inco Limited.

anode than with the metal anode, with the anode potential for the sulfide anode being sufficiently large so that some water is decomposed to gaseous oxygen and free acid. This results in an acidic anolyte, pH 1.9, deficient in nickel, and nickel can be deposited faster from the purified electrolyte at the cathode than can enter the electrolyte at the anode. This situation is corrected during anolyte purification by adding basic nickel hydroxide or nickel carbonate both to increase the amount of nickel in solution and to neutralize the excess acid to pH 4. The purified electrolyte pumped to the cathode compartment will contain 60 grams per litre of nickel.

The cathodes, analyzing +99.95% nickel and weighing 140 pounds (64 kg), are removed from the cell after 10 days. They are then sheared to appropriate size and sent to market. The anodes, after use with three or four cathodes, are sluiced to remove the adhering sulfur and precious metal sludge and are returned to the converter for remelting. The washed solids from the anode bag are heated to 275°F (135°C) to melt the 97% elemental sulfur, which is filtered off leaving the undissolved sulfides and precious metals. These separated products are treated to recover their contained values.

The current efficiency of the nickel ion deposition at the cathode is very good, 99.5%, at a current density of 20 amperes per square foot (0.09 m^2) of cathode surface. This is slightly higher than the 16 amperes current density used with nickel metal anodes.

3. Softened Lead Anodes, as processed at Oroya, are electrolytically refined by the *Betts process* to produce bismuth free lead of unusually high purity, 99.999%.

Anodes of softened lead, 36 by 24 by 1 inches (90 × 60 × 2.5 cm) and weighing 350 pounds (159.1 kg), are cast and placed in cells with thin, 10 pound (4.5 kg) pure sheet lead cathodes. The anodes and cathodes are placed 50 to 60 in a cell and spaced $1\frac{5}{8}$ inches (4.1 cm) face to face. Closer spacing than this is conductive to touching and short circuiting and is avoided.

The electrical connections are of the Walker multiple system, with the tanks arranged in double sets to take advantage of common busbars. The tanks themselves are of conventional concrete lined with asphalt construction and similar to those used for copper refining. A tank holding 50 electrodes, anodes and cathodes, will measure some 9 by $2\frac{1}{2}$ by 4 feet (2.74 × 0.76 × 1.22 m). A few plants have the tanks arranged in a cascade, so that the tanks are at different elevations and the electrolyte flows by gravity from the higher to the lower levels. If the tanks are all on one level, then pumps provide circulation by moving the solution from tank to tank, withdrawing it from the bottom of one tank and feeding it into the top of the next.

The electrolyte is a solution of lead fluosilicate, $PbSiF_6$, and 8 to 15% total hydrofluosilic acid, H_2SiF_6, with 3 to 5% of the hydrofluosilic acid free. The electrolyte also contains 5 to 10% lead. Glue added to the electrolyte helps to ensure a solid cathode deposit and is consumed at the rate of $1\frac{1}{2}$ pound glue (0.68 kg) per ton cathode lead deposited. The current density is 16 amperes per square foot (0.09 m^2) cathode area at a current efficiency of 90 to 93% and a cell voltage of 0.5 volt.

As a result of the electrolysis, lead, tin, zinc, and iron go into solution in the anolyte; of these only tin is quite close to lead in the electromotive series and will also deposit at the cathode. However the prior operation of softening eliminated the tin, and there will be

Table 1.34. Comparative Refined Nickel Analyses

	Carbonyl Process	Matte Anodes	Reduced Metal Anodes
%Ni	+99.95	99.27	99.73
%Co	Nil	0.70	0.25
Total	99.98	99.97	99.98

no more than a trace left to be deposited. Copper, bismuth, the precious metals, cadmium, arsenic, and antimony are all insoluble and remain attached to the anode in the form of a mud or sponge. After four days the remaining 25 to 35% of the anode is removed from the cell, its insoluble mud layer washed and scraped off, and the cleaned anode scrap then taken to the anode furnace for remelting. The short, four day cycle is brought about by the formation of neutral lead fluosilicate in the insoluble anode mud layer which increases current resistance and limits anode efficiency. There is also the danger of the thickening mud layer sliding off and either causing short circuits or becoming entangled with the depositing cathode lead.

The anode mud is filtered, with the filtrate being returned to the tank house for reuse, and the solid residues is treated especially to recover the contained precious metals, bismuth, antimony, arsenic, and copper. The high-purity lead cathodes, which have increased in weight to 150 pounds (68.2 kg), are melted and cast into pigs for market.

CHEMICAL REFINING

This is a method of refining carried out in cases where the impure metal compound which has been produced by conventional furnace smelting can be most conveniently chemically treated to give a high-purity refined metal product and recover the impurities as by-products, by processes which include either volatilization and then deposition from the decomposed metallic vapor or selective leaching and precipitation. Both of these processes can be carried out within narrow tolerance limits so that it is possible to separate the desired metal and its impurities with considerable accuracy and also to separate in some instances the impurity by-products from each other.

The volatilization process requires that the metal to be refined be in a state which will easily volatilize and form a metallic vapor with the gaseous compound that is being supplied for this reaction. The impurities in the metal do not combine to form any metallic vapor and are excluded for reacting. The gaseous compound supplied for the volatilization reaction is now decomposed away from its combined metallic vapor, which separates into pure metallic vapor to be deposited as refined metal. The gaseous reaction compound is reused as the absorbing vehicle for the metallic vapor in the preceding volatilization stage.

In the same manner selective dissolution of the desired values in leach solutions and selective precipitation from these solutions separate the metal being refined from the associated impurities found with it, with the impure metal compound being pretreated to be in a suitable condition to be easily soluble in the leaching solution provided.

Consequently with chemical refining a gaseous compound is the vehicle that selectively combines with and removes the metal being purified in one instance, while in the second case a leach solution is the solute for this same type of selective collection, with the final refined metal being extracted from vapor in one and from leach liquid in the other.

TYPES OF CHEMICAL REFINING

There are two distinct types of chemical refining, one being that of vapometallurgical refining, such as the carbonyl process for nickel, where a reactive gas is supplied to combine selectively with the desired metal component of the impure metal being refined. The other type is that of selectively precipitating the desired metals as compounds from a leach solution and then chemically separating them from each other, as in the cobalt processes.

Both methods of chemical refining require specific pretreatments to have the impure metal in a condition where it will readily react with the gas supplied in one case and react with the leach solution in the other. These pretreatments consist of conventional smelting and roasting operations.

Vapometallurgical Refining is used for the treatment of nickel sulfide matte which is given the preliminary pyrometallurgical treatments of roasting the sulfide to an oxide and then reducing the oxide to semipure metal with hydrogen-rich water gas in tall, gas-tight reducers. The impure metallic granules resulting are passed through volatilizers and are here brought into contact with the reactive gas, carbon monoxide, at a slightly elevated temperature and atmospheric pressure, to combine as metallic carbonyl and be carried out in the departing gas stream. When heated to a higher temperature the carbonyl decomposes into carbon monoxide and practically pure metal which deposits on the surface of metal pellets that are added to the decomposer units.

Leaching and Precipitation methods of chemical refining also require preliminary treatments to convert the material into a suitable form for processing.

Two different compound forms are treated in this manner. One is an *oxide* which is given the initial pretreatment of electric furnace smelting to form an alloy containing most of the metal to be refined. The alloy is then acid leached, and the metal and by-product impurities are selectively chemically precipitated from the pregnant solution. The metal is precipitated as the hydrate which is heated to convert it to the oxide, and then the oxide is reduced with carbon to the final refined metallic condition.

Arsenides are also chemically refined, and here the pretreatment consists of blast furnace smelting to drive off most of the arsenic, roasting the resultant matte and speiss to remove the remaining sulfur and arsenic, and finally pugging the roasted calcines with sulfuric acid to form water-soluble sulfates of the desired metals. The pugged sulfates are water leached and treated in the same manner as was the oxide, with the desired metal and the by-product impurities being selectively chemically precipitated from the leach solution. The metal is precipitated as the hydrate and, as before, is heated to convert to the oxide. Then the oxide is reduced with carbon to the metallic state.

CHEMICAL REFINING PROCESSES

1. Nickel Carbonyl Process, or the *Mond process*, named for its developer, Ludwig Mond, used by the International Nickel Company, takes advantage of the fact that if impure nickel metal is brought into contact with carbon monoxide gas between 100°F (38°C) and 200°F (93°C) the metal will react with the gas to form volatile nickel carbonyl which is carried out of the reaction chamber in the gas stream:

$$Ni + 4CO = Ni(CO)_4$$

This reaction is quite readily reversible, and if the carbonyl gas is passed through a tower containing nickel shot heated in the temperature range of 300 to 600°F (150 to 316°C), the nickel carbonyl will decompose into metallic nickel, which deposits on the shot causing it to grow.

$$Ni(CO)_4 = Ni + 4CO$$

The mild conditions of low temperature and atmospheric pressure used in the process take only the nickel into the volatile carbonyl phase, so that the operation is highly selective and produces a very pure product, 99.95% nickel. This compares with refined electrolytic nickel from nickel metal anodes which analyzes 99.73% nickel with 0.25% cobalt and with refined electrolytic nickel from nickel matte anodes which analyzes 99.27% nickel and 0.70% cobalt.

Preliminary fire refining converts nickel matte into the impure nickel metal used in the carbonyl process. This is accomplished by roasting the sulfide matte to oxide in fluid bed roasters and then reducing the nickel oxide to metal in reducers using water gas analyzing 51% hydrogen and 40% carbon monoxide as the reducing agent. The impure metal granules that result are the feed material for the carbonyl chemical refining process.

The multiple-hearth volatilizing units are gas-tight cylindrical columns, similar in design to multiple-hearth roasters, 40 feet high and 6 feet in diameter (12.2 × 1.83 m), with 21 hearths. The impure nickel metal granules are fed in at the top of the volatilizer and travel downward from hearth to hearth, while the carbon monoxide gas is introduced at the bottom of the volatilizer and travels upward. The nickel carbonyl formed joins the rising gas stream and flows out the top. The carbonyl formation reaction produces heat, and to keep this controlled and around the 140°F (60°C) needed for active carbonylation, cooling water is used to remove the excess heat. This water can flow down the outside of the volatilizer cooling externally or can be circulated through the hollow hearths to cool internally.

Only partial extraction of nickel as carbonyl takes place in one pass of the impure granules through a volatilizer; and to achieve the 95% extraction required, it is necessary to repeat the process through a total of eight volatilizers, which takes four days. The capacity of a battery of eight volatilizers is 20 tons crude nickel granules per day.

The residue granules discharged from the eighth volatilizer contain copper, iron, cobalt, and precious metals, in addition to some remaining nickel. These are given further treatment to separate and recover the remaining values.

The exit gases from the eight volatilizers contain varying amounts of nickel carbonyl,

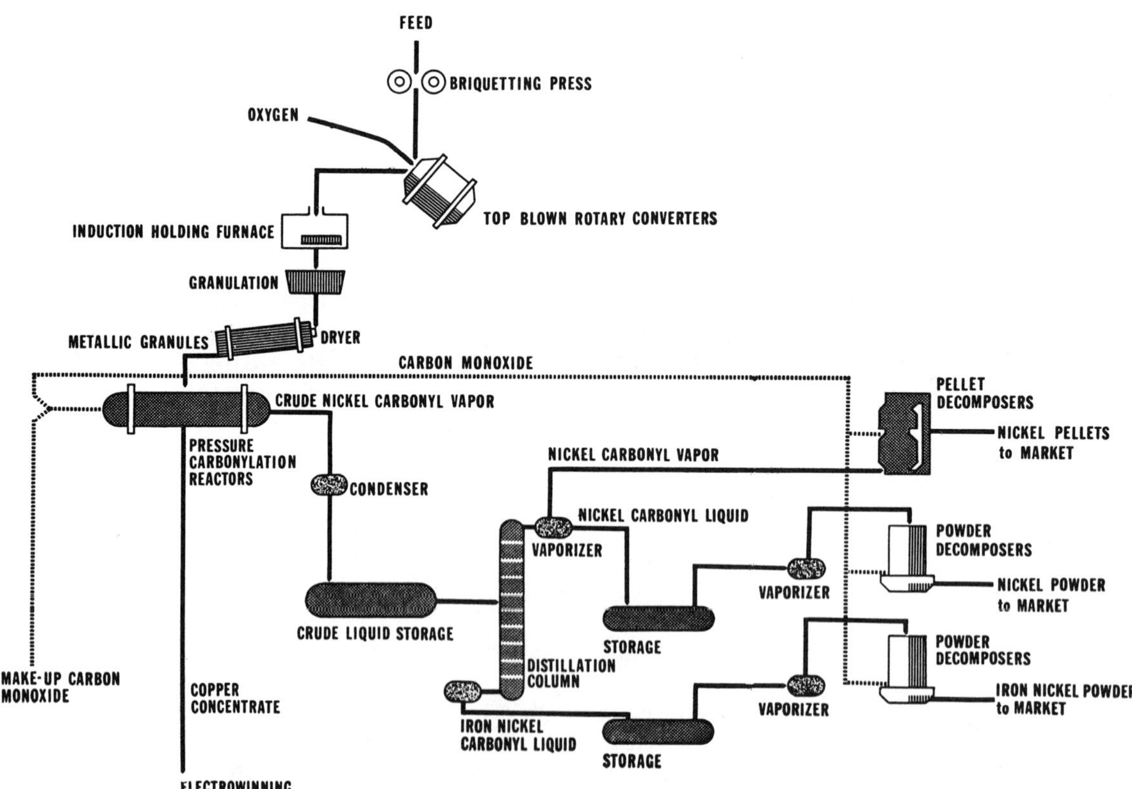

Figure 1.76. Nickel carbonyl refining process flow sheet. *Source:* Highly automated nickel refinery will use unique process, *Industrial Heating*, Vol. 39, No. 8, 1972, p. 1416.

from 15% by volume from the first to 0.5% by volume from the eighth. The combined gas from the battery will average 8% and is sent to the pellet decomposer.

This unit consists of a preheating chamber 20 feet high and 4 feet in diameter (6.1 × 1.22 m) through which nickel pellets flow by gravity and then on down, at 450°F (232°C), into a 10 foot high and 4 foot diameter (3.05 × 1.22 m) decomposer vessel, where the descending hot pellets meet a rising stream of carbonyl gas. The heat from the pellets is transferred to the carbonyl gas which decomposes and deposits its contained nickel on the surface of the pellets. The pellets run out of the bottom of the decomposer by gravity and are lifted by bucket elevator to make another pass down through the preheater and decomposer.

Nickel powder is added periodically to grow new pellets, and about three months is required to grow a pellet in the decomposer to the $\frac{5}{16}$ inch size (7.8 mm) required for the market. The repeated nickel deposition on the pellets as they cycle through the decomposer gives them an onionlike structure as they grow in size, and the movement of the pellets as they pass down by gravity prevents any pellets from adhering and sticking together.

The exhaust gas from the decomposer, which is carbon monoxide with a very small amount of undecomposed nickel carbonyl, is returned to the volatilizers.

2a. Cobalt Speiss as produced by the blast furnace treatment of arsenical cobalt is the result of smelting concentrates containing nickel, copper, and iron. The speiss is then ground to minus 80 mesh and roasted at 1300°F (705°C) in order to remove most of the arsenic and sulfur as volatile gases and to oxidize the iron.

The roasted speiss, which contains 23% cobalt, 21% iron, 9% nickel, 2.5% copper, 11% arsenic as arsenate, and 8.5% sulfur, is mixed in a pug mill with water and sulfuric acid to convert the cobalt, nickel, copper, and iron to water-soluble sulfates. This sulfated mixture goes on to a small water-filled leach tank stirred by a mechanical agitator to dissolve the soluble sulfates and from there to a thickener. The solids from the thickener underflow containing the insoluble, unsulfated portion of the roasted speiss are dried and returned to the blast furnace, while the thickener overflow containing the soluble sulfate values goes on to chemical processing.

The solution is treated in mechanically agitated tanks first with sodium chlorate to oxidize the iron, then with lime to raise the pH to 3.25 to precipitate it, but not above this, as copper will then also come out of solution and precipitate. The iron, arsenic, and sulfur come out of solution as ferric arsenate ($FeAsO_4$), ferric hydroxide (Fe_2O_3), and calcium sulfate ($CaSO_4$). The precipitated solids are removed as thickener underflow and are discarded. The thickener overflow still containing the cobalt, nickel, and copper is now put through cementation tanks where the copper is precipitated on scrap iron and removed.

The solution coming from the copper circuit contains 1.5% cobalt + nickel, 0.01% copper, and 0.2 to 0.3% iron, at a pH of 2.9. This is batch treated in 70 ton neutralizer tanks by heating and agitating with steam lances to 158°F (70°C), adding sodium chlorate to oxidize all the iron to ferric hydroxide, and then adding lime to raise the pH to 5.4 to precipitate all the iron and copper from the solution. These precipitates are allowed to settle, then removed and vacuum filtered, and finally returned to the initial iron removal circuit.

The clean solution in the neutralizer tanks is decanted off and filtered to remove any fine precipitate still remaining. Any recovered solids are added to the settled neutralizer tank residues being returned to the iron removal circuit. The filtered solution, containing cobalt and nickel, is pumped to the cobalt precipitation tanks, acidified with sulfuric acid to pH 2.6, and sodium hypochlorite, NaClO, is added to precipitate cobaltic hydroxide, $Co(OH)_3$.

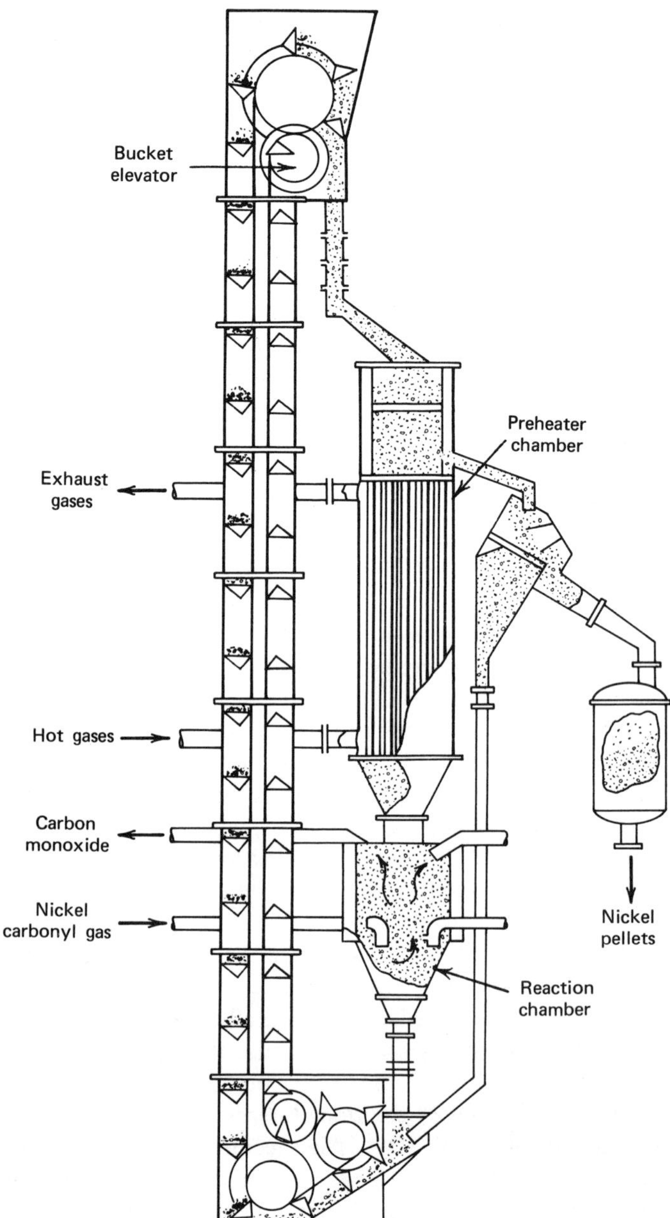

Figure 1.77. Nickel carbonyl pellet decomposer. *Source:* J. R. Boldt, Jr., and P. Queneau, *The Winning of Nickel*, Methuen and Company Ltd., London, 1967. Copyright Inco Limited, 1967, p. 378.

This final hydroxide precipitate is separated in a plate-and-frame filter press and the press cake mixed with sodium carbonate and dried for 8 hours at 1382°F (750°C). The dry precipitate, which at this stage is about 70% cobalt as Co_2O_3, is taken to cold-water leach tanks to wash out the calcium oxide along with the sulfur as sodium sulfate. The sodium carbonate which was added assists in this removal by transforming the calcium sulfate to more soluble sodium sulfate, and it also precipitates any soluble cobalt sulfate which went into solution.

The cobalt-free solution is pumped off and discarded, while the cobalt oxide residue, when sulfur free, 4000 pounds (1818 kg) to a batch, is charged to rotary kilns for 8 hours. Charcoal is added at the rate of $\frac{1}{3}$ to $\frac{1}{2}$ pound (0.15 to 0.23 kg) of charcoal per pound (0.45 kg) of cobalt metal, the kilns are heated to 1830°F (1000°C), and the cobalt oxide is reduced by carbon to grain metal. This grain metal is melted in electric furnaces and granulated in water to give cobalt shot analyzing 98.6% cobalt.

Nickel is still present in the neutralizer tank solution after most of the cobaltic hydroxide has been precipitated with sodium hypochlorite, and this solution is first reprecipitated with additional sodium hypochlorite to strip out any remaining cobalt. Then after treatment in a clarifying press the clear solution goes to the nickel plant. Here the pH is raised to 8.6 to precipitate nickel hydroxide, which is sold without further processing.

2b. Cobalt White Alloy from Katanga is one of the products of electric furnace smelting of cobalt oxide ores and concentrates rich in copper. The oxides are reduced in a conventional steel making-type three-electrode electric furnace with coke and give two alloys which are easily separated because of differences in specific gravity. These are the heavier red alloy, 89% copper and 4.5% cobalt, which is taken to a copper refinery; and the lighter white alloy, 42% cobalt, 15% copper, and 34% iron, which is treated by chemical refining to give refined cobalt metal.

The white alloy is heated to red heat, in which temperature range it is most brittle, and is crushed to powder in a hammer mill. This powder is leached in dilute 20% sulfuric acid which dissolves the cobalt and iron but not the copper, which is mostly left as a residue and shipped to a copper smelter for treatment. The pregnant solution is neutralized with returned impure cobalt carbonate and then filtered. At this stage the solution contains 98% of the cobalt in the white alloy and has iron and copper contents of 10 and 3 grams per litre, respectively. The dissolved iron is oxidized with air to the ferric state. Then the pH is raised to 3.5 with lime to precipitate most of the ferric hydrate, which is removed and discarded. More lime is now added to raise the pH to 5.5, at which level the remaining iron, the copper, and some 10 to 14% of the cobalt in solution precipitates. This residue is removed and recycled back through the initial leaching operation to recover its values.

The filtered, purified solution is pumped to tanks with mechanical agitators, and sodium carbonate is added to raise the pH and precipitate pure cobalt carbonate, which is removed. A second precipitation is made by adding additional sodium carbonate to raise the pH to 8.4, and this removes all of the remaining cobalt in solution but gives an impure product. This impure cobalt carbonate is used to neutralize the sulfuric acid in the initial leaching operation.

The pure cobalt carbonate is roasted to CoO in rotary, oil-fired furnaces at 1435°F (780°C) to give a 75% cobalt product. This oxide is compressed into small cylinders $\frac{1}{2}$ inch in diameter (12.5 mm) by $\frac{1}{2}$ inch long (12.5 mm) and fed with charcoal into vertical reduction furnaces with carborundum walls. The furnaces are heated to 1830°F (1000°C) and charcoal reduction of the cobalt oxide takes place. After reduction is complete the metal cylinders, which are +99% pure cobalt, are separated from the excess charcoal used in the furnace and are polished and barreled for the market.

NONREACTIVE METALS, PYROMETALLURGICAL EXTRACTIVE PROCESSES

The individual nature of ores, and to a lesser degree local conditions, makes it hardly ever possible to have rigid, standard processing flow sheets. No one metallurgical project is exactly like another, and consequently each one must be adapted to a special, specific case. So while the general operations for treatment are the same, their combinations will differ in accordance with the individuality of the material being processed.

The combinations of processes using pyrometallurgical methods to extract and refine copper, nickel, lead, and cobalt from their various natural metallic compounds are summarized in Table 1.35.

Table 1.35. Nonreactive Metals, Pyrometallurgical Extractive Processes

Copper

Sulfide

A. Beneficiate
Partially oxidizing roast
Smelt to matte
Convert matte to blister copper
Fire refine

B. Beneficiate
Partially oxidizing roast
Smelt to matte
Convert matte to blister copper
Fire refine to anodes
Electrolytically refine

C. Beneficiate
Smelt to matte
Convert matte to blister copper
Fire refine to anodes
Electrolytically refine

Nickel

Oxide

A. Sinter or briquette
Smelt to matte with gypsum or pyrite
Convert to white metal stage
Roast to oxide
Electric furnace reduction to metal

Garnierite

Beneficiate
Sinter
Smelt to matte with gypsum
Convert matte to white metal stage
Roast to oxide
Retort reduction to metal

Sulfide

A. Beneficiate
Partially oxidizing roast
Smelt to matte
Convert matte to white metal stage
Slow cooling
Separation of sulfides by flotation
Cast nickel sulfide anodes
Electrolytically refine

B. Beneficiate
Partially oxidizing roast

Smelt to matte
Convert matte to white metal stage
Slow cooling
Separation of sulfides by flotation
Oxidizing roast
Furnace reduction of oxide to metal
anodes
Electrolytically refine

C. Beneficiate
Smelt to matte
Convert matte to white metal stage
Slow cooling
Separation of sulfides by flotation
Oxidizing roast
Furnace reduction of oxide to metal
anodes
Electrolytically refine

D. Beneficiate
Partially oxidizing roast
Smelt to matte
Convert matte to white metal stage
Oxidizing roast
Acid leach of impurities
Furnace reduction of oxide to metal
anodes
Electrolytically refine

E. Beneficiate
Partially oxidizing roast
Smelt to matte
Convert matte to white metal stage
Oxidizing roast
Acid leach of impurities
Water gas reduction to metal
Form nickel carbonyl with carbon
monoxide
Decompose carbonyl to metal

F. Beneficiate
Partially oxidizing roast
Smelt to matte
Convert matte to metal with oxygen
Form nickel carbonyl with carbon
monoxide
Decompose carbonyl to metal

Lead

Sulfide (With Silver)

A. Beneficiate
   Oxidizing double sintering
   Smelt to metal
   Dross, slag fuming
   Softening
   Desilverize
   Dezinc
   Debismuthize

B. Beneficiate
   Oxidizing double sintering
   Smelt to metal
   Dross, slag fuming
   Softening
   Electrolytically refine
   Silver from anode slimes

C. Beneficiate
   Flash smelt to metal
   Convert
   Desilverize

D. Beneficiate
   Flash smelt to metal
   Refine

E. Beneficiate
   Pelletize
   Convert
   Refine

Cobalt

Oxide

Beneficiate
Smelt to alloys, red and white
White alloys sulfuric acid leach
Precipitate cobalt carbonate
Calcine carbonate to oxide
Furnace reduction of oxide to metal

Arsenide

Beneficiate
Smelt to speiss
Roast speiss to oxide
Sulfate oxide with sulfuric acid
Water leach
Precipitate cobalt hydroxide
Calcine to oxide
Furnace reduction of oxide to metal

REFERENCES

Anderson, J. N. and P. E. Queneau, Eds., Pyrometallurgical Processes in Nonferrous Metallurgy, *Metallurgical Society Conference AIME*, Vol. 39, Gordon and Breach, New York, 1965.

Biswas, A. K. and W. G. Davenport, *Extractive Metallurgy of Copper*, Pergamon, New York, 1976.

Boldt, J. R., Jr., *The Winning of Nickel*, Methuen, London, 1967.

Bunshah, R. F., Ed., *Techniques of Metals Research*, Vol. 1, Pt. 2, Techniques of Materials Preparation and Handling, Interscience, New York, 1965.

The Staff of Batelle Memorial Institute, *Cobalt Monograph*, Centre d'Information du Cobalt, Columbus, Ohio, 1960.

Cotterill, C. H. and J. M. Cigan, Eds., *AIME World Symposium on Mining and Metallurgy of Lead and Zinc*, Vol. 2, Extractive Metallurgy of Lead and Zinc, Port City Press, Baltimore, 1970.

Dennis, W. H., *Metallurgy of the Nonferrous Metals*, Pitman, New York, 1961.

Diaz, Carlos, Ed., *The Future of Copper Pyrometallurgy*, Chilean Institute of Mining Engineers, Santiago, Chile, 1973.

Ehlrich, R. P., Ed., *Copper Metallurgy*, Extractive Metallurgy Division Symposium on Copper Metallurgy, AIME, 1970.

Habashi, F., *Principles of Extractive Metallurgy*, Vol. 1, Gordon and Breach, New York, 1969.

Hampel, C. A., Ed., *Rare Metals Handbook*, Chapman and Hall, London, 1956.

Harbison, E. J. and J. A. Davidson, *Oxygen in Copper Smelting*, Air Products and Chemicals, Inc., Allentown, Pennsylvania, 1972.

Henrie, T. A. and D. H. Baker, Jr., Eds., *Electrometallurgy*, Extractive Metallurgy Division Symposium on Electrometallurgy, AIME, New York, 1968.

Herbert, I. C., New Copper Extraction Processes, *J. Metall.*, Vol. 26, No. 8, 1974.

Howard-White, F. B., *Nickel, an Historical Review*, D. Van Nostrand, New York, 1963.

Jones, H. R., Pollution control in the non-ferrous metals industry, *Pollution Control Review*, No. 13, Noyes Data Corp., Park Ridge, New Jersey, 1972.

Jones, M. J., Ed., *Advances in Extractive Metallurgy and Refining*, The Institution of Mining and Metallurgy, London, 1968, 1972.

Annual Review Issue, *J. Metall.*, Vol. 28, No. 3, 1976.

Annual Review Issue, *J. Metall.*, Vol. 29, No. 3, 1977.

Annual Review Issue, *J. Metall.*, Vol. 30, No. 4, 1978.

Lurgi Manual, Lurgi Gesellschaften, Frankfurt am Main, 1961.

Melcher, G., E. Müller and H. Weigel, KIVCET Process, *J. Metall.*, Vol. 28, No. 7, 1976.

Mineral Facts and Problems, U.S. Bureau of Mines Bulletin 667, U.S. Department of the Interior, 1975.

Oroya Metallurgical Operations, Cerro De Pasco Corp. (now Centromin Peru), 1972.

Pehlke, R. D., *Unit Processes of Extractive Metallurgy*, Elsevier, New York, 1972.

Queneau, P., Ed., *Extractive Metallurgy of Copper, Nickel and Cobalt*, Interscience, New York, 1961.

Robeitte, A. G. E., *Electric Smelting Processes*, Wiley, New York, 1973.

Rosenbaum, J. B., Minerals Extraction and Processing: New Developments, in *Materials: Renewable and Nonrenewable Resources*, Abelson, P. H. and A. L. Hammon, Eds., American Association for the Advancement of Science, Washington, D.C., 1976, pp. 113–117.

Rosenqvist, T., *Principles of Extractive Metallurgy*, McGraw-Hill, New York, 1974.

Ryan, W., Ed., *Non-ferrous Extractive Metallurgy in the United Kingdom*, The Institute of Mining and Metallurgy, London, 1968.

Sohn, H. Y. and M. E. Wadsworth, Coordinators, *Rate Processes of Extractive Metallurgy*, The Metallurgical Society of AIME, New York, 1976.

Szekely, J. and N. J. Themelis, *Rate Phenomena in Process Metallurgy*, Wiley, New York, 1971.

Themelis, N. J., G. McKerrow, P. Tarassoff and G. Hallett, Noranda Process, *J. Metall.*, Vol. 24, No. 4, 1972.

Thomas, R., Ed., *Engineering and Mining Journal Operating Handbook of Mineral Processing*, Vol. 1, McGraw-Hill, New York, 1977.

Queneau, P., C. E. O'Neill, A. Illis and J. Warner, Top Blown Rotary Converter, *J. Metall.*, Vol. 21, No. 7, 1969.

World Minerals Availability, 1975–2000, Vol. 2, Outlook for Mineral Processing, Nodule Mining, and Transportation, Stanford Research Institute, SRI Project ECC 3742, Stanford, California, 1976.

Yannopoulos, J. C. and J. C. Agarwal, Eds., *Extractive Metallurgy of Copper*, Vol. 1, Pyrometallurgy and Electrolytic Refining; Vol. 2, Hydrometallurgy and Electrowinning, Metallurgical Society of AIME, New York, 1976.

Young, R. S., *Cobalt, Its Chemistry, Metallurgy, and Uses*, Reinhold, New York, 1960.

2. Nonreactive Metals, Hydrometallurgical Treatments

Wet chemical processing is a newer development that has appeared to supplement the older pyrometallurgical processes used for the extraction and recovery of metals. This increasing development of wet metallurgy has been greatly influenced by the fact that the ever growing demand for metals makes it necessary to use additional ores for this metal source, some of which, for technical or economic reasons, are better treated by hydrometallurgical rather than pyrometallurgical methods.

The common metals recovered by wet metallurgical processes are copper, nickel, cobalt, gold, and silver. With individual variations the most usual general processing will be first to roast to change the metallics into compounds soluble in a leaching solution. The roasted metals are then leached into solution and separated from the insoluble residue by decanting or filtration, and the solution is purified if necessary and then treated chemically or electrolytically to precipitate metals or metal salts. In some instances a final refining operation additionally purifies the metal product.

These process treatments are grouped into their major categories in Table 2.1.

ROASTING

Roasting base metal ores or concentrates as a preliminary to leaching takes the form of oxidizing sulfides to oxides or sulfates or reducing silicates and oxides to a crude metallic state. This is done with the purpose of changing the metallics into compound forms that are more soluble in the leaching solution, to volatilize certain soluble impurities that could contaminate the pregnant solution, and to make the metallic compounds porous to be more readily attacked by the leaching solvent.

As roasting is a relatively expensive operation, it can only be used on high-grade material that can justify the expensive treatment costs. However the effect of an oxidizing or sulfatizing roasting of low-grade sulfide ores can be obtained by their natural weathering over an extended period of time.

Table 2.1. Nonreactive Metals, Hydrometallurgical Treatments

Roasting Methods

1.   Nickel laterites high in magnesia—hearth
2.   Cobalt sulfide concentrate—hearth, fluid-bed
3.   Copper sulfide concentrate—fluid-bed

Leaching Methods

1a.   Copper oxide ores and concentrate—tank percolation, tank agitation
1b.   Copper sulfide ores and concentrates—in place, heap, tank percolation, pressure
1c.   Native copper and copper carbonate concentrates—tank percolation
2a.   Nickel laterites high magnesia roasted calcines—tank agitation
2b.   Nickel laterites low magnesia ore—pressure
2c.   Nickel sulfide concentrate—pressure
2d.   Nickel sulfide matte—tank agitation
3a.   Cobalt sulfide roasted calcines—tank agitation
3b.   Cobalt oxide concentrate—tank agitation
4.   Gold and silver ores and concentrates—tank agitation, heap

Precipitation Methods

1a.   Copper sulfate solution—electrolytic, cementation (tank and cone, cells and tanks, launders), solvent extraction—electrolytic
1b.   Copper ammoniacal solutions—stills, solvent extraction—electrolytic
1c.   Copper chloride solutions—vacuum or refrigeration
2a.   Nickel ammoniacal solution—stills, autoclave
2b.   Nickel sulfate solution—electrolytic
3.   Cobalt sulfate solution—chemical
4.   Gold and silver cyanide solution—chemical replacement, carbon adsorption—electrolytic

Refining Methods

1a.   Copper electrolytic cathodes—fire
1b.   Cement copper—fire, electrolytic
1c.   Copper sulfide percipitate—fire
1d.   Copper oxide precipitate—fire
2a.   Nickel oxide precipitate—fire
2b.   Nickel sulfide precipitate—chemical reduction
3.   Cobalt hydroxide precipitate—electrolytic
4.   Gold and silver—fire, electrolytic, chemical

The roasting reactions used here are the same as have been discussed previously for the pyrometallurgical treatment of the nonreactive metals.

TYPES OF ROASTING

The types of roasters used for the preliminary treatments of ores and concentrates to be leached are multiple-hearth roasters and fluid-bed roasters, the operations of which have been previously described.

In one case a reducing roast is required to reduce an oxide and silicate to an impure metallic state which can be dissolved by ammonia. These roasters are very large, $22\frac{1}{2}$ feet in diameter (6.86 m) and 17 hearths high, supplied at the bottom with hot producer gas

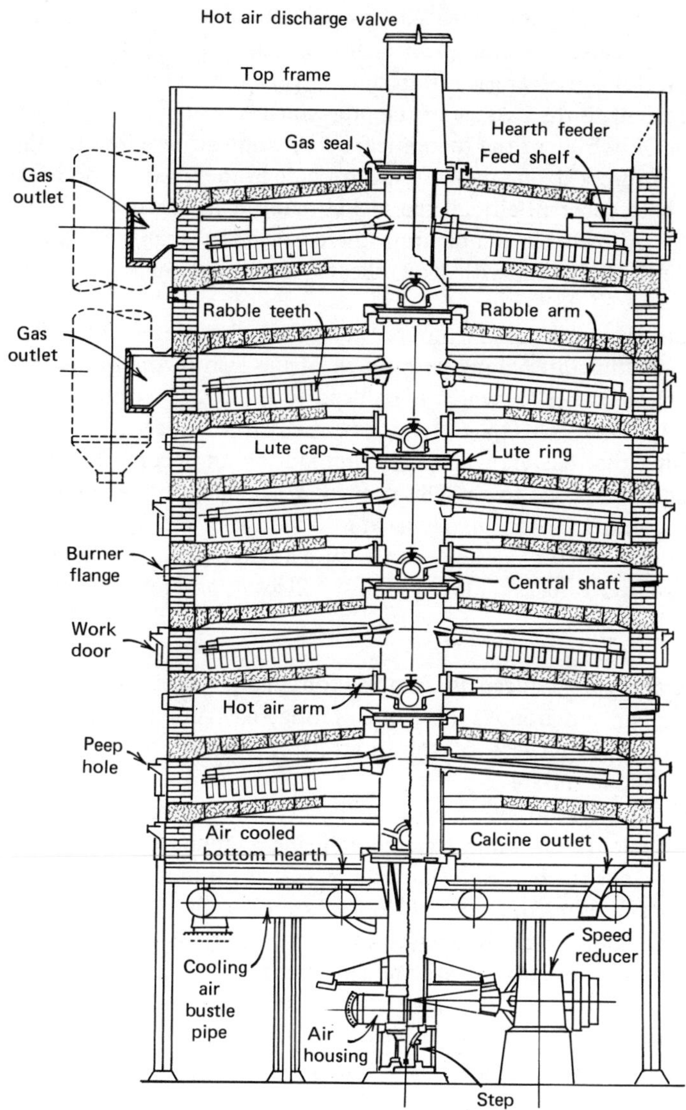

Figure 2.1. Nichols Herreshoff multiple-hearth roaster. *Source:* Courtesy of Nichols Engineering and Research Corp.

made by burning anthracite coal. To ensure a strong reducing atmosphere in the roaster only one half of the theoretical air requirement is provided for the combustion of the producer gas.

One other preliminary roast used is the treatment of sulfides, in order to oxidize them to water-soluble sulfates. Eight-hearth roasters are used for this operation, with special attention to the control of the temperature and admission of air to ensure that the product will not be oxidized beyond the desired sulfate form.

The third roast is an oxidizing roast of a sulfide concentrate so that its values will be soluble in an acid, spent electrolyte leach solution.

ROASTING PROCESSES

1. Nickel Laterites, 1.5% nickel and high in magnesia (8%), are given a reducing roast to reduce the nickel silicate and oxide, $(Mg, Fe, Ni)_6Si_4O_{12}(OH)_8$ and $(Fe, Ni)O(OH) \cdot nH_2O$, to the metallic state, which will be soluble in ammonia, and at the same time to reduce a minimum amount of iron which will also be soluble.

The ore is pretreated by drying and grinding to 90% minus 200 mesh, after which it is transferred to the top-charging hearths of multiple-hearth roasters, $22\frac{1}{2}$ feet in diameter (6.86 m) and 17 hearths high. Hot producer gas is introduced at the bottom of the roaster and flows upward, countercurrent to the passage of the ore, which is being rabbled downward from hearth to hearth. Only 50% of the theoretical combustion air required for the producer gas is admitted to ensure a strong reducing atmosphere in the roaster, and heating is kept at a low rate to have substantially all the nickel reduced to metal before the ore reaches $1400°F$ ($760°C$). Above this temperature there is a phase change with any nickel still remaining as oxide combining with iron silicate, and it is then much more difficult to reduce.

The residence time in the roaster is brief, $1\frac{1}{2}$ hours, and during this period the nickel present is reduced to the metallic state, along with much of the iron in the limonithic fraction of the ore, $(Fe,Ni)O(OH) \cdot nH_2O$, which is reduced to Fe_3O_4. The iron content of the serpentine fraction, $(Mg,Fe,Ni)_6Si_4O_{10}(OH)_8$, is essentially not reduced.

The hot, reduced ore leaving the roasters will quickly reoxidize and become insoluble to the ammonia leach solution if exposed to air. This necessitates cooling the ore down to $300°F$ ($150°C$) in rotary coolers, 9 feet in diameter and 80 feet long (2.74×24.4 m), partially immersed in a water bath for cooling, and with a nonoxidizing atmosphere. Flue dust passing out in the roaster gases is removed in dust collectors and returned to the roasters.

2. **Cobalt Sulfide Concentrate**, as produced in Zambia, containing 3 to 5% cobalt and much larger amounts of copper and iron, is given a sulfating roast in an eight-hearth mechanically rabbled roaster. The purpose of this roast is to convert the maximum amount of cobalt to a water-soluble sulfate.

Special conditions of roasting are required to convert the maximum amount of cobalt to the water-soluble form while at the same time solubilizing a minimum amount of copper and iron. This is accomplished by controlling the gas flow for temperature variation on the different hearths and admitting cold air to the lower hearths. The top hearth is essentially for drying, with ignition of the concentrate by oil burners on the second hearth. The actual roasting is in three stages, with the first stage having the cobalt, iron, and copper sulfides converted to oxides at a temperature of not over $1110°F$ ($600°C$), which minimizes the formation of insoluble cobalt ferrites. In the second stage the cobalt and copper oxides are largely converted to sulfates by reacting with SO_3 at 580 to $620°F$ (304 to $327°C$):

$$CoO + SO_3 = CoSO_4$$

In the third stage decomposition of the cuprous sulfate to insoluble oxide and further sulfating of cobalt oxide by SO_3 take place. The gas temperature for these last reactions is kept at about $1110°F$ ($600°C$) on hearths 4 to 7, and that of the bottom hearth, 8, is lowered to about $930°F$ ($500°C$). Retention time in the roaster is about 13 hours, and capacity is 40 tons of feed per day.

A somewhat similar overall treatment is practiced in Finland, where the Outokumpu Oy Cobalt Plant processes cobaltiferrous pyrite concentrate containing some nickel, copper, and zinc. Here too a selective sulfatizing roast is given to the concentrate to convert the valuable metals (cobalt, copper, nickel, zinc) to water or dilute acid-soluble sulfates, while most of the iron remains as insoluble oxide.

The concentrate, along with metal bearing dead roasted calcine returned from the domestic pulp and paper industry, is screened, and all coarse lumps are crushed and then fed to a fluid-bed roaster. The feed, both concentrate and calcine, is quite fine, on the order of 50 to 60% minus 200 mesh.

The roaster has a long, rectangular shape, 24.6 feet (7.5 m) high, with four compartments, each having a grate area of about 13 by 13 feet (4 × 4 m). The feed ratio of concentrate to calcine is on the order of one part concentrate to three or four parts calcine, and the sulfur content of the concentrate will average 48% as compared to 2.5% for the calcine. The concentrate contains about 0.7% cobalt, and the calcine, 0.80%.

Green, dry concentrate is added to each of the four compartments by a rotating disc feeder, at a rate that is sufficient to maintain the correct gas atmosphere and temperature. The temperature is kept at about 1256°F (680°C) and can be raised by increasing the concentrate feed rate or lowered by spraying water in each compartment. Dry calcine is fed to the first compartment of the roaster by a conveyor, and then proceeds through the roaster by means of openings in the walls between compartments.

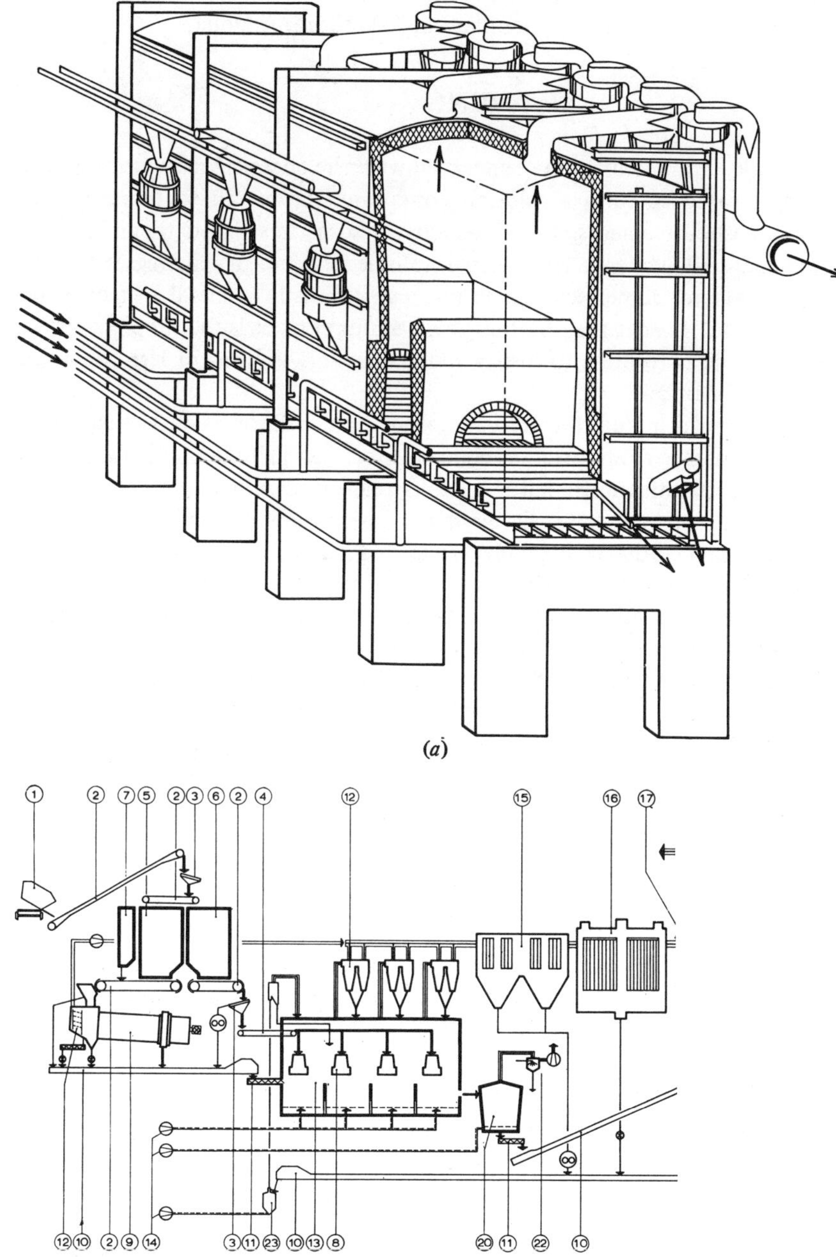

(a)

(b)

Figure 2.2. Four compartment-type fluid-bed roaster for cobalt sulfatizing roast: **1,** railway wagon; **2,** belt conveyor; **3,** vibrating screen; **4,** moving belt conveyor; **5,** calcine bin; **6,** concentrate bin; **7,** sodium sulfate bin; **8,** disc feeder; **9,** rotary kiln; **10,** redler conveyor; **11,** screw feeder; **12,** cyclone; **13,** fluid bed roaster; **14,** rotary piston compressor; **15,** waste heat boiler; **16,** electrostatic precipitator. *Soure:* M. Palperi and O. Aaltonen, *J. Metall.*, Vol. 23, No. 2, 1971, pp. 36, 37. (a)—cutaway view of interior; (b)—cross sectional view.

The height of the fluidized bed is about $6\frac{1}{2}$ to $8\frac{1}{2}$ feet (2 to 2.5 m), and all of the product is taken out of the fourth compartment of the roaster. Air for the sulfating reactions and for fluidizing is supplied by four compressors, with one for each compartment.

The sulfatized product from the roaster is first cooled in fluidized-bed coolers and then sent over high-intensity, dry magnetic separators. The incompletely sulfatized material is magnetic and is recycled back to the roaster, while the sulfatized nonmagnetics pass on to the leaching and precipitation operations that follow.

3. Copper Sulfide Concentrate, 325 mesh, containing 25% copper is fed as a slurry into a fluid-bed roaster to convert the sulfide to oxide soluble in H_2SO_4-spent electrolyte. The SO_2 in the roaster gases is processed to make sulfuric acid, and the insoluble silicious iron residue is furnace reduced to make sponge iron pellets. This type of operation is carried out at the El Paso Natural Gas copper plant at Lakeshore, Arizona.

LEACHING

Leaching processes in hydrometallurgy are concerned with the chemical dissolution of the raw materials being treated to form a solution containing the metals to be recovered. This leaching of the desired elements is done selectively so as to separate them from the bulk of unwanted material which is left as an insoluble residue. In most cases the ore being leached is one that for some reason does not respond easily to metal extraction by pyrometallurgical treatment, or it may be that the leaching process is the simplest one to use, or that the ore grade is too low to justify more than this slower but less expensive hydrometallurgical treatment.

Many reagents are known, in addition to water alone, that will form aqueous solutions capable of dissolving metals from ores or concentrates and include acids, bases, and salts. There are several qualifications that a leaching reagent should satisfy, the reagent cost being one of the most important. Other qualifications are the ability to dissolve selectively the desired metal, the ease of procuring and handling, the ability to be recovered and reused, and the extent to which it corrodes process equipment. Sulfuric acid is probably the most widely used solvent, with its aggressiveness, cheapness, and general availability balancing its nonselectivity. Nitric and hydrochloric acids are also fine solvents but are restricted in use because of their heavy attack on equipment as well as their nonselectivity and higher cost. Ammonia is used extensively in certain applications, and its high cost requires that it be recovered, regenerated, and reused in the leaching circuit. It is a highly selective solvent for nickel, copper, and cobalt in particular and does not attack steel equipment.

Ferrous sulfate ($FeSO_4$) and sulfuric acid gradually form when water containing dissolved air contacts broken rock surfaces containing pyrite (FeS_2):

$$2FeS_2 + 7O_2 + 2H_2O = 2FeSO_4 + 2H_2SO_4$$

In the presence of additional oxygen, the ferrous sulfate is slowly oxidized to ferric sulfate which helps in the dissolution of a number of copper sulfides, by converting them to soluble sulfates, even though it is not the principal solvent itself:

$$Cu_2S + Fe_2(SO_4)_3 + 2O_2 = 2CuSO_4 + 2FeSO_4$$

Ferric chloride processes are also under study at the present time.

Several strains of oxidizing bacteria appear to accelerate the leaching rate very considerably when sulfuric acid is used on copper sulfide ores. The effect is to accelerate the oxidation of sulfide minerals to form acid and soluble copper and iron sulfates, due to the reoxidation of ferrous to ferric ions and the possible direct attack by these bacteria on the metal sulfides. Some of the bacteria that are effective are *Thiobacillus thiooxidans*, which obtains energy by oxidizing sulfur to form sulfates, and *Thiobacillus ferrooxidans*,

Ferrobacillus ferrooxidans, and *Ferrobacillus sulphooxidans* which are effective in oxidizing ferrous iron to ferric iron:

$$4FeSO_4 + O_2 + 2H_2SO_4 = 2Fe_2(SO_4)_3 + 2H_2O$$

These bacterial strains complement each other in that the *Thiobacillus thiooxidans* produces acid which keeps the ferric iron produced by the other bacteria in solution. This forms an acidified ferric sulfate solution which is a powerful solvent for copper minerals:

$$2S + 3O_2 + 2H_2O = 2H_2SO_4$$

Leach solutions containing only acid can dissolve most oxide or oxidized minerals, but a much larger amount of acid must be added to the leach solution than when pyrite is also present. Nearly all oxide and sulfide minerals will dissolve in dilute sulfuric acid and ferric sulfate, though the dissolution rate in some cases is quite slow, as low as 1% per month. The addition of chlorine or sodium hypochlorite to the conventional sulfuric acid leaching solution speeds the dissolution rate considerably.

Gaseous reagents dissolved in leach solutions are often used in hydrometallurgy, the principal one being oxygen. The amount of oxygen in solution that can dissolve in the aqueous phase will depend on its pressure in contact with the solution; and increasing the pressure of the reagent gas will increase its solubility. The temperature of the overall system also has a bearing, with increasing temperature initially lowering gas solubilities. Solubilities are often at a minimum from 165°F to 212°F (74°C to 100°C), and above this temperature gas solubilities rise sharply if the heated system is under pressure to prevent vaporization of the aqueous phase. The composition of the solution also has an effect on gaseous solubility, with large concentrations of dissolved salts causing a decrease. The purpose of dissolved oxygen is to oxidize the desired metals and any metallic iron present and to put them in a condition where the desired metal oxide is more readily reactive with the leaching solution and the iron oxide is less reactive.

Selective leaching is desired to the greatest degree obtainable in order to dissolve the sought after metals into the leach solution while leaving the unwanted portion as an insoluble residue. This difference can be enhanced in a number of ways, one of which is to choose a leach liquor in which the desired metals have a much faster rate of solubility than do the undesired elements. This condition can sometimes be improved by thermal pretreatment where a sulfide can be given an oxidizing roast to produce a water-soluble sulfate, or an oxide can be given a controlled reduction roast to produce a calcine with reduced metal which will dissolve easily in the leaching solution but still retain most of the iron as oxides, which are less readily soluble. Concentration of the chosen solvent also has a bearing on selectivity, as does the temperature, both of these being adjusted to take advantage of any inherent solubility differences between valuable metals and waste rock in a particular leaching situation.

Economically it is important to establish leaching conditions that will dissolve the desired elements as rapidly as possible; and reagent concentration, temperature, and agitation are the principal controlling variables that influence leaching rate. The minerals must also be ground finely enough to be exposed to and in thorough contact with the leach solution, and it should have a low content of acid-consuming gangue and basic minerals if an acid leaching solution is used. Other variables having an effect on leaching rates are the pulp density of the leach slurry, the availability of agitation, and the contact time allowed between solids and liquid.

TYPES OF LEACHING

Many types of materials are satisfactory feeds for leaching operations, including low-grade and high-grade ores, concentrates, roasted calcines, and matte.

The leaching of low-grade ores must be done on a very large scale for a sufficient amount of metallics to be extracted to make the operation profitable. Large amounts of solvent are required to treat the large bulk of ore being processed, and the solvent must be low priced to have the operation profitable. With high-grade ores, concentrates, calcines, and matte, the valuable metal has been collected into a smaller bulk and the ratio of waste material to values is lower, so that the amount of reagents required for this smaller mass will be less than for low-grade ore, and the cost of the leaching reagents is not as important a factor.

In all ores and concentrates the mineralogical composition of the metallic minerals must be soluble in the reagents used, and the waste rock should be substantially insoluble. The same is true of the metallic compounds formed after roasting and smelting prior to leaching. Contact of the metallics with the leach solution, and size after crushing to expose these metallic surfaces, have an enormous bearing on the speed of dissolution. A piece of rock containing a metal compound can only be leached if the metal compound is exposed to the leaching solution. Consequently a copper sulfide ore ground to minus 60 mesh can be leached in 6 hours, compared to five days for the same ore crushed to $\frac{1}{4}$ inch size (6.25 mm), and five years if crushed to only 6 inch size (15 cm).

The rate of leaching decreases as the metal extraction comes closer to completion. This slowing down may be due to several things, such as a decrease in the concentration of the metal still to go into solution, a decrease in the strength of the reagent, especially if a batch leach is being performed, and finally low diffusion rates through the semistagnant layers of leach liquor which collect in the tiny pockets, formed as the metallics dissolve, in the solids being leached.

Leaching is done in a variety of ways. The method used is governed by the expense for treatment that can be borne, the difficulty of dissolving the desired metallics, and the amount of material being treated. Both atmospheric and pressure leaching is commonly used. The usual procedures include underground (in-place) leaching, dump or heap leaching, tank leaching at atmospheric pressure, and leaching in autoclaves at high and low pressures.

Underground (In-Place) Leaching is used in areas surrounded by tight and impermeable rock to dissolve values, most often copper, from low-grade ore deposits and from the worked-out stopes and support pillars left from conventional mining methods for high grade ore deposits. In all cases the ore must be broken down fine enough to expose the mineral values to the leaching solution. This is often done by block-caving mining methods in low-grade ore bodies. In higher-grade deposits the support pillars in worked-out areas will be blasted down and broken up. The treatment cost of this underground leaching is the lowest of any method of metal extraction, and low-grade ores can be worked at a profit that cannot be processed by any other method. The operation is comparatively slow, proceeding for years in some cases, but the capital expenditure and equipment costs are also at a minimum.

Worked-out stopes and caved areas are flooded with barren leach solution or are permitted to fill with acidic mine waters formed from the action of dissolved air in the water reacting with pyrite and forming ferric sulfates and sulfuric acid. In some cases the broken ore is first allowed to weather, or oxidize, in the stopes before being flooded with leach solution, and this oxidized mineral is then more easily soluble.

The pregnant solutions produced by underground leaching during a five or six month contact period are pumped to storage ponds on the surface, where a retention of the solution removes slimes and solids picked up from mill tailings used for mine backfill and various residues left from mining. The solution is then pumped to precipitation tanks to remove the metallic values, after which it is returned to the underground workings to dissolve more of the desired metallics.

Heap or Dump Leaching at one time was mostly used to recover the values from discarded waste rock with a metal value below that of milling grade or because an ore was

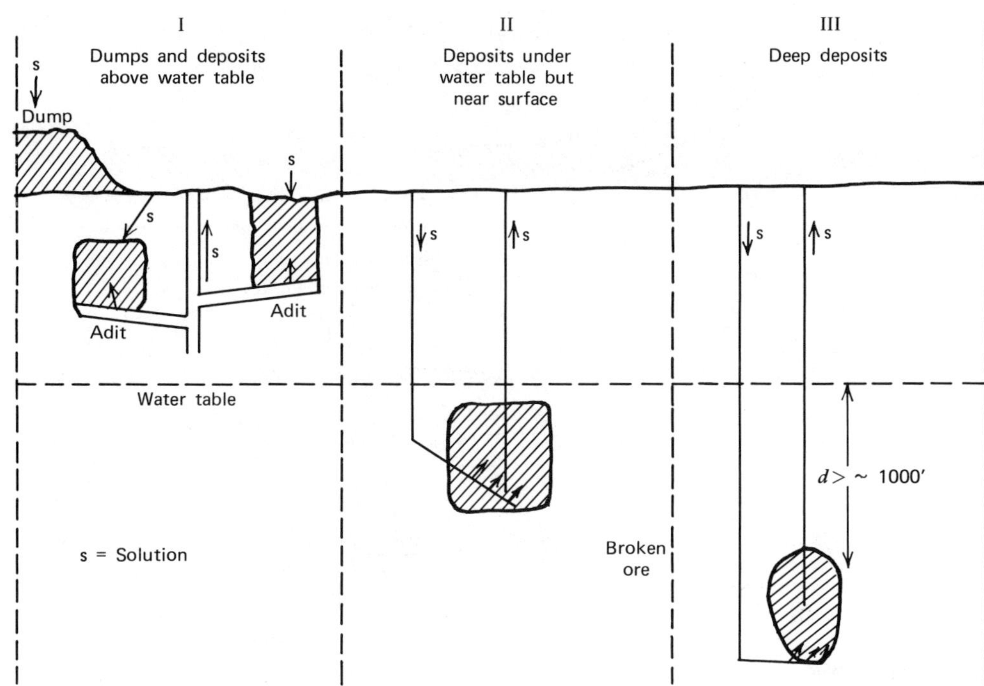

Figure 2.3. Leaching in place. *Source:* Rate processes of extractive metallurgy–hydrometallurgy, H. Y. Sohn and M. E. Wadsworth, Coordinators, Met. Soc. of AIME, 1976, p. 98.

not easily amenable to flotation treatment. Now however, as several large ore deposits have been discovered that are of relatively low grade, heap leaching has become more the accepted method of treating these types of ores and processing them on a large scale.

In heap leaching, waterproof pads, sloped for drainage and collection of the leaching liquor, are made of concrete, tamped clay, or plastic sheet, and the ore is heaped on these. A layer of coarser, 4 to 8 inch (10 to 20 cm), chunks should be put down first to assist later drainage and should be put down carefully so as not to puncture the waterproof layer of clay or plastic sheet, if these are used. Culverts are built in under the heaps to facilitate drainage of the leach liquor to a main launder leading to a storage pond below the dump. This pond must also be lined with clay or plastic film to prevent seepage and loss of the pregnant solution. Ventilating flues are sometimes built into the heaps, especially if sulfides are being leached, to assist in air circulation and oxidation of the ore to a soluble condition.

The leaching solutions pumped to and sprayed over the heaps to trickle down through them and drain off to the storage pond can consist of mine waters containing ferric sulfate and sulfuric acid, natural rainwater, water to which sulfuric acid has been added, and returned barren leach solution after precipitation of its values. All that can be accomplished by a single application of solution is to wash off the soluble salts on the surface of an ore chunk; an excessive volume of solution is undesirable and unnecessary as it simply means that a more dilute pregnant solution results.

The pregnant solution is pumped from the storage pond to precipitation processing where the metal values are removed, and the barren solution is then recycled back to be used over again as leaching liquor spray.

Tank Leaching is done in two very different forms. One is on a large scale in which several thousand tons of porous ore are treated at one time in large concrete tanks by a series of leaching and washing operations. The second system of tank leaching is done on a smaller scale in containers where agitation is used to circulate and aerate the solids and solution. The ore in this second case is not porous and must be ground to a sufficiently fine size to expose its values to the leaching solution.

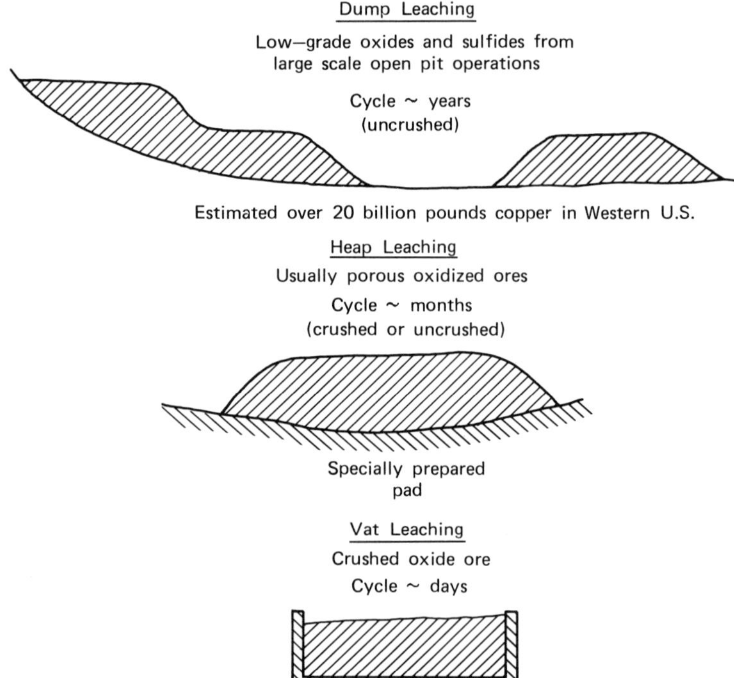

Figure 2.4. Dump, heap, and vat surface leaching methods used for copper ores. *Source:* Rate processes of extractive metallurgy—hydrometallurgy, H. Y. Sohn and M. E. Wadsworth, Coordinators, Met. Soc. of AIME, 1976, p. 97.

1. Percolation or Sand Leaching is used for porous higher-grade ores than are treated by heap leaching, those containing 1 to 2% of metal in a soluble form that can be treated in a relatively coarse size, $\frac{3}{8}$ inch (9.4 mm), and be leached completely over a period of a few days.

The tanks used for leaching are concrete, lined with lead or asphalt to be acid resistant, and are equipped with false filter bottoms to facilitate the addition and withdrawal of leach solution and wash water. The tanks are quite large, 150 by 110 by 17 feet (45.7 × 33.5 × 5.2 m), and hold 10,000 to 12,000 tons of ore to be leached, which is added to tanks by conveyor belts and removed after leaching by bucket excavators.

Percolation leaching is essentially a batch operation, as the ore must be charged and the residue removed while the tank is out of operation. This necessitates having several extra tanks in the circuit as their large size takes some time for charging and digging out, and the whole leaching operation cannot be closed down while waiting for a tank to be prepared for the circuit.

The principle of leaching is a somewhat complex countercurrent flow of solution over several stages so that the first leach on fresh ore is made with weak (spent) solution that has already been in contact with ore for an extended period of time and is next to be withdrawn as final pregnant solution from the leaching circuit and pumped to the precipitation circuit to remove its contained values. The final leach on ore that has been in contact with solutions of varying strength for several days is made with fresh (strongest) leach solution. After this final leach the remaining insoluble solids are washed, dug out of the tank, and discarded. This system utilizes the weakest (spent) solution coming into contact with fresh ore containing the most easily soluble compounds, in order to use effectively the last remaining leach liquor strength. On the other hand, fresh (strongest) leach solution contacts the least soluble portion of the ore, remaining after several days of soaking in solutions of gradually increasing strength, and this fresh solution will dissolve the least soluble metal values still remaining. The countercurrent flow of leach solution has the solution pumped from tank to tank after residence in each of a day, so that in each batch leaching operation the solution, as its strength is used up through several leachings and becomes weaker, is in contact with ore that has been leached for a shorter

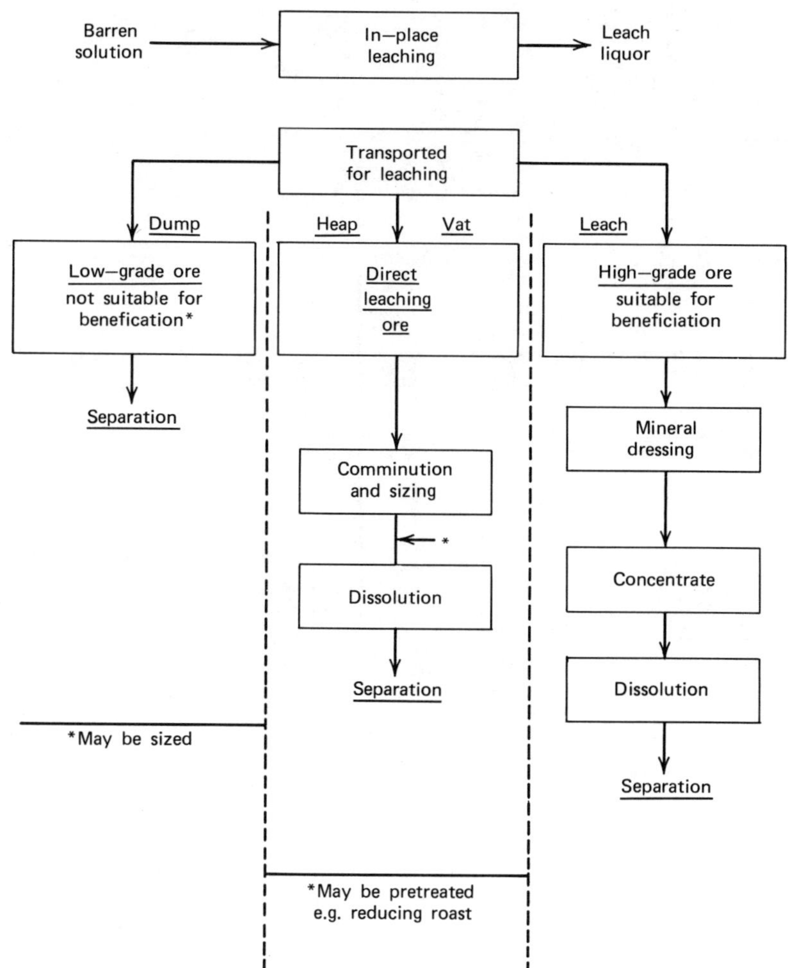

Figure 2.5. Hydrometallurgical processing for various ore grades. *Source:* Rate processes of extractive metallurgy—hydrometallurgy, H. Y. Sohn and M. E. Wadsworth, Co-ordinators, Met. Soc. of AIME, 1976, p. 2.

and shorter number of days. The liquid may be added at the tank bottom for upward percolation or at the top for downward percolation.

The barren solution after precipitation of the values is returned as fresh (strongest) solution to the leaching circuit, and at this point will have any necessary makeup water or acid added to it. The leaching circuit runs from five to eight days, with metal extraction from 87 to 93%.

2. Agitation Leaching is used on higher-grade ores where the tonnage to be treated is smaller than that leached by the percolation method, and the desired mineral is so fine grained or so well disseminated that extensive crushing and grinding is necessary to liberate it and expose it to the leaching solvent.

Either bubbling compressed air or mechanical agitation are two methods used to keep the pulp in circulation until dissolution is complete, the contact time of solids and liquor being a matter of hours rather than the days of the percolation process. The mechanical agitators are simply impellers lowered into the tank, while the air-agitated tanks are often Pachuca tanks which have a central pipe that behaves as an air lift. With these, a compressed-air line opening at the bottom of the pipe pushes solution and pulp up to the pipe top, from which it overflows and spreads out into the tank, to drop back down to the bottom and be again picked up by the air lift. The Pachuca tank is an extremely popular, simple, and efficient design with no moving parts to get out of order and performs admirably in circulating the whole tank charge. A cone-bottomed Pachuca will be

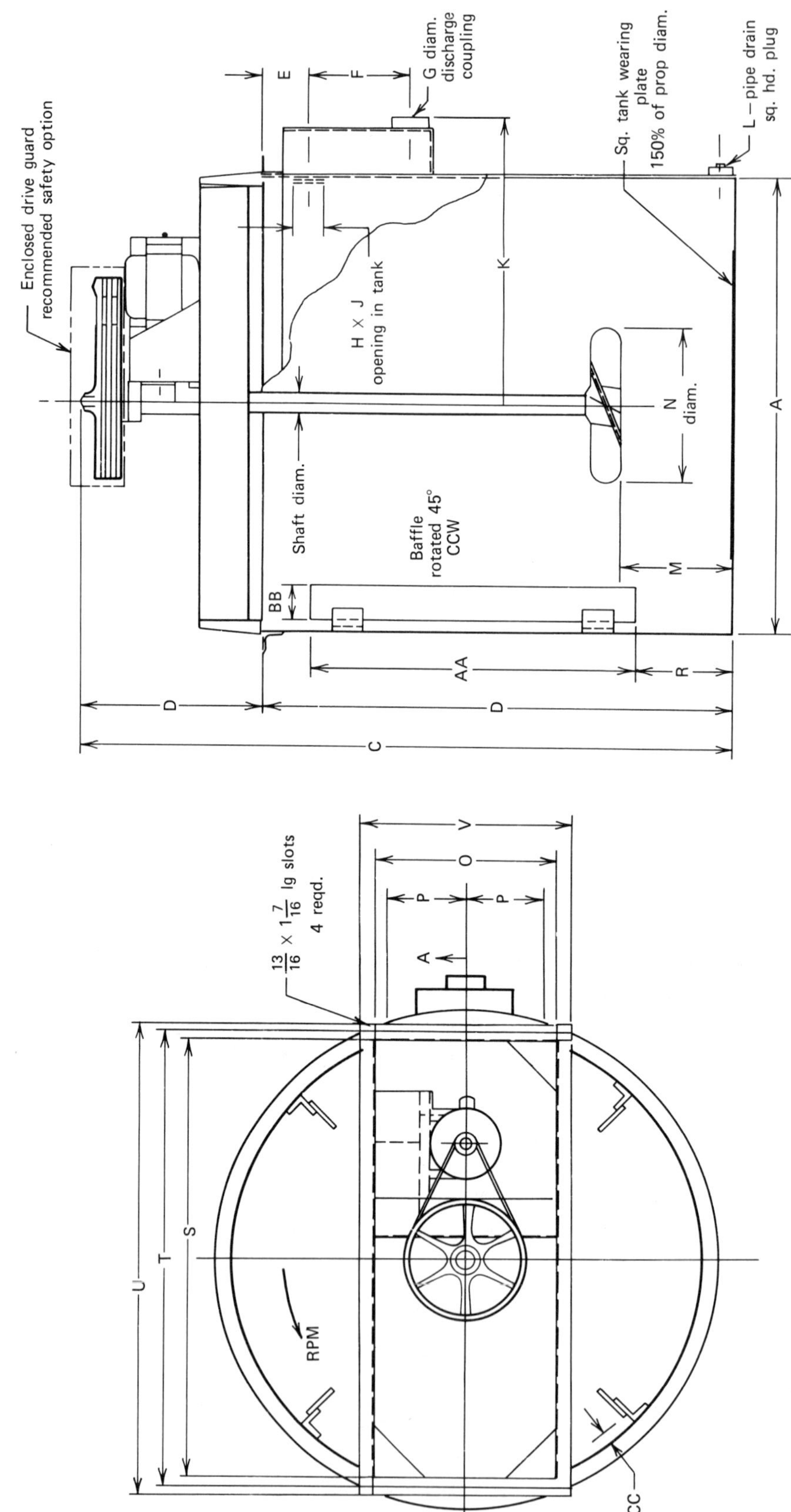

Figure 2.6. Agitation-type leach tank. *Source:* Courtesy of Denver Equipment Division of the Joy Industrial Equipment Company.

40 feet high and 10 feet in diameter (12.2 × 3.05 m), with a 10 inch central stand pipe (25 cm). Construction is of wood or steel; and, where necessary for corrosion protection, inert corrosion-proof linings are installed. Similarly agitators for mechanically agitated tanks must be rubber covered or made of some corrosion resistant material if solution corrosion is a problem. In some instances where extreme agitation is not required, thickeners are used for the leaching tanks.

The agitation type of leaching operation can be either a single batch process or can be a countercurrent continuous-flow operation utilizing several tanks similar to the system used in percolation leaching and followed by washing of the undissolved slimes. These slimes are often washed in thickeners and then filtered before discarding. The pregnant solution clarified by decantation and filtering is taken to the precipitation circuit, and the barren leach liquor is often returned to the leaching circuit for reuse.

Pressure Leaching is done for two purposes—to speed the dissolution of all values into the leach solution and to improve the solubility rate of solids that are at best only slowly soluble at atmospheric pressure. The closed autoclaves used for pressure leaching also permit higher temperatures than are possible with open tanks, and this too speeds the dissolution rate.

Gaseous reagents such as oxygen are often important for the speedy dissolution of metallics, and the amount of these gases which can be kept in the leach solution will depend on the pressure of the gas in the autoclave above the solution surface, with a greater amount being in solution the higher the gas pressure and subsequently the greater the rate of solids dissolution. Materials such as sulfides, which are relatively insoluble under normal, open-tank leaching conditions, become soluble if a high enough pressure can be maintained to force considerable amounts of oxygen into the leach solution. This does away with the preliminary treatment required for sulfides in which they must first be given an oxidizing roast to convert the sulfides to soluble sulfates or oxides before leaching.

Autoclaves for high-pressure leaching are made of metal for strength and of corrosion-resistant stainless steels or titanium to withstand the severe attack imposed by the leach solution at high temperature and pressure. The autoclaves are also often glass-, lead-, or brick-lined to prevent equipment corrosion and usually have some type of agitators. Frequently there are also built-in heating or cooling coils for the provision of adding or removing heat from the leach solution inside the vessel during the process.

Autoclaves are found in both vertical and horizontal styles, with both types being several times longer (four to five times) than the diameter. The horizontal autoclave type used on sulfide concentrates is divided into four compartments, with a mechanically driven impeller in each compartment to keep the solids in suspension. Cooling coils are also provided to maintain the leaching reaction at the optimum temperature.

Leaching is carried out in several reaction vessels in series, and in about 90 minutes 95% of the metal content of the solids will be taken into solution. The pregnant solution is separated from the insoluble residue in thickeners and goes on to the precipitation plant to recover the contained values.

LEACHING PROCESSES

1a. Copper Oxide ores and concentrates are treated by heap, tank percolation, and agitation types of leaching, all using atmospheric pressure. Some of the ore bodies contain sulfides as well as oxides, and when the overall grade is high enough to warrant such treatment, the sulfides will be concentrated and removed by mineral beneficiation, leaving the not so easily concentrated oxides to be recovered from the gangue rock by leaching. In a newer type of processing, as is done at the Hecla-El Paso Plant at Lakeshore, Arizona, the sulfide portion of the ore is roasted and sulfuric acid is produced from the SO_2 in the roaster gases. This H_2SO_4 is then used to leach the oxidized portion of the ore.

Tank Percolation Leaching of copper oxide ore is carried out in large-scale operations with a countercurrent sulfuric acid leach followed by countercurrent water washing, as is done at Chuquicamata in Chile.

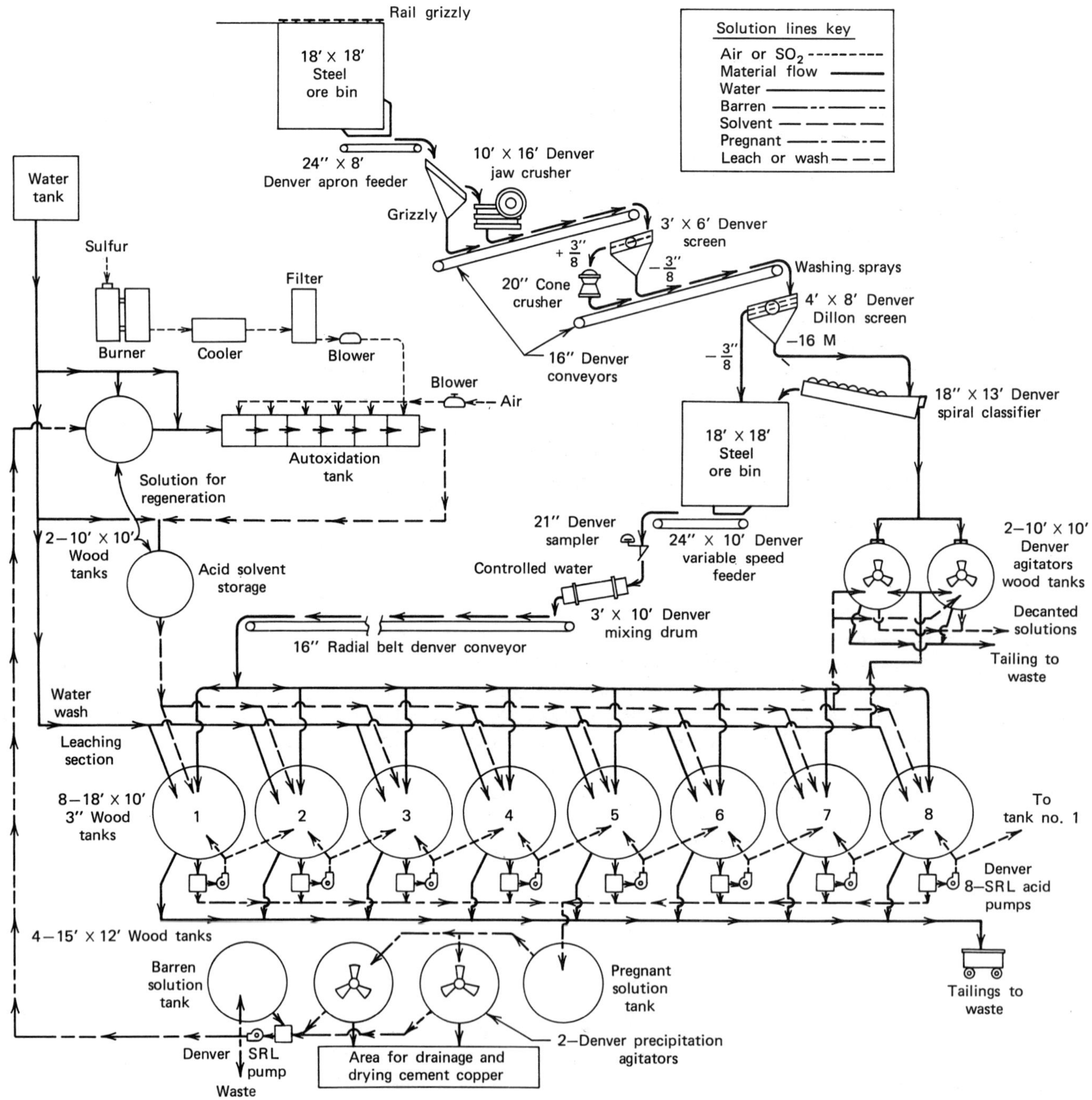

Figure 2.7. Small-scale leaching of copper ores flow sheet. *Source:* Denver Equipment Co., Modern Mineral Processing Flowsheets, 1962, p. 38.

The ore is crushed by successive coarse and fine crushing plants down to $\frac{1}{2}$ to $\frac{3}{8}$ inch (12.5 to 9.4 mm) maximum size, after which the fines, which have been found to be detrimental to uniform material bedding for leaching, can be removed in a washing plant for separate recovery treatment. The ore is charged into a leaching vat in amounts of from 7500 to 11,000 tons per vat, and the vat dimensions will be as large as 150 feet long, 110 feet wide, and 18 feet deep (45.7 X 33.5 X 5.5 m). The vats are constructed of concrete lined with acid-proof brick or mastic and have wooden filter bottoms. The ore is not bedded when charging a vat but is loaded at one location from top to bottom before the loading bridge is advanced some 12 to 15 inches (30 to 37.5 cm), and loading proceeds at a second location. A vat holding 7500 tons or ore takes 15 hours to charge.

Leaching of the ore, which analyzes 1.4 to 2.1% copper, is done either by batch percolation or countercurrent percolation, with both accomplishing the same degree of leaching and metal extraction of over 90%. Both types of leaching work on the principle of leaching fresh ore with weak acid, leaching partly dissolved ore with strong acid, and then washing the insoluble residues with water before they are discarded. Percolation in either case can be accomplished with the solution added either from the bottom or the top of the tank, and the total leaching time will be the same, five or six days. Several extra vats are needed in addition to those actually being used for leaching at any one time, as these extra tanks will be temporarily out of the circuit while they are charged with fresh ore and have the residues washed and the residues dug out for disposal. Each of these operations takes a day apiece to accomplish.

With batch percolation the ore is submerged in the treatment solution and allowed to soak for the prescribed length of time needed, from 8 to 24 hours. Then the solution is pumped to the next tank where the soaking treatment is repeated. The first three soakings with fresh ore produced solutions containing respectively 42, 32, and 22 gpl copper, which are blended together and sent to the precipitation circuit for copper recovery. Two further soakings with strong solution used on the now partly dissolved ore will give solutions containing 19 gpl copper, and these solutions, which still contain considerable usable acid, are pumped to vats 1 and 2 to be used for the first and second leach solutions. The insoluble residue remaining after the fifth soaking is given several water washes, the successive washes decreasing in copper content from 9 gpl in the first wash to 1 gpl

Figure 2.8. Copper oxide ore leaching operation at Chuquicamata. *Source:* The Anaconda Company.

after several washes. These wash solutions are combined and pumped off to be stripped of their copper content, while the washed residues are dug out of the vat and disposed of.

Countercurrent percolation differs from batch percolation in that instead of merely soaking the ore in the treatment solution, the leach solution is both circulated in each vat and also at the same time slowly advanced from vat to vat. The circulation rate in a vat is 2.5 tons solution per 24 hours per ton ore, and the advance rate is 0.7 ton solution per 24 hours per ton ore.

Six vats are used, with the ore being in contact with leach solution for six days before it is washed and dug out. Fresh, strong acid is pumped to the tank which has ore for its sixth day of leaching and is then advanced day by day until after six days the solution, containing 42 gpl copper, is removed from the vat containing fresh ore leached for only one day and is sent to the precipitation plant to recover the contained copper. The insoluble residues are washed several times, with the wash water going to a copper stripping circuit, and the residues are then dug out of the vat and discarded.

The fresh strong acid used for leaching contains 75% sulfuric acid, and after contact with the ore the final pregnant solution which is removed to go to the precipitation circuit will have only 14% sulfuric acid remaining. These corrosive solutions are transferred in lead, plastic, or wood stave pipes.

Tank Agitation Leaching is done in preference to percolation leaching in certain specialized cases. These include situations where the mineral values are very finely disseminated and finer crushing is necessary to expose these fine-grained metallics to the leach solution. Another situation is the condition where the ore contains mixed oxides and sulfides, with the sulfides being removed by froth flotation and the finely ground oxide ore and gangue remaining being too small sized for effective percolation leaching.

Finally some oxide ores contain carbonates which release carbon dioxide in contact with the leaching solutions, and this released gas seriously interferes with either upward or downward percolation.

Leaching is carried out in Pachuca tanks, or tanks with mechanical or combined air lift and mechanical agitation. In all cases the exposed equipment surfaces are covered with lead, acid-proof brick, rubber, or plastic to prevent rapid corrosion by the leaching solution. The retention time of ore in the agitator section is $4\frac{1}{2}$ to 5 hours, and metal extraction is well over 90%. Acidity of the leaching solution is held to 9 to 11 gpl sulfuric acid, which is adequate to take the acid-soluble copper into solution while dissolving a minimum amount of iron, which is a contaminant.

The Twin Buttes, Arizona, operation of the Anamax Mining Company (owned jointly by Amax, Inc., and the Anaconda Company) will treat at capacity 10,000 tpd oxide ore carrying about 1% acid–soluble copper. The ore, ground to 95% minus 48 mesh, is acid leached for 5 hours in five rubber-lined, mechanically agitated tanks 30 feet in diameter by 31 feet high (9 $\times$ 9.3 m) arranged in a cascade, and will yield 36,000 tons per year of copper.   Up to 250 pounds (112.5 kg) sulfuric acid is used per ton or ore, with the acid being supplied from a nearby smelter.

Leached slurry at 50% solids is washed countercurrently with return raffinate (from the SIX circuit which follows) in a series of four 400 foot diameter (120 m) thickeners. Solids (underflow) advance from thickeners no. 1 through no. 4 and then on to the tailings pond, while the SIX raffinate (overflow) advances from thickeners no. 4 to no. 1, from where the pregnant solution passes on to pH adjustment and then the SIX concentration circuit.

The slurry leaves the pH adjustment vessel at about 10% solids and is clarified in a two-stage 400 feet in diameter (120 m) thickening system followed by six pressure sand filters 12 feet in diameter (3.6 m), which give a final solution containing about 4 to 10 ppm suspended solids. Filtration of the suspended solids gives better mixer–settler efficiencies during solvent extraction, less organic entrainment and loss, decreased iron transfer by suspended solids, and less sludge buildup at the organic-aqueous interface.

Heap Leaching is becoming more important and is now done on a relatively large scale to treat low-grade ores, such as the 0.4 to 0.45% copper ore mined by the Bluebird plant of Ranchers Exploration and Development Corporation in Arizona.

A new leach area is readied by grading, sloping, and then compacting the natural surface with standard road building equipment. A total of 21 or 22 heaps are built up by 20 foot (6.1 m) lifts and are rotated as necessary through leaching, drying out, additions of new ore, and repiping, so that 11 or 12 of the heaps will be undergoing active leaching at any one time. The average life of a heap being leached is about 120 days.

The surface area of the leach dump has a network of 2 inch (5 cm) PVC pipes on 8 foot (2.44 m) centers, and these pipes are equipped with small needle valves which drip leach solution at the rate of 200 gpm per heap, totaling some 2400 gpm.

The leach solution for the first 60 days of a new leach cycle is acidified raffinate (the aqueous off-solution from the final solvent extraction vessel, to which makeup acid is added) carrying 20 gpl H_2SO_4 and 0.2 gpl copper. Straight raffinate, without added H_2SO_4, of 5 to 7 gpl H_2SO_4 content is used for the second 60 days of each cycle. The solution percolates down through the heaps and drains to a surge sump, from where it is piped to a storage pond. The pregnant solution contains about 1.9 gpl copper and 3 to 4 gpl H_2SO_4 and is drawn from the pond to be stripped of copper values by solvent extractions.

1b. Copper Sulfides are leached in a variety of manners, by in-place, heap, and tank percolation leaching at atmospheric pressures and in agitated autoclaves at high and low pressures. In the atmospheric leachings the sulfides are oxidized to soluble sulfates and oxides which are readily dissolved in the aqueous sulfuric acid and ferric sulfate leach solutions or can be leached in the sulfide form by ferric chloride. With the high-pressure autoclaves a strong aqueous ammonia solution takes the copper into solution. Copper, being one of the few metals that form complex ammonium ions, can be selectively recovered in this manner. The low-pressure leach is carried out at near-atmospheric pressure in an oxygen–ammonia–ammonium sulfate system and is the newest of the leaching processes.

In-Place Leaching is the least expensive method of leaching and is done without removing the ore from underground workings. Undeveloped ore bodies can be broken down by block caving, low grade ore can be broken down in stopes from which high-grade ore has been removed, and high-grade support pillars can be broken down after the rest of the mine is worked out. In all cases a piece of rock containing copper can only be leached if it is porous enough to have the leach solution penetrate it, and 4 inches (10 cm) is probably the maximum size that can be treated in this manner. The surrounding rock should also be nonporous so that the leach solution will not soak into it and be lost.

The copper sulfide mineral surfaces which are exposed are naturally oxidized to water-soluble copper sulfate by alternate contact with air and water. The leach solutions accumulate in drainage tunnels under the ore body or in the lower areas of the mine and are pumped to the surface for precipitation treatment to remove their values. They are then pumped back into the underground workings to dissolve more metallics, and the cycle is repeated.

Bacterial oxidation assists in oxidizing the sulfide ores, and the formation of ferric sulfate from any pyrite present also assists in oxidizing copper sulfide to a water-soluble form:

$$Cu_2S + Fe_2(SO_4)_3 + 2O_2 = 2CuSO_4 + 2FeSO_4$$

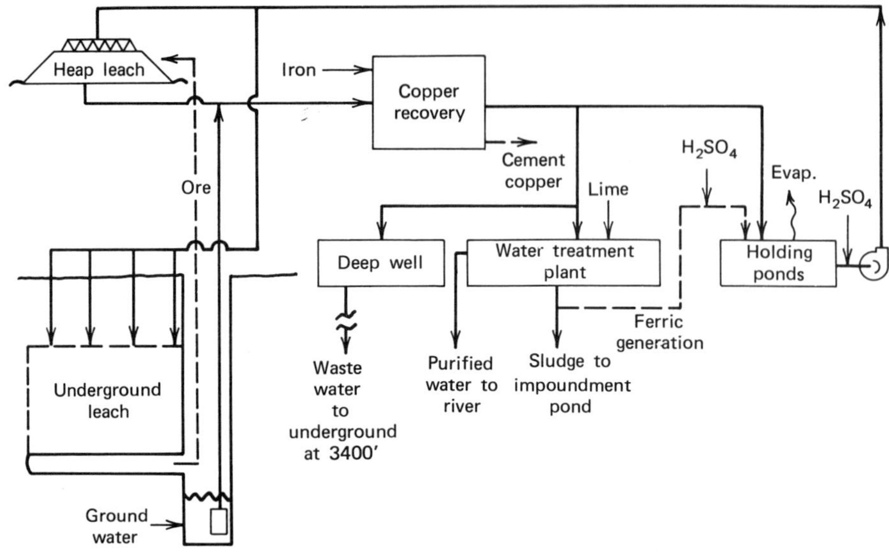

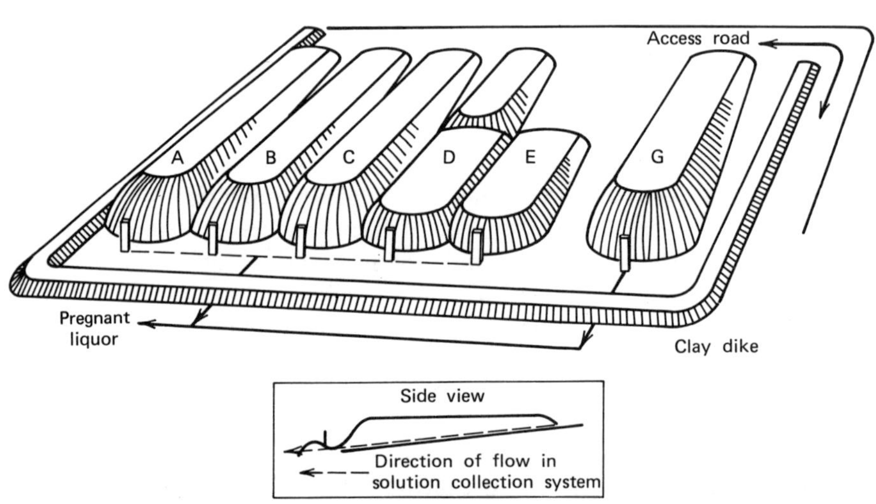

Figure 2.9. Heap leaching circuits. Heap and in-place underground leaching. *Source:* J. C. Yannopoulos and J. C. Agarwal, Eds., *Extractive Metallurgy of Copper*, Met. Soc. of AIME, Vol. 2, 1976, pp. 859, 863.

Copper ores containing no pyrite must have sulfuric acid added to the leaching water in amounts ranging up to five times the weight of the copper dissolved. The rate of dissolution is slow, often as low as only 1% per month, but the capital cost is also low and makes it economically possible to treat ores of less than 0.5% copper content.

Heap Leaching is similar to in-place leaching in that it depends on the natural oxidation of the sulfide minerals by continual contact with air and water. Oxides, sulfides, and their mixed ores which are removed in mining high-grade ores but are too low grade themselves to warrant further expensive beneficiation and pyrometallurgical treatment are treated by heap leaching, with the process consisting of first piling the ore in an area which has drainage to a prepared sump. Leach solutions containing sulfuric acid are then sprayed over the piles and allowed to drain down through them, dissolving the oxides and sulfates that have oxidized from the copper sulfide. The pregnant solution collects in the sump and is pumped from here to the precipitation circuit to remove the contained copper,

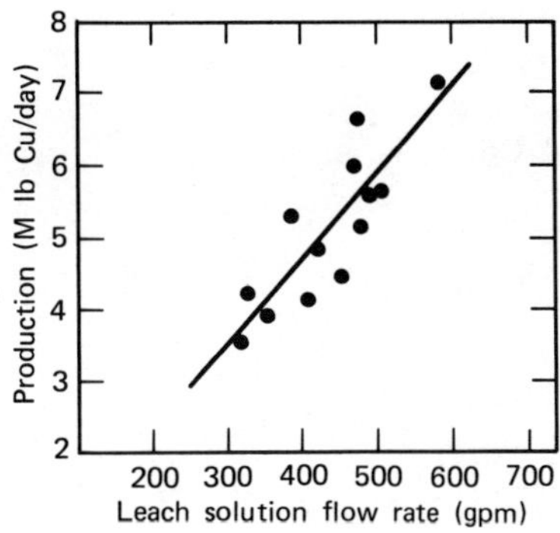

Figure 2.10. Solution flow versus production, Mountain City, in-place leaching. *Source:* J. C. Yannopoulos and J. C. Agarwal, Eds., *Extractive Metallurgy of Copper,* Met. Soc. of AIME, Vol. 2, 1976, pp. 859.

after which it is run back to be sprayed again over the oxidizing ore heaps and the process repeated.

Ores that contain on the order of 1% copper can be economically treated by heap leaching, with the pregnant solutions going to precipitation averaging 1.5 gpl copper. The process is quite slow, and ore 4 to 6 inches (10 to 15 cm) in size may take several years to achieve a suitable extraction.

Tank Percolation Leaching is used on a mixed oxide and sulfide ore, leaching them together by using an aqueous solution that contains sulfuric acid to dissolve the copper oxide and ferric sulfate to oxidize the copper sulfide to the soluble copper sulfate that can also be taken into solution.

The general processing is similar to the percolation leaching done with coarse copper oxide ores. Large, lead-lined, concrete vats 175 by 68 by 19 feet (53.4 × 20.7 × 5.8 m) in size are filled with approximately 10,000 tons ore each and are leached for eight days. Countercurrent leaching is also used here, so that fresh strong solution comes in contact with ore undergoing its eighth and last day of soaking. Pregnant solution, for its last day of contact before being drained off to the precipitation plant, will be in a vat with fresh ore having its first day of soaking.

Percolation is upward, with the leach solution entering the vat through a lead pipe just above the tank bottom and below the wooden filter bottom. Overflow from the vat is into a launder which feeds a pump serving the next vat in the series.

After leaching is completed in the eight day cycle, the residues are given 10 washes requiring three days, and this wash water is pumped to a precipitation circuit to have its contained copper stripped. The residues after washing are dug out and discarded.

The mixed oxide–sulfide ore contains 0.6% oxide and 0.7% sulfide, for a total copper content of 1.3%. The ore is crushed to $\frac{3}{8}$ inch (9.4 mm) size before leaching and washed to remove the fine slimes, which are treated separately in an agitation leaching operation.

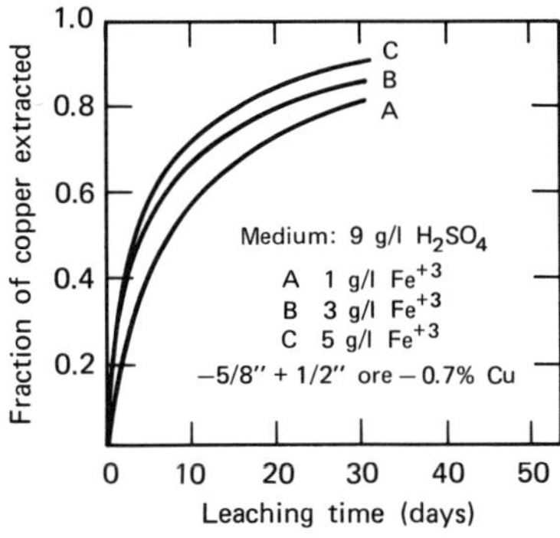

Figure 2.11. Effects of ferric sulfate concentration on leaching recovery and rates. *Source:* J. C. Yannopoulos and J. C. Agarwal, Eds., *Extractive Metallurgy of Copper,* Met. Soc. of AIME, Vol. 2, 1976, p. 863.

**Table 2.2. Copper Extracted Versus
Leaching Time—Heap and In-Place
Underground Leaching, Mountain City,
Nevada (du Pont de Nemours & Co.):
Heap Leaching**

Heap	%Cu Extracted	Days Operated
A	45.5	373
B	25.8	262
C	27.3	94
D	26.4	72
E	61.3	60
G	17.7	77

Source: J. C. Yannopoulos and J. C. Agarwal, Eds., *Extractive Metallurgy of Copper*,
Vol. 2, The Metallurgical Society of AIME,
1976, p. 861.

The extraction of the oxide copper is 97% and 79% for the sulfide copper, giving a total of 87%. The ferric sulfate required in solution will vary with the amount of copper sulfide to be oxidized, with general practice holding the average strength in the range of 7.5 gpl ferric sulfate.

Tank Leaching is also used for mixed oxide and sulfide copper ores, where the two ores are beneficiated and separated as oxide and sulfide concentrates to be individually leached. This type of treatment is carried out extensively at large successful commercial installations in Zambia and Zaire, with 300,000 tons of copper annually being produced in Zaire alone, and at the new plant of the Hecla Mining Company—El Paso Natural Gas Company's Lakeshore, Arizona, operation. This plant is scheduled to produce 65,000 tons of copper per year, of which 35,000 tons will be from leaching oxide ore and 30,000 tons from the sulfide ore.

After beneficiation in a 9150 ton per day sulfide mill, the 325 mesh sulfide concentrate, containing about 25% copper, is thickened and/or filtered to control the density of the slurry feed to two fluid-bed roasters, which have space rates of 1.5 to 2 feet per second (0.45 to 0.6 m per second). After roasting, the calcines are leached in a multistage tank leaching operation with spent electrolyte from the electrowinning precipitation circuit which follows, and the pregnant solution produced will carry about 55 gpl copper and 14 gpl H_2SO_4.

The SO_2 gas from the roasting is sent to an acid treatment plant to make H_2SO_4 for the tank leaching of the oxide concentrate, while the insoluble siliceous iron residue remaining from this roasted calcine leach is furnace reduced to make sponge iron pellets for the cementation precipitation of copper from the copper oxide pregnant solution.

High-Pressure Leaching of copper sulfide concentrates, the *Sherritt Gordon process*, is done in combination with the treatment of nickel, copper, and cobalt sulfide flotation concentrates, of which copper is not the chief constituent and analyzes 2% of the total, as compared to 10% nickel content.

Leaching is carried out in two stages, each stage using four compartment autoclaves 44 feet long by 11 feet in diameter (13.4 × 3.35 m), with the conditions of the first stage being 185°F (85°C) and 120 psi (827 kPa) and the second stage 175°F (80°C) and 130 psi (896 kPa). One third of the leaching time is in the first stage and two thirds in the

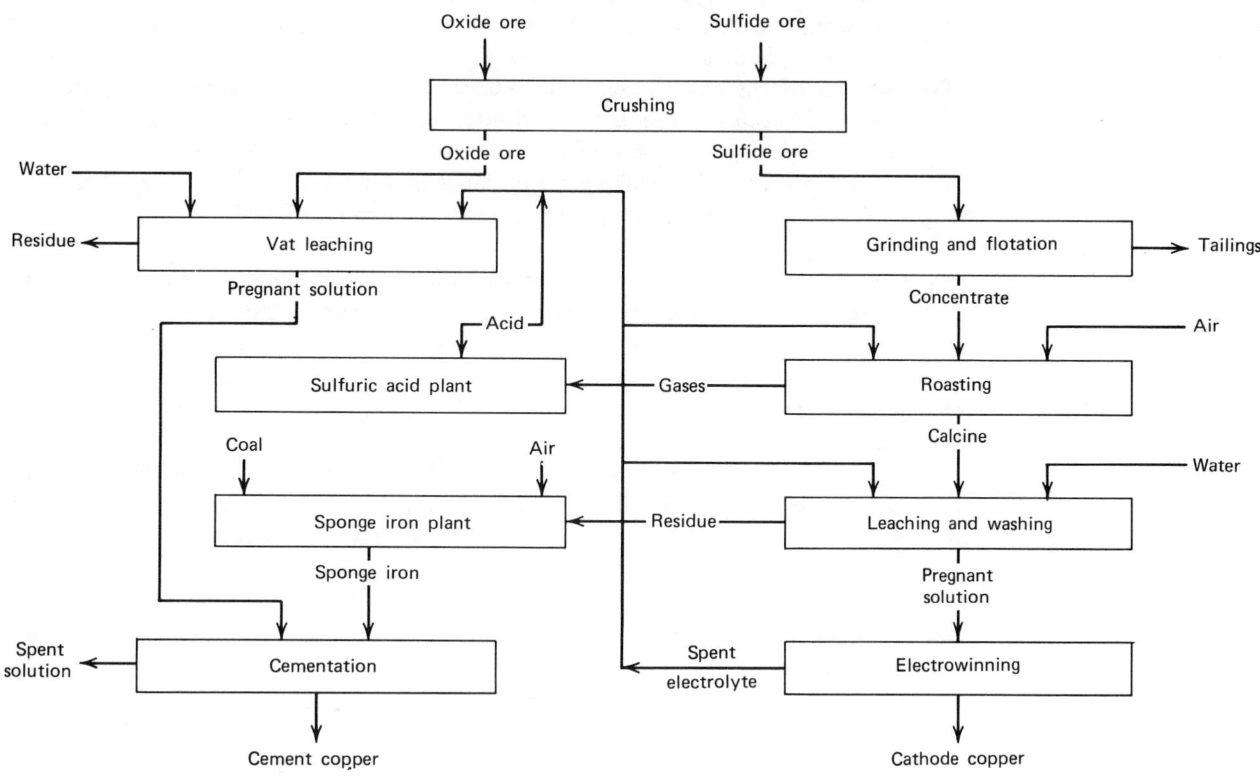

Figure 2.12. Combined copper sulfide and copper oxide leaching process. *Source: J. Metall.*, Vol. 27, No. 2, 1975, p. 19.

second stage. The leaching reaction is exothermic, and the autoclaves are equipped with cooling coils to carry away the excess heat generated, as well as agitators to keep the fine solids in suspension and promote the best leaching action.

The leaching operation involves a reaction of the sulfide minerals with dissolved oxygen, ammonia, and water that converts them to soluble ammines which go into solution. This pregnant solution is taken and treated in a precipitation circuit to recover the 7 gpl copper it contains.

Low-Pressure Leaching of copper sulfid concentrate slurry is carried out in the Anaconda Company's *Arbiter process.* A plant rated at 36,000 tons copper per year was started up in September 1974 and closed and put on standby in June 1975. After a lengthy shutdown for winterizing and other modifications, it was put back into operation in 1976.

The Arbiter process leaches a copper sulfide concentrate slurry in an oxygen-ammonia-ammonium sulfate system at $102°F$ $(70°C)$ under about $\frac{1}{3}$ atmosphere pressure [5 pounds per square inch (kPa per sq cm)]. Copper goes into solution as copper amine sulfate $(CuNH_4SO_4)$, and iron pyrite is insoluble. As originally planned, leaching is followed first by solvent extraction stripping of the solution and then by electrolytic precipitation to recover the copper. However a more recent process innovation is the possible elimination of these two steps to recover copper directly as a cuprous ammonium sulfite precipitate, by treating the leach solution with SO_2. This modification could reduce capital cost requirements by one third.

Ferric Chloride Leaching of copper iron sulfides ($CuFeS_2$) is one of the newest processes, and of these the *Cymet process* of Cyprus Mines Corporation shows promise. A copper sulfide concentrate, 50% minus 200 mesh, is countercurrent leached in ferric chloride solution at atmospheric pressure to produce cuprous chloride in the pregnant solution.

Precipitation of the cuprous chloride is carried out by either vacuum or refrigeration techniques and produce cuprous chloride cyrstals. These are reduced to nearly pure copper by hydrogen in a fluid-bed reactor and are removed as copper pellets.

A second ferric chloride leaching operation is underway at Duval's CLEAR plant at Sieritta, Arizona, and is to have a projected production rate of 32,500 tpy.

1c. Native Copper and Copper Carbonates are treated by ammoniacal leaching to extract their values. Coarser native copper is removed by gravity concentration, after which classification separates the finer particles into plus 100 mesh and minus 100 mesh fractions. The minus 100 mesh copper is extracted by flotation, while the plus 100 mesh portion is leached with cupric ammonium carbonate in batch downward-percolation leaching processes.

The leach solution for copper carbonates is ammonia and ammonium carbonate, and for native copper it is ammonia and ammonium carbonate plus the addition of dissolved oxygen. The native copper first dissolves to cuprous ammonium carbonate,

$$Cu + Cu(NH_3)_4CO_3 = Cu_2(NH_3)_4CO_3$$

and this is then oxidized to cupric ammonium carbonate by reacting with oxygen and ammonium carbonate:

$$2Cu_2(NH_3)_4CO_3 + 4(NH_4)_2CO_3 + O_2 = 4Cu(NH_3)_4CO_3 + 4H_2O + 2CO_2$$

The actual solvent is the cupric ammonium carbonate, which is reduced back to the cuprous compound by the copper being dissolved. The air oxidation of this cuprous compound provides the required amount of cupric ammonium carbonate necessarily present in order to have leaching efficiency.

The actual leaching is carried out on batches of 1000 tons of solids in steel tanks 54 feet in diameter and 12 feet high (16.46×3.66 m), which are enclosed to contain the ammonia vapors. After a 12 hour soak the pregnant solution, which contains 5% copper, is washed with steam to drive off the ammonia and is then sent on to the precipitation circuit to recover the contained copper. Extraction is in the range of 80%.

2a. Nickel Laterites, while widespread throughout the world and making up a considerable proportion of the known nickel reserves, are a low-grade, complex ore, not too well adapted to pyrometallurgical treatment due to the lack of sulfur for matte formation. Leaching is the accepted processing treatment and is carried out by two quite different methods, differentiation depending on whether or not there is a high percentage of basic magnesia in the ore. If magnesia is high, it makes acid leaching economically impossible as enough acid would have to be supplied to neutralize the magnesia as well as take the nickel into solution. High magnesia ores then dictate an ammoniacal basic leach, which is carried out at atmospheric pressure and ambient temperature.

Low magnesia ores, on the other hand, can be leached with sulfuric acid as there is no problem involved of wasting large amounts of acid to neutralize basic gangue in the ore. This leach is carried out at high temperature and pressure to speed the overall dissolution process.

The lateritic ores are most widely distributed in regions with warm climates and heavy rainfall where weathering of serpentinized ferromagnesian rocks has taken place over long periods of time. During this weathering, percolating ground waters have dissolved out some of the silica and magnesia from the parent serpentine rock and have left the less soluble oxides of iron, cobalt, nickel, chromium, and aluminum concentrated in the residue remaining.

High-Magnesia Laterites have been treated for the greater length of time by the *Nicaro process*, and the ore going to this leaching process contains 1.4% nickel, 8% magnesia, and 14% silica. The ore, $H_2(NiMg)SiO_4 \cdot nH_2O$, is dried, ground to 90% minus 200 mesh, and given a roast to reduce the nickel oxide to nickel metal which is compatible with ammoniacal leaching.

The roasting process is quite critical for several reasons. One is that a sufficiently low heating rate and high reducing atmosphere must be maintained in order to reduce the nickel to metal below 1400°F (760°C). Above this temperature an exothermic phase change takes place in which any remaining nickel oxide is substituted for magnesium oxide in the olivine which now forms from the serpentine and makes nickel oxide more difficult to reduce. The ratio of reducing to oxidizing constituents must be held as close to 1:1 as possible on the lower hearths to obtain satisfactory nickel oxide-to-metal reduction and ferric iron reduction to magnetite:

$$NiO + H_2 = Ni + H_2O$$

$$3Fe_2O_3 + H_2 = 2Fe_3O_4 + H_2O$$

The reduced nickel metal is actually in the form of a nickel–iron solid-solution alloy rather than as nickel metal alone.

To prevent reoxidation of the fine reduced particles as they are discharged hot from the roaster, the discharge is directed into coolers with nonoxidizing atmospheres. These coolers are steel tubes 80 feet long and 9 feet in diameter (24.4 × 2.74 m) slowly rotating in a water bath, with the reduced metal which enters at over 1400°F (760°C) being discharged at 300°F (149°C).

The cooled solids are added to an ammonia–ammonium carbonate solution (6.5% ammonia) making a 20% solids pulp density slurry, which is leached in a series of compartmented, concrete impeller-aerated tanks. The nickel–iron alloy oxidizes to nickel ions and ferrous ions which pass into solution, with the ferrous ions then further oxidizing to the ferric state and precipitating out as a gelatinous mass:

$$FeNi + O_2 + 8NH_3 + 3CO_2 + H_2O = Ni(NH_3)_6 + Fe^{2+} + 2NH_4 + 3CO_3$$

$$4Fe^{2+} + O_2 + 2H_2O + 8OH = 4Fe(OH)_3$$

The discharge from each leach tank flows to a leaching thickener, from which the overflow, an ammoniacal ammonium carbonate solution containing 12 gpl nickel, goes to the nickel recovery circuit.

The pulp underflow from this leaching thickener is washed with water containing ammonia in a series of four thickeners. Washing is countercurrent, with the underflow traveling from thickeners 1 through 4, while the wash water is added to thickener 4 and the overflow is then pumped back from 4 through 1. The final wash liquid, which is the overflow from thickener 1, is added to the overflow from the leaching thickener going to the nickel recovery circuit. The final washed residues, which are the underflow from washing thickener 4, are sent to a plant for ammonia removal and recovery.

Low-Magnesia Laterites, which analyze 1.35% nickel, 1.66% magnesia, and 3.7% silica, are selectively leached with sulfuric acid at high temperature and pressure to recover the nickel and cobalt. This is the *Moa Bay process* (Freeport Nickel). If the ore is leached in an aqueous sulfuric acid solution at atmospheric pressure and ambient temperature the considerable amount of iron present in the ore (68% iron oxide) will readily go into solution with the nickel and cobalt. However with the same acid concentration but higher temperature, 450° to 500°F (232° to 260°C), and pressure of 610 psi (4203 kPa), all three metals have a lesser tendency to dissolve. The solubility of the iron is most affected and only minor amounts will go into solution under these conditions, while better than 95% of the nickel and cobalt will still dissolve. The relatively lower content of basic magnesia in the ore permits the economical use of sulfuric acid as the leaching solvent, without large amounts of acid being wasted in merely neutralizing a high basic content in the ore.

The mined ore is screened and washed to remove and reject the material coarser than 20 mesh, and the minus 20 mesh portion is made into a slurry of about 25% solids and pumped to storage thickeners at the leaching plant. These thickeners are 325 feet in diameter (99.1 m) by 32 feet deep (9.75 m), and one thickener can furnish ore slurry storage for three to five days of full plant operation. The ore slurry is settled to 45% solids in the thickeners to provide feed to the leaching operation.

Agitated tank preheaters with steam absorption first raise the slurry temperature to 170°F (77°C). Then the slurry is heated to the reaction temperature of 450°F (232°C) with steam injection at 610 psi (4203 kPa) in heated towers, and from here flows by gravity to a series of four autoclaves. The autoclaves are 50 feet high and 10 feet in diameter (15.2 × 3.05 m) and are lined with lead and acid-proof brick. Agitation is accomplished by injecting 610 psi (4203 kPa) steam through a 16 inch diameter (40 cm) titanium centrally supported draft tube, and circulation is induced on the Pachuca principle. Sulfuric acid at 98% concentration is fed with plunger pumps into the first reaction vessel of the series, and combining there with the 45% slurry from the leaching thickener, the mixture flows by gravity through the series of four autoclaves for a combined leaching time of between 1 and 2 hours.

The leached slurry which overflows the fourth stage reactor passes through a heat exchanger where it is cooled to below 275°F (135°C) and then, still under pressure, through special flow-control chokes into a flash tank where it is reduced to atmospheric pressure. The leached slurry goes on to thickeners, where asphalt linings and acid-proof brick construction hold down equipment corrosion, to separate the pregnant solution from the insoluble residues. The pregnant solution overflow from the first thickener in the six-stage washing circuit goes to precipitation and contains 5.95 gpl nickel, which means 96% of the available metal is being extracted. The residues are washed through five more countercurrent washing thickeners before being discarded with a remaining nickel content of only 0.06%.

2b. Nickel Sulfides are leached in two quite different forms—as a flotation concentrate in an ammoniacal leach in autoclaves at high temperature and high pressure and as a converter matte which is finely ground and then leached with a sulfuric acid solution in agitated tanks at atmospheric pressure and ambient temperature. In both instances the nickel is taken into solution along with the copper and cobalt also present, and these are then all selectively precipitated.

Flotation Concentrate, in the *Sherritt Gordon process*, is leached in a strong aqueous ammonia solution in the ranges of 160 to 190°F (71 to 88°C) and 100 to 150 psi (690 to 1034 kPa), with the leach solution oxygenated with air. The nickel, which analyzes 10%, reacts with the ammonia, water, and dissolved oxygen to form soluble ammines:

$$NiS + 2O_2 + 6NH_3 = Ni(NH_3)_6SO_4$$

Copper and cobalt sulfides present in lesser amounts will also go into solution in the same manner as the nickel, while iron sulfide forms an insoluble ferric oxide and is left as an insoluble residue with the SiO_2:

$$4FeS + 9O_2 + 8NH_3 + 4H_2O = 2Fe_2O_3 + 4(NH_4)_2SO_4$$

The concentrate is mixed into a slurry with solution returned from the second stage of a two-stage leaching operation and is pumped to the first stage of leaching. This is carried out in a four-compartment autoclave with adjustable weirs over which the slurry flows from compartment to compartment. Impellers in each compartment keep the solids in suspension, and cooling coils remove heat from the exothermic leaching reactions to keep the temperature at about $185°F$ ($85°C$), along with the pressure of 120 psi (827 kPa). One third of the total leaching time of 8 hours is used up in this first stage.

The slurry, after this first leach, is pumped through water-cooled heat exchangers, which reduce the temperature to $90°F$ ($32°C$), and then on to a thickener. The gas escaping from the process also goes through a cooler, then to an adsorption column to recover the ammonia for recycling. The thickener overflow goes to a precipitation circuit, while the underflow is filtered and the solids repulped into a slurry with fresh ammonia solution and pumped to the second-stage leaching autoclaves. These autoclaves are the same for both leaching stages, 44 feet long and 11 feet in diameter (13.4×3.35 m), with four compartments having mechanical agitation in each and cooling coils to adjust the heat. The conditions for the second leach are $175°F$ ($80°C$) at 130 psi (896 kPa) for two thirds of the 8 hour total leaching time. The leached slurry from this second stage also passes through a cooler and on to a thickener, from which the overflow solution goes back to the first leaching stage to be pulped with sulfide concentrate and makes up the first-stage leach slurry. The insoluble solids of Fe_2O_3 and SiO_2 from the thickener go through a wash circuit in which they are filtered, washed, repulped, then filtered, washed, and repulped again, a total of three times before discarding. The wash solution is returned to the second leaching-stage thickener.

The pregnant solution, which is taken off from the first-stage leaching thickener and sent to the precipitation circuit, analyzes 50 gpl nickel, 1 gpl cobalt, 10 gpl copper, 180 gpl ammonium sulfate, 10 gpl sulfur as thiosulfate and polythionates, and 100 gpl free ammonia.

Converter Matte at the Outokumpu Harjavalta Smelter is a low-sulfur composition analyzing 63% nickel, 28% copper, and 7% sulfur. It is crushed and ground to 90% minus 270 mesh and sent to a two-stage leaching circuit where it is leached in mechanical and air-agitated tanks with a sulfuric acid solution.

The ground matte is filtered and repulped to the desired slurry concentration before going to the first stage of leaching, which consists of three 1200 cubic foot (34 m^3) agitated leach tanks in series. The leached slurry from the third tank in the series passes on to a cyclone-classifier circuit which separates the pregnant solution from the remaining solids. The pregnant solution is sent to a precipitation circuit, while the remaining solids are further leached in a second-stage series of two agitated tanks.

A series of cyclic reactions takes place in these first three tanks, with copper cementing out as nickel dissolves. Then in the second-stage, two-tank series the remaining solids of preleached matte and cement copper have most of their remaining values dissolved in a sulfuric acid solution containing 40 gpl free acid. The leached slurry from this stage goes to a thickener, with the overflow pregnant solution passed on to a precipitation circuit. The insoluble residues are filtered and water washed, with the collected wash water then being added to the pregnant solution and the final matte residues returned to the smelter.

3a. Cobalt Sulfides, as in Zambia, are found in combination with larger amounts of associated copper sulfides. As these ores analyze only 0.1% cobalt with 3% copper, it is necessary first to beneficiate them to improve the cobalt content to 3 to 5%. The cobalt

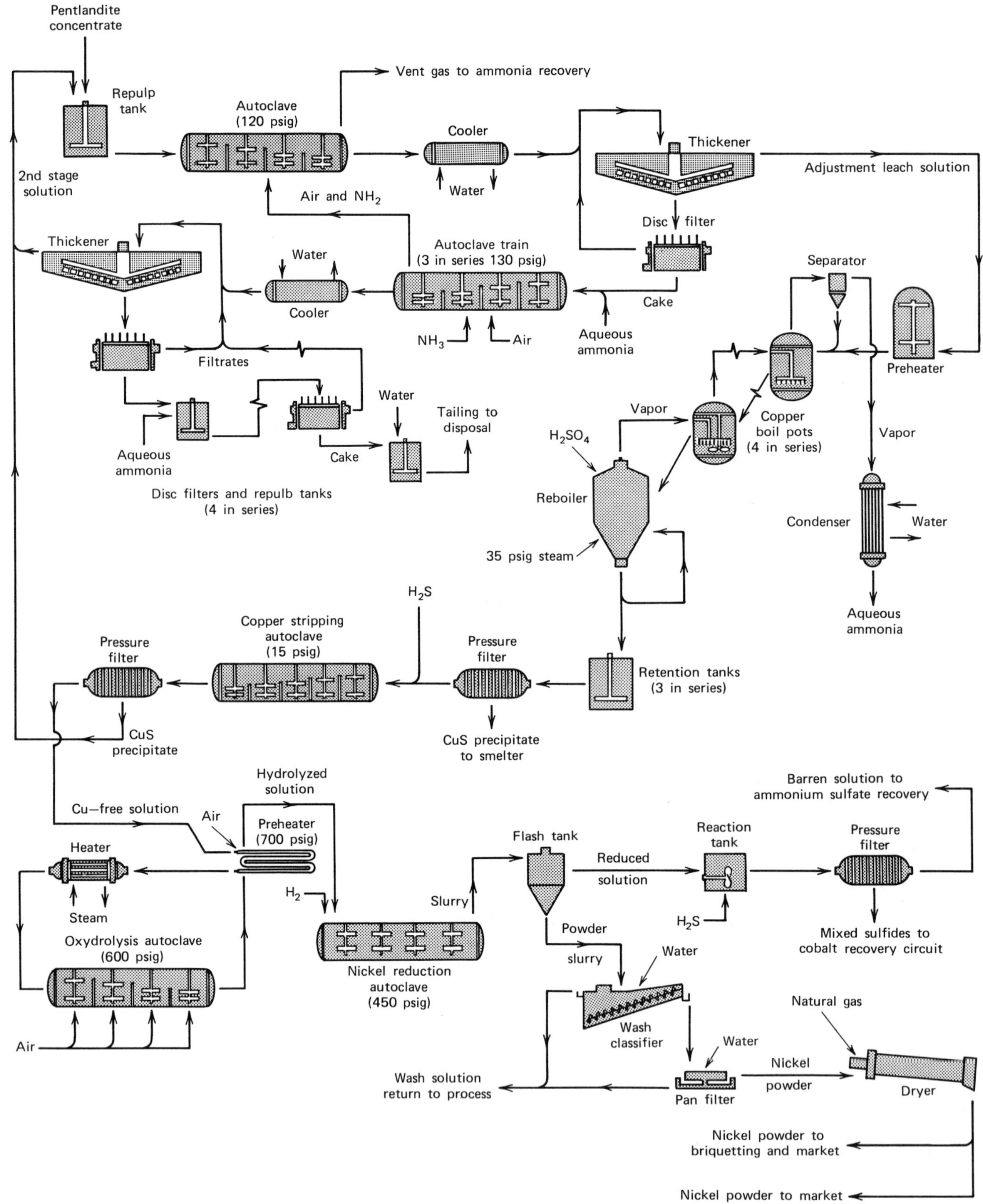

Figure 2.13. Sherritt–Gordon's sulfide leach process. *Source:* J. R. Boldt, Jr., and P. Queneau, *The Winning of Nickel*, Methuen and Company, Ltd., London, 1967. Copyright Inco Limited, 1967, p. 303.

concentrate will also contain a varying but still much larger percentage of copper (25 to 30%).

The cobalt sulfide concentrate is given a sulfating roast to convert the cobalt sulfides to water-soluble sulfates which can be leached with hot water. This roasting is done in eight-hearth, mechanically rabbled multiple-hearth roasters that have a capacity of 40 tons of feed per day and a retention time of 13 hours.

The water solubility of the cobalt sulfate after roasting is expected to be 80 to 85% of the available cobalt, and this is batch leached in heated water in mechanically agitated tanks at atmospheric pressure. The tanks are filled with hot water at 158°F (70°C) and weighed and roasted calcines are added to give a pulp specific gravity of 1.3 to 1.5. Agitation is continued for 1 hour, and the temperature, which the heat of reaction raises to 176°F (80°C), is maintained at this level by steam coils.

Leached calcines are filtered, repulped with water, and refiltered with a water spray wash on the filter cake. The combined filtrates contain 10 to 25 gpl cobalt and are sent to the precipitation plant, while the high-copper filter cake is shipped to the copper smelter.

The *Finnish sulfatized cobalt calcines* produced in a fluid-bed roaster at Outokumpu Oy are water leached in a countercurrent leaching and washing system consisting of two thickeners in series.

The overflow from thickener 2 is used to quench the sulfatized calcines from the roaster, and the resulting slurry is pumped into thickener 1. The overflow from thickener 1 is the pregnant solution pumped to the precipitation circuit, while the underflow from thickener 1 is repulped with filter solution and fed to thickener 2. The underflow from thickener 2 is filtered, with this filtrate used to repulp the thickener 1 underflow, and

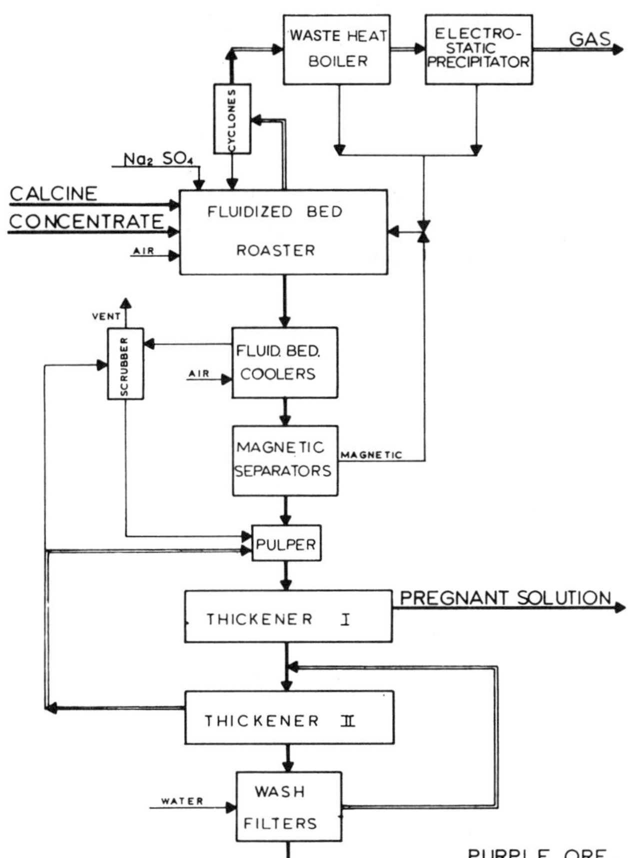

Figure 2.14. Cobalt sulfides roasted to sulfates and leached in water. Outokumpu Oy, Finland. *Source:* M. Palperi and O. Aaltonen, *J. Metall.*, Vol. 23, No. 2, 1971, p. 36.

washed. This leach residue filter cake is designated as "purple ore" and contains as little as 0.01% of the soluble metal. It is dried in a rotary kiln and storage piled.

The pregnant solution will have a typical analysis of 20 gpl cobalt, 9 to 10 gpl nickel, 8 to 9 gpl copper, 11 to 13 gpl zinc, and 7 to 9 gpl iron.

3b. Cobalt Oxides are found both as combined copper and cobalt oxides, and as more complex mixed ores where both oxides and sulfides of copper and cobalt are present. The general treatment of both types of ores is the same; after preliminary treatments they are leached in sulfuric acid solutions at atmospheric pressure.

The oxide ores are concentrated and then leached in a sulfuric acid solution using air agitated Pachuca-type vats. The copper oxide dissolves readily, but cobalt oxide is only slightly soluble unless ferrous sulfate is added as a reducing agent, in which case the cobalt oxide is then almost completely dissolved. After thickening and filtering to remove the insoluble residue, the pregnant solution is sent to the precipitation plant to selectively remove the dissolved cobalt and copper.

The mixed oxide and sulfide ore is also concentrated and then given a sulfating roast in fluo-solid roasters to convert the sulfides in the concentrate to soluble sulfates or oxides. Leaching is then carried out in sulfuric acid solutions in a manner similar to that used to treat the oxide concentrate.

4. Silver and Gold are quite often found in association with one another to some degree—from the case where almost all gold ores contain some silver at least in a minor degree to that of both being valuable by-products of many copper and lead sulfide ore deposits. Silver and gold ores, which are mined and treated primarily for the extraction of these precious metals, can be leached at atmospheric pressure in dilute aqueous sodium or potassium cyanide solutions, which readily dissolve the metals if oxidizing conditions are maintained:

$$4Ag + 8NaCN + O_2 + 2H_2O = 4Na[Ag(CN)_2] + 4NaOH$$

$$4Au + 8NaCN + O_2 + 2H_2O = 4Na[Au(CN)_2] + 4NaOH$$

Most gold ores are still treated by cyanide leaching, but it is no longer a widely used method for silver extraction as many of these ores are now treated by flotation which recovers about 97% of the silver and the flotation concentrate is then smelted to separate the silver as a final metal product ready to be refined.

The ores are crushed and ground finely enough to expose the metallic particles to the action of the leaching solution, and this fine grinding in most cases will be continued until at least 60% of the material is minus 200 mesh. Grinding is also usually carried out in a cyanide solution, and this early contact begins the dissolution of the metals in the leach solution. Lime is added to the grinding mills in amounts of 1 to 2 pounds (0.454 to 0.908 kg) per ton of ore treated to act as both a neutralizing agent for any acidity the ore may have that would consume large amounts of cyanide and also as a settling agent for the small particles in solution after fine grinding. The sulfides of copper, iron, antimony, and arsenic are the acidic contaminants of an ore that enter the leach solution and deprive the silver and gold of both cyanide and oxygen unless treated for protective alkalinity with lime. These are known as cyanicides.

Fine grinding is followed by thickening in conventional thickeners, with the clear cyanide solution overflow either being sent back to the fine grinding circuit as a source of cyanide or sent to the precipitation plant to recover the silver and gold it contains. The thickened underflow, having a solution-to-solid ratio of 1:1, is made up to required cyanide solution strength and pumped to the center column air lift-type agitators to be leached. Cheaper sodium cyanide is invariably used as the reagent for leaching.

The strength of the dissolving solution is usually about 1 pound (0.454 kg) of cyanide per ton of water for gold (equivalent to 0.05% cyanide solution) and at least double this for silver. A stronger solution and longer leaching time are required for silver for two reasons—one that silver is more difficult to take into solution, and the other that the quantity of silver to be dissolved is generally much larger than the amount of gold available in the ore. Air lift-type tanks from 22 to 60 feet high (6.6 to 18.3 m) and 8 to 30 feet in diameter (2.44 to 9.0 m), generally of steel construction, with usually three or four, but as many as seven, connected in series, ensure both thorough agitation and contact of particles and leach solution, in addition to providing compressed air which in

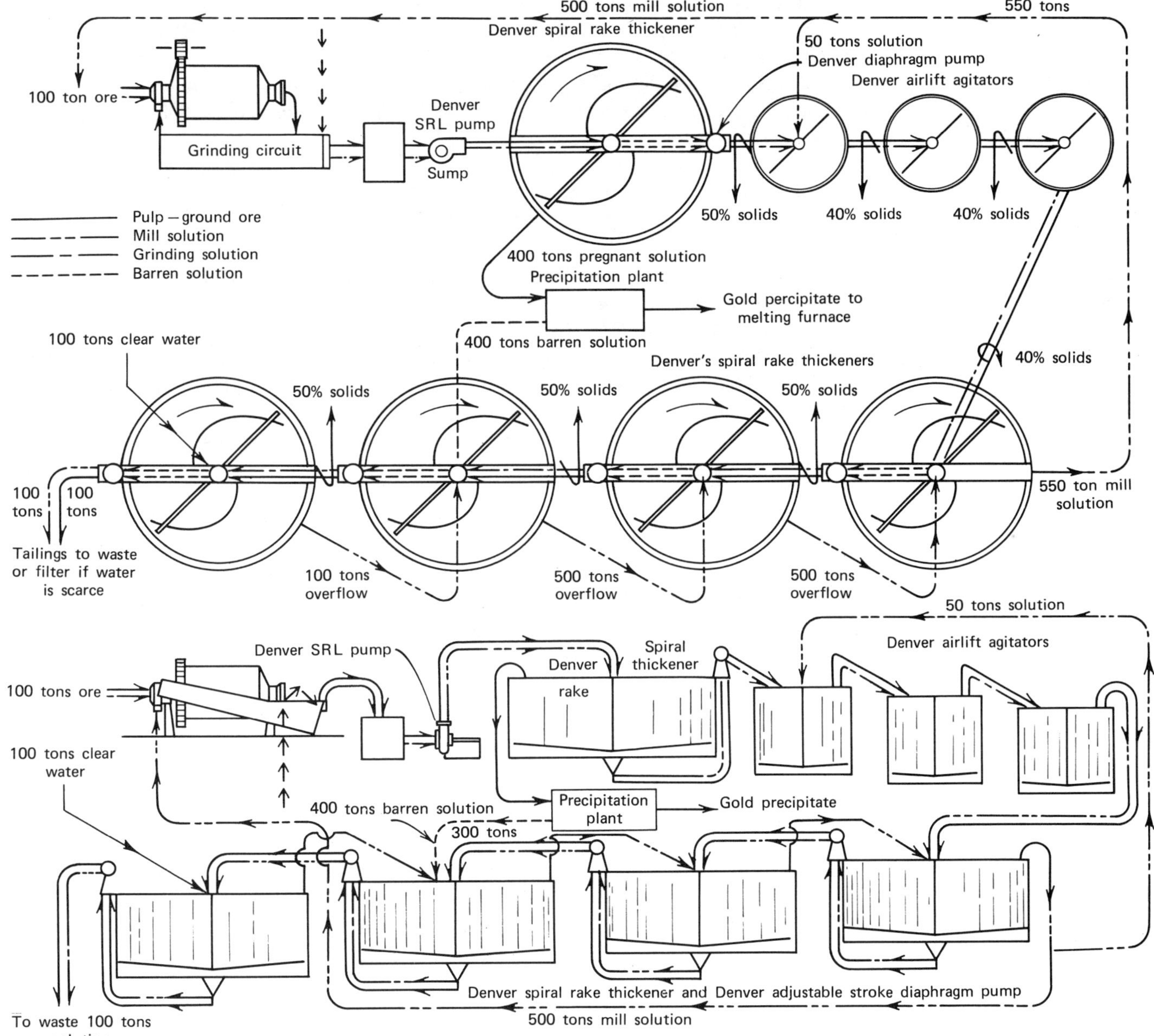

Figure 2.15. One hundred ton per day cyanide leaching plant. *Source:* Denver Equipment Co., Modern Mineral Processing Flowsheets, 1962, p. 60.

addition to giving agitation supplies the necessary oxygen required for dissolving the precious metals. The tanks in series, with the pulp traveling from one tank to the next, ensure that all the pulp is aerated sufficiently for the best metal extraction. The time spent in leaching will vary from 20 to 40 hours, and metal recovery will approach +90%. The pregnant solution is filtered to remove the insoluble residue and then pumped on to the precipitation circuit to recover the dissolved metals.

Heap Leaching is a new innovation in the treatment of low-grade gold ores, and the simplicity of the system gives a respectable production rate at minimum capital cost.

The leaching is done on permanent watertight pads, with subsequent evacuation and disposal of the insoluble tailings. The cycle of heap emplacement, leaching, washing, and tailings removal will require about one month. The lime–cyanide leach solution will be 0.5% NaCN at a pH of 11, and reagent consumption is 1 pound (0.45 kg) cyanide and 2 pounds (0.91 kg) lime per ton of ore treated.

The Smokey Valley Mining Company of Round Mountain, Nevada, of which Copper Range Company is a major contributor, is the innovator of this latest form of gold ore leaching.

PRECIPITATION

The pregnant solution containing the dissolved metal values from the leaching process is treated in a number of ways to precipitate the dissolved metal content and recover it in solid form. In some cases the leach liquor must first be purified to remove secondary metals which also went into solution during leaching and which, if not first selectively removed, will also precipitate with the valuable metal product and contaminate it. In other cases direct selective recovery of the valuable metal is possible from the solution as it comes from the leaching circuit, without the necessity of preliminary purification.

There are two general methods of precipitation—by electrolytic deposition using insoluble anodes and by chemical precipitation. Electrolytic precipitation is done at atmospheric pressure, while chemical precipitation uses both atmospheric conditions and high pressure and temperature.

A new additional method, which is fast maturing as a major development, is a combination of solvent extraction followed by electrolytic precipitation. The solvent extraction functions as a highly versatile preliminary mechanism for purifying leach solutions and concentrating metal values into smaller, more manageable solution volumes for precipitation. Another new method, somewhat similar to solvent extraction, uses an adsorption circuit with activated carbon to strip the pregnant solution, followed by electrolytic precipitation.

Electrolytic deposition gives a very pure deposit that requires only a minimum of further refining and regenerates the solvent so that it can be reused again in the leaching circuit after electrolytic deposition has removed most of the dissolved metal. It is also a selective method of precipitation in some cases. These are where the electromotive series is such that the voltage applied to the electrolyte in the cell (the pregnant leach solution) is sufficient to decompose the valuable metal compound in solution and have this metal deposit at the cathode, but with the applied voltage not sufficiently high enough also to decompose the other contaminating metal compounds in the solution. These therefore will remain in solution and do not deposit at the cathode. If the decomposition voltage needed for the valuable metal compounds will also decompose contaminating metal compounds, the leach solution must first be purified to remove these secondary metals before electrolytic precipitation can be applied.

Chemical precipitation, which is the formation of a solid compound no longer soluble in the pregnant solution, can be accomplished in a number of ways. Some of these processes also first require purification of contaminant dissolved metal compounds before chemical precipitation can be applied to recover the valuable metal compound, while in other treatments selective precipitation is possible without prior purification. Generally the product of chemical precipitation is of not too high a purity and needs considerable further refining to have a pure final metal product. However the treatment costs are low, very much less than those for electrolytic precipitation, and as a consequence pregnant solutions of quite low metal values can be economically treated. In some instances the stripped solution after precipitation can also be regenerated and sent back to the leaching circuit as fresh leach liquor.

Cementation is one of the most commonly used methods of removing values from leach solutions by a displacement reaction in which a more active metal reduces the ions of the valuable metal to the metallic state and itself enters the solution as a replacement. Hydrogen displacement of metal values from solution is also used, with the necessary conditions of elevated temperature and pressure and the addition of metallic nuclei on which the deposited metal can grow.

Precipitation can also be produced through removal of a complexing agent such as ammonia, which can be volatilized on heating the pregnant solution with steam and so free the metal ions combined with it to react with other ions still in solution and form an insoluble compound which precipitates.

Addition agents such as hydrogen sulfide can be added to form an insoluble precipitate with a valuable metal, or the oxidation state of a metal in solution can be modified to cause hydrolysis and have a metal precipitate as a hydroxide. Manipulation of the solution temperature and pH is also used to achieve the selective crystallization of metallic compounds from pregnant solutions.

TYPES OF PRECIPITATION

Precipitation is the middle stage of hydrometallurgical extraction and is the processing step to remove the dissolved metal as a solid product from the dissolving leach solution, with this precipitated metal then usually being refined to its final, pure, commercially usable form.

The material being leached is not all dissolved, and there will be an insoluble residue left, frequently waste rock high in SiO_2, which has no part in the precipitation operation where the metal values that have selectively gone into solution during leaching are now being recovered. An essentially clear pregnant solution is required for precipitation, and this is separated from the insoluble residues by decantation or by filtering, or the two combined. After precipitation the precipitate is separated from the barren solution and dewatered by filtering.

In many cases of precipitation processing the pregnant solution is not completely stripped of its metal content. But after partial precipitation the solution is rejuvenated as a leach solution and is reused in the leaching circuit to dissolve another batch of the valuable metal from a load of fresh feed material in the leaching tanks.

Of the various methods of precipitation it can generally be said that the individual method chosen in each case is the method which is best suited to recover the metal from that particular pregnant solution. There are however additional considerations to be made such as the scale of the leaching operation and the resulting tonnage of pregnant solution to be treated, the percent of valuable metal in solution, whether it is high or low, and finally the ease with which valuable metal can be precipitated from the particular pregnant solution and the purity of the precipitate required from the process.

There are two main categories of precipitation. One is electrolysis, which depends on the fact that when direct current is passed through a metallic solution decomposition occurs and metal will deposit at the cathode. The other is chemical, where the addition of a reagent to the solution will throw out the metal as a solid precipitate. Because of the cost of electricity for the electrolytic process, only solutions containing relatively large amounts of metal can be treated in this manner, with the high operating costs being balanced by the relatively quick processing time and large tonnage of metal extracted. To expedite these factors, several of the newest hydrometallurgical plants are using solvent extraction or activated carbon adsorption as a preliminary step before electrolytic precipitation, both to purify the pregnant solution and also to concentrate the metal values into smaller solution volumes for treatment. This form of precipitation is done at atmospheric pressure and ambient or slightly elevated temperature.

The second main category, chemical precipitation, is used in a larger variety of processings than is electrolytic precipitation. As it can be a cheap, but efficient, ion replacement method, it can be used economically with large volumes of quite dilute pregnant solution. In other instances direct combination with an added reagent will selectively produce the valuable metal as a precipitated compound, or a reagent which alters the pH will give the condition to cause precipitation. In some instances metals that are only precipitated with difficulty can be chemically precipitated by adding a reagent when the pregnant solution is at high temperature and pressure in an autoclave, while increases in the temperature alone can volatilize complexing compounds and so release metal ions to precipitate.

The purity of precipitates from electrolytic treatment is usually such that only a minimum of further refining is necessary, while with chemical precipitation there is a range from pure metal powder products in final form to quite impure precipitates that still require considerable further refining.

Electrolytic Precipitation depends on the clarified pregnant leach solution, which in effect is the electrolyte containing the valuable metal to be recovered, decomposing when a high enough direct current voltage, the so-called decomposition voltage, is passed through it. There is a definite decomposition voltage for each metal, which varies according to the position of the metal in the electromotive series. If several metals are in solution, the one with the lowest decomposition voltage will deposit at the cathode first, with the other metals staying in solution until the voltage is raised sufficiently to decompose them as well. For example, the decomposition voltage of an aqueous solution of copper sulfate is 1.49 volt, and for zinc sulfate it is 2.35 volt, which means that at a voltage of 1.49 volt direct current, only the copper will deposit from a solution of mixed copper and zinc sulfates.

Power consumption is the product of the amperes used per hour multiplied by the total voltage applied to the circuit, with the power costs being a considerable portion of the total plant costs of operations. Consequently any reduction in power usage is important, and tests are made constantly to check for any electrode short circuiting or current leakage. The amount of metal deposited is a function of time and amperage according to Faraday's Law, which states that an equivalent weight of metal (the atomic weight divided by the valence) in grams will be deposited by 96,500 coulombs (amperes $\times$ seconds) of electricity, or that 500 ampere days of electricity will deposit one equivalent weight of metal in pounds.

The amount of metal that theoretically is to be deposited by calculation is never reached, as cell efficiency is never 100% because of current leakage, short circuits, and polarization due to gas bubbles collecting on the anode or cathode which raise the cell resistance. The cell efficiency obtained can be calculated by comparing the actual weight of metal which has deposited with the weight that should have deposited according to Faraday's law.

Cell temperature is important in that, as the solution temperature is raised, its resistance, and with this the decomposition voltage, is lowered. There is a limit to the degree of electrolyte heating that can be applied, however, and excessive heating costs will more than use up any savings in electricity. Sometimes too undesirable side effects show up with excessive temperature and can cause uneven cathode growth, the unwanted deposition of impurities, or even vaporization of the electrolyte.

Purification of impurity metals, which were dissolved in the leach circuit along with the valuable metal, must be taken care of before electrolysis if there is a possibility that these impurities also deposit at the cathode along with the valuable metal being electrolytically precipitated, and so contaminate it. These impurities are usually removed by selective chemical precipitation followed by filtering, and the purified solution is then pumped to the electrolytic cell. With the impurities removed before electrolysis there is no buildup of these contaminants in the cell, and spent electrolyte, after the valuable metal has been removed, is simply used over and over again as the solvent in the leaching stage.

The electrical connections in the cell are the multiple system, similar to that used for electrolytic refining, except that now there is one more anode than cathode in the complement of electrodes in a cell. The anodes are insoluble, generally of lead or stainless steel, and are used only to carry current into the cell. The cathodes, on which the precipitated metal comes out of solution, can be either thin sheets of the same metal as is being deposited or a different, inert metal from which the deposits are periodically stripped. In one instance a wad of steel wool is used to provide the maximum cathode surface area for deposition. The electrodes, both anodes and cathodes, are usually about 3 feet square (0.9 m^2) and set in the tanks with about 2 inch spacings (5 cm) between each. The cells are commonly of wood or concrete, with lead, plastic, or asphalt linings to protect against corrosion by the usually highly acidic electrolyte. Fiberglass cells are also

Table 2.3. Decomposition Voltages[a]

Electrolyte	Decomposition Voltage (V)	Electrolyte	Decomposition Voltage (V)
Zinc sulfate	2.55	Ammonium hydroxide	1.74
Zinc bromide	1.80	Sulfuric acid	1.67
Nickel sulfate	2.09	Nitric acid	1.69
Nickel chloride	1.85	Phosphoric acid	1.70
Lead nitrate	1.52	Monochloracetic acid	1.72
Cadmium sulfate	2.03	Dichloracetic acid	1.66
Cadmium nitrate	1.98	Perchloric acid	1.65
Cobalt sulfate	1.92	Acetic acid	1.57
Cobalt chloride	1.78	Trichloracetic acid	1.51
Copper sulfate	1.49	Hydrochloric acid	1.31
Silver sulfate	0.80	Azoic acid	1.29
Silver nitrate	0.70	Oxalic acid	0.95
Sodium hydroxide	1.69	Hydrobromic acid	0.94
Potassium hydroxide	1.67	Hydriodic acid	0.52

Source: H. J. Creighton, *Principles and Applications of Electrochemistry*, Vol. 1, Wiley, New York, 1943, p. 241.

[a]The decomposition potentials vary, not only for the solutions of the different electrolytes, but also because of changes in concentration, electrode material, cell size, temperature, and the presence of agents that may affect the cell function.

now being used. A tank of typical dimensions, 11 feet long, $3\frac{1}{2}$ feet wide, and $3\frac{1}{2}$ feet deep (3.35 × 1.07 × 1.07 m), will hold 30 anodes and 29 cathodes.

Chemical Precipitation, in general terms, is accomplished by adding a reagent, which can be a metal powder, a gas, a chemical compound, or heat as steam, in different sets of circumstances, to react and throw the valuable metal out of solution as a solid precipitate. In some cases the precipitate will be a relatively pure metal powder, while in others it is a compound that still needs considerable treatment to reach the pure metal state. Most chemical precipitation is done at atmospheric pressure, but there are cases where high temperature and pressure, as obtained in autoclaves, are necessary to cause the precipitating reactions to occur.

1. Cementation of a valuable metal from a pregnant leach solution depends on a displacement reaction in which a more active metal reduces the ions of the precipitating metal to the metallic state, which then comes out of solution, while the ions of the more active metal enter the solution in replacement.

Solutions both high and low in metallic values can be treated by cementation; and while many methods have been tried as alternatives, this ancient method of precipitation is very widely used in a great number of different extractive applications, from recovery of metal values in very dilute solutions to stripping of values from high grade solutions, and purification of contaminants before precipitation by another method.

Dilute solutions, such as one from heap leaching of a copper ore and assaying only $\frac{1}{2}$ to $2\frac{1}{2}$ gpl copper, are treated on a large scale with iron replacement metal in the form of light gauge scrap iron and steel, steel cans, or sponge iron, placed in tanks or troughs and the solution allowed to flow over it. Replacement takes place with copper coming out of solution and the iron being dissolved.

The cement copper formed is periodically washed off the surface of the iron scrap and the wash water filtered to collect this precipitated metal. The barren solution is pumped back to the leaching circuit to be reused as leach liquor. The whole operation is quite inexpensive with the operational equipment being quite limited and the scrap iron for precipitation being the major expense.

Solutions with high metal content can also be treated by cementation, both for purification and for extraction of the valuable metals. Low sulfur nickel-copper matte analyzing 63% nickel and 28% copper can be taken into solution and then have the copper selectively precipitate, as nickel powder is added to replace the copper in solution. This is carried out in mechanically agitated tanks, and after filtering out the cemented copper the solution, still containing the nickel from the matte, is pumped to cells to precipitate this metal electrolytically.

Gold and silver are stripped from their leach solution by replacement with fine zinc powder. The powder is fed into the intake of a pump feeding a filter press with leach solution, and the precipitation reaction is rapid enough that the press is filled with precipitate, which is separated in this way from the now barren solution.

2. Hydrogen Precipitation depends on reduction to metal of the metallic compounds in solution, which will result in metal powder as the end product, and one of high enough purity to need no further refining.

High temperature and pressure are required, as are nuclei of the metal to provide a base on which precipitation can readily take place. Autoclaves similar to those used for leaching are employed on a batch basis, with successive loads of solution depositing metal precipitate on the nuclei, which are growing in size, and then the barren solution being withdrawn.

Ammoniacal nickel solutions are heated to 400°F (204°C) at 450 psi (3100 kPa) in a reducing hydrogen atmosphere, in the presence of fine nickel powder for nucleation. Several batches of solution are reduced and deposit their metal content on the nuclei until the powder formed grows to sufficient size. This powder is then withdrawn to complete the precipitation cycle for the autoclave, which is then charged with fresh nuclei and fresh solution for a new batch cycle.

3. Hydrogen Sulfide will form a metallic sulfide compound with the valuable metal which is no longer soluble in the leach solution. Autoclaves with strong mechanical agitation are used and the higher the temperature the faster will be the precipitation rate. Gaseous high-purity hydrogen sulfide is injected to maintain pressure in the autoclave at 150 psi (1034 kPa), along with preheated solution at 245°F (118°C). The sulfide combines readily with such metals as nickel, copper, lead, and zinc to form metal sulfides. Seeding of the pregnant solution with recycled sulfide precipitate helps to improve metal recovery and control particle size produced. These too are batch processes.

4. Steam Heat added to a pregnant solution, can in some cases as with ammonia, remove a complexing agent by volatilization, and release the metal ions which had been held by the ammonia so that they can react with other ions in the solution and form an insoluble precipitate.

Cylindrical, multi-tray, steam heated towers, with the pregnant solution flowing downward, countercurrent to the flow of rising steam, are used in this type of precipitation. As the descending solution is gradually depleted of ammonia the metal will begin to precipitate and form a new and insoluble metallic compound. The vapors passing off the top of the towers are condensed and scrubbed to recover the ammonia, while the slurry from the bottom is thickened and filtered to separate the precipitate. These towers can be operated on a continuous basis.

5. Chemical Compound Additions and changes in pH will form new insoluble compounds, or remove compounds from solution which are soluble in acid mixtures but not in basic mixtures.

Iron is commonly precipitated by oxidizing, either by aeration or with a chemical oxidizer, to convert ferrous iron to ferric iron, and then be precipitated as the hydroxide. Iron, copper, cobalt, and nickel are precipitated selectively as hydroxides in aerated, air-agitated, solutions by raising the pH with milk of lime. Iron precipitates at pH 3.5, copper at 5.8, cobalt at 8.3, and finally nickel at 9.4. These precipitations are carried out in batch operations, and the solid precipitates formed are each removed by filtering before the next precipitation is made at a progressively higher pH.

PRECIPITATION PROCESSES

1a. Copper Oxide pregnant leach solution, resulting from the leaching and washing performed on ores, has the copper precipitated electrolytically from the high metal content pregnant solutions and by cementation from the lower metal value leach solutions and the wash solution used to clean insoluble residues before they are discarded.

High-Value Pregnant Solutions are sometimes sent through a purification process before electrolysis, especially if there is acid-soluble iron in the material being leached that will be dissolved and remain in solution at the end of the leach cycle. The iron content should not be above 1.2 gpl; to accomplish this, the ferrous iron, which makes up 90% of the iron in solution, is oxidized to ferric iron by adding manganese dioxide and then precipitated as ferric hydroxide. This is carried out by raising the pH to 2.2 by partly neutralizing the acid solution with additions of copper hydroxide filter cake. The precipitated iron is removed by thickening.

The control of soluble iron in solution has had a considerable effect on electrolytic cell performance, increasing current efficiency quite substantially with corresponding related increases in productivity and reduction in power consumption. Iron concentration of over 4 gpl will only allow current efficiencies of below 80%, with the current efficiency progressively rising to above 90% as the iron concentration falls to the 1 to $1\frac{1}{2}$ gpl range. Iron removal by purification also has an effect on the preceding leaching operation, for if purification is not carried out leaching must necessarily be done at a lower acid concentration to minimize the amount of iron taken into solution, and this lower acid concentration will impair the copper extraction as an unwanted side effect.

The pregnant solution going to the precipitation circuit will contain from 30 to 50 gpl copper in solution, of which approximately half will be deposited at the cathodes on its cycle through the electrolytic cells, with the partially depleted electrolyte, which has now increased in sulfuric acid enrichment some 35 gpl as the copper was removed, then being pumped back to the leaching circuit to be reused there as fresh dissolving liquor:

$$CuSO_4 = Cu^{2+} + SO_4{}^{2-}$$

$$SO_4 + H_2O = H_2SO_4 + \tfrac{1}{2}O_2$$

The leach solution, after clarification by thickening and filtering, is stored in lead-lined concrete tanks, prior to being pumped through lead or plastic pipe to the precipitation plant. The electrolytic cells are arranged in units and sections, with a typical arrangement of six cells to a section and 24 sections to a unit. Each cell has 41 anodes and 40 cathodes connected in a multiple system, with cell dimensions of 15 feet long, 4 feet wide, and 4 feet deep (4.57 × 1.22 × 1.22 m). Direct current is supplied from rectifiers to give a current density at the cathode of 16 amperes per square foot (0.09 m^2) at full load. The sections of each unit and the cells of each section are connected in series electrically.

The cells are of reinforced concrete lined with antimonial lead sheet, asphalt mastic, or fiberglass-reinforced polyester; polyethylene has largely replaced lead in all the pipelines of diameter under 4 inches (10 cm).

Starting sheet cathodes are of pure copper, 3 feet by 3 feet (0.09 × 0.09 m), $\frac{1}{4}$ inch thick (6.25 mm), and weighing 11 or 12 pounds (5 or 5.45 kg) each. These are supported in the cell by two welded-on sheet copper loops through which can be placed a copper hanger rod with its contact end resting on a cell bus bar and the other end on a rubber insulator. The anodes of antimonial lead are the same size as the cathode starting sheets, are made with cast-in support bars for suspension and electrical contacts, and weight 300 to 500 pounds (136 to 227 kg). Porcelain insulators at the corners and center prevent their touching the cathodes and short circuiting them.

The clarified, and frequently purified, leach solution is mixed with concentrated sulfuric acid, and sometimes a portion of spent electrolyte, to form the feed solution to the electrolytic cells. This higher acidity permits operation at lower cell voltage, and the resulting final mixture contains 45 to 55 gpl acid and 30 to 50 gpl copper. Flow of feed solution can be individually to each cell or to parallel tanks in a section which are arranged in a cascade and is initially by gravity flow from a storage tank.

A six day cathode cycle is common, and this would have the cathodes removed after six days to be replaced with fresh copper starting sheets. The lead anodes are left in position as they are inert in the cell and used only to carry current. The cathodes, 99.96% copper, are melted and cast into commercial shapes for market.

Low-Value Pregnant Solutions until quite recently were treated only by cementation to precipitate and recover their copper content. However there is now an alternative method of processing in that the dilute values in leach solutions of only 2 gpl copper can be concentrated to 40 to 45 gpl by solvent extraction. This solution strength now being the equal of a high-value pregnant solution, precipitation can be by electrowinning (electrolytic precipitation).

1. Cementation is an inexpensive treatment method which makes economically feasible the extraction of small copper contents from low-value pregnant leach solutions or insoluble-residue wash solutions that may contain as little as $\frac{1}{2}$ to $2\frac{1}{2}$ grams per litre of copper.

Scrap iron is still almost universally used for cementing copper from dilute solutions and represents the largest cost item of the process. Most light-gauge sheet steel, which has the largest reactive surface for its weight, is lacquered or soldered and must be burned to remove this and then shredded, which preparation adds to the metal expense. Heavy scrap is less efficient because of its relatively small surface and frequent high carbon content. Sponge iron is being used instead of scrap iron in some plants and has been coming into use on an increasing scale. Sponge consumption is similar to that of scrap, based on the content of metallic iron in both materials.

The scrap consumption varies considerably from plant to plant, depending largely on the ferric iron content of the pregnant solutions. All ferric iron must be reduced to the ferrous state, and much of any free acid must be used up before copper precipitation can take place. The overall reactions are

$$CuSO_4 + Fe = Cu + FeSO_4$$

$$H_2SO_4 + Fe = H_2 + FeSO_4$$

$$Fe_2(SO_4)_3 + Fe = 3FeSO_4$$

all taking place simultaneously. The cementation of the copper is the only profitable reaction, with the other two representing a necessary operating loss of iron; so that while theoretically one pound (0.454 kg) of copper should be replaced in solution by 0.89 pound (0.404 kg) of iron, in practice a good working figure for solutions containing ferric iron is two pounds (0.908 kg) of scrap per pound (0.454 kg) of copper precipitated.

Pyrrhotite has been found in tests to be a very efficient reducer of ferric iron, and exposure of pregnant solutions to lump pyrrhotite for a few hours has reduced ferric to ferrous iron, with an accompanying reduction in scrap consumed during cementation of some 35%

$$Fe_7S_8 + 32H_2O + 31Fe_2(SO_4)_3 = 69FeSO_4 + 32H_2SO_4$$

The purity of the cement copper being precipitated depends to a considerable degree on the cleanliness of the scrap iron being used, and dirty, oxidized scrap will contaminate the cement copper. The purity of the copper dictates where it can be added to the process flow sheet, with badly contaminated material being returned to the smelter reverberatory furnace or converter, and high purity precipitate being melted in the anode furnace to produce electrolytic anodes. The best iron scrap goes to the final, scavenging sections of the cementation plant to strip the copper still in solution to the lowest possible figure, which will leave some 60 mg per litre unprecipitated. This stripped solution is recycled to the leaching circuit to be reused as high-acid leach solution.

Several types of equipment are used for cementation, some quite ancient and others just recently developed. Of the methods commonly in use, cones, flat launders, V-troughs, tanks, and cells all operate on the same basic principle, and only the form of equipment used differs.

a. TANK-AND-CONE PRECIPITATORS, developed by Kennecott Copper, are one of the most recent and efficient precipitators, giving a 90 to 95% copper product, using less scrap for each pound of copper precipitated, and recovering 99% of the copper precipi-

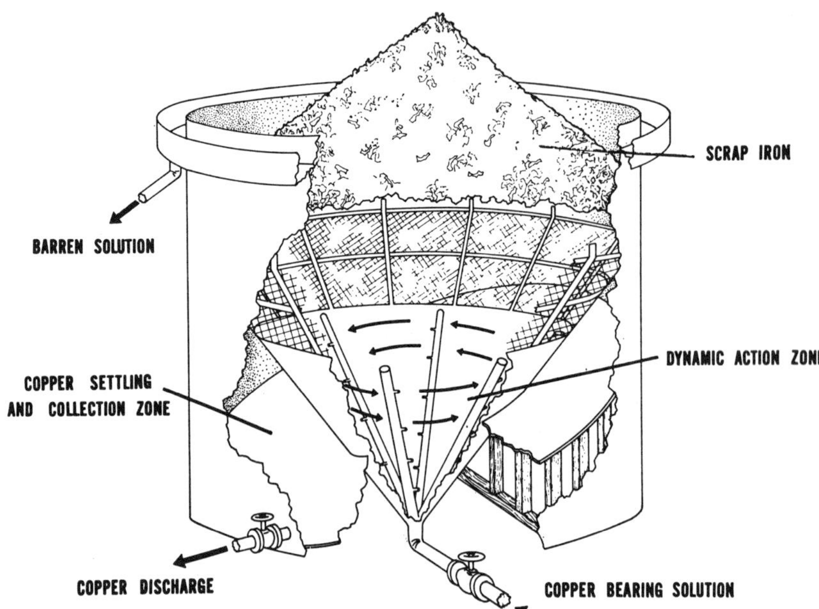

Figure 2.16. Cone-type cement copper precipitator. *Source: J. Metall.*, Vol. 18, No. 10, 1966, p. 1139.

tated, and recovering 99% of the copper from solutions containing 0.4 to 4.0 gpl, in a high capacity system featuring automatic controls and mechanized material handling.

The amount of scrap iron consumed is decreased, and the capacity of cement copper output is increased due to the fact that of the three chemical reactions important to copper cementation, it has been found that one,

$$CuSO_4 + Fe = Cu + FeSO_4$$

has a much faster reaction rate if a high surface area of iron is used and that this reaction essentially will be completed before the other two reactions, which both consume large amounts of iron,

$$H_2SO_4 + Fe = H_2 + FeSO_4$$

Figure 2.17. Top view of cone-type copper precipitators showing conveyor system and charged cones. *Source:* Courtesy of Kennecott Copper Corp. Photograph by Don Green.

and

$$Fe_2(SO_4)_3 + Fe = 3FeSO_4$$

have proceeded any great distance toward their own equilibria.

The precipitator consists of a 14 foot diameter wooden tank (4.27 m) 24 feet high (7.32 m), which contains at the bottom an inverted stainless steel cone 11 feet high (3.35 m). Mounted on top of the cone is a conical screen, $4\frac{3}{4}$ feet high (1.45 m) and 1 inch mesh (2.5 cm), which continues the taper of the cylindrical steel cone until it meets the wall of the wooden tank, about two thirds of the way up the tank.

The cone is filled with shredded, detinned, delacquered scrap iron, and pregnant solution is pumped into the bottom. Cementation takes place, and the high-velocity swirling motion of the solution in the cone washes the precipitated copper from the surface of the scrap iron and carries it upward to where it falls through the screen at the top of the cone and accumulates at the bottom of the wooden tank to be removed later. The partially decopperized solution overflows the top of the tank and is pumped to the bottom of a second precipitator where the action is repeated. Spent liquor from this second precipitator is returned as leach solution to the leaching operation.

A cone will process about 2500 gpm (11,250 lpm) of leach solution.

b.   CELLS AND TANKS are used in a series of several (six to ten) placed so that there is a drop in elevation to allow a gravity flow of solution through the cells. The cells are of acid-proofed wood or concrete, 8 feet wide, 5 feet deep, and 12 feet long (2.44 × 1.52 × 3.66 m), with a screen grid holding the scrap iron off the cell bottom to permit good solution circulation. The precipitated copper is recovered by taking a cell out of the circuit and washing the adhering copper off the scrap iron with high-pressure hoses down through the bottom screen and out through a valve into a settling tank or vacuum filter. Periodically the settled copper is dug out of the settling tank, dried on a pad, and shipped to a smelter. This product is some 82% copper.

The first two cells in the circuit always receive the highest-value pregnant solution and will precipitate about 60% of the total copper production. The next pair of cells remove 20% more of the total recovery, and the remaining cells remove the rest. The cells that are most active must be washed more frequently and have scrap iron added at shorter intervals.

Figure 2.18. Side view of cone-type copper precipitators. *Source:* Courtesy of Kennecott Copper Corp. Photograph by Don Green.

c. LAUNDERS are one of the oldest pieces of equipment used for precipitation and consist of flat-bottomed acid-proofed concrete or wooden troughs 32 inches high by 32 inches wide (80 × 80 cm) and up to 1600 feet long (488 m). A wooden lattice with $\frac{1}{4}$ inch (6.25 mm) openings is suspended 17 inches (42.5 cm) above the bottom of the launder, and scrap iron is piled on this. Pregnant solution filling the launder then slowly flows through by gravity, and as much as 97 to 98% of the copper content may be cemented out. Periodically the launder is drained and the cement copper hosed off the scrap iron, down through the wooden lattice and out the end of the launder, where it is settled from the wash water and collected. This cement copper is returned to the smelter, while the decopperized pregnant solution is recycled back to the leaching circuit.

New V-trough precipitators use sponge iron to precipitate the major portion of copper in solution, followed by scrap iron to cement out the remaining copper from the partially decopperized solution. The V-trough precipitator design minimizes the caking and blinding problems formerly encountered when using fine sponge iron.

2. Solvent Extraction and Electrowinning is the new alternate choice to cementation for the treatment of low-value pregnant solutions and is based on relatively new solvent ion exchange processing. Variously called SIX (solvent ion exchange) or LIX (liquid ion exchange), the technique which was first used in a commercial copper circuit in 1968 is becoming a major production method and in 1977 was producing over 200,000 tpy of high-purity copper.

Since 1968, nine SIX-type copper plants have started production, with all of the plants now in operation using General Mills LIX series extractants. The largest of these operations using a SIX circuit are the Nchanga Consolidated Mines Chingola plant in Zambia, with a 90,000 tpy capacity, and Codelco's 36,000 tpy operation in Chile, which was scheduled to go on stream in 1977. Numerous other new SIX plants, in addition to those already in operation, are being planned for Zambia, Zaire, Peru, Europe, Japan, and the United States.

The first United States plant to utilize SIX electrowinning in a commercial copper circuit was the 7000 tpy Bluebird plant of the Ranchers Exploration and Development Corporation at Miami, Arizona. The largest in North America is the new 10,000 tpd copper oxide processing plant of the Anamax Mining Company at Twin Buttes, Arizona, which was brought into production in 1975.

Table 2.4.   Recent Ion Exchange Plants for Copper Extraction

Company	Location	Yearly Capacity (tons)	Start-Up	Extractant
Ranchers Exploration and Development	Arizona	7000	1968	12% LIX 64N in 80% kerosene
Cyprus Bagdad	Arizona	7000	1970	LIX
Nchanga Consolidated Mines	Zambia	90,000	1973	LIX 64N
Anaconda (Arbiter)	Montana	36,000	1974	LIX 65N
Cyprus Johnson	Arizona	5000	1975	LIX
Anamax	Arizona	30,000	1975	14% LIX 64N in kerosene
Cities Service	Arizona	7000	1976	LIX
Minero Peru	Peru	30,000	1977	LIX
Codelco	Chile	36,000	1977	LIX

The solvent extraction–electrowinning operations at these United States plants are generally alike, and the solvent extraction process that is carried out is essentially similar to that which is described in detail for the processing of uranium pregnant solutions in the chapter on the hydrometallurgical treatments of the reactive metals. The recovery of the metal values by this liquid-liquid method of ion exchange is normally done at a very economical rate with a minimum of capital investment, and after concentration by solvent extraction the metal bearing solution is sent to the tank house to remove its copper content by electrowinning.

To optimize solvent extraction, the pH of the Anamax tank leached pregnant solution is adjusted from 1.5 to 2.5 by adding unreacted ground slurry from the grinding circuit. Then the solution is clarified to about 4 to 10 ppm of suspended solids and sent to the extraction circuit.

Initially a selective organic solvent, in this case 14% LIX 64N in 86% kerosene (the "organic"), immiscible with water, is intimately contacted countercurrently in mixer units with the copper pregnant leach solution, and the copper passes from the aqueous leach solution into the "organic" solvent. The immiscibility of the organic solvent with

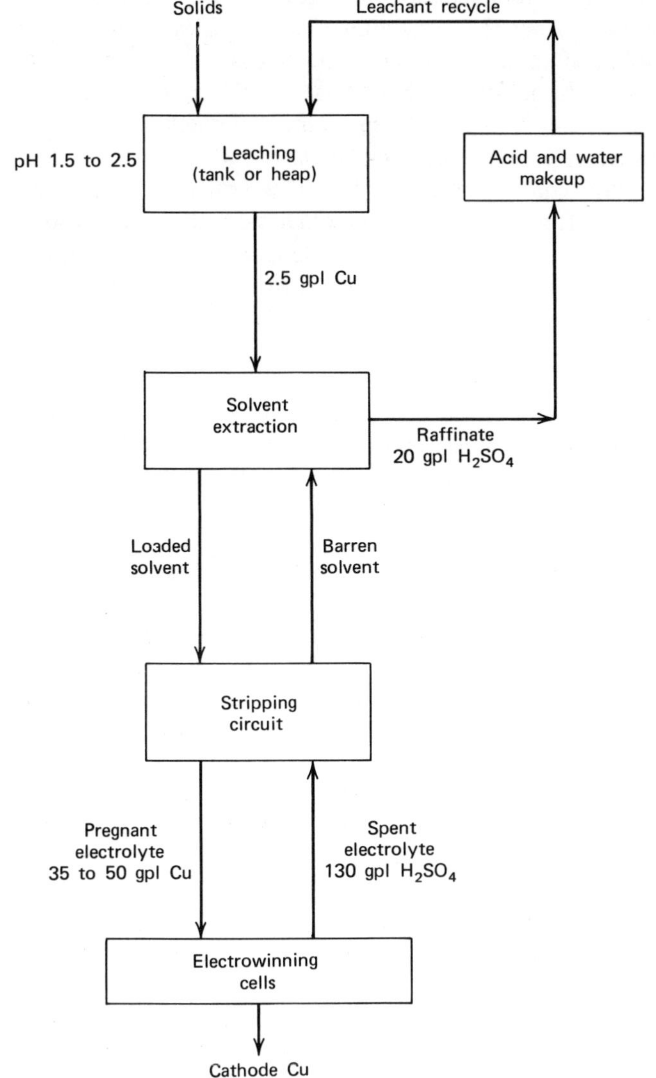

Figure 2.19. Copper leach processing–solvent extraction followed by electrowinning.

water permits a separation of the two phases (water and organic in kerosene) in the settler units, and the copper values are now concentrated into a much smaller volume of organic solution. This operation is known as the extraction circuit. The barren solution left after extraction is called the "raffinate."

The loaded solvent solution is next sent on to the stripping circuit to recover the copper values by stripping them from the organic extraction liquor with spent electrolyte and are now in a much more concentrated form suitable for treatment by electrolytic precipitation. The stripped organic is returned to be used in the extraction circuit.

The Ranchers' Bluebird plant collects pregnant leach solution from the leached heaps into a butyl rubber-lined pond, and this pond serves a five-vessel liquid solvent extraction exchange train in which three vessels are on extraction and two on stripping, with all having a counterflow of organic and aqueous phases. The much larger Twin Buttes plant splits the pregnant liquor into two streams, each of 3560 gpm, to go on to two parallel extraction trains of mixer–settlers, which have four extraction and two stripping mixer–settlers per train.

The pregnant solutions contain about 2.5 gpl copper for the Anamax Twin Buttes plant, and after passing through the respective solvent extraction's extraction and stripping circuits, the concentration of copper has increased to 50 gpl.

The raffinate, the barren solution from the extraction circuit, carries about 20 gpl H_2SO_4, along with a small amount of copper (0.08 gpl), and this is pumped to a surge tank where makeup water is added to adjust for small losses in the system, plus any makeup acid that may be required, and is then recycled to the countercurrent washing thickeners which are washing the slurry from the leaching tanks.

The electrowinning plant is quite conventional and generally similar in design and operation to that already mentioned for processing copper oxide sulfuric acid leach solutions. Copper is removed from the strong electrolyte (50 gpl copper, 90 gpl H_2SO_4), and acid is regenerated to provide a high-acid (130 gpl H_2SO_4) spent electrolyte for return to the stripping circuit mixer–settlers:

$$2CuSO_4 + 2H_2O = 2Cu + 2H_2SO_4 + O_2$$

The electrolytic process takes place in PVC-lined concrete cells, with wooden floors, so that any leakage in the PVC liner is easily spotted from the floor below. The anodes are sheet lead with 10% antimony, the copper cathodes starting sheets are 3 feet wide by 4 feet high (0.9 $\times$ 1.2 m), and the current density is 15 amperes per square foot (0.09 m^2).

The Twin Buttes plant has four electrolytic circuits in its tankhouse, these being (i) the starting sheet circuit with 36 cells, (ii) the first commercial circuit with 36 cells, and (iii) and (iv) the second and third commercial circuits each with 72 cells. Each circuit is arranged in a cascade and is equipped with its own recirculation tank, pumps, and head tank. Incoming strong electrolyte displaces solution from one cell to the next, and overflow piping in the recirculation tanks is arranged so that fresh feed to any circuit must pass through the cells before it can overflow to the next circuit.

Strong electrolyte from the stripping circuit must be virtually free of entrained organic material because of the adverse effect that this organic can have on electrowinning. Consequently the electrolyte is cleaned in a coalescer before circulating to the tankhouse cells.

The Twin Buttes coalescer is a two-compartment unit with a 186,000 gallon (837,000 litres) first-compartment feed reservoir, in which entrained organic material surfaces and is decanted off. A bottom outlet in the reservoir delivers strong electrolyte to a second compartment, where it is pumped through two coalescing units. The organic that collects in the dome of each of these coalescing units is tapped at intervals and returned to the mixer–settlers. Cleaned electrolyte discharges from the coalescers into the tankhouse feed reservoir for processing in the electrolytic cells.

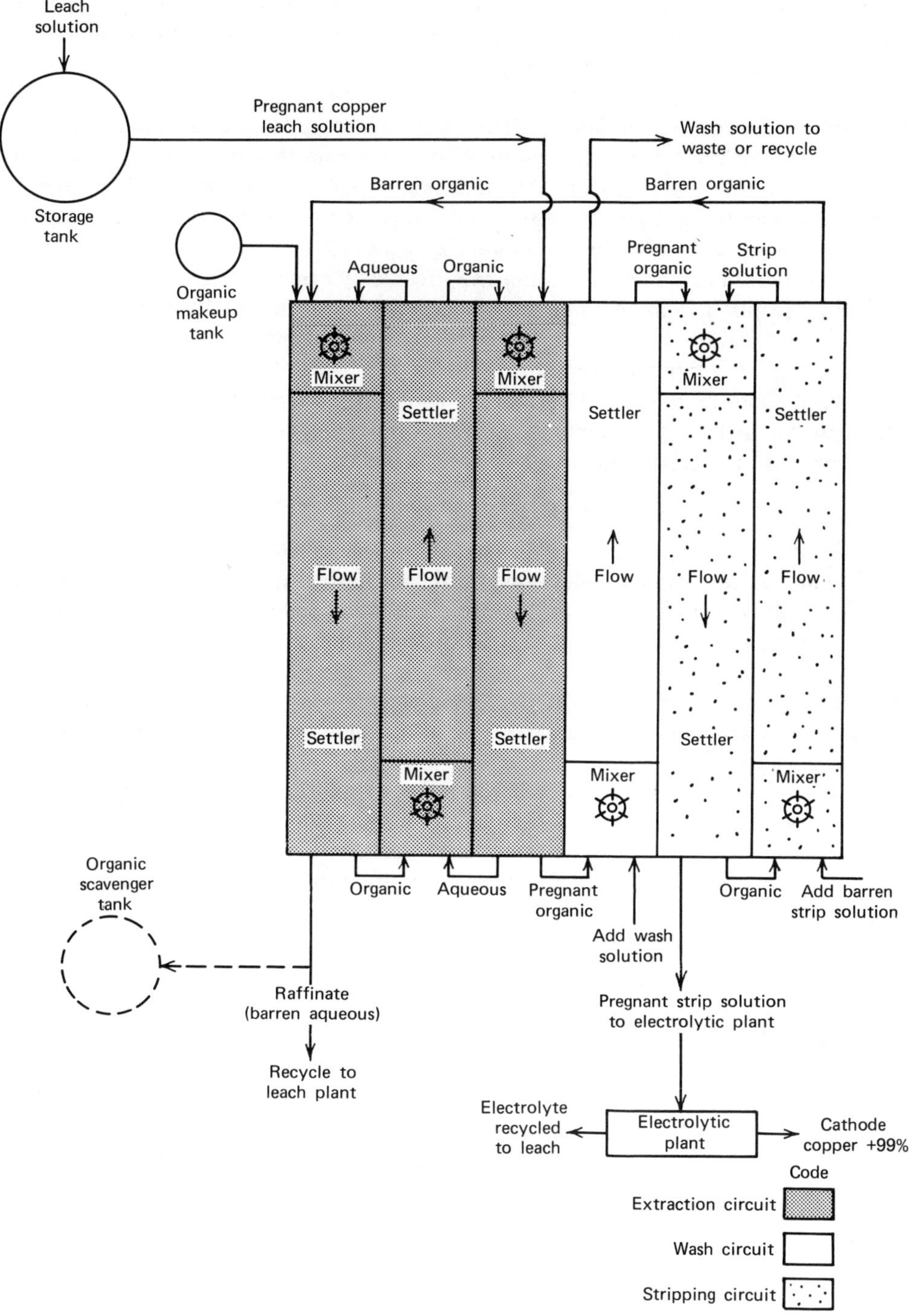

Figure 2.20. Solvent extraction of copper from leach liquor. *Source:* Denver Equipment Co. Deco Trefoil, Bull., M7-F99, Nov.–Dec., 1965, p. 21.

The new Miami, Arizona, plant of the Cities Service Company takes a lower-grade leach solution (only 0.85 gpl) into its SIX circuit, and after processing the liquor in an operation which has six extraction and four organic stripping vessels, the solution is upgraded to 35 gpl of copper and sent on to a 60 cell electrowinning plant. Total capacity output of this plant is rated at 1 million pounds cathode copper per month.

In the *Arbiter* ammoniacal leach process, solvent extraction is also carried out, using LIX 64N in a high flash-point carrier of distillate, and yields an electrolyte of copper sulfate for electrowinning, with the precipitation circuit returning an acidified spent electrolyte to the solvent extraction stripping stage. The residues from the ammonia leach circuit are washed with raffinate and then processed by froth flotation to recover small amounts of undissolved copper and precious metals.

1b. Copper Sulfide leach solutions are for the most part treated in exactly the same manner as are copper oxide leach solutions, for the reason that in atmospheric leaching the sulfides are first oxidized to soluble sulfates and oxides which are readily dissolved in the leaching solvents and give pregnant solutions similar to those resulting from copper oxide leaching. These high-value solutions are treated by electrical precipitation, while the low-value solutions and wash solutions are treated by cementation.

The one exception from an oxide precipitation method is the precipitation of the pregnant solution resulting from the ammoniacal pressure leaching of mixed nickel, copper, and cobalt sulfide flotation concentrate.

This leach solution contains 7 gpl copper and 150 gpl ammonium sulfate; and with heating to 250°F (121°C), ammonia is driven off leaving sulfur ions to combine with cupric ions and form copper sulfide which precipitates:

$$Cu^{2+} + S_2O_3^{2-} + H_2O = CuS + 2H^+ + SO_4^{2-}$$

The precipitation operation is carried out in closed pot stills, with heating to the required temperature provided by injecting steam. Mechanical agitators keep the precipitating copper sulfide in suspension until the reaction is complete, and it is then filtered and washed in pressure filters and shipped to a copper smelter to recover the copper content.

1c. Native Copper and Copper Carbonates are precipitated from ammoniacal leach solutions as an adjunct to the operation of driving off the ammonia and recovering it for reuse in the leaching circuit.

The cuprous ammonium carbonate in the pregnant solution is broken down to release the ammonia, which is recovered separately, by heating with steam to above the boiling point in a batch operation. The pregnant leach solution is treated in steel cylinders, 10 feet in diameter (3.05 m) and 16 feet high (4.88 m) with dome tops and cone bottoms, until the ammonia level is reduced to less than 0.01%. By this time the ammonia has been removed and the copper has all been changed to, and precipitated as, black copper oxide:

$$Cu_2(NH_3)_4CO_3 + H_2O = 2CuO + 4(NH_3) + CO_2 + H_2O$$

The charge in the steel reaction vessel is now pushed out by steam pressure through lines connected to the cone bottom to a filter where the precipitated copper oxide is recovered, to be dried and shipped to market. Extraction is on the order of 80%.

2a. Nickel Laterites are leached in two widely different methods. One is an ammoniacal leach at atmospheric pressure which produces a basic nickel carbonate when precipitated from the leach solution. The other leaching method is in sulfuric acid at high temperature and pressure, which gives nickel sulfide as the precipitated value.

High-Magnesia Laterites are treated by a basic ammoniacal leach, the *Nicaro process*, as the large amount of basic waste makes it economically unrealistic to waste excessive amounts of leaching acid to neutralize this high magnesia content.

The clarified pregnant solution from the leaching circuit thickener still contains some ore fines and traces of iron. This iron is precipitated by briefly agitating the solution with air in aerating tanks to convert the ferrous iron to ferric hydroxide, which is filtered off, along with the ore fines, in a filter press.

The clear liquid remaining is pumped to a parallel row of stripping stills, cylindrical multitray steel towers 50 feet high and 7 feet in diameter (15.2 × 2.13 m), in which the solution flows downward. The stripping steam, which carries off the ammonia to a reagent recovery system, flows upward. As the descending solution gradually loses its complexing ammonia, freed nickel begins to precipitate:

$$Ni(NH_3)_2 = Ni + 2NH_3$$

The nickel ions produced combine with hydroxyl and carbonate ions in solution to form insoluble nickel carbonate:

$$5Ni + 6OH + 2CO_3 = 3Ni(OH)_2 \cdot 2NiCO_3$$

The slurry containing this nickel carbonate is separated from the remaining liquid by thickening and vacuum filtering, and the resulting cake is dried and calcined to nickel oxide in a rotary kiln:

$$3Ni(OH)_2 \cdot 2NiCO_3 = 5NiO + 3H_2O + 2CO_2$$

This oil-fired kiln, 12 feet in diameter (3.66 m) and 160 feet long (48.8 m), is heated to 2450°F (1343°C) at its hottest point 18 feet (5.49 m) from the feed end of the kiln and produces a calcine about 76.5% nickel. A cyclone dust collector and an electrostatic precipitator collect the dust in the gases leaving the kiln and recycles it back into the kiln feed.

Low-Magnesia Laterites are treated by a sulfuric acid leach, the *Moa Bay process*, at high temperature and pressure. An acid leach can be used as there is no large amount of basic waste to react with the acid and enormously increase the amount of leaching solvent required.

The leach slurry is thickened after having been brought down to atmospheric pressure. The thickener overflow, after pretreatment with hydrogen sulfide gas to reduce ferric to ferrous iron and flocculate any fine ore solids, goes to the neutralization circuit. Here, in a series of agitator-equipped wooden tanks, the pregnant sulfuric acid solution is neutralized with a slurry of coral mud analyzing over 90% calcium carbonate. The reaction is rapid, and $CaSO_4 \cdot 2H_2O$ is precipitated in an amount equivalent to the acid available. This precipitate is removed as thickener underflow to be discarded as tailings, while the clear thickener overflow is further treated to precipitate the nickel and cobalt contents.

The liquor is preheated to 170°F (77°C) by steam in the first stage of heating. It is then heated in a second stage to 245°F (118°C), again with steam, and pumped into horizontal cylindrical autoclave precipitators at 150 psi (1034 kPa). The autoclaves are lined with acid-proof brick and divided into three compartments, each with an impeller for agitation. High-purity gaseous H_2S is injected to maintain the 150 psi (1032 kPa) pressure.

In the precipitation, nickel and cobalt combine with hydrogen sulfide to form sulfides and sulfuric acid, with 99% of the available nickel in solution and 98% of the cobalt reacting:

$$NiSO_4 + H_2S = NiS + H_2SO_4$$

The batch reaction ultimately stops, and the resulting slurry is blown into a flash tank to release the pressure. The H_2S gas released is recovered and recycled, while the liquid carrying the precipitates of nickel and cobalt flows to a 60 foot diameter thickener (18.3 m), with the precipitate underflow from this thickener passing on to two further hot-water washing thickeners before finally being discharged as a 65% solids slurry.

Ninty-nine percent of the nickel in the leach solution is recovered as precipitate, for an overall recovery of 95% of the nickel available in the ore.

2b. Nickel Sulfide leach solutions are the result of two quite different processes, and as a consequence their pregnant solutions receive quite different treatments to precipitate the contained values. In one case a sulfide flotation concentrate is pressure leached in ammonia, and then the pregnant solution is reduced with hydrogen gas to recover nickel metal as a powder. In the other case a nickel matte is acid leached to obtain a solution to be used for electrolytic precipitation.

Flotation Concentrate, as treated by Sherritt Gordon, after an ammoniacal pressure leach to take the nickel, cobalt, and copper in the concentrate into solution has the copper and ammonia removed from the resulting pregnant solution by heating it in pot stills. The ammonia is driven off as a gas upon heating, leaving free sulfur ions from ammonium sulfate to combine with copper as copper sulfide and precipitate as such, then to be removed by filtering.

The filtrate from the copper stripping circuit analyzes 46 gpl nickel and 0.9 gpl cobalt and is pumped on to the metals recovery plant, where a preliminary operation preceding nickel precipitation eliminates unsaturated S_2O_3 and S_3O_6 (thionates and thiosulfates) and ammonium sulfamate, $NH_4SO_3NH_2$, which could give sulfur contamination of the nickel powder endproduct. These are eliminated by oxidizing the unsaturates and hydrolyzing the sulfamate to form the more stable sulfate ion. This is accomplished by heating the solution and holding it at an elevated temperature and pressure in an autoclave under oxidizing conditions. The overall process is called oxydrolysis.

The oxydrolysis autoclave is a 3000 gallon capacity (13,650 litres) four-compartment unit, equipped with heating coils and an agitator in each compartment. The solution before reaching the autoclave is first heated in a preheater to 250°F (121°C) and then heated in a heat exchanger to 425°F (219°C) before going into the autoclave where the temperature is raised still further to 475°F (246°C) and air admitted to maintain an oxidizing pressure of 600 psi (4134 kPa).

The thionates and thiosulfates react earliest, beginning to combine to form sulfates with oxygen, water, and ammonia in the preheating stages, and finish reacting in the autoclave:

$$S_2O_3^{2-} + S_3O_6^{2-} + 4O_2 + 3H_2O + 6NH_3 = 6NH_4^+ + 5SO_4^{2-}$$

The sulfamate starts to hydrolize rapidly in the autoclave at about 450°F (232°C) and forms ammonium sulfate:

$$NH_4SO_3NH_2 + H_2O = 2NH_4^+ + SO_4^{2-}$$

A retention time of 20 minutes in the autoclave is sufficient to bring about the required reactions, and when this is complete the solution is discharged and cooled to 400°F (204°C) by running it through the feed solution preheater.

The conditioned nickel-bearing solution is pumped to an autoclave feed tank, which is maintained at 400°F (204°C) and 220 psi (1516 kPa) and is metered in batches from there into the precipitation autoclaves, which operate at 400°F (204°C) and 350 to 450 psi of hydrogen (2412 to 3100 kPa). Fine nickel particles in the autoclaves act as seed for nickel precipitation. The nickel, reduced by hydrogen from the ammoniacal leach solution, deposits on these nuclei:

$$Ni(NH_3)_2SO_4 + H_2 = Ni + (NH_4)_2SO_4$$

After reduction is completed with a charge of leach solution, the autoclave agitators are shut off, the nickel powder is allowed to settle, and the spent solution is withdrawn. A new batch of solution is then charged from the feed tank and the operation is repeated. This cycle of successive batch precipitations is continued until the nickel powder has been built up to the required amount (50 batches); and at this time the entire nickel charge is withdrawn to flash tanks as a 95% solids slurry. The powder is washed, filtered, and dried and can be shipped as is or can be briquetted.

Overall nickel recovery is on the order of 90 to 95%, and nickel purity is 99.9%.

The discarded spent solution still contains some nickel and most of the cobalt (0.8 gpl of each), and these are recovered by precipitating with hydrogen sulfide at atmospheric pressure in a secondary circuit.

Matte of low sulfur composition, as treated at the Outokumpu Harjavalta plant, is sulfuric acid leached to dissolve the nickel, copper, and cobalt, with the resulting pregnant solution purified by first cementing and then electrodepositing the copper. Then the cobalt is removed as hydroxide by precipitation with black nickel hydroxide:

$$Ni(OH)_3 + Co = Ni + Co(OH)_3$$

The now purified nickel solution, of 75 gpl, is preheated to 145°F (63°C) and has the pH adjusted to 3.2. Additions are made of sodium sulfate and boric acid, and the electrolyte is pumped to the cathode compartments of the electrolytic cells. A conventional nickel cell is used, with a tightly woven terylene bag around the nickel starting sheet cathode, and a $\frac{1}{2}$ inch (12.5 mm) solution head is maintained in this bag above the solution level in the anode compartment. Anodes are of rolled lead sheet. The average current density in the cell is 17 amperes per square foot (0.09 m^2), and the cell voltage is 3.5 volts.

The nickel contained in the electrolyte deposits 25 gpl of its nickel content at the cathode before flowing through the diaphragm terylene bag into the anode compartment. The anolyte solution also now contains 40 gpl free acid, as hydrogen ions have formed at the anode in chemically equal amounts to the nickel deposited at the cathode, and this high acid solution is run off and sent back to be used in the second stage of the preceding leaching circuit.

Finished cathodes average +99.95% nickel and weigh 165 pounds (75 kg).

3a. Cobalt Sulfides in Zambia are found in combination with larger amounts of copper sulfides. They are beneficiated to separate the two sulfides, after which the cobalt sulfide portion is concentrated by mineral beneficiation, roasted to the soluble sulfate form, and leached with warm water.

The pregnant solution contains 10 to 25 gpl cobalt, 7 to 10 gpl copper, and 0.3 gpl iron. Purification to remove the copper and iron is carried out first, by precipitation as hydrates. This is accomplished by adding milk of lime to the solution, along with vigorous air agitation in Pachuca tanks and raising the pH so that at the end of the operation it is 6.4 to 6.6. The precipitate is removed by thickening, which leaves a solution practically free of copper and iron.

The cobalt in the purified solution is now precipitated in agitated tanks as the hydroxide, $Co(OH)_2$, at pH 8.3 with milk of lime. The hydroxide pulp is thickened and filtered to remove the spent leach solution.

3b. Cobalt Oxides (Zaire) are taken into solution along with much larger amounts of copper oxides by dilute sulfuric acid leaching.

The pregnant solution is treated by electrolysis to recover the copper, which is the main metallic constituent, leaving the spent electrolyte containing 6 to 7 gpl cobalt. Some iron and copper are also still present, and these are first removed by selective precipitation with milk of lime, at pH 3.5 for the iron and pH 5.8 for the copper. If necessary, further copper removal can be carried out by cementing any still remaining copper on cobalt shot at a pH of 2.5.

The cobalt is finally precipitated as hydroxide from the purified solution, which contains about 15% cobalt, by milk of lime at pH 8.2. The cobalt is separated by filtering from the barren solution.

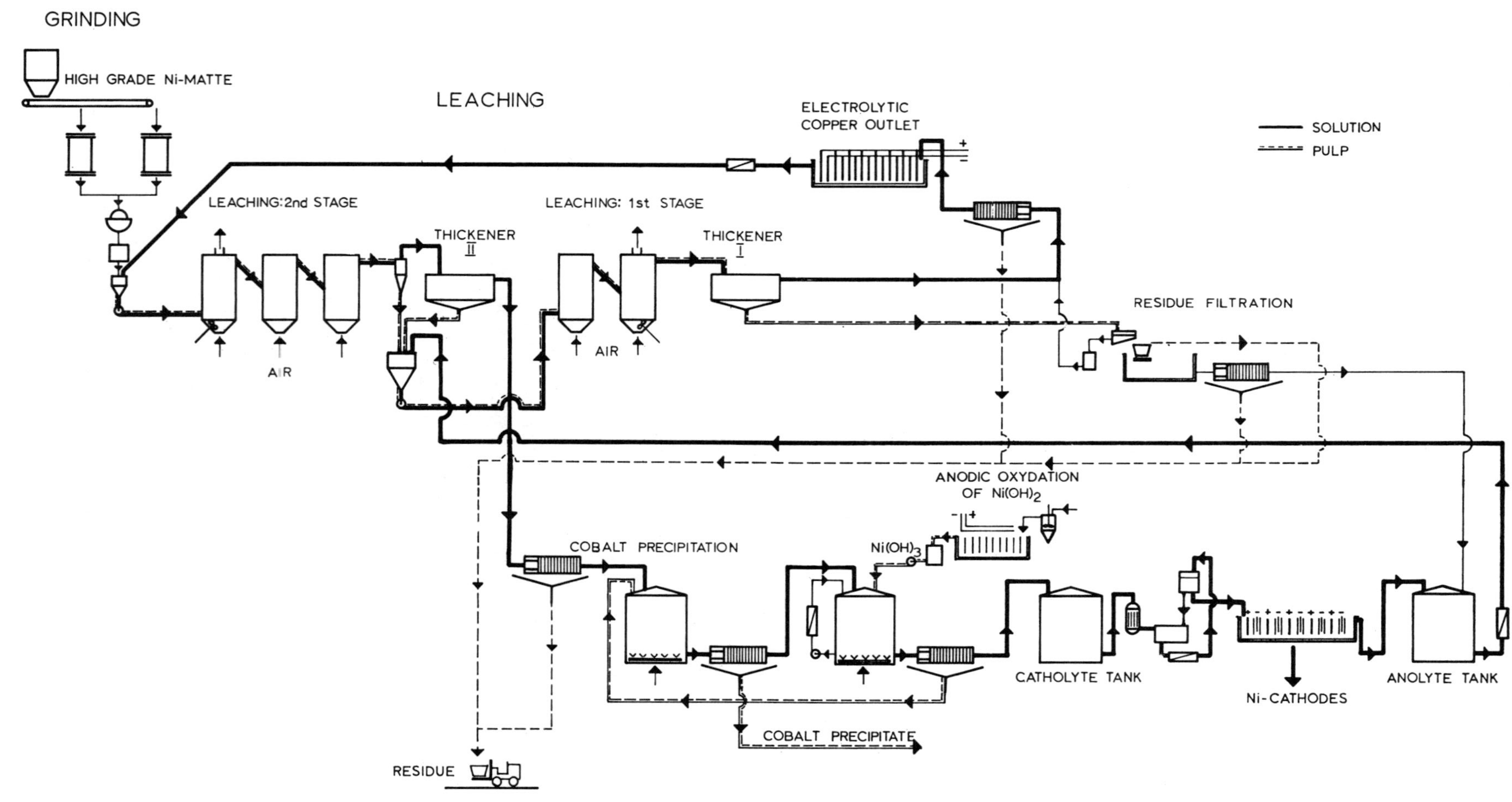

Figure 2.21. Electrolyte purification–nickel electrowinning flow sheet of the Harjavalta Nickel Refinery. *Source: Can. Inst. Min. Metall. Bull.,* Vol. 57, No. 626, 1964, p. 656.

4. Silver and Gold are leached in solution of sodium or potassium cyanide, with the pregnant solution then being filtered so as to be absolutely clear for precipitation.

Gold and silver are recovered from solution by the addition of zinc dust:

$$2NaAu(CN)_2 + Zn = Na_2Zn(CN)_4 + 2Au$$

$$2NaAg(CN)_2 + Zn = Na_2Zn(CN)_4 + 2Ag$$

There is also some tendency for Zn to unite with excess NaCN, which causes a larger consumption of zinc dust than is theoretically required to liberate the gold and silver present:

$$Zn + 4NaCN + 2H_2O = Na_2ZnCN + 2NaOH + H_2$$

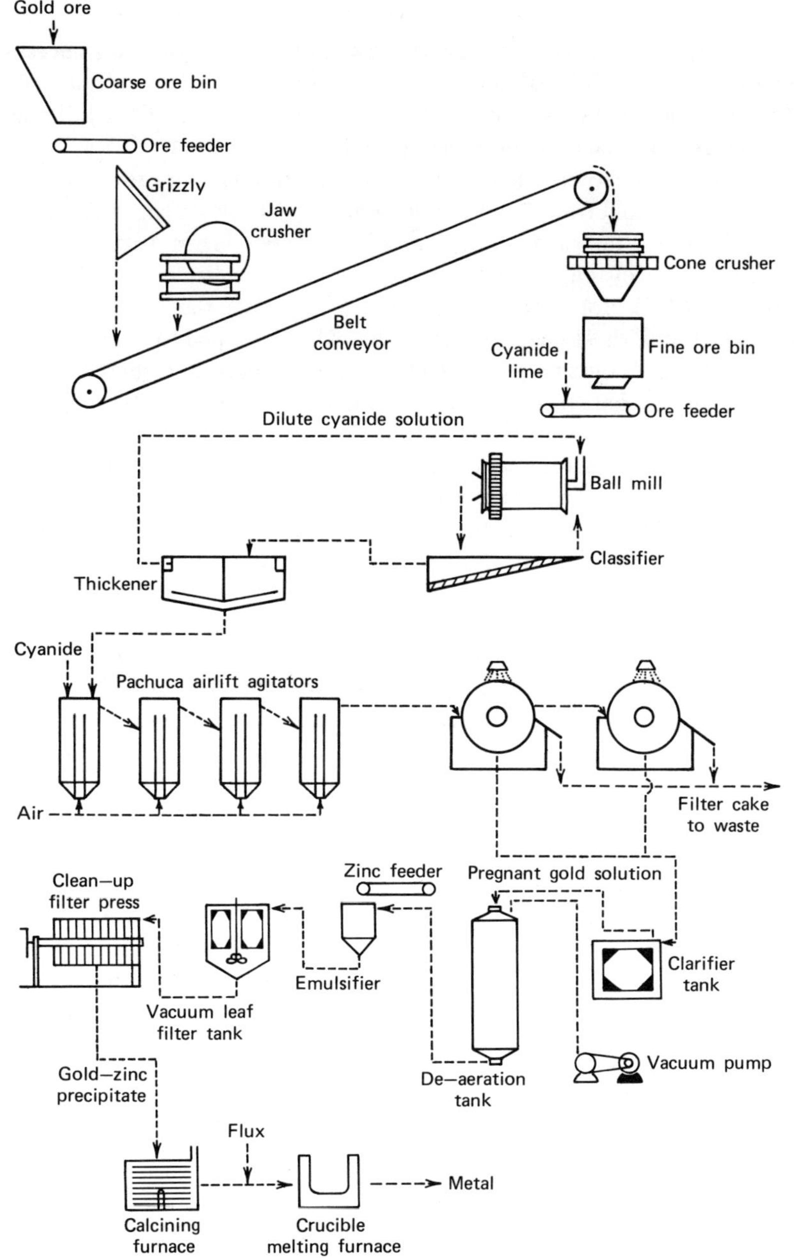

Figure 2.22. Leaching and precipitation of gold flow sheet. *Source:* W. H. Dennis, *Metallurgy in the Service of Man*, MacDonald and Janc's Publishers Ltd., London, 1961, p. 333.

It is essential that no oxygen be present in the solution to be precipitated, for any oxygen at this stage will combine with the fine zinc powder to form zinc hydroxide and zinc ferrocyanide, which interfere with the precipitation reaction and give a low-grade product. Removal of oxygen is accomplished by subjecting the solution, contained in a tank, to a vacuum, with the reduced pressure causing the oxygen (air) to bubble to the surface and be sucked away.

Zinc dust is then added to the deaerated solution, with intimate contact being made by adding the dust at the feed end of the suction pump. Precipitation is quite rapid, and the barren solution containing the particles of gold and silver in suspension, along with excess zinc dust, is pushed by the suction pump into a plate-and-frame press, where the solids are retained and the barren solution run off to be discarded.

Carbon-in-Pulp Recovery The Homestake Mining Company, the largest United States gold producer, was prohibited in 1970 by the U.S. Water Pollution Control Administration from further use of mercury in its plant because of downstream pollution. This necessitated a drastic change in procedural treatment, for until then, 60% of the gold had been recovered from the grinding circuit by amalgamation with mercury and the additional 35% from conventional cyanide leaching and precipitation.

The revised processing finally chosen was cyanide leaching followed by a carbon-in-pulp adsorption circuit to strip the gold from the pregnant solution. This new plant was started up in March, 1973, and reached full capacity to treat 2250 tons per day in August, 1973.

After leaching with sodium cyanide in a series of seven 30 foot diameter by 22 foot deep (9.0 × 6.6 m) air lift tanks for a total contact time of 20 hours, the gold in solution averages 0.06 ounce per ton with 91% recovery, and the pulp is pumped to the number 1 adsorption tank where it first contacts the activated carbon. Coconut carbon 6 to 16 mesh size, is used because it is harder and more resistant to breakage and abrasion than other carbons.

The adsorption circuit consists of four center airlift agitator tanks 18 feet in diameter by 16 feet deep (5.4 × 4.8 m). The leach pulp is moved through the tanks from number 1 to number 4, and the activated carbon is moved counter to the pulp flow. Complex ions of gold and silver cyanides are absorbed from the solution by the activated carbon.

Dissolved values are lowered as the pulp flows from tank to tank, the pulp flowing from tank 4 being the final tailing going on to waste. As the carbon is advanced from tank to tank, in countercurrent movement, the gold load on the carbon increases, and it is held in tank 1 until the load builds up to 300 to 400 ounces of gold per ton of carbon.

In the desorption circuit the carbon is first washed and screened over 24 mesh screens to clean and separate the loaded carbon from insoluble fines, and then the gold and silver are stripped from the carbon by a process in which the equilibrium between adsorption and desorption are reversed and gold goes into solution to be deposited on steel wool cathodes. Stripping is done by a solution of hot caustic cyanide at 190°F (88°C), with solution strength maintained at 1% sodium hydroxide and 0.2% sodium cyanide, in a stainless steel cone bottomed tank which holds about 1 ton of carbon.

When the carbon is sufficiently stripped of gold (below 5 ounces of gold per ton) the pregnant solution flows out through three electrolytic cells in series, where the gold and silver is deposited at the cathodes. Each cell consists of a fiberglass tank, a stainless steel screen anode, and a cathode consisting of about 10 pounds (4.5 kg) of medium-coarse steel wool, which provides a maximum surface area for deposition.

The electrolytic precipitation cells are connected in series, and the pregnant solution from the desorption circuit carrying the gold and silver flows from cell 1 through cell 2 and on to cell 3, from where the stripped solution is reheated and recycled back to the desorption tanks to be reused.

The cathodes in the electrolytic cells are switched to have optimum recovery and

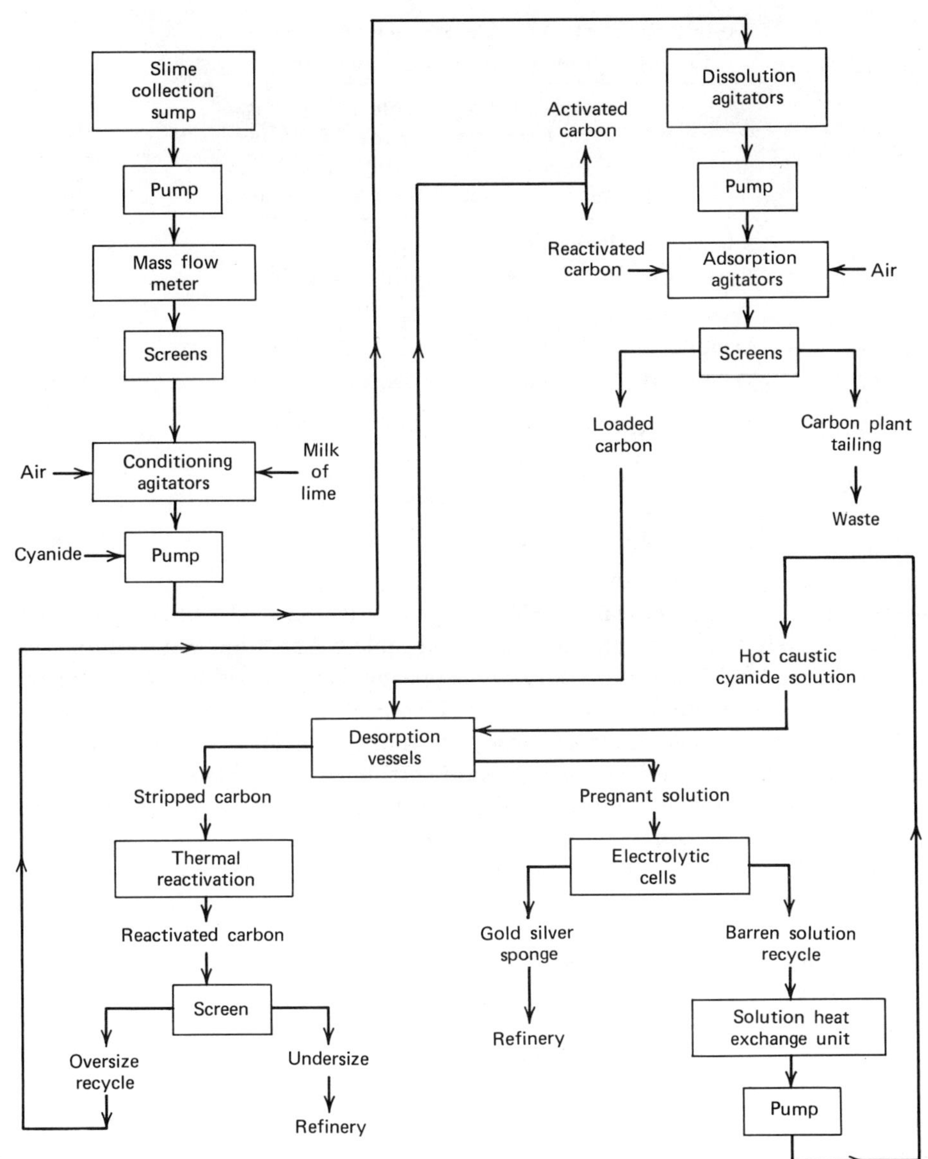

Figure 2.23. Flow sheet for Homestake Mining Company's carbon-in-pulp plant for gold recovery at Lead, South Dakota. *Source: World Mining*, Vol. 27, No. 11, 1974, p. 45.

deposition of the precious metals. The cathode in cell 1 is left in place until it builds up a load of about 1000 ounces of gold and 200 ounces of silver; then it is removed and sent to the refinery. The cathode from cell 2 then is moved to cell 1, the cathode from cell 3 to cell 2, and a new clean steel wool cathode is placed in cell 3. This ensures always having a heavily loaded cathode in the number 1 position for removal to the refinery and a very lightly loaded cathode in cell 3 for scavenging.

The stripped carbon after desorption is reactivated and then returned to the adsorption circuit. Reactivation is carried out by heating the carbon for $\frac{1}{2}$ hour at $1100°F$ ($593°C$) in a small gas-fired stainless steel rotary kiln, 12 feet 7 inches long by 20 inches in diameter (3.8×0.5 m). The carbon is dewatered to about 40% moisture before being fed to the kiln, and the feed end of the kiln is sealed so that the steam generated is exhausted from a flue at the discharge end of the kiln and prevents any air from entering. This reactivation drives off most of the fouling agents (ferrous iron–sulfur, sodium hydroxide, and organic compounds) that have been picked up by the carbon and restores its activity for reuse.

Table 2.5. Typical Profile of Gold Adsorption Circuit at Homestake Mining Company

Location	Gold Loading on Carbon (ounces gold per ton carbon)	Solution Values Tank Outflows (ounces gold per ton solution)
Adsorption plant feed		0.064
Tank no. 1	375.0	0.024
Tank no. 2	150.0	0.009
Tank no. 3	85.0	0.002
Tank no. 4	20.0	0.0005

Source: K. B. Hall, *World Mining*, Vol. 27, No. 11, 1974, p. 45.

REFINING

Refining is the last stage in the treatment of metals that have been extracted by a leaching process to selectively take them into solution, to precipitate them from that solution, and collect them in the precipitated form. Finally the precipitate is purified to the pure metal.

Not all precipitates require refining. Some are marketed as the chemical compounds in which they precipitate. Others are in an already high-purity metallic condition that can be used without further treatment. The only processing required in these two cases will be to dry the chemical compounds before shipping and to melt and cast, or briquette and sinter, the pure metal into convenient industrial shapes.

Metal precipitates are of higher metallic content than their counterpart chemical compound precipitates, and because of this often require less involved refining processes. Fire refining, chemical refining, and electrolytic deposition from an impure anode to a pure metal cathode are the types of processes most generally used.

TYPES OF REFINING

As refining is the final processing step of hydrometallurgical extraction, it is the one in which the metallic precipitate is treated to give the desired metal in a pure, commercially usable form.

Precipitates are collected in two quite different forms, which require quite different refining treatments. The more common of these forms is to have the precipitate as a *chemical compound*, sometimes relatively complex, which contains the desired metal as only a part of the compound. The other form is as *metal*, which can be quite pure and used as is without further treatment, or it may be contaminated and must be processed to remove the contaminants.

The metallic compound precipitates are oxides, sulfides, and hydroxides. The oxides are pyrometallurgically reduced with carbon to the metallic state and used in that condition, or they can be leached with sulfuric acid to give an electrolyte from which the metal can be deposited at a cathode. Sulfides are either returned to the smelter furnace or converter stages for pyrometallurgical recovery, or they can be leached in an autoclave with high temperature and pressure, and the metallic compound that is put into solution can then be reduced out as a pure metal precipitate by a gas such as hydrogen. Hydroxides are calcined to oxides and then reduced to metal with carbon in furnacing operations, or they can be dissolved in an electrolyte and electrolytically deposited from it.

The contaminated metal precipitate can be given a furnace treatment to drive off the volatile impurities and remove most of the others as a slag. This partially refined metal can then be cast into anodes and electrolytically refined, or it can be given a chemical treatment to dissolve the few impurities remaining and leave a pure metal residue. If the metal is quite badly contaminated, as happens in cementation, it may have to be sent back to the smelting or converting operations in order to recover the valuable metal from the impure precipitate.

High-purity metal precipitates are usually given a minimum of fire refining before casting into desired industrial shapes, but in some cases are marketed as precipitated, without further refining.

REFINING PROCESSES

1a. Copper Oxide leach solutions produce two different precipitates, with the high-copper pregnant solutions being treated by electrical precipitation and depositing high-purity copper on a cathode, while the low-copper pregnant solutions have their copper content cemented out by replacement of the copper ions in solution by iron. The refining procedures of these two precipitates will be different, with the cathode copper already in the pure state and not needing much further refining, while the cement copper can be either relatively pure or badly contaminated by the iron oxide and other impurities that are removed with the cement copper from the precipitation units. The percent purity determines where the cement copper can be returned to the process stream. Badly contaminated product is returned to the smelter melting furnace, and high-purity cement copper is added to the refinery anode furnace.

Cathode Copper from the electrolytic precipitation of high-value pregnant solutions is 99.96% pure. As such, some is sold without further processing for alloying. However most of the cathode copper is remelted and cast into wire bars, billets, ingots, and slabs. During the melting for this casting, advantage is taken when the metal is liquid to effect some slight further fire refining and to adjust the oxygen content to 0.03% for tough-pitch copper.

Small reverberatory-type furnaces are used, with a typical model having a silica or magnesite brick bottom, magnesite side walls to the slag line, and silica or chrome magnesite above this and a suspended magnesite roof. Washed cathodes are charged to the furnace, along with available refined scrap, and are melted, which takes from 8 to 10 hours. Flapping is now carried out by inserting iron pipes into the copper and blowing in air until button samples show the required block grain on fracture, indicating that oxygen content of the bath is at 0.9%. The bath is then skimmed clean and a protective layer of charcoal spread over it.

Poling with green hardwood logs for 2 hours reduces the oxygen content to 0.03%, the tough-pitch state, as again shown by fracturing button samples. The furnace charge is then tapped and cast into molds.

Cement Copper purity depends to a considerable degree on the quality of the scrap iron and steel being used, and dirty, rusted scrap will have this dirt and iron oxide carried off with the cement copper and lower its grade. Consequently cement copper from different plant operations comes in varying grades from 50% copper or below, to as high as 95% copper.

The purity of the cement copper determines where it can be recycled back into the overall sequence of processes to recover the contained copper, with high-purity cement copper being added to the fire refining anode furnace, intermediate purity to the smelter converter, and low purity to the smelter melting furnace.

1b. Copper Sulfide leach solutions, where the sulfides have been oxidized before leaching, are identical to the previously mentioned methods of treatment for copper oxide precipitate refining. Cathode copper is melted, fire refined, and cast into industrial shapes, while cement copper is returned to whatever process operation in the flowsheet its purity warrents.

The ammoniacal leach of mixed copper, nickel, and cobalt sulfide flotation concentrates produces a copper sulfide precipitate, and this too is returned to the smelter and added to the smelting furnace or the converters. Large tonnage of low-grade material will be put in the smelting furnace, while small tonnage of high grade can be added to the converters.

1c. Native Copper and Copper Carbonates precipitated from an ammoniacal leach are in the form of black copper oxide, CuO. This is dried and shipped to market as is, or it can be reduced to metallic copper with carbon in a smelting operation.

2a. Nickel Laterites have two leaching treatment methods, governed by the amounts of associated basic material, with a calcined nickel oxide as the final precipitate from an ammoniacal leach of a high-magnesia content ore, and a 65% solids slurry of nickel sulfide in water as the precipitate from a pressurized sulfuric acid leach of a low-magnesia content ore.

The nickel oxide from the ammoniacal leach is mixed with coal and partially reduced to sinter containing 88% nickel, while the nickel sulfide slurry from the acid leach is redissolved in sulfuric acid, the iron, aluminum, copper, zinc, and lead impurities are precipitated and removed, and the nickel in solution is then reduced to nickel metal powder with hydrogen.

High-Magnesia Laterites, *Nicaro process* have as a final product from the precipitation operation a calcined nickel oxide, 76.5% nickel, produced from calcining nickel carbonate. This oxide is upgraded to 88% nickel by partial reduction and sold to the steel industry.

Reduction consists of putting the calcined nickel oxide through a hammer mill to break up any lumps and then pelletizing it with a mixture of fine anthracite coal and returned collected fines from dust collectors. The pellets are spread on a traveling grate-type sintering machine 7 feet wide and 42 feet long (2.13 X 12.8 m) and ignited. The grate on the machine moves at a rate of 8 inches per minute (20 cm) and cyclone and bag house dust collectors remove the fines sucked off through the windbox. These fines are recycled back to the pelletizing operation.

The sinter cake is broken up in rolls crushers, screened to the sizes desired, and shipped to the steel plants as alloying material.

Low-Magnesia Laterites (*Moa Bay*) produce as a precipitation endproduct a 65% nickel sulfide solids slurry containing 55% nickel, 36% sulfur, and 6% cobalt, with lesser amounts of iron, aluminum, chromium, copper, lead, and zinc.

The sulfide slurry is pumped into impeller-agitated, lead-lined steel pressure vessels operating at 350°F (177°C) and 700 psi air pressure (4823 kPa), and the sulfide solids are oxidized by the oxygen in the air to form water-soluble sulfates. The oxidation reaction produces sufficient heat to keep the system at the desired operating temperature. The slurry is discharged from the autoclave through a flash tank to relieve the high pressure and then into a small thickener. The thickener underflow contains the excess sulfides that did not have a chance to react in the autoclave, and these are returned to join the stream of fresh slurry being fed to this first operation.

The thickener overflow contains 50 gpl nickel, 5 gpl cobalt, and lesser amounts of iron, aluminum, chromium, copper, lead, and zinc and is purified in two stages. By adjusting the solution pH to 5.3 with ammonia and aerating it, iron, chromium, and aluminum precipitate as hydroxides and are filtered off. This first purification is carried out in tanks at 180°F (82°C) and atmospheric pressure. The remaining liquor is then adjusted with sulfuric acid to pH 1.5 in a rubber-lined reactor, and hydrogen sulfide is pumped in to precipitate the copper, lead, and zinc as sulfides. These are removed by filtering.

The purified liquid, now containing only nickel and cobalt in solution, is pumped through heat exchangers to preheat it to the reaction temperature of 375°F (190°C) and then into autoclaves at 750 psi (5168 kPa) hydrogen pressure. A seed powder of fine nickel is used as a precipitation catalyst on the first batch of liquor in the autoclave. The deposited metal powder is built up and densified by carrying out a further series of reductions with fresh batches of liquor before the powder is finally removed. Ammonia is continually added to neutralize excess acid and maintain the solution pH at 1.8:

$$NiSO_4 + 2NH_3 + H_2 = Ni + (NH_4)_2 SO_4$$

The spent liquor, which still contains essentially all of the cobalt, is evaporated, and the metallic compound crystals are treated to recover the values in them.

The nickel powder in the autoclave, after sufficient particle size buildup over several densifications, is blown out into a flash tank, washed in spiral classifiers, filtered, and dried. The resulting powder is 99.8% nickel.

2b. Nickel Sulfides produce precipitates of nickel metal powder from the ammoniacal leaching of sulfide flotation concentrate and nickel metal cathodes from the sulfuric acid leaching of low-sulfur matte. Both of these products are high-purity metals, on the order of 99.9% nickel, and are not further refined.

3a. Cobalt Sulfides (Zambia, Canada, Russia) roasted to soluble sulfates and leached in warm water are precipitated as cobalt hydroxide after the pregnant solution has been purified to remove copper and iron. The cobalt hydroxide is redissolved in spent electrolyte and deposited in electrolytic cells as cathode cobalt of 99.8% purity, which is either marketed as broken cathode metal or is melted and shotted.

The electrolyte is prepared by repulping the cobalt hydroxide precipitate with spent electrolyte and adjusting the pH with sulfuric acid to 5.9. This solution contains about 20 to 30 gpl cobalt as well as some insoluble calcium sulfate. This latter is removed by filtering so as to have a clarified electrolyte going to the cells.

The electrolysis is carried out in lead-lined cells with lead anodes and steel cathodes at a current density of 17 amperes per square foot (0.09 m^2) of cathode surface. The electrolyte is heated to 140°F (60°C), and the steel cathodes are treated with a derusting solution to minimize deposited metal adherence and facilitate ease of stripping. Five-day cathode cycles give a 50 pound (22.7 kg) cobalt deposition, 99.8% pure, which is stripped by hand and broken into small chunks in a rolls crusher for shipping to market.

The spent electrolyte is recycled back to dissolve the cobalt hydroxide from the precipitation circuit.

3b. Cobalt Oxides (Zaire) having been taken into solution by leaching with sulfuric acid are precipitated as cobalt hydroxide after preliminary purification steps to precipitate the iron and copper that are also in the pregnant solution.

In an electrolytic precipitation, which in many ways is quite similar to that just described for cobalt sulfides, the hydroxide precipitate is dissolved in spent electrolyte, and cobalt is deposited at the cathode of an electrolytic cell. Some zinc and sulfur are deposited with the cobalt. These are removed by melting the cathodes in an electric furnace and removing the final impurities, after which the 99% cobalt is granulated for shipment.

The cake of cobalt hydroxide precipitate, after filtering, is repulped with acidic spent electrolyte at pH 3. This precipitate, containing 18% cobalt, 45% calcium sulfate, and some nickel, zinc, and manganese, is partly dissolved in the spent electrolyte, leaving an excess of the cobalt hydroxide still in suspension, along with calcium sulfate. The solids content can reach 40 gpl in this electrolyte pulp, which analyzes 15 to 20 gpl cobalt.

Electrolytic deposition of the cobalt is possible only in a neutral solution, which is fulfilled by the excess of cobalt hydroxide precipitate in the electrolytic acting as a neutralizing agent for the acid generated at the anode and keeping the electrolyte pulp at about pH 7.0. The possible harmful effects of solids in the electrolyte is avoided by intense agitation in the cell, which is obtained by blowing air through especially designed anodes.

Electrolysis in the cells is carried out using 13 lead anodes and 12 mild steel cathodes, 3.2 inches (8 cm) apart, at a current density of 40 amperes per square foot (0.09 m^2). Current efficiency is rated at 85 to 88%. The cathode cobalt is periodically stripped and analyzes 92 to 94% cobalt. Some metallic cobalt is also recovered by electromagnets from the sludge of MnO_2 and $CaSO_4$ in the bottom of the tank, where nodules and needles fell during cathode deposition. The cobalt at this stage also contains 3 to 4% zinc and a little entrained basic sulfate.

The pulp from the cobalt electrolytic cells (the spent electrolyte) goes to Pachuca tanks where sulfuric acid is added to dissolve the excess cobalt hydroxide precipitate used for cell neutralization. This acid solution at pH 3.0 is run out of the Pachuca into thickeners for clarification and then back to the repulping section ahead of the electrolytic cells to dissolve fresh cobalt hydroxide precipitate. Some fresh sulfuric acid is added as makeup solution before the repulping.

The cobalt cathodes are refined in a magnesite-lined electric arc furnace at 2912°F (1600°C), where most of the zinc is removed by volatilization at the beginning of the process, and the remainder is oxidized and removed by agitating the bath. The sulfur is removed in a high-lime slag, which also combines with any remaining manganese. The molten bath of cobalt metal is then deoxidized by poling and granulated by pouring out of the furnace into a stream of water. The endproduct is a high-purity cobalt +99%, with particularly low sulfur and carbon contents.

4. Silver and Gold are leached in sodium or potassium cyanide solutions, from which the metal values are precipitated as metallics by replacement of their ions in solution with zinc metal ions. The resulting precipitate is first melted and fire refined, with the metal produced cast into anodes for electrolytic refining. Silver, which is generally in the larger amount, is removed first by electrodeposition in cathode compartmented cells; and the insoluble anode sludge, which contains the gold, is either chemically refined or melted and cast into anodes for treatment in electrolytic refining cells.

The cyanide leach precipitate is dried and smelted in small oil- or gas-fired furnaces with borax, silica sand, and sodium carbonate fluxes to volatilize the excess zinc dust remaining from the precipitation circuit and to remove some of the other impurities in a slag. The resulting doré bullion, about 950 fine, is cast into anodes for electrolytic refining of the silver.

Compartmented cells are used for silver refining with the impure anodes placed in cloth compartments. The principal difference in the two types of cells is that in one, the Thum cell, the anodes are horizontal, while in the Moebius cell the electrodes are vertical. Both cells are quite small.

The Thum Cell is a glazed earthenware dish, 4 feet long, 2 feet wide, and 1 foot deep ($1.22 \times 0.61 \times 0.3$ m), filled with a neutral electrolyte of silver nitrate solution made by dissolving silver in nitric acid and containing 60 to 80 gpl silver nitrate. The bottom of each cell is fitted with a block of carbon which is the cathode, while the silver, cast into anodes 1 foot square (0.09 m^2) and $\frac{1}{2}$ inch thick (12.5 mm), is placed four or five anodes deep in a wooden tray with a slotted bottom, lined with canvas, and positioned across the top of one end of the earthenware cell. Silver "candlestick" electrodes with a broad base are placed on the pile of anodes and on the cathode. Then direct current is passed through the cell. The cell voltage is 3 to 3.5 volts, with a cathode current density of 10 amperes per square foot (0.09 m^2).

Under the action of the passing current, silver is dissolved from the anode, which corrodes away in four days, and deposited as loose crystals on the cathode in the bottom of the cell at the rate of 330 ounces per day. The gold, platinum, and other insolubles are left as a black mud on the cloth in the anode tray and are periodically removed. The silver crystals on the cathode are removed with a scraper, melted under charcoal, and cast into bars which assay +999 fine.

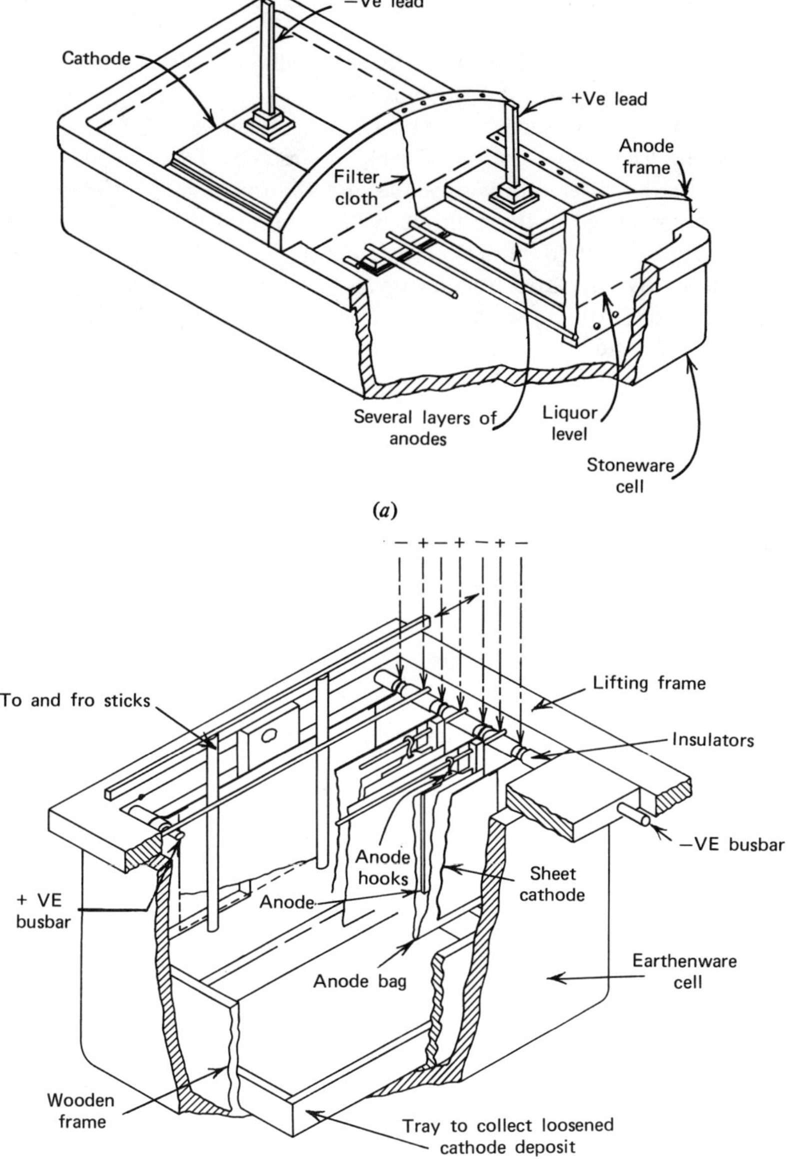

Figure 2.24. Silver refining cells: (*a*) Thum cell; (*b*) Moebius cell. *Source:* A. E. Richards, Refining of gold and silver, in *The Refining of Non-ferrous Metals*, Copyright The Institution of Mining and Metallurgy, London, 1950, pp. 107, 108.

The Moebius Cell is also made of glazed earthenware or plastic, and one small cell holds four anodes and five cathodes. The anodes are cast doré bullion 17 inches high by $8\frac{1}{2}$ inches wide by 1 inch thick (43 × 21 × 2.5 cm), while the cathodes are of titanium or stainless steel and measure 20 inches by 15 inches by $\frac{1}{16}$ inch (50 × 37.5 × 0.15 cm). Both anodes and cathodes are suspended vertically in the cell, with the anodes enclosed in canvas bags to collect the insoluble anode mud which contains the gold.

Anodes and cathodes are connected to busbars in the multiple system, and as current is passed silver leaves the anode and deposits at the cathode. Cell voltage is 2.7 volts, and cathode current density is 28 amperes per square foot (0.09 m²). The electrolyte is maintained at about 90°F (32°C) and is a solution of silver nitrate containing 150 gpl AgCl at a pH 1.0 to 1.5.

The anode bags are removed and cleaned of their gold-bearing sludge twice a week, and silver must be removed regularly from the cathodes so that the deposit does not grow out to touch the anode and short-circuit the cell. Silver scraped off the cathodes collects in wooden trays at the bottom of the cell and will analyze 999 to 999.8 fine.

The Wohlwill Process is used to refine electrolytically gold bullion of +94% gold content, such as the anode slimes from the Thum and Moebius cells.

The cells are 16 inches long by 11 inches wide by 12 inches deep (40 × 27.5 × 30 cm) and are made of porcelain. These hold the tiny 8 inch by 4 inch by $\frac{1}{2}$ inch anodes (20 × 10 × 1.25 cm) cast from the silver refining residue and the rolled gold cathodes. The anodes and cathodes are connected in the multiple system, with a 1 volt cell voltage and 80 to 100 amperes per square foot (0.09 m²) current density. The electrolyte is heated to 167°F (75°C) and is gently circulated by blowing air through it. It is an aqueous solution of hydrochlorauric acid, $HAuCl_4$, made from a solution of gold chloride containing 4% gold and 5% hydrochloric acid.

As electrical current is passed, anode gold goes into solution and deposits at the cath-

Table 2.6. Moebius Cell Operating Data

Anodes

17 × $8\frac{1}{2}$ × 1 in. (43 × 21 × 2.5 cm) doré bullion, four per cell, each anode surrounded by a cloth bag

Life, 60 hr

Scrap, 15% of anode

Cathodes

20 × 15 × $\frac{1}{16}$ in. (50 × 37.5 × 0.15 cm) titanium or stainless steel, five per cell

Electrolyte

Silver nitrate, 150 gpl AgCl

pH 1.0 to 1.5

90°F (32°C)

Voltage

2.7 V per cell

Current

Cathode current density 28 A/ft² (0.09 m²)

Circuit System

Multiple

Figure 2.25. Silver refining room with over 100 electrochemical cells. *Source:* Photograph courtesy of Handy and Harmon; all rights reserved.

Figure 2.26. Electrolytic cells for parting silver and gold at Port Pirie Refinery of the Broken Hill Associated Smelters Pty. Ltd. *Source:* The Silver Institute Letter, Vol. 8, No. 5, 1978, p. 1. Photograph by Val Foreman.

ode, giving a high-purity bullion of +999 fine. The cathodes are removed after 36 hours, melted, and cast into bars.

Chemical Refining is also used to refine gold. In one of these methods the gold slimes are boiled with a solution of sulfuric acid and potassium nitrate, which will dissolve silver and the platinum group metals and leave the gold as a residue of high purity. The gold bullion is melted and cast into bars.

NONREACTIVE METALS, HYDROMETALLURGICAL EXTRACTIVE PROCESSES

There is less general variation in the treatment of ores and concentrates by hydrometallurgical extractive processes than by pyrometallurgical methods, with the treatment sequence following the same pattern to a greater degree for the major metals processed in this manner. However there are still some individual differences in processing so that a particular operation can be used which will best suit the extractive operation to the material being treated, sometimes because of physical properties and other times because of desirable chemical reactions.

The combinations of the processes using hydrometallurgical methods to extract and refine copper, nickel, cobalt, gold, and silver from their commonest natural metallic compounds are given in Table 2.7

Table 2.7. Nonreactive Metals, Hydrometallurgical Extractive Processes

Copper

Oxide

A. Tank percolation leach in H_2SO_4 solution
Electrolytic precipitation
Fire refine cathode copper

B. Tank agitation leach in H_2SO_4 solution
Electrolytic precipitation
Fire refine cathode copper

C. Heap leach in acidified raffinate
Solvent extraction
Electrolytic precipitation

D. Tank leach in H_2SO_4 solution, countercurrent wash in raffinate
Solvent extraction
Electrolytic precipitation

Sulfide

A. In-place leach in H_2SO_4 solution
Cementation precipitation
Return cement copper to smelter reverberatory or anode furnace

B. Heap leach in H_2SO_4 solution
Cementation precipitation
Return cement copper to smelter reverberatory or anode furnace

C. Tank percolation leach in H_2SO_4 solution
Electrolytic precipitation
Fire refine cathode copper

D. Beneficiate
Pressure leach in ammoniacal solution
Precipitation in stills as copper sulfide
Return to smelter reverberatory or converters

E. Beneficiate
Pressure leach in oxygen-ammonia-ammonium sulfate solution
Solvent extraction
Electrolytic precipitation

F. Beneficiate
Tank leach in ferric chloride
Electrolytic precipitation

Table 2.7. *(Continued)*

Native and Carbonates

    Beneficiate
    Tank percolation leach in ammoniacal solution
    Precipitate in stills as copper oxide

Nickel

Laterites

A. High MgO ore
    Beneficiate
    Reducing roast to crude metal
    Tank agitation leach in ammoniacal solution
    Precipitation in stills as nickel carbonate
    Calcine nickel carbonate to nickel oxide
    Fire refine nickel oxide to metal

B. Low MgO ore
    Beneficiate
    Pressure leach in H_2SO_4
    Neutralize leach solution
    Precipitate in autoclave as nickel sulfide
    Oxidation of sulfide to nickel sulfate in autoclave
    Pressure leach of nickel sulfate in water
    Precipitation in autoclave with hydrogen to nickel powder

Sulfide

A. Beneficiate
    Pressure leach in ammoniacal solution
    Precipitation in autoclave with hydrogen to nickel powder

B. Beneficiate
    Smelt to matte
    Tank agitation leach in H_2SO_4 solution
    Electrolytic precipitation

Cobalt

Sulfide

    Beneficiate
    Roast to cobalt sulfate
    Tank agitation leach in water
    Chemical precipitation of copper and iron
    Chemical precipitation of cobalt hydroxide
    Tank agitation leach in spent electrolyte
    Electrolytic precipitation as cathode cobalt

Oxide

    Beneficiate
    Tank agitation leach in H_2SO_4 solution
    Chemical precipitation of copper and iron

Table 2.7. (*Continued*)

Chemical precipitation of cobalt hydroxide
Tank agitation leach in neutralized spent electrolyte
Electrolytic precipitation
Fire refine cathode cobalt

Gold and Silver

Native

A. Beneficiate
Tank agitation leach in cyanide solution
Replacement precipitation as metal
Fire refine to doré bullion anodes
Electrolytic refining of silver
Melt and cast gold anodes from anode sludge
Electrolytic refining of gold

B. Beneficiate
Tank agitation leach in cyanide solution
Replacement precipitation as metal
Fire refine to doré bullion anodes
Electrolytic refining of silver
Chemical gold refining of anode sludge

C. Heap leach in cyanide solution
Replacement precipitation
Fire refine to doré bullion anodes
Electrolytic refine

D. Beneficiate
Tank agitation leach in cyanide solution
Adsorption on activated charcoal
Strip in desorption circuit
Electrolytic precipitation
Electrolytic refining of gold and silver

REFERENCES

Biswas, A. K. and W. G. Davenport, *Extractive Metallurgy of Copper*, Pergamon, New York, 1976.

Boldt, J. R., Jr., *The Winning of Nickel*, Methuen, London, 1967.

Burkin, A. R., *The Chemistry of Hydrometallurgical Processes*, D. Van Nostrand, New York, 1966.

Carbon in pulp to recover gold. *World Mining*, Vol. 27, No. 11, 1974.

The Staff of Batelle Memorial Institute, *Cobalt Monograph*, Centre d'Information du Cobalt, Columbus, Ohio, 1960.

Dennis, W. H., *Metallurgy of the Non-ferrous Metals*, Pitman, New York, 1961.

Ehrlich, R. P., Ed., *Copper Metallurgy*, Extractive Metallurgy Division Symposium on Copper Metallurgy, AIME, New York, 1970.

Griffith, W. A. et al., Development of the roast-leach-electrowin process for Lakeshore, *J. Metall.*, Vol. 27, No. 2, 1975, p. 19.

Habashi, F., *Principles of Extractive Metallurgy*, Vol. 2, Gordon and Breach, New York, 1970.

Hall, K. B., *World Mining*, Vol. 27, No. 11, 1974.

Hampel, C. A., Ed., *Rare Metals Handbook*, Chapman and Hall, London, 1956.

Harjavalta Nickel Refinery, *Can. Inst. Min. Metall. Bull.*, Vol. 57, No. 626, 1964.

Herbert I. C., New Copper Extraction Processes, *J. Metall.*, Vol. 26, No. 8, 1974.

Hydrometallurgy Applied to Copper Ores, Denver Equipment Division, Joy Manufacturing Co., Bul. G3-B146, Deco Trefoil, Denver, 1969.

Inco Pressure Carbonyl Process, *J. Metall.*, Vol. 21, No. 7, 1969.

Jones, M. J., Ed., *Advances in Extractive Metallurgy and Refining*, The Institution of Mining and Metallurgy, London, 1972.

Annual Review Issue, *J. Metall.*, Vol. 28, No. 3, 1976.

Annual Review Issue, *J. Metall.*, Vol. 29, No. 3, 1977.

Miller, J. D. A., Ed., *Microbial Aspects of Metallurgy*, Elsevier, New York, 1972.

Mineral Facts and Problems, U.S. Bureau of Mines Bul. 667, U.S. Department of the Interior, Washington, D.C., 1975.

New Copper Extraction Processes, *J. Metall.*, Vol. 26, No. 8, 1974.

Palperi, M. and O. Aaltonen, Sulfatizing Roasting and Leaching of Cobalt Ores at Outokumpu Oy, *J. Metall.*, Vol. 23, No. 2, 1971.

Pehlke, R. D., *Unit Processes of Extractive Metallurgy*, Elsevier, New York, 1972.

Queneau, P., Ed., *Extractive Metallurgy of Copper, Nickel and Cobalt*, Interscience, New York, 1961.

Refining of Non-ferrous Metals, The Institution of Mining and Metallurgy, London, 1950.

Roast–Leach–Electrowin Process, *J. Metall.*, Vol. 27, No. 2, 1975.

Rosenbaum, J. B., Minerals Extraction and Processing: New Developments, in *Materials: Renewable and Nonrenewable Resources*, Abelson, P. H. and A. L. Hammon, Eds., American Association for the Advancement of Science, Washington, D.C., 1976.

Rosenqvist, T., *Principles of Extractive Metallurgy*, McGraw-Hill, New York, 1974.

Sohn, H. Y. and M. E. Wadsworth, Coordinators, *Rate Processes of Extractive Metallurgy*, The Metallurgical Society of AIME, New York, 1976.

Spedden, H. R., Cone Type Copper Precipitators, *J. Metall.*, Vol. 18, No. 10, 1966.

Thomas, R., Ed., *Engineering and Mining Journal Operating Handbook of Mineral Processing*, Vol. 1, McGraw-Hill, New York, 1977.

Wadsworth, M. E. and F. T. Davis, Eds., Unit Processes in Hydrometallurgy, *Metallurgical Society Conference AIME*, Vol. 24, Gordon and Breach, New York, 1964.

World Minerals Availability, 1975-2000, Vol. 2, Outlook for Mineral Processing, Nodule Mining, and Transportation, Stanford Research Institute, SRI Project ECC 3742, Stanford, California, 1976.

Yannopoulos, J. C. and J. C. Agarwal, Eds., *Extractive Metallurgy of Copper*, Vol. 2, Hydrometallurgy and Electrowinning, Metallurgical Society of AIME, New York, 1976.

Young, R. S., *Cobalt, Its Chemistry, Metallurgy and Uses*, Reinhold, New York, 1960.

Van Arsdale, G. D., *Hydrometallurgy of Base Metals*, McGraw-Hill, New York, 1953.

3. Reactive Metals, Pyrometallurgical Treatments

Pyrometallurgical treatments have been used for many years to process such familiar reactive metals as aluminum and zinc and also are the common processing procedures used to extract the newer reactive metals, magnesium and titanium. In all the pyrometallurgical processing of reactive metals the operation is carried out with no contact with air, so that there is no reaction between the heated metallic material and the oxygen in the air. This exclusion of air is accomplished in a variety of ways—by using sealed retorts or retorts with a positive internal pressure, by using vacuum furnaces or furnaces filled with inert gas, or by having the extractive reactions carried out under a protective bath of molten salts.

The processes used are combinations of first roasting, to convert the metallic compounds into more easily treated oxides or chlorides, and then smelting in retorts or furnaces from which air is excluded. Many of these metals are in their final, commercially usable condition after smelting, while others can be still further purified by fire and electrolytic refining.

These processing treatments are grouped together into their major categories in Table 3.1.

ROASTING

Roasting of reactive base metal ores or concentrates, prior to smelting, is carried out with the intention of changing the initial compounds into those more easily treated in the metal processing operations to follow, and also to transform fines into lumps that can be treated in the blast furnace. Sulfides are roasted to oxides, which are then reduced to volatile metal vapor in distillation processes. Carbonates are calcined to oxides and then converted to chlorides by a chloridizing roast. Metallic oxides are also given a chloridizing roast; the chlorides produced in both these cases are reduced to metal in the subsequent smelting processes.

Table 3.1.   Reactive Metals, Pyrometallurgical Treatments

Roasting Methods

1.   Zinc sulfide concentrate—flash, fluid-bed, hearth, sintering
2.   Magnesium carbonate—electric furnace chlorinator
3.   Titanium oxide—electric furnace chlorinator

Smelting Methods

1.   Zinc sulfide roasted calcines—horizontal retort, vertical retort, electrothermic furnace, sealed blast furnace
2a. Magnesium chloride—molten salt electrolysis
2b. Magnesium oxide—vacuum retort
3.   Titanium tetrachloride—reaction vessel
4a. Aluminum oxide—molten salt electrolysis
4b. Aluminum trichloride—molten salt electrolysis

Fire Refining Methods

1.   Zinc metal—fractional distillation, liquation

Electrolytic Refining Methods

1.   Aluminum metal—molten salt electrolysis

The roasting reactions of the oxidizing roast for sulfides are the same as have been previously discussed for the pyrometallurgical treatment of nonreactive metals, where the metal sulfide heated to an elevated temperature combines with oxygen in the air to break down into a metal oxide and sulfur dioxide gas:

$$MS + 3O = MO + SO_2$$

The calcining of carbonates to oxides uses heat to break down the carbonates into metal oxide and carbon dioxide gas:

$$MCO_3 = MO + CO_2$$

As the reactions involved are reversible, a certain minimum temperature must be maintained to ensure a reasonable reaction rate and also to ensure that the dissociation pressure of the decomposing compounds is greater than the back pressure of the gas being evolved. The product gas, CO_2, must be constantly removed either by gas flow out of the calciner or by a slight vacuum on the vessel, to ensure that it does not build up sufficiently in concentration to reverse the reaction. The critical temperature to have this process proceed at commercially important tonnages is quite high, usually in excess of 1832°F (1000°C). As it is expensive to supply such large amounts of heat, all precautions are taken to waste as little heat as possible.

The material to be calcined is ground rather finely and sometimes predried, so that the reaction will take place rapidly and completely. The optimum size of feed is quite important; large particles require much longer heating time, and even then they may not be completely decomposed through to their centers. Too much fine material is also a handicap, as it can pack to restrict the outward flow of CO_2 gas produced, which builds up a back pressure causing the calcining reaction to reverse. Fines can also be lost by being carried out in the escaping gas stream, or also require extensive dust collecting equipment to recover them.

Chloridizing roasts of metallic oxides to chlorides take place in a type of chlorinator, with the metallic oxide first being well mixed with carbon, briquetted and coked, and then reacted with chlorine, either as liquid chlorine or in solid form as sodium chloride or ammonium chloride, to form the metallic chloride:

$$MO + 2Cl + C = MCl_2 + CO$$

The chlorides produced are in liquid or vapor form. These are run out as liquid from the chlorinators to produce high boiling point chlorides or pass off as vapor which is caught and condensed from chlorinators that have low boiling point products.

TYPES OF ROASTING

The types of roasters used for the preliminary conversion of reactive metallic ores and concentrates are flash roasters, fluid-bed roasters, multiple-hearth roasters, and sintering machines for the oxidation roasting of sulfides. Kilns are used for calcining carbonates to oxides, and chlorinators are used to convert oxides to chlorides.

Fluid-Bed Roasters, Multiple-Hearth Roasters, and Sintering Machines have been discussed in the section dealing with the oxidation roasting of nonreactive metal sulphides, and the principle of operation carried out is the same for both types of metallic compounds, reactive and nonreactive.

Flash Roasting is an additional method to those already covered above and is a rather specialized type used where a quite complete oxidation roast is required and sulfur below 1% in the calcines is desirable. This roaster was developed from the conventional multiple-hearth roaster, with the realization that much of the actual roasting was accomplished as the particles dropped between hearths and during this short time interval had optimum gas–solid contact, with the particles at this stage surrounded on all sides by heated air. Flash roasters can be constructed by removing several hearths from a multiple-hearth roaster, so that a large, open combustion chamber is available for fine, dry concentrate to be blown in at the top and roast as it drops to the bottom of the open chamber.

There are two main designs of flash roasters, both with the same basic construction of a circular steel shell lined with refractory brick, refractory hearths on which there are rotating air-cooled rabbles driven from a central rotating shaft, and drop holes at the inner and outer circumferences of each successive hearth.

The feed material must be both dry and fine to have the flash roaster operate successfully, and the difference in design characteristics of the commonly used types is in the placement of the drying hearths and combustion chamber in the roaster. *One type* has the open combustion chamber at the top of the roaster with four hearths below this. Wet flotation concentrate is first dried by being continuously fed to the bottom two hearths and there dried by hot gases withdrawn from the upper combustion chamber. The dried concentrate is raked from the bottom hearth into an air-swept Marcy ball mill to break up any agglomerations and, if necessary, to grind the concentrate still finer before feeding it to the combustion chamber of the roaster. An air classifier followed by a cyclone separator remove the solids from the air stream coming from the ball mill, and these solids collected in a dry feed bin are blown through a burner into the top of the roaster combustion chamber. Ignition of the sulfide takes place near the burner nozzle and continues as the fine particles drop down through the open combustion chamber to settle on the top hearth below. The roasted calcines are next raked through drop holes to the second hearth and are then discharged from the roaster. The hearths are used in pairs with no connection between the pairs; the top two hearths are employed to collect and discharge the roasted calcines, while the bottom two hearths are used solely for drying wet concentrate prior to roasting.

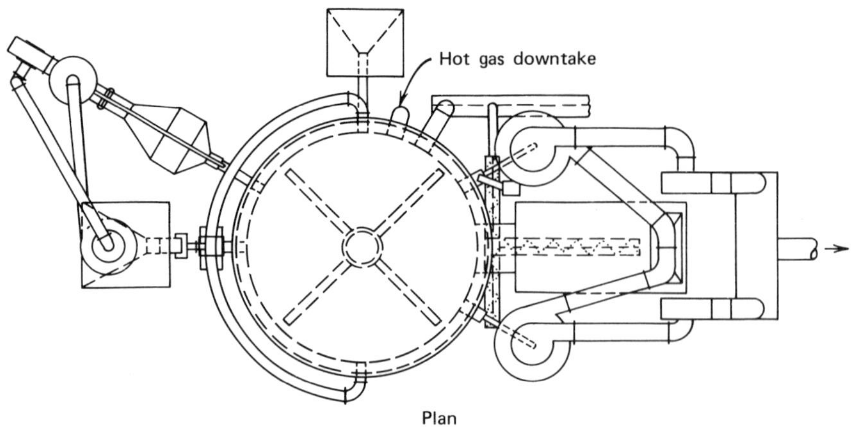

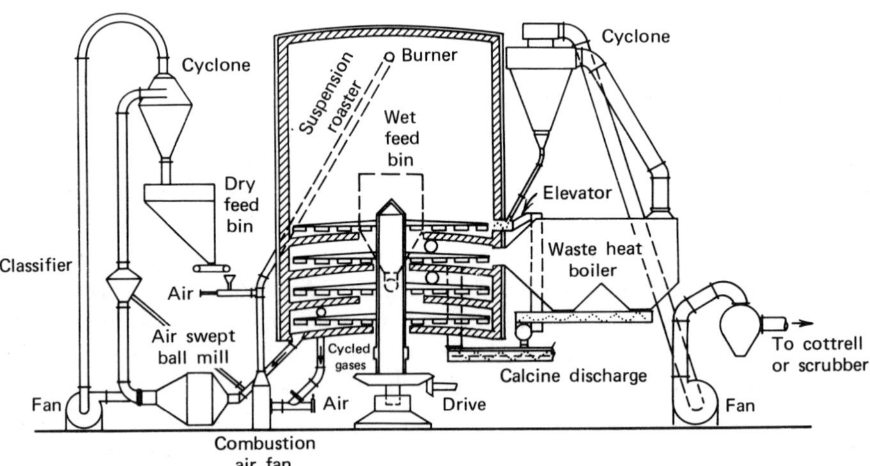

Figure 3.1. Flash-type roaster with drying hearth below combustion chamber. *Source:* R. E. Eyre, *Can. Min. Metall. Bull.*, Vol. 54, No. 592, 1961, p. 591.

The second type of flash roaster differs in having the open combustion chamber in the center of the roaster, with one or two drying hearths above and two or three collecting hearths below this chamber. Wet concentrate (3 to 4% moisture) is fed to the top drier hearths; and as it is rabbled across these, it is dried by hot gases from the combustion chamber. The dried concentrate then passes out into an air-swept ball mill, is collected after grinding and agglomeration breakup, and blown into the top of the combustion chamber through a burner. Ignition again takes place close to the burner nozzle, and the particles roast as they settle downward in the chamber to be collected on, and discharged from, the bottom hearths.

In both roaster types about 60% of the roasted calcines settles on the collecting hearths, with the other 40% carried off by the gas stream leaving the combustion chamber. This 40% is removed by cyclone or electrostatic dust collectors. This dust has a low sulfur content similar to the settled calcines and is also a finished product ready for the next processing step.

Roasting temperature can be controlled in both types of roasters by controlling the concentrate feed rate. This is blown through the feed burners, along with sufficient air for combustion, into the top of the hot roasting chamber and gives fast ignition of the sulfide contained in the dry concentrate feed, just as though a powdered fuel were being

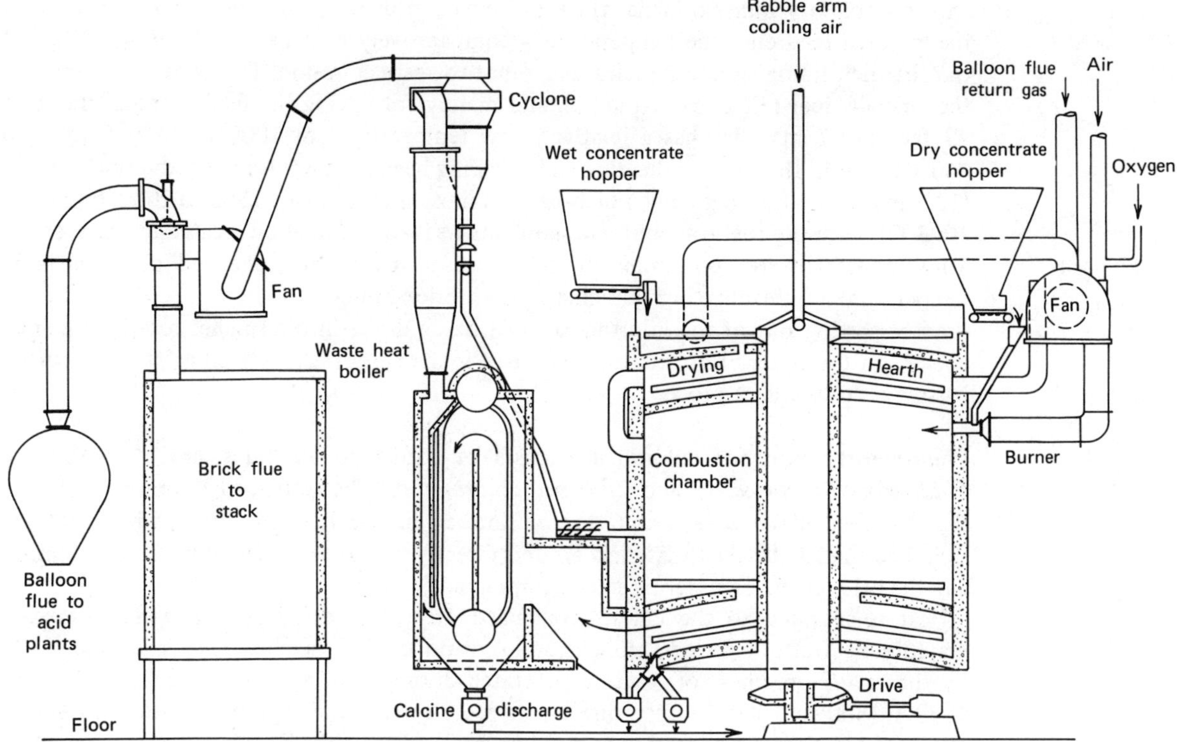

Figure 3.2. Flash-type roaster with drying hearth above combustion chamber. *Source:* R. E. Eyre, *Can. Min. Metall. Bull.*, Vol. 54, No. 592, 1961, p. 589.

used. The exothermic heat of oxidation is usually sufficient to maintain a heat of 1832°F (1000°C) required for a dead roast. In the event of too high a temperature being produced, cool gas can be drawn from the drying chamber and returned to the combustion chamber to help keep the temperature under control as well as to recover dust which is produced in the drying operation.

Calcination of carbonates to oxides can be carried out in the conventional *multiple-hearth roaster*, or more frequently in *rotary kilns*. In either case, as this is an endothermic reaction, external fuel must provide the heat required for the reaction to proceed at a reasonable rate, and this is accomplished through utilizing gas, oil, or pulverized coal burners.

Multiple-hearth roasters used for calcining are similar in design to those already discussed.

Modern rotary kilns are designed with a mild steel circular shell lined with refractory brick to resist the mechanical and chemical action of the kiln charge, as well as to stand up to the elevated temperature at the burner end. Kilns are set in a slope of $\frac{1}{4}$ to $\frac{3}{4}$ inch per foot from the feed end (6.25 to 18.75 mm per 0.3 m). A typical speed of rotation is 1 to $1\frac{1}{2}$ rpm.

The kiln shell is designed to be a rigid support for its lining and charge. To accomplish this, riding rings are fastened at intervals on reinforced sections of the shell to transmit the weight of the kiln to support rollers mounted on a concrete foundation. The rotary drive is moved by a girth gear mounted around the shell near the center of the kiln and is powered by an electric motor working through a speed reducer.

A typical kiln would be about 125 feet long and 10 feet in diameter (38.11 × 3.05 m), with a refractory lining 6 inches thick (15 cm) and divided into three zones. In zone 1, the elevated feed end, the temperature is comparatively low, below 800°F (427°C), and the firebrick lining should be hard and dense to resist abrasion. This zone would include the first 40 feet (12.2 m) of the kiln and mainly for drying the feed. Zone 2, the next 45 feet (13.7 m), also has a relatively low temperature, of 1000 to 1400°F (538 to 760°C), and in this zone preheating and calcining begins. The third zone, the final 40 feet (12.2 m), is the burning zone which has the most intense heat, 1500 to 2000°F (816 to 1093°C), with the fuel-fired burners supplying heat for calcination located in this section. This imposes severe conditions on the refractories for this zone and demands high-temperature, nonreactive high-alumina or silica brick linings.

After passing out of the kiln the calcined oxide drops into a smaller rotating inclined cooler, on the order of 40 feet long by 4 feet in diameter (12.2 × 1.22 m), to cool before passing on to the next process.

Chlorinators, used for chloridizing roasts of oxides to chlorides, are of two types, *fluidized-bed chlorinators* and *static-bed chlorinators*. The static-bed chlorinator requires that the feed material be briquetted or sintered before it can be used, while with the fluid-bed model the feed materials are fed directly to the bed without prior combination and the briquetting or sintering step is eliminated.

Heat is supplied to the chlorinators partly by electrical resistance passing between graphite electrodes inserted through the lower portion of the chlorinator shell and partly by the exothermic heat of reaction generated during chlorination, to reach the 1472 to 1832°F (800 to 1000°C) operating temperature range. One type of chlorinator has two sets of three electrodes inserted in the bottom fourth of the shell, one set at the bottom and the other one quarter of the height up from the bottom. The resistance to the electrical current flowing through the briquetted charge between sets of electrodes generates the required heat. The other type of chlorinator has only one set of electrodes inserted through the shell one third of the distance up from the bottom and projecting into a bed of carbon resistor blocks. Current flow between electrodes through these blocks causes resistance heating, and the temperature is raised to the required level.

Both chlorinators are tall, on the order of 20 feet (6.1 m), circular, vertical steel shells, refractory lined with a protective brick layer to prevent chlorine attack on the steel shell.

The top of each chlorinator has a sealed hopper for periodic charging of baked briquettes, consisting of calcined oxide, coal (coke), and a binder for the static bed unit, or separate lots of calcined oxide and coal (coke) for the fluid-bed type. The charge completely fills the vessel in the unit heated by two sets of electrodes and forms a bed on top of the hot resistor blocks in the other type. The resistor block layer and the feed material bed above it are of approximately equal depth, each about two fifths of the height of the reactor opening.

Gaseous or liquid chlorine is added from several inlets at the bottom of either chlorinator, and is preheated before it rises to contact the charge. The rate of addition is on the order of $\frac{1}{2}$ to 1 pound of Cl_2 per pound of metallic element in the charge (227 g to 454 g Cl_2 added per 454 g of metallic element).

In the chlorinator the oxide in the charge is first reduced to a carbide by the coal or coke also charged, and then this carbide reacts with the chlorine to form a metallic chloride. The metallic chloride can be removed from the chlorinator as a liquid tapped from the bottom of the vessel, if it has a boiling point higher than the reaction temperature of the process. If the chloride produced has a lower boiling point than the reaction temperature, then it will pass out as a vapor through an outlet at the top of the chlorinator. This chloride vapor first passes through a dust collector to reduce the amount of solids suspended in the gas stream, and from there it passes to a condenser. The condensers are vertical towers into which is pumped a spray of the same cold liquid metallic chloride, condensing the metallic chloride vapor in the gas stream and letting it run into a settling tank where any insoluble matter and sludge are separated out as a bottom layer below the condensed liquid.

ROASTING PROCESSES

1. Zinc Sulfide Concentrates are roasted by a variety of methods, all with the intention of reducing the sulfur content by oxidation to below 1%, in which form the zinc oxide product can be reduced by carbon in a retort or blast furnace to zinc metal. The more completely the oxidizing roast is carried out, the better, as any zinc sulfide remaining in the calcines cannot be reduced by carbon during retorting and will be lost as unreacted, discarded residue. Any zinc sulfate formed during roasting is also lost, as this will reduce to zinc sulfide in the retort and go out as unreacted residues in this form. With the atomic weight of zinc being twice that of sulfur, 65 compared to 32, every pound (454 g) of sulfur in ZnS will combine with over 2 pounds (908 g) of zinc which is retained unreduced in the retort residue. This makes dead roasting of zinc sulfide to zinc oxide very meaningful.

Calcines treated in the zinc blast furnace are not as sensitive to sulfur content, and in this case the hot tuyere air blown in will combine with any remaining sulfur to form SO_2 gas, which goes off in this form.

The methods used for oxidizing roasting of zinc sulfide concentrates are flash roasting, fluid-bed roasting, multiple-hearth roasting, and sintering, used singly or in combination. All are designed to produce material of the physical form and mechanical strength appropriate for the process that follows.

Flash Roasting is carried out in a modification of the conventional multiple-hearth roaster, where fine flotation concentrate (85% minus 200 mesh) containing 30% sulfur can be given an oxidizing roast and produce a zinc oxide with less than 1% sulfur content.

The wet filter cake (3 to 4% moisture) product of the preceding beneficiation process is first dried by passing it through one or two drying hearths, over which hot gases are drawn from the roasting operation. These drying hearths may be either above or below the roasting chamber, depending on the particular roaster design.

After drying, the concentrate is rabbled out at the periphery of the bottom drying hearth and slides down a chute into a 9 foot diameter (2.74 m) air-swept Marcy ball mill. Oversize particles and agglomerates are broken up in the ball mill, and when of sufficiently fine size and weight, these particles are sucked out of the ball mill to an air classifier, and from here the particles too coarse are returned to the ball mill and the finer particles are blown on to a 9 foot cyclone dust collector (2.74 m). The cyclone collector removes the solids from the air stream and drops them into a dry feed bin, ready to be charged to the combustion chamber of the flash roaster.

The combustion chamber is 20 to 25 feet in height (6.1 $\times$ 7.62 m) and can be placed either as the top two thirds of a roaster, with all hearths below, or can be located as the center section with the drying hearth above and the final roasting and discharge hearths below. In either case, the dry concentrate from the feed bin is blown horizontally through a burner nozzle, along with the air for combustion, located at the top of the combustion chamber. The fine sulfide particles will ignite in front of the burner and continue to burn in suspension while falling vertically through the height of the chamber. Cool cycled gas from the drying hearths is sucked into this burner and mixed with atmospheric air in order both to help control the considerable exothermic heat of the reaction,

$$ZnS + \tfrac{3}{2}O_2 \longrightarrow ZnO + SO_2 \quad \Delta H^\circ = -111{,}000 \text{ calories per mole}$$

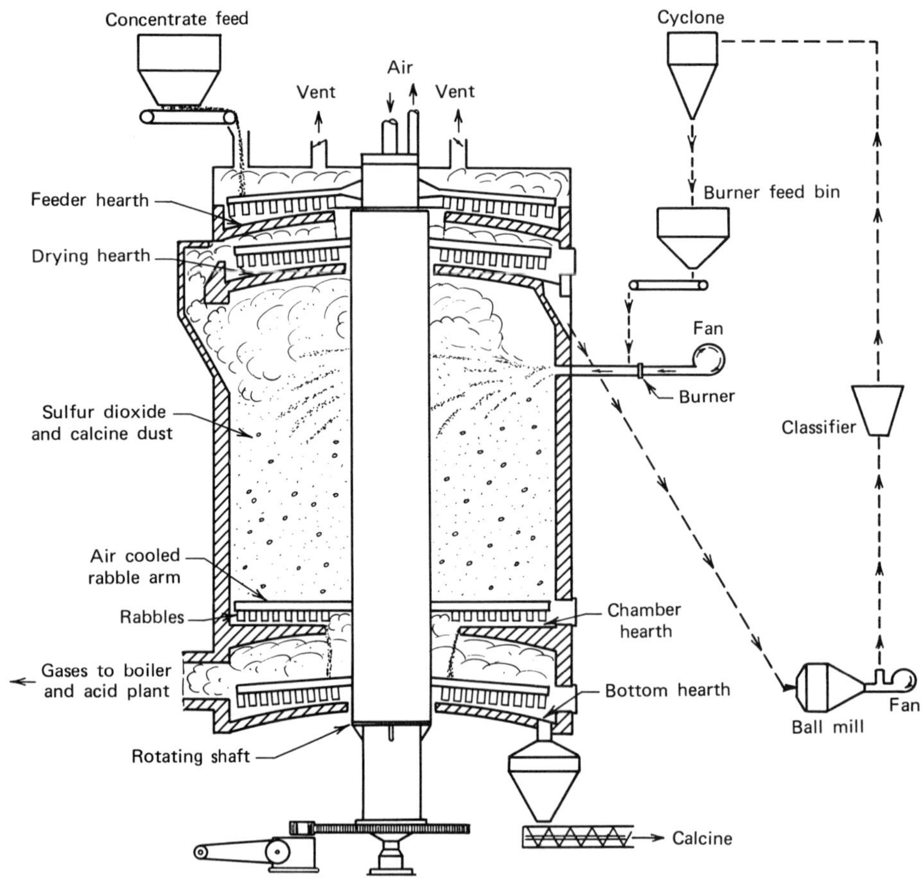

Figure 3.3. Zinc sulfide concentrate flash roaster. *Source:* Courtesy of St. Joe Minerals Corporation.

Table 3.2. Typical Zinc Roasting Operations

Type of Roaster	Operating Temperature, (°F)	Feed Capacity (tons/day)	Dust in Offgas (% feed)	SO_2 in Offgas (%)
Multiple-hearth	1200–1350	50–120	5–15	4.5–6.5
Multiple-hearth[a]	1600–1650	250	5–15	4.5–6.5
Fluid bed[b] (Dorr-Oliver)	1640	140–225	70–80	7–8
Fluid bed[a] (Dorr-Oliver)	1650	240–350	75–85	10–12
Fluid bed (Lurgi)	1700	240	50	9–10
Suspension (Flash)	1800	120–350	50	8–12
Fluid column	1900	225	17–18	11–12

Source: U.S. Bureau of Mines Information Circular 8629, 1974.

Dead roast, except where noted otherwise.

[a] First stage is a partial roast in multiple-hearth, second stage is a dry-feed dead roast in Dorr-Oliver fluid bed.

[b] Slurry feed.

212

and to recover dust produced in the drying operation. Because of this exothermic heat produced, the reaction is highly autogenous and proceeds at over 1832°F (1000°C).

Approximately 60% of the roasted calcines settle through the combustion chamber to collect on the bottom collecting and discharge hearths, and it is while being rabbled across these two or three hearths and finally out through a chute to a calcine bin that final desulfurization takes place. A 30% S concentrate will flash down to 8% S in the combustion chamber and be further reduced to below 1% S on these last lower hearths.

Because of the degree of fineness of the concentrate feed, necessary for both speed and completeness of the oxidation reaction, a considerable portion (40%) of dry feed is carried out of the combustion chamber in the escaping stream of hot gas and does not collect on the finishing hearths below. The hot gases are usually led off the roaster from the bottom collecting hearth to try and ensure that as much as possible of the finer calcines will collect on the hearth and go on directly to the calcine bin. The gases go first to waste heat boilers to recover part of their sensible heat, and a fan sucks them from here to a large cyclone dust collector where the larger solids are removed and returned to the upper most collecting hearth at the bottom of the roaster combustion chamber. The gas stream from the cyclone dust collector is next blown through the cyclone fan to an electrostatic precipitator, or to a wet scrubber tower, to remove the finer dust particles still present. These solids are low-sulfur calcines which can be added directly to the calcine bin as finished product. The clean gas, high in SO_2, is then piped to the acid recovery plant for sulfuric acid manufacture.

Fluid Bed Roasters use pelletized feed (minus 3 to plus 14 mesh) at roasting temperatures above 1742°F (950°C). The roasters used are rectangular in shape, $15\frac{1}{2}$ feet high and 20 feet long (4.73 × 6.1 m), and tapered with a cross-sectional dimension of $6\frac{1}{4}$ feet at the top and $3\frac{1}{4}$ feet at the bottom (1.91 × 0.99 m). Fluid-bed depth is 6 feet (1.83 m), with the feed input opening being $6\frac{1}{2}$ feet (1.98 m) from the bottom of the roaster and the

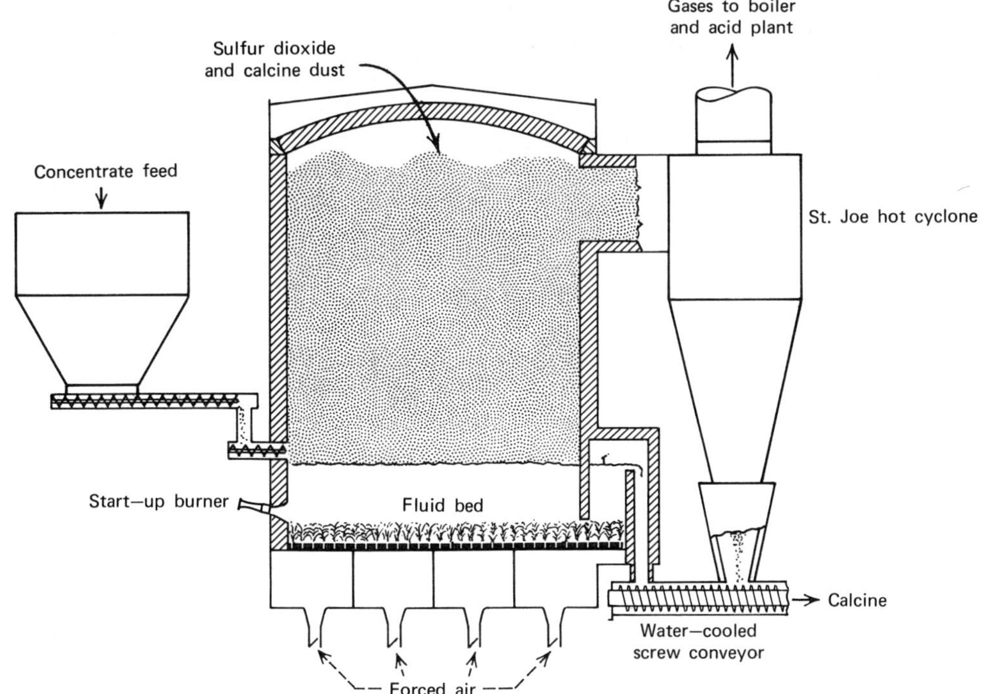

Figure 3.4. Zinc sulfide concentrate fluid-bed roaster. *Source:* Courtesy of St. Joe Minerals Corporation.

calcine discharge port 8 inches (20 cm) from the bottom. Air injection into the bed is at the rate of 14,000 to 17,000 cubic feet per minute (398 to 482 cubic meters per minute), with the lower blowing rates being used with the higher temperatures.

The coarse calcine, in which the sulfur content has been reduced from 33% S in the concentrate feed to only 0.5% S, flows from the roaster into a calcine cooler and from there to a storage bin. This consists of some 55% of the total calcine product. The hot gases escaping from a gas port at the top of the roaster go first to a waste heat boiler to recover part of their sensible heat, then to a cyclone dust collector where 38% of the total calcines are removed. Half of this cyclone product is added to the overflow calcines in the calcine cooler, and the other half is returned to the pelletizing plant as a binding ingredient. The sulfur content of these final cyclone calcines is slightly over 1% S.

The roaster gases still contain 7% of the total calcines as fine dust, and this is removed in a second stage dust collector, usually an electrostatic precipitator, after which the dust-free gas is allowed to pass out of the roasting circuit. This dust is also added to the calcine storage bin and is very low in sulfur, about 0.1% S. The combined sulfur content of all three calcines, from the roaster overflow, the cyclone collector, and the electrostatic precipitator, is on the order of 0.75% S.

Multiple-Hearth Roasters are used both for complete roasting operations and as the first stage in a two-part roast using sintering as the second and final step.

Roasters with twelve roasting hearths and a top drying hearth are used, and the exothermic heat from the sulfur in the feed is sufficient to reach a reaction temperature of 1292 to 1472°F (700 to 800°C) that will roast the feed material from 30% S down to 8% S. However a dead roast with a very low sulfur calcine product requires a high finishing temperature to drive off the remaining sulfur and decompose any zinc sulfate that has formed, and this necessitates a final roasting temperature of 1832°F (1000°C). Oil or gas fired burners on the lower hearths raise the temperature to this level.

In some instances multiple-hearth roasting is completed at the 8% S calcine level, and a second roast on the sintering machine reduces the sulfur content in the final sintered calcines to below 1% S. This finishing sintering roast keeps the SO_2 gas escaping from the multiple-hearth roaster from being diluted and contaminated by the burner products of combustion if it is to be used for sulfuric acid manufacture. It also decreases the maintenance and repair costs by having the multiple-hearth roasting operation done at a lower temperature.

Sintering is carried out both as the second stage of a two-stage roast, where a multiple-hearth roast is used as the first step, and also as a complete roasting process by itself. The sinter product is ideal feed material for blast furnace smelting of the zinc oxide product; when crushed to $\frac{1}{4}$ inch (6.25 mm) size it makes excellent feed material for retort distillation.

The most common method of sintering is the mixing of crushed return sinter with raw concentrate to be roasted, as well as mixing CaO fluxing agents if the sinter is to be smelted in the zinc blast furnace and roasting this mixture. The raw concentrates, returned sinter, and fluxing agents are moistened to produce a satisfactory consistency of feed, and a layer is spread over a grate of a continuously operating sintering machine. This layer is then ignited from the top by a gas or oil burner, and air is sucked down through the burden to continue the sintering as the grate moves toward the discharge end of the machine. Combustion is completed by the time the sinter is discharged, and the sinter is graded for size. Sinter of +1 inch size (+2.5 cm) is taken to storage bins, or part is crushed and returned for the addition to raw concentrate. The −1 inch sinter (−2.5 cm) is all returned to be added to raw concentrate feed to the sintering operation.

Raw concentrate, sintered without dilution by low-sulfur sintered fines, will roast too

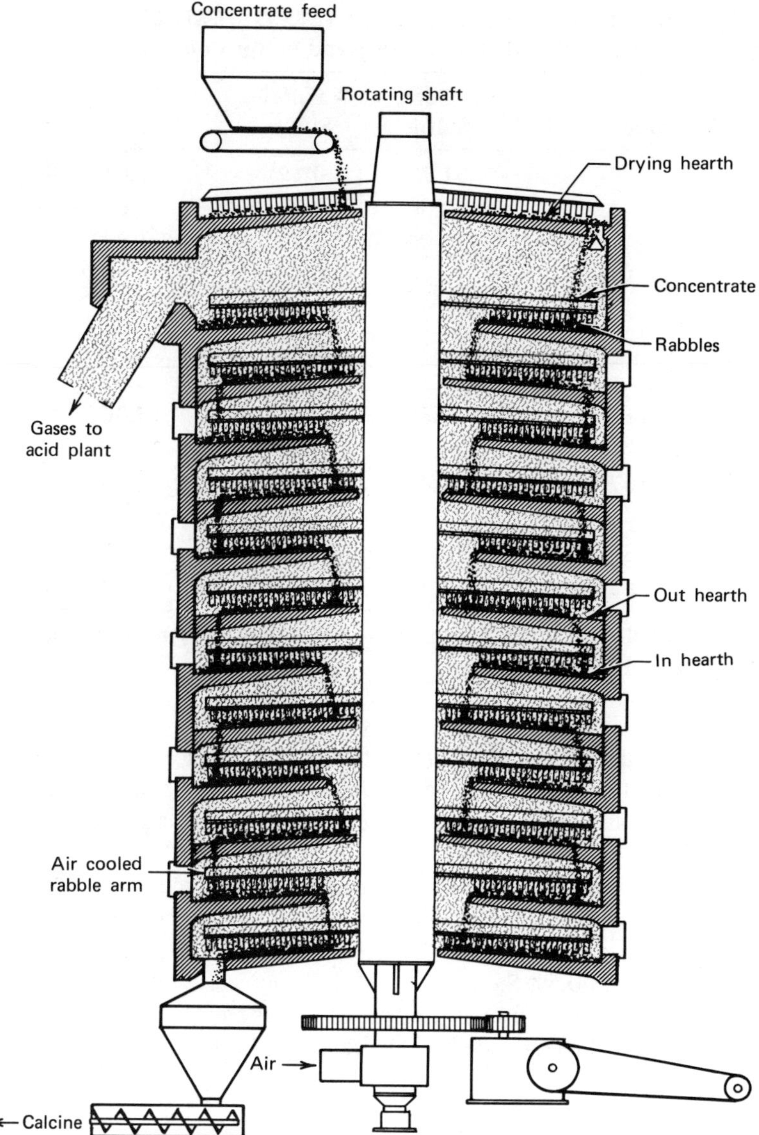

Figure 3.5. Zinc sulfide concentrate multiple-hearth roaster. *Source:* Courtesy of St. Joe Minerals Corporation.

fiercely, and there is danger of the exothermic heat of oxidation melting the charge, fusing it together and forming liquid compounds which will stop the roasting action. Mixing in of presintered fines in proportions of four or five parts sinter to one part raw concentrate keeps the amount of the sulfide fuel in the mixture at a low level, at 5 to 6% S, and permits close, smoothly controlled roasting. The sintered calcines produced are below the 1% S level desired, and as the level of sulfur in the feed mixture is comparatively low, excessively high sintering temperatures that would melt the charge are avoided.

Sintering is also used as the second stage of a two-part roast begun in a multiple-hearth or fluid-bed roaster. The roast may leave as much as 8% S in the calcines, which can be successfully sintered without additional fuel or only a small amount (5%) mixed in; or the calcines can be much lower than 8% S, in which case there is insufficient heat from sulfur oxidation alone for sintering, and 10 or 11% of finely ground coke or coal must be added as additional fuel.

The different proportions of the charge are carefully prepared and then mixed with water to give a uniform feed of proper density. The preroasted calcines are screened to

Table 3.3.   Zinc Multiple Hearth Roasting, ASARCO MEXICANA, Nueva Rosita Plant: Hearth Temperature and Sulfur Content

Hearth Number	Temperature (°K)	% Total Sulfur	% Sulfate Sulfur	SO_4/Total S
Feed	—	31.77	0.19	0.61
1	—	22.00	0.36	1.63
2	—	17.30	0.32	1.85
3	1103	16.90	0.52	3.07
4	—	16.20	0.47	2.91
5	1113	12.50	0.56	4.48
6	—	8.58	0.74	8.62
7	1123	6.38	0.98	15.36
8	—	4.63	1.00	21.60
9	1073	3.45	1.25	36.23
10	—	3.08	1.67	54.22
11	888	2.27	1.90	83.70
12	888	2.71	2.34	86.34

Source: C. H. Cotterill and J. M. Cigan, Eds., AIME World Symposium of Lead and Zinc, Vol. 2, 1970, p. 471.

remove any lumps, these lumps being crushed and returned, and the coke (or coal) is ball milled to be of sufficiently fine size. From 35 to 70% of the sinter being produced is recycled back and mixed in and gives an open charge for proper air penetration during sintering.

The sinter cake discharges from the end of the machine to a grizzly, and the fines dropping through the grizzly are returned as diluent to the charge mixer. The final sinter product is crushed and screened to $+\frac{3}{16}$ to -1 inch size ($+4.68$ to -25 mm), and sent to storage bins. It will average about 0.5% S.

The usual size of sintering machines used in zinc plants ranges from $3\frac{1}{2}$ feet wide by 45 feet long (1.06 × 13.5 m) up to 12 feet wide by 168 feet long (3.6 × 50.4 m). The hearth

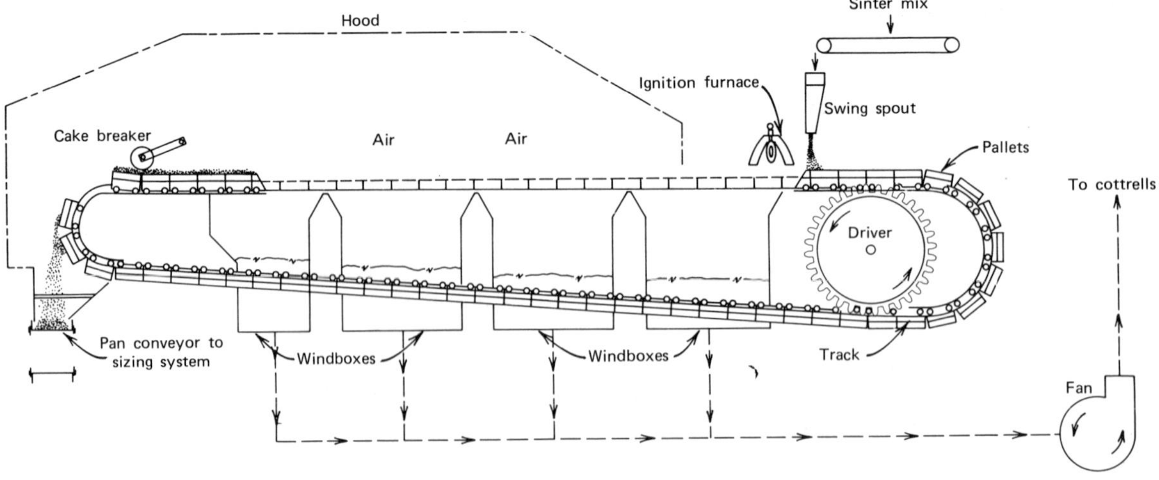

Figure 3.6.   Zinc sulfide concentrate roasted and agglomerated on downdraft sintering machine. *Source:* Courtesy of St. Joe Minerals Corporation.

area can be subdivided to enable separate removal of the first roasting gases richest in SO_2, which can be treated to produce sulfuric acid.

Sintering operations produce considerable fume of lead and cadmium, which is recovered in baghouses and electrostatic precipitators. Lead and cadmium chlorides volatilize at the sintering temperature, $PbCl_2$ at 1742°F (950°C) and $CdCl_2$ at 1760°F (960°C), and can be encouraged to form by wetting the sinter feed with a 1% solution of sodium chloride. The precipitated fume is sent to lead or cadmium plants for treatment.

Downdraft machines are widely used in the zinc industry. The downdraft is produced by sectionalized wind boxes installed beneath the line of travel of the pallets.

2. Magnesium Carbonate is first calcined to magnesium oxide in oil or gas fired rotating kilns at 1650°F (900°C) to drive off carbon dioxide and convert the raw magnesite ($MgCO_3$) to magnesia (MgO).

The magnesium oxide, which was ground in ball mills to 98% minus 100 mesh before calcining, is mixed with fine coal (200 mesh) and briquetted, using liquid magnesium chloride as a binder. The briquettes are coked in rotary kilns at a temperature just sufficient to cement the mixture without charring it and are then charged to the chlorinators.

Chlorinators are tall, refractory-lined circular steel shells $24\frac{1}{2}$ feet exterior height by $7\frac{1}{2}$ feet interior diameter (7.47 X 2.29 m), with a sealed charging hopper at the top to prevent gases produced from escaping. Heat is maintained at 1562°F (850°C) by electrical resistance through the briquetted charge between two sets of three 15 inch diameter (37.5 cm) graphite electrodes. One set of electrodes is inserted 7 feet (2.3 m) up from the interior furnace bottom and the other set, 8 inches (20 cm) from the furnace bottom. Magnesia briquettes are loaded in through the top-charging hopper in 300 pound batches (136.2 kg) and keep the chlorinator full of charge. Gaseous chlorine is fed through three equispaced inlet pipes close to the bottom of the vessel and is preheated as it rises through the electrically heated zone to contact and react with the briquetted charge.

The reaction that takes place transforms the magnesium oxide to magnesium chloride,

$$MgO + 2Cl + C = MgCl_2 + CO$$

with the magnesium chloride, which collects as a watery liquid in the bottom of the chlorinator, being drained out about once an hour and carried off to the next processing step.

The gases from the reaction are passed out through a gas outlet port at the top of the chlorinator, and contain some magnesium chloride as well as hydrochloric acid, carbon

Table 3.4. Typical Zinc Sintering Operations

	Operation		
	1	2	3
New feed material	Calcine	Calcine	Concentrate
Total charge capacity (tons/day)	240–300	400–450	550–600
Machine size (ft)	3.5 X 45	6 X 97	12 X 168
Fuel added to feed (%)	6–7	10–11	0–2
Total sulfur in new feed (%)	8	2	31
Recycle (% new feed)	35–75	40–70	80
Operating temperature (°F)	1900	1900	1900
Dust in offgas (% feed)	5	5–7	5–10
Offgas SO_2 content (%)	1.5–2.0	0	1.7–2.4

Source: U.S. Bureau of Mines Information Circular 8629, 1974.

monoxide and dioxide, chlorine, and hydrocarbons. The gases are scrubbed in towers from which the magnesium chloride and hydrochloric acid are recovered. This acid solution is neutralized with magnesium oxide to form the magnesium chloride used as the binder in the briquetting process. The solution has a strength of about $2\frac{1}{2}$ pounds (1.14 kg) of magnesium chloride per gallon (4.55 litres).

The briquettes are practically all used up during the chlorination reaction, except for a small residue left of silica, alumina, and iron oxide. This residue gradually accumulates, and every three or four weeks it is necessary to shut down the chlorinator to clean it out, and also to repair or replace the electrodes, as may be required.

Dolomite, $MgCa(CO_3)_2$, containing 13% magnesium as compared to 29% magnesium in $MgCO_3$, is also used as a source of magnesium metal and is calcined to $MgCa(O)_2$ prior to a distillation processing to remove the magnesium as metal. Dolomite, 22% MgO and 30% CaO, is crushed in jaw or gyratory crushers, washed, and ground in ball mills. The ground material is then heated in rotating kilns at 1800 to 2000°F (982 to 1093°C) to drive off carbon dioxide and produce calcined dolomite.

3. Titanium Oxide in the form of rutile (TiO_2) or ilmenite ($FeO \cdot TiO_2$) is converted to titanium tetrachloride by chlorination in the presence of carbon. Rutile is the preferred feed material, even though ilmenite is the far more plentiful domestic ore. This is due to the fact that rutile is about 95% TiO_2, as compared to 40 to 60% TiO_2 for the ilmenite, and also because of the economic and engineering problems arising from iron oxide in the ilmenite also chloridizing and then having to be separated from the resulting titanium tetrachloride. Both static-bed and fluid-bed chlorinators are used, with American plants preferring the static-bed type.

The Static Bed Chlorinator uses sintered feed, which is prepared by mixing measured amounts of fine rutile, ground coke (or coal), and coal tar in a pug mill, after which it is dumped on, and then spread out into, a uniform layer on the moving table of a circular kiln. The table carries the mixture into the hot, burner-heated zones of the kiln, and in this first kiln zone air is admitted to burn the volatile material given off by the coal tar. Air is limited in the subsequent zones to restrict the oxidation of the carbon in the mixture. But despite this there is some carbon loss, and extra coke is formulated into the mix to retain sufficient carbon to react with the titanium oxide during chlorination. The last section of the kiln is unheated and serves as a cooling zone, after which the finished sinter is scraped off the table and transferred to storage bins prior to being charged to the chlorinator.

The Fluid-Bed Chlorinator has some advantages during operation in that the sintering or briquetting step is eliminated and ground coke and rutile can be fed into it directly. Also the continuous overflow of reaction residues does away with the periodic shutdowns necessary to clean out the static-bed type of unit.

Carbon resistor blocks, heated by graphite electrodes projecting through the side walls, fill the bottom two fifths of the static-bed chlorinator, and coked sinter is fed intermittently through a sealed feed chamber at the top to form a bed above the resistor blocks. The sinter bed depth is approximately equal to the resistor block depth, each being about two fifths of the chlorinator interior opening, and each is about 8 feet (2.44 m) deep.

Gaseous chlorine is fed into the bottom of the vessel through inlet pipes and is preheated by its passage upward through the hot carbon resistor blocks before reaching the sinter and reacting. Chlorine reaction efficiency is 75 to 85%:

$$2TiO_2 + 3C + 4Cl_2 = 2TiCl_4 + 2CO + CO_2$$

Gaseous titanium tetrachloride is evolved, along with carbon monoxide, carbon dioxide, and the excess unreacted chlorine. If pure TiO_2 and carbon are used in the charge there should, theoretically at least, be no residue and all of the sinter should be converted to gases. In practice however there is silica impurity in the rutile, ash from the tar binder, and bits of unreacted residue, all of which build up in the chlorinator so that it must be shut down periodically and cleaned out.

The temperature in the chlorinator is maintained at 1472 to 1832°F (800 to 1000°C) by heat from the electrical resistance of the carbon resistor blocks and from the exothermic heat of the chlorination reaction. This extra electrical resistance heat is required during the reaction to bring the reactor charge up to the required reaction temperature and to balance the heat losses during the operation.

The chlorinators are tall, circular, brick-lined steel cylinders, 20 feet high and 13 feet in diameter (6.1 × 3.96 m). The electrodes are inserted about 6 feet (1.83 m) from the bottom of the furnace, and the chlorine inlets project through the shell at the bottom of the vessel, underneath the carbon resistor blocks. Charging is carried out intermittently through a sealed hopper at the top of the chlorinator, and just below and to one side of the charge opening is a port carrying off the gases of reaction. The rate of chlorination is 40 to 50 pounds (18.2 to 22.7 kg) of $TiCl_4$ per hour per square foot (0.09 m^2) of chlorinator bed area.

The carbon tetrachloride, which boils at 277°F (136°C), passes out of the chlorinator as a gas, and after first removing most suspended unreacted rutile and carbon dust solids from the gas in a cyclone dust collector, the cleaned gas goes on to a $TiCl_4$ condenser. This is a vertical tower into which cold, liquid circulation titanium tetrachloride is sprayed, cooling the $TiCl_4$ in the chlorinator gas, and dropping it out as liquid droplets. The condensed liquid tetrachloride is run to a settling tank, where any sludge and insoluble matter also removed by $TiCl_4$ spray will settle out. This leaves a clear liquid layer of $TiCl_4$ above the settled solids. The gas, cleaned of metallic chlorides, is passed out of the condenser for disposal.

Most of the metallic impurities in rutile are as easily chlorinated as is titanium, including iron chloride, vanadium oxychloride, silicon tetrachloride, tin tetrachloride, and several other metallic chlorides. The bulk of these chlorides are insoluble in titanium tetrachloride; and as they have higher boiling points, they can be separated by fractional distillation quite easily. Vanadium oxychloride is a special case as it boils at 262°F (127°C), which is very close to the boiling point of 277°F (136°C) for $TiCl_4$. It is not possible to design a distillation column to make this separation, but the $VOCl_3$ can be selectively precipitated chemically by hydrogen sulfide and is removed in this manner quite successfully in a large-scale operation.

The purity of the $TiCl_4$ produced will be on the order of 99.98%, and it is stored in

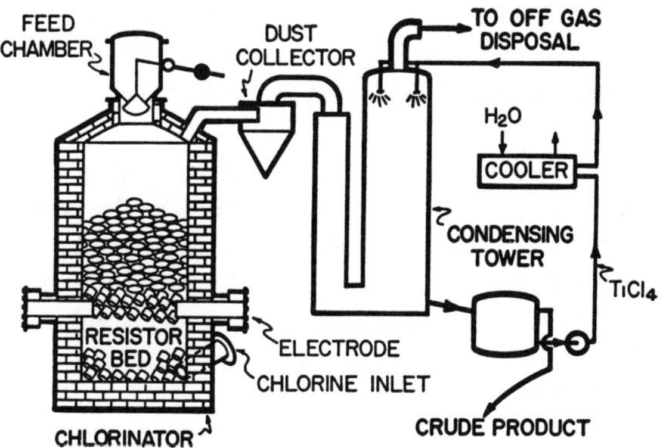

Figure 3.7. Titanium oxide chlorination. *Source:* R. L. Powell, *Chem. Eng. Prog.*, Vol. 50, No. 11, 1954, p. 579.

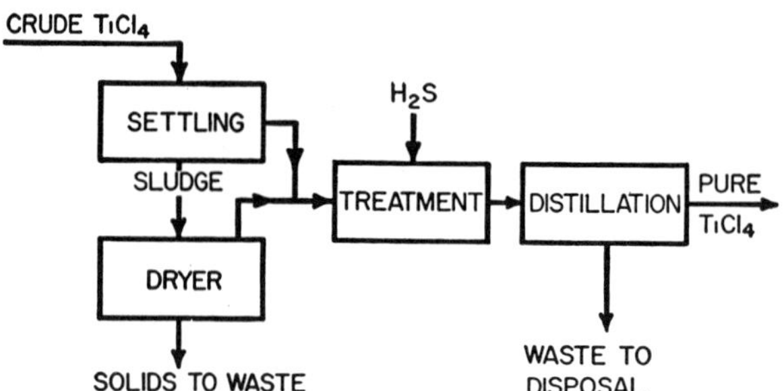

Figure 3.8. Purification of titanium tetrachloride. *Source:* R. L. Powell, *Chem. Eng. Prog.*, Vol. 50, No. 11, 1954, p. 579.

mild steel containers in the presence of helium or argon, which prevents hydrolysis or the adsorption of atmospheric oxygen or nitrogen.

SMELTING

Smelting is a process carried out with the intention of liberating the metallic element from the compound in which it is contained in the smelting charge and in most cases liberating it in a pure enough form to be commercially useable without further refining.

The charge to the smelting reactor can be both solid and liquid, depending on the process. In several instances it is a combination of purified compounds put together for inter-reaction rather than the extraction of a small amount of metallics from a large amount of associated waste material, as is the common smelting processing step for the treatment of nonreactive metals. However there are a few instances where this nonreactive type of processing is used for reactive metals as well.

Some of the processes are batch operations, and in these a solid charge is added. The metallic product is removed as a vapor which is condensed to liquid metal, while the now barren remaining part of the charge is removed and discarded. Other processes are continuous operations in which the reacting components are added either as solids or as liquids to be absorbed in molten salt baths and interreact there, with a steady flow of both charge to the reaction vessel and products removed from it after reaction.

The metallic values are recovered from the smelting process as either liquid or liquid and then solidified metallics, while the waste products are left as solid unmelted residues, liquid slag, gases, or liquid compounds that have solidified.

TYPES OF SMELTING

There are three main types of reactive metal smelting—reduction smelting in a retort or sealed blast furnace, electrolytic dissociation from liquid salt electrolyte, and inter-reaction of two purified liquid metallic compounds at an elevated temperature.

With Reduction Smelting the solid charge, which has been preheated to carry sensible heat to an endothermic reaction, is in the form of an oxide that can be reduced by carbon or ferrosilicon to liberate its metallic value. The reduced metallic value has a low boiling point and is removed as a metallic vapor which is then condensed to liquid metal. In retorting, the waste material in the charge is not melted and is removed as a solid to be discarded; while with the blast furnace and its higher temperature, the waste is melted and run off as a liquid slag.

Retorting can be either batch or continuous, while the blast furnace is a continuous process. Both are endothermic reactions.

Electrolytic Dissociation has either a liquid metallic chloride compound or a solid oxide powder dissolved in a molten electrolyte. Current is then passed to dissociate the metallic compound, and the released metal collects at the cathode, while gas is given off at the anode. Heat must be supplied to keep the electrolyte liquid and elevated to the required reaction temperature. These electrolytic cells are continuous operation units.

Interaction of Pure Liquid Compounds at an elevated temperature is an exothermic reaction between a pure liquid metallic chloride and a pure liquid metal, in order to reduce the metallic element of the chloride. This is a batch process. On completion of the reaction the liquid charge is cooled to solidify before removal and separation of the metallic value formed from the waste residue.

SMELTING PROCESSES

1. Zinc Sulfide Roasted Calcines are reduced from the oxide form by carbon (as carbon monoxide) to give zinc metal. This operation is carried out by a variety of retorting and smelting operations, some quite recent and others very ancient. These processes are the horizontal retort, the vertical retort, the electrothermic furnace, and the blast furnace, with the horizontal retort being a batch operation and the others all continuous processes.

In all these methods, coal or coke is used as the reducing agent and zinc is released as a metallic vapor that must be condensed to metal. Other impurity metals with vapor pressures close to that of zinc are also removed with the zinc and can be removed later by redistillation if a serious contaminant.

The retorts and blast furnace are designed so that the zinc vapor does not contact air or oxygen to reoxidize while still at a high temperature and must be condensed and cooled before being exposed to air.

The reduction reaction is endothermic, with the overall process written as

$$ZnO + C = Zn + CO \qquad \Delta H^\circ = +57{,}000 \text{ calories per mole}$$

and the intermediate reactions being

$$ZnO + CO = Zn + CO_2$$

$$CO_2 + C = 2CO$$

To provide the necessary heat for the reaction to proceed [temperatures reaching 2372 to 2462°F (1300 to 1350°C)], external heating of the retorts must be provided and fuel charged to the blast furnace. Heat is supplied by natural gas to the furnaces surrounding the horizontal and vertical retorts, by electrical energy to the electrothermic furnace, and by burning coke in the blast furnace. The charges are also preheated for the vertical retort, electrothermic furnace, and blast furnace; and the tuyere air to the blast furnace is heated in addition. All of these preheated charges carry sensible heat into the reaction vessel to aid in reaching the fairly high reaction temperature.

As the boiling point of zinc, 1664°F (907°C), is lower than the temperature at which carbon will begin to reduce zinc oxide, 2048°F (1120°C), this means that zinc will be liberated as a vapor rather than as a liquid metal. In this condition it is quite susceptible to oxidation by air and water vapor and by carbon dioxide, as the reduction reaction is reversible:

$$Zn + CO_2 = ZnO + CO$$

The use of excess carbon limits the amount of carbon dioxide and consequently the amount of reoxidation taking place, by having this excess carbon combine with any carbon dioxide to form carbon monoxide:

$$C + CO_2 = 2CO$$

Also, keeping the gases in the retort or furnace at a high temperature, approaching 1832°F (1000°C), until they pass into the condenser also restricts oxide formation, as this is above the zinc reoxidation temperature.

However in spite of these precautions it is not possible to eliminate completely the reoxidation of some zinc. Depending on the efficiency of the process, between 3 and 15% of the zinc metal produced can be lost as solid zinc particles coated with oxide. This is referred to as "blue powder," and it can be recycled through the distillation process to recover its metal content.

Horizontal Retorts are the oldest zinc-producing method, having been developed in Europe at the beginning of the nineteenth century, and still provide a considerable proportion of the world's zinc. At one time figures placed this as high as 37% of the world's production, but it has been steadily decreasing, largely replaced by leaching and electrolytic processing which now produces 78% overall of world zinc production. For instance there are no longer any operational horizontal zinc plants in the United States.

Rows of small refractory retorts are placed in a horizontal position inside a gas-fired regenerative furnace, which used the heat from the exit gases to preheat the air necessary for combustion. Several hundred retorts can be positioned in one furnace, with banks being four or five retorts high and a hundred or more retorts long.

The retorts are made on hydraulic presses from mixtures of raw and burnt clay in various proportions, a 50–50 mix being fairly common. The green clay retorts, containing 12 to 15% water, are taken to a drying chamber where the temperature is gradually increased to 158°F (70°C) over a period of several weeks. Next the retorts go to the baking ovens where the temperature is slowly raised over 48 hours to 2462°F (1350°C). From here they are taken as needed to replace cracked or broken retorts in the operating furnace. Retorts of silicon carbide and fire clay mixed with silicon carbide are treated in the same way and have partially replaced fire clay. The silicon carbide increases the cost but has greater resistance to thermal shock, greater thermal conductivity, and greater resistance to mechanical and flame abrasion, all of which give a superior retort with a four to five month life, compared to the 40 to 50 day life of the fire clay retort.

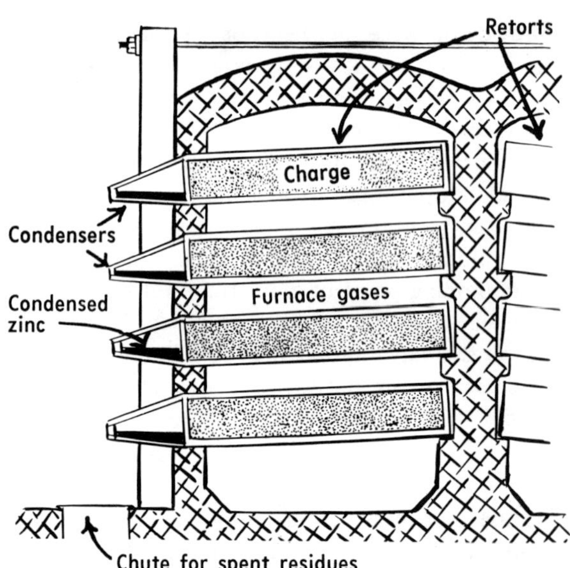

Figure 3.9.   Zinc horizontal retorts. *Source:* The Carborundum Company.

**Table 3.5.   Zinc Horizontal Retort—Clean-Out and Charging Machines,
ASARCO MEXICANA, Nueva Rosita Plant**

Clean-Out Machines

All motors except for the water pump, are 440 V, 60 Hz, 1180 rpm, Reliance equipment

Each machine uses one 1.49 kW traction motor, one 3.73 kW auger advance and regress motor with fluid drive, one 746 W hydraulic oil pump motor, one 1.49 kW water pump motor, and two 2.24 kW auger motors (right and left)

Augers at 120 rpm advance and regress at 14 m/min

Dodge No. 80, 25.4 mm pitch, riveted chain is used throughout

Complete machine with water load weighs 8 148 kg

Charging Machines

Charge belt motors are 440 V, 69 Hz, 1165 rpm, 5.6 kW made by Westinghouse

Elevator and conveyor belt gear motors are 746 W, 440 V, 60 Hz, 1800 rpm U.S. Motors with a 14.4 : 1 ratio, or 125 rpm output

Charge belt pulleys are driven by three "B" section Vee belts (6 belts, 3 each pulley)

Charge belt speed 704 m/min

Elevator belt speed 49 m/min

Conveyor belt speed 38 m/min

Machine weight 2500 kg

Source: Cotterill, C. H. and J. M. Cigan, Eds., AIME World Symposium of Lead and Zinc, Vol. 2, 1970, p. 489.

Condensers, which fit into the open end of the retort and project out of the hot furnace, are not subjected to the same high-temperature conditions as the retorts and consequently can be made of cheaper materials. Raw clay, often mixed with crushed broken retorts, is formed into a condenser shape by forcing a mandrel into a clay billet held in a cone-shaped mold, which is then dried and baked. A sheet iron cannister called a prolong is sometimes fitted to the outer end of the condenser to further enhance its condensing and collecting ability. As condensers have a much shorter life than do retorts, only six to seven days, their cost must be held to a minimum.

The premixed charge of solid zinc oxide and fine anthracite coal is charged into an empty retort by a mechanical charging machine. The retort capacity of $2\frac{1}{2}$ cubic feet (0.07 m^3) will take approximately 150 pounds (68.1 kg) of charge, this made up of 60 to 70% of zinc oxide powder and 30 to 40% fine coal by weight. The condenser is next fitted to the open end of the retort and fastened in place with moist clay. Then the prolong, if used, is attached to the protruding end of the condenser.

The charging machines consist of two grooved pulleys into which the charge falls through chutes from an overhead hopper. These pulleys, traveling at a high rate of speed, at about 2100 to 2300 feet (640 to 700 m) per minute, sling the charge into the retort and pack it from back to mouth fairly evenly. A lever adjustment, controlled by the operator, raises or lowers the charging machine position to fill retorts in the various horizontal bank levels.

The reduction reaction in the retort is that the carbon in the coal will reduce the oxygen in the zinc oxide, ultimately to produce metallic zinc and carbon monoxide:

$$ZnO + C = Zn + CO \quad \Delta H^\circ = +57{,}000 \text{ calories per mole}$$

As this is an endothermic reaction, heat must be provided to reach the required temperature of 2372°F (1300°C) where the reaction proceeds at a fairly rapid rate. This is accomplished by the retorts being enclosed in a furnace and heated externally.

As distillation proceeds, a flame of escaping CO burns to CO_2 at the mouth of the condenser, while the zinc vapor distilled from the retort reaction collects in the condenser as liquid zinc. The condenser temperature is rather critical and must be maintained within a range of 788 to 932°F (420 to 500°C). If the temperature is below the melting point of zinc at 788°F (420°C), the zinc will solidify and become oxidized as blue powder; while if the temperature is much above 932°F (500°C), the zinc vapor will not all condense and is lost to the atmosphere.

Both 24 and 48 hour retort cycles are used, with the actual retorting reaction taking up all but $4\frac{1}{2}$ hours of the cycle, this shorter time being used for charging and removing spent residues from the retorts. With the 24 hour cycle three tappings of metal are made, and with the 48 hour cycle there are five tappings. In each case the last draw of metal is made just before removal of the spent residue at the end of the cycle. The drawing of the metal is a lengthy procedure requiring some 4 hours for each draw in a 400 retort operation.

The zinc metal is run by gravity from the condensers into a kettle carried on an overhead rail in front of the retorts, with a scraper used to drag out the last portions. Blue powder, zinc dust, and other residues are also scraped out at this time and as they float on the liquid zinc surface are easily removed by skimming with a perforated shovel.

The first draw can be cast into ingots and sold directly, but the last draw is practically always off-grade due to high iron or lead content; the quantity of these metals distilling off with the zinc increases as the retorting temperature rises during the cycle. This off-grade product must be either redistilled or sold at a discount.

In a 48 hour, five-draw cycle the total percent of zinc removed in each subsequent draw would be 25, 26, 23, 16, and 10%. Plant recovery is rated as averaging between 88 and 90%.

At the end of the cycle, after the last metal drawing has been made, the condensers are removed, and water-cooled augers rotating at 120 rpm bore out the still solid, spent residue from the retorts. The residue is then either treated to recover any remaining metal content, or is discarded.

The retorts are now recharged and the reduction cycle repeated.

Vertical Retorts, developed by the New Jersey Zinc Company, differ from the older horizontal retorts in that while still remaining a retorting process, the vertical retort is much larger and is a continuous operation. The labor required for its operation is consequently much less, and because of its superior silicon carbide construction the retorts last two to three years before requiring replacement. After this time the retorts begin to leak zinc vapor through the refractory joints into the heating chamber, with a subsequent loss of zinc metal production.

The retorts, which are 40 feet high, 7 feet long, and 1 foot across ($12.2 \times 2.13 \times 0.3$ m), receive a charge of 3000 pounds (1362 kg) of coked briquettes every 27 minutes and will produce 10 tons zinc metal every 24 hours. The briquettes are a mixture of 60% zinc oxide and 25% fine bituminous coal, with a binder of 8 to 9% clay and 1% sulfite liquor. These components are thoroughly mixed together in a pug mill and then put through a roll-briquetting press. These first briquettes are fed to a second roll-briquetting press, which serves to give a further densification of the briquetted feed prior to coking. The final briquettes are pillow shaped and weigh slightly over 1 pound (454 g).

It is necessary to coke the green briquettes in order for them to be strong enough to withstand the rough treatment of handling and charging into the vertical retort, and also to maintain their shape during passage through the retort and while being removed from it and carried off as spent residue. The charging of hot briquettes from the coking furnace also carries sensible heat to the retort as an aid in supplying heat for the endothermic reaction.

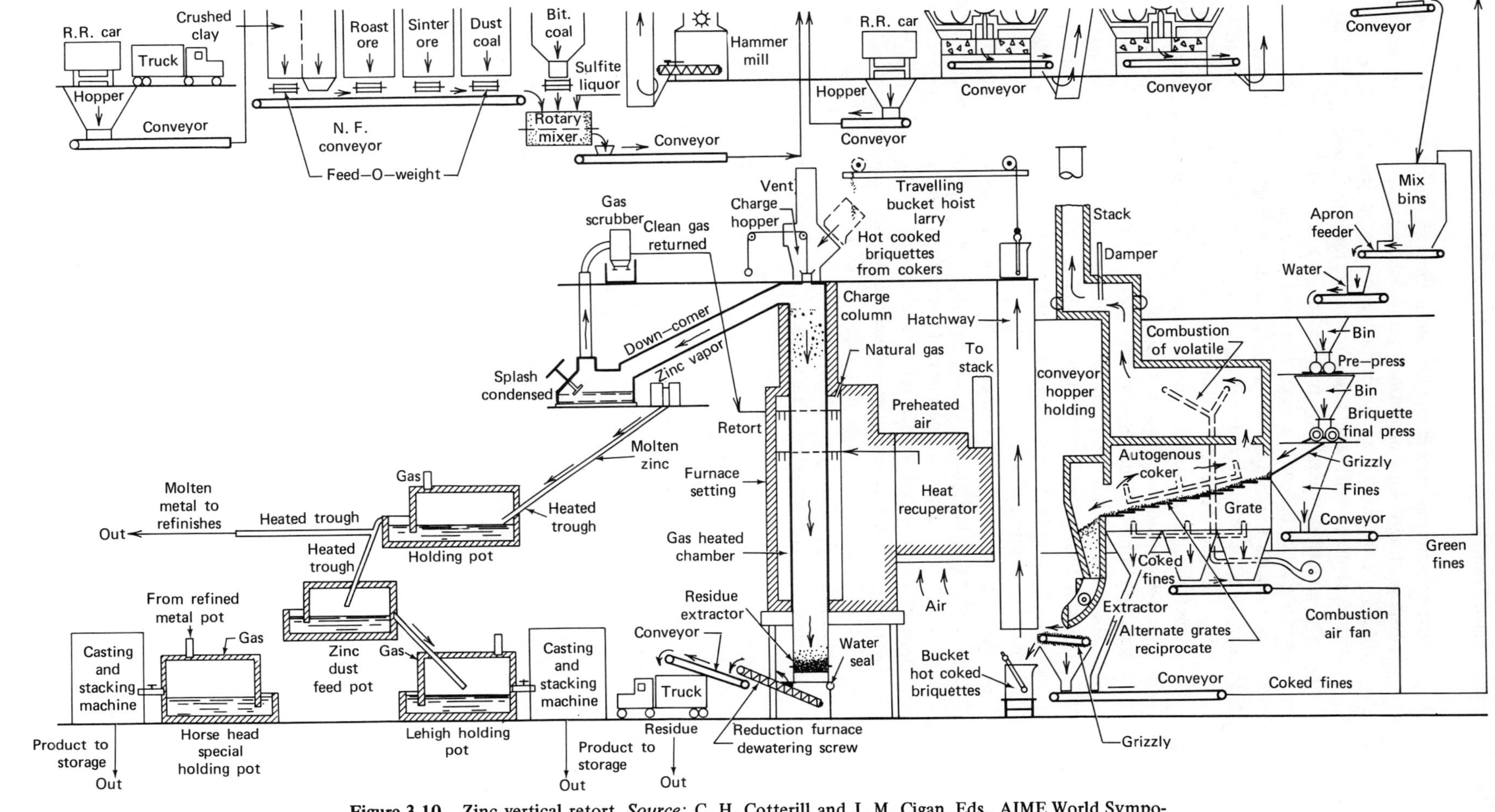

Figure 3.10. Zinc vertical retort. *Source:* C. H. Cotterill and J. M. Cigan, Eds., AIME World Symposium of Lead and Zinc, Vol. 2, 1970, p. 527.

The bituminous coal in the briquettes provides volatiles as fuel to support a thermally self-supporting coking operation. Temperatures in the coker are controlled not to exceed 1652°F (900°C), so that the briquettes on discharge are largely volatile free but have not been heated above the zinc reduction temperature of 2048°F (1120°C).

The green briquettes drop in a continuous flow from the second roll-briquetting press over a grizzly to remove fines and then into the coker, which is essentially an upright furnace with a series of downwardly inclined step grates. Alternate grates have a slow reciprocating motion which gives a forward motion to the bed of briquettes, and the briquette feed rate and grate motion speed provide a thin layer of uniformly distributed briquettes throughout the whole coker. Combustion air is provided through the grates from below, and this combines with the distilled volatiles from the coal to maintain the necessary combustion temperature.

The coker has a holding chamber at its bottom with a roll-discharge mechanism, and on a regular time schedule 3000 pounds (1362 kg) of briquettes are run out into a boxlike steel "coke bucket" and hoisted to the retort charging floor to be dumped hot into a retort for reduction.

The upper 5 or 6 feet (1.52 or 1.83 m) of the retort projects above the heating chamber which surrounds the retort, and it is into this upper portion that the hot, coked briquettes are charged. The internal pressure in the retort is quite low, so a simple, round tapered cap covered with a few shovelfulls of blue powder is adequate to prevent gas leakage from the charge port between chargings. The retort is kept filled from top to bottom with briquettes, and it takes 24 hours for a freshly charged briquette to pass down through and out of a retort.

Heating chambers 35 feet high (10.67 m) enclose the two long (7 feet, 2.13 m) walls of the retort, and heat from them is transferred through the silicon carbide retort walls to give the heat required for the reduction reaction of zinc oxide by carbon to take place and zinc metal vapor to be released. Fuel for these heating chambers is 70% natural gas and 30% scrubbed CO by-product gas recovered from the reaction in the retort.

The gaseous products from the retort reaction consist of 40% zinc vapor and 45% CO, and the remainder of hydrogen, nitrogen, and CO_2. These gases exit close to the top of the retort into a downward-sloping, rectangular-shaped refractory flue, which leads them into a splash condenser.

The splash condenser is an enclosed, elongated refractory chamber holding a bath of liquid zinc at a fixed level by continuous overflow and maintaining a relatively constant temperature at about 932°F (500°C) by an immersed water-cooled coil, which is submerged or raised by thermocouple control. A graphite impeller 14 inches (35 cm) in diameter rotating at 400 rpm is dipped into the zinc bath and throws up a shower of liquid zinc droplets which entirely curtains the end of the flue out of which the retort gases must pass. These zinc droplets combine with and condense the zinc vapor in the gas and add it to the liquid metal pool in the condenser. There is relatively heavy wear on the graphite impellers; their life is about three weeks before they need replacement.

The condenser must be kept within a temperature range where it is cool enough to condense the zinc vapor, but not so cool that the zinc vapor solidifies and blue powder is formed. The rapidity of condensation is an aid in minimizing blue powder formation, and it can be kept down to 3 to 5% of the zinc produced. Part of this powder floats on the condenser surface from where it is skimmed off, and part proceeds on with the CO gases passing through the condenser to a scrubber tower. Here the gases are cleaned of blue powder and then recycled back as fuel in the retort furnaces. The collected blue powder is dried and used as a sealant over the charge caps on the retorts.

The molten zinc overflows the condenser into a launder and flows to a holding furnace from which some of the metal is cast into slabs for market, and part is refined to lower the cadmium and lead contents. Zinc recovery is on the order of 97%.

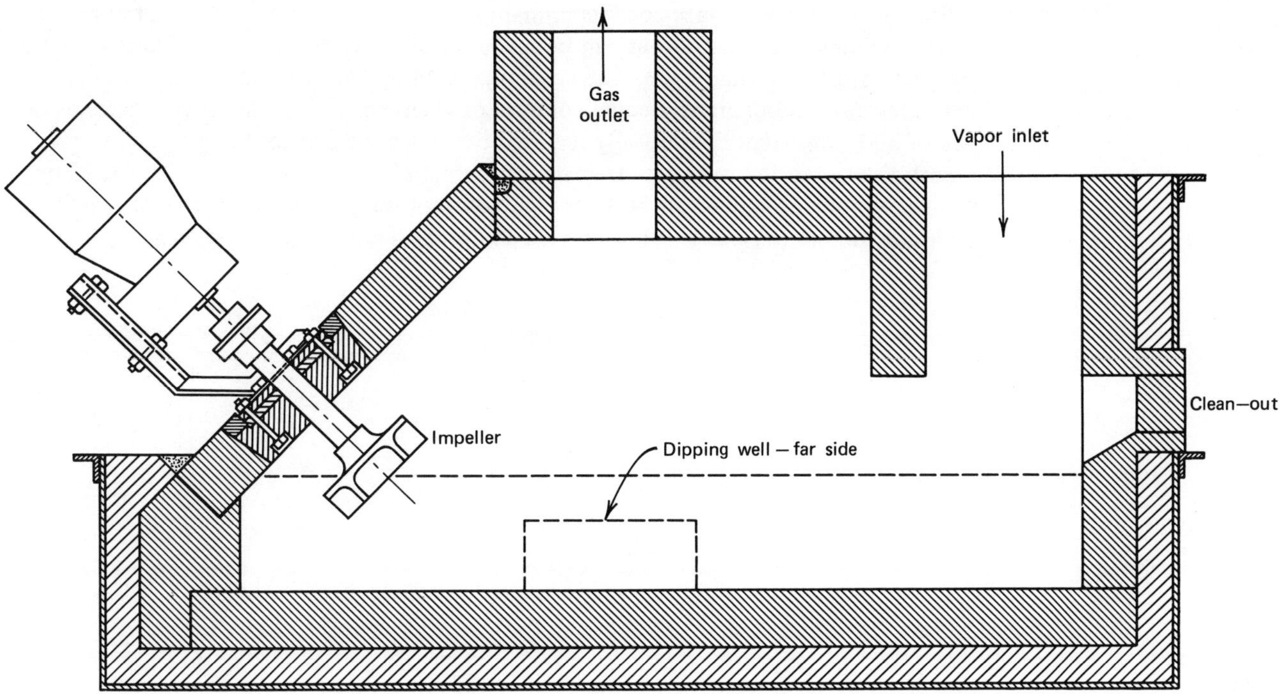

Figure 3.11. Zinc vertical-retort splash condenser. *Source:* C. H. Cotterill and J. M. Cigans, Eds.. AIME World Symposium of Lead and Zinc, Vol. 2, 1970, p. 540.

As the briquettes pass down through the retort they retain their shape due to their inherent strength but lose 60% of their weight as the reaction of reduction takes place and the zinc and carbon in their constituents is removed. The residue briquettes contain 25 to 30% carbon, 3% zinc, and other trace metals present in the feed ore. A slowly rotating drum extractor called a spider withdraws the spent briquettes from the bottom of the retort and drops them into a closed water bath, from which they are continuously removed by a screw conveyor. The rate of withdrawal by the rotating drum is synchronized to maintain a retention time of the briquettes in the retort to give the 97% zinc extraction required.

The spent briquettes are sometimes used for such purposes as fuel for heating boilers because of their still relatively high carbon content or can merely be discarded as waste to the dump.

The Electrothermic Furnace, a development of St. Joe Minerals Corporation, is also a continuous process with a zinc production of close to 100 tons per day at an efficiency of 92% recovery. The major difference in the construction of this unit and the vertical retort is that heat is generated internally by resistance to current flow by the charge in the furnace, between sets of graphite electrodes inserted through the top and bottom furnace walls.

The furnace is circular, 45 feet high and $7\frac{1}{2}$ feet in diameter (13.7 × 2.28 m), and constructed of a series of short sections of high and superduty firebrick. Steel water jackets enclose the lower half, the main smelting zone, of the furnace below the vapor ring. This annular vapor ring is about in the middle of the furnace, and through it the distilled zinc vapor and furnace gases pass to the condenser. The vapor ring and condenser are both lined with silicon carbide bricks.

The charge to the furnace consists primarily of zinc oxide sinter and coke, but as much as 25% of the total zinc output can be made up of other zinc-bearing materials such as granules, dross, screenings, etc. The coke rate is 44% of the weight of the sinter, which translates into approximately equal volumes of each, with a stochiometric excess of carbon of 300% over that theoretically required to reduce the zinc in the sinter.

Both the sinter and coke are sized to approximately 1 inch by $\frac{1}{4}$ inch (25 × 6.25 mm) and drawn out of individual bins through constant-weight feeders to a rotary, gas-fired preheater, where the charge is both mixed and preheated. This preheater is 5 feet in diam-

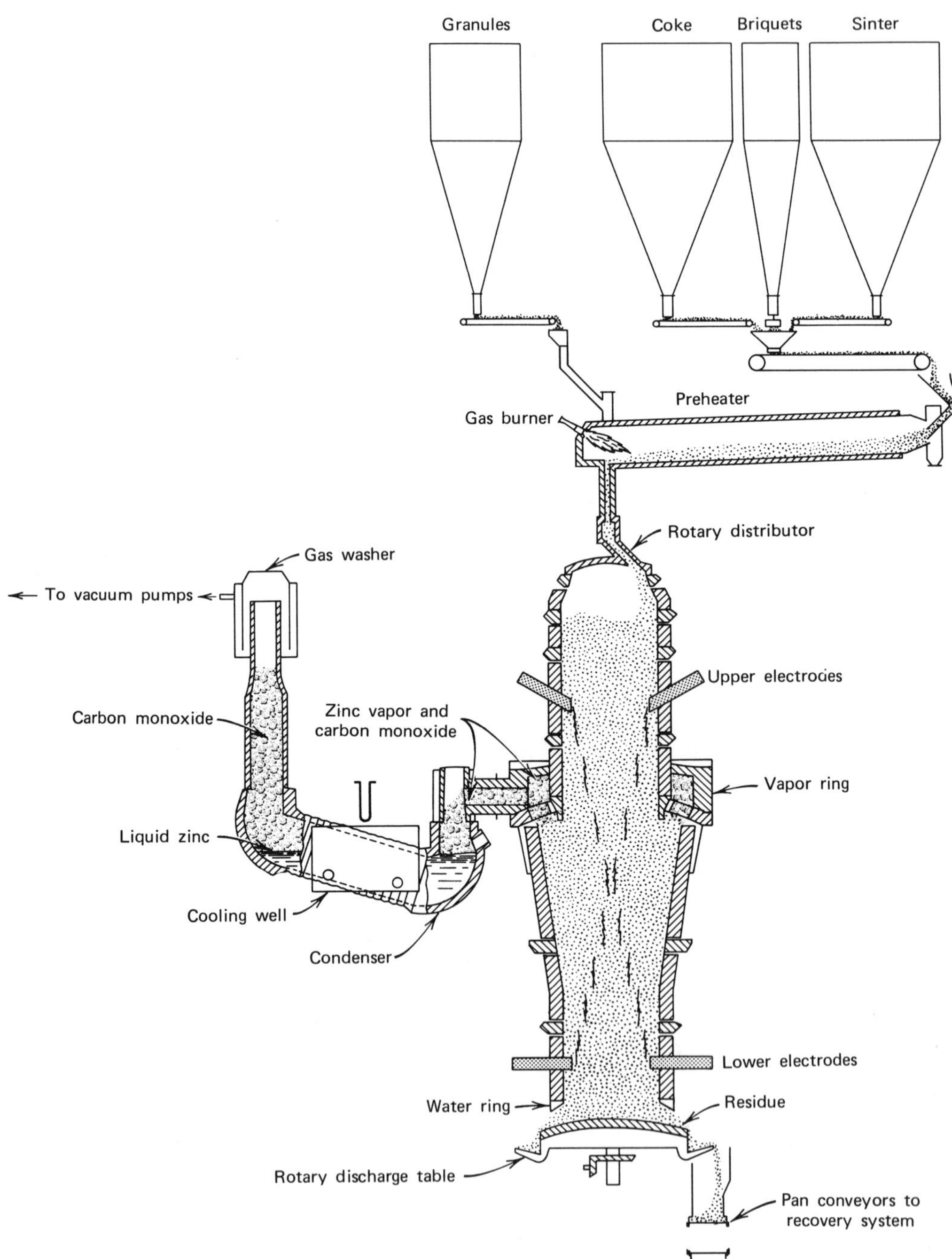

Figure 3.12. Electrothermic zinc furnace. *Source:* Courtesy of St. Joe Minerals Corporation.

eter by 29 feet long (1.52 × 8.84 m), is fired by CO gas recovered from the furnace reaction, and discharges the now well-mixed charge at about 1382°F (750°C).

A rotating distributor slowly turning at 0.4 rpm continuously spreads the hot feed in the top of the furnace and at the same time seals the top of the shaft to prevent the escape of any gases or metal vapor. The furnace is kept full of charge, and the time required for fresh charge to pass down through and out of the furnace is about 22 hours.

Heating in the furnace is accomplished by the resistance of the charge to current flow between eight equispaced, pairs of graphite electrodes; one of the sets of pairs is situated near the top of the shaft, and the other set of the pairs is close to the bottom. The electrodes are 8 inches (20 cm) in diameter and $29\frac{1}{2}$ feet apart (9.0 m), with the top electrodes angling downward into the furnace at 30° for a distance of 19 inches (47.5 cm). The bottom electrodes protrude at right angles into the furnace for 6 inches (15 cm). The top electrodes are consumed at the rate of about $1\frac{1}{2}$ inches (3.75 cm) per day and need repositioning every few days, while the lower electrodes have much less consumption and as a consequence do not require repositioning. Total electrode consumption amounts to about $2\frac{1}{2}$ pounds (1.14 kg) per ton of zinc metal produced.

The total power to a furnace will be 10,000 kW at 200 to 230 volts, and resistance heating from this will give a temperature in the center of the furnace at the vapor ring elevation of 2192 to 2552°F (1200 to 1400°C) and a temperature of 2372°F (1300°C) at the lower electrode elevation.

The gases that escape out the vapor ring and on to the condenser leave the furnace at 1562°F (850°C), and are composed of 45% zinc vapor and 45% CO and the remainder nitrogen, hydrogen, and CO_2.

The condenser, which is situated just below the vapor ring, is U-shaped with a vertical inlet and outlet. The bottom part of the U is the metal condenser pool, and it is into this liquid zinc pool that the condenser, operating under a vacuum of 10 to 12 inches of mercury (25 to 30 cm), draws the furnace gases to have the zinc vapor condense here and add to the pool volume. The condenser is made of boiler plate lined with silicon carbide refractory brick, and the bottom of the U is 19 feet long (5.8 m), has a cross-sectional area of 22 square feet (2.04 m^2), and is inclined upward, away from the furnace, at about 22° to the horizontal. Water-cooled pipe loops, in at attached cooling well, control the temperature of the metal in the condenser at 896 to 932°F (480 to 500°C), and the condenser along with the attached cooling well holds 45 tons of zinc metal.

This zinc is tapped periodically and cast into slabs for market, or else goes to the refinery to be processed into higher-grade zinc.

The furnace gases pass through the condenser and into a high-velocity water scrubber where the entrained solids and blue powder are removed. This slurry is treated to recover the solids which are briquetted and returned to the furnace. The clean gas, which is 80% CO, is piped around the plant for fuel.

The spent residue from the bottom of the furnace consists of coke, sinter residue, and some liquid slag. However the excess of coke keeps the charge from being stuck together by the liquid slag, and it is not difficult to handle.

The bottom of the furnace wall is supported by a water-cooled steel ring, and 12 inches (30 cm) below this is a refractory covered discharge table rotating at 0.03 rpm. This slowly rotating table continuously removes some of the spent residue, which is resting on it, out of the bottom of the furnace. But most of the spent charge is extracted by periodically ramming a heavy bar with an arrowhead point into the exposed residue, in the 12 inch (30 cm) gap between furnace bottom and rotating table, and then withdrawing it. This drags out the residues, which fall into a pan conveyor and are carried off for disposal or further treatment. There are two poker bar machines to a furnace, one on each side, and their frequency of operation is controlled by a cobalt-60 gamma-ray source charge height detector in the upper shaft of the furnace. This maintains a retention time of charge in the furnace to give the required zinc extraction of 93%.

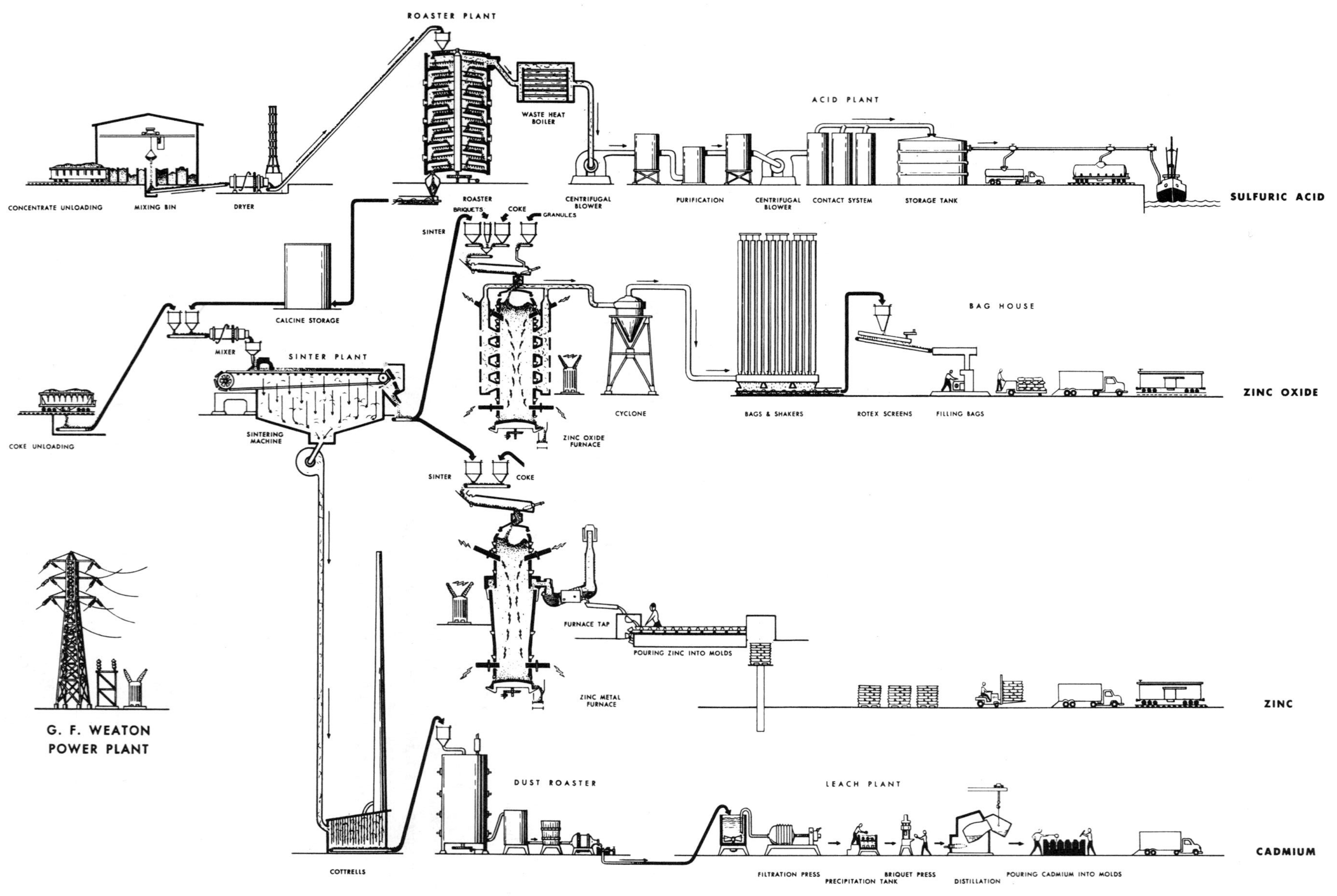

Figure 3.13. Josephtown electrothermic zinc smelter. *Source:* Courtesy of St. Joe Minerals Corporation.

Table 3.6.   Grades and Chemical Composition of Slab Zinc[a]

| | Composition (%) | | | |
	Lead, maximum	Iron, maximum	Cadmium, maximum	Zinc, minimum, by difference
Special High-Grade[b]	0.003	0.003	0.003	99.99
High-Grade	0.07	0.02	0.03	99.90
Intermediate	0.20	0.03	0.40	99.50
Brass Special[c]	0.60	0.03	0.50	99.00
Prime Western	1.60	0.05	0.50	98.00

Source: U.S. Bureau of Mines Information Circular 8629, 1974.

[a]Prime Western zinc was the specification first established for use in hot-dip galvanizing. Brass Special and Intermediate were largely used in alloying with copper to form brass. High-Grade, and later Special High-Grade, were established with the advent of alloys containing small amounts of aluminum, to be used in the die casting of intricate functional parts cast to very close dimensional tolerances.

[b]Tin in Special High-Grade shall not exceed 0.001%.

[c]When specified for use in the manufacture of rolled zinc or brass, aluminum shall not exceed 0.0005%.

The Zinc Blast Furnace process for producing zinc is the newest of the four pyrometallurgical processes, having been first commercially introduced in 1950 by Imperial Smelting Processes Limited. It is also much the largest production unit, with a daily output as high as 300 tons, as compared to 0.06 tons for the horizontal retort, 10 tons for the vertical retort, and 100 tons for the electrothermic furnace. There are now 12 zinc blast furnace plants scattered around the world, and from these comes 14% of world zinc production.

Normal blast furnace practice is carried out in that carbonaceous material is burned in close association with zinc oxide calcines to reduce the oxide to metal. As in the other zinc distillation processes, the zinc is released as a metal vapor which must be condensed.

The charge to the furnace consists of hot lump sinter direct from the sintering machines, preheated coke, and a small addition of lime for slag making.

The discharged sinter from the sintering machine passes first through a pronged breaker rotating at 3 rpm and then to a spiked roll crusher at 80 rpm, which takes the maximum size down to slightly over 3 inches (7.5 cm). Fines are screened out on a double-deck screen with 100 mm and 20 mm screens and are recycled back to the sintering machine. The preferred treatment for the lump sinter, to have it reach the furnace at the highest temperature, is to discharge it hot into 600 ton capacity insulated storage bins within a few minutes of leaving the sinter machine, or it can be quenched, stored in large bins of several thousand tons capacity, and withdrawn for furnace feed as required. A typical sinter will analyze approximately 42% zinc and 20% lead.

Coke, sized between 2 and 3 inches (5.0 and 7.5 cm) is put through a coke preheater to raise its temperature to 1472°F (800°C). This preheater is a refractory-lined steel shaft containing a column of coke and is heated by fuel gas from the blast furnace. A control valve on the exit stack keeps the preheater at a slight positive pressure to prevent air entering. This hot coke preheating improves the overall thermal efficiency of the blast furnace process.

Accurate proportioning of the feed (coke, sinter, and hard burned lime) is vital to proper furnace operation. To accomplish this a comparatively high degree of automation of materials handling is used. Each coke preheater, sinter storage bin, and flux storage bin

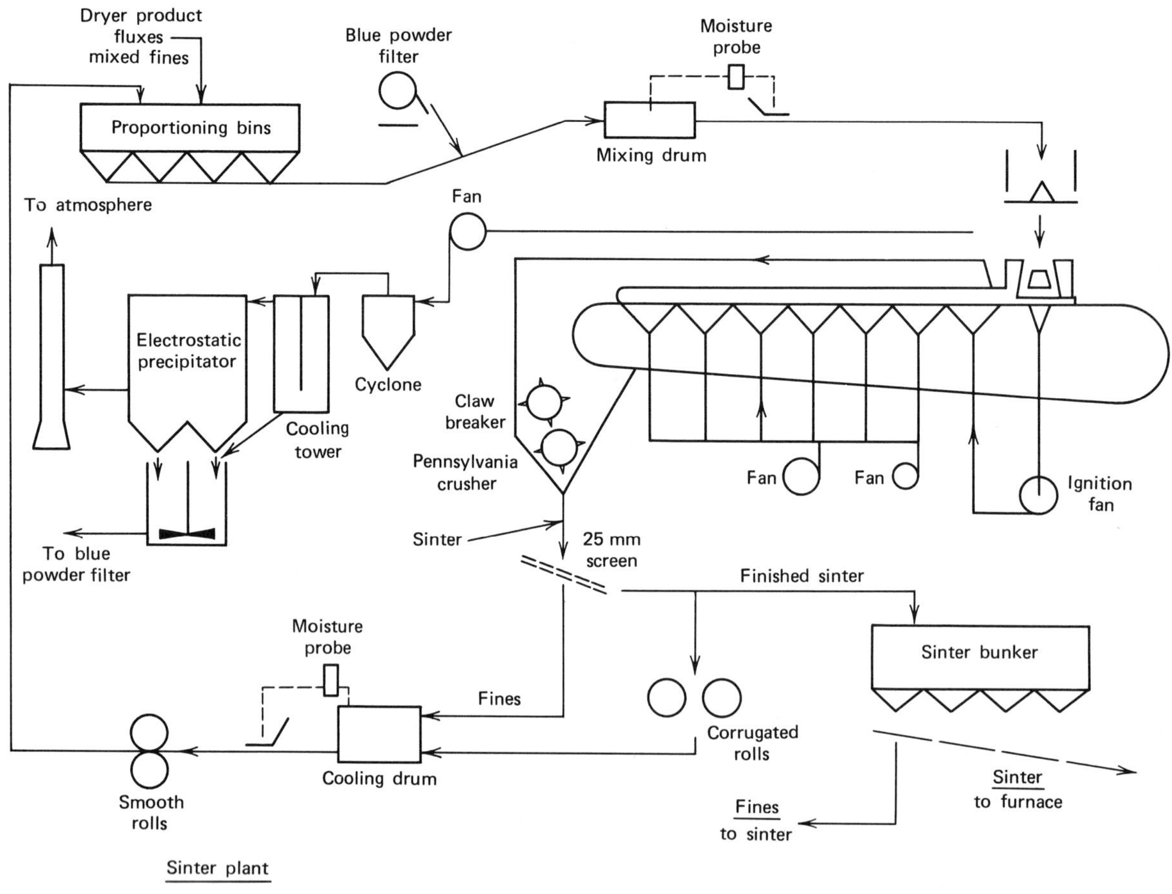

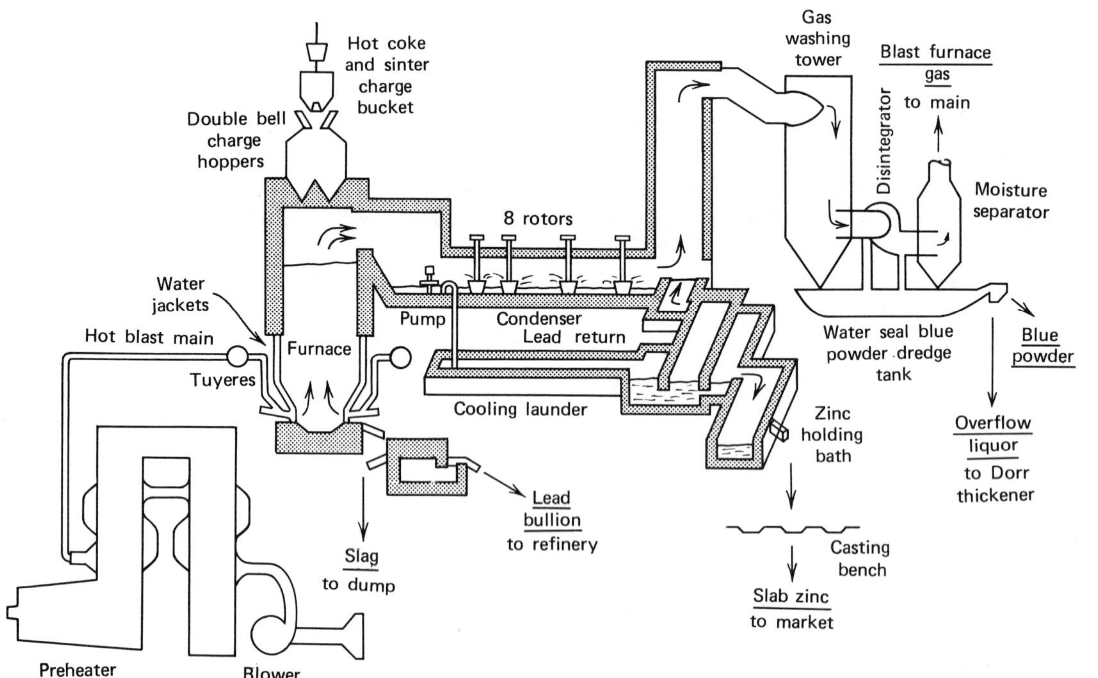

Figure 3.14. Imperial smelting furnace. *Source:* C. H. Cotterill and J. M. Cigan, Eds., AIME World Symposium of Lead and Zinc, Vol. 2, 1970, p. 687.

has a weight hopper which enables the precise amount of each component to be weighed out automatically. A transfer car brings two charge buckets beneath the weigh hoppers to collect the feed and moves from hopper to hopper until the proportioned charge has been gathered. The car then leaves the loading area and passes to a crane, which raises the buckets and travels along an overhead track to a point above two double-bell and hopper charging units in the furnace roof, where they discharge their contents into the furnace shaft. The weight ratio of coke to zinc in the feed is on the order of 0.8 coke to 1.0 zinc.

The furnace is rectangular in shape with rounded ends and is generally the approximate size of a lead blast furnace, with the shaft area varying between 185 and 292 square feet (17.1 and 27.0 m^2). The upper part of the furnace is a firebrick-lined steel shaft, while the lower part is constructed of hollow water cooled steel jackets, or in one case of $1\frac{3}{8}$ inch steel plates (3.43 cm) cooled by external water trickling, as is used in the steel industry. Two stoves are used to heat the tuyere air to 1742°F (950°C), operating on a 40 minute "on blast" cycle and a 30 minute firing cycle. Blast furnace gas or fuel oil is used for heating. These units contain 200 tons of checker bricks and are 85 feet tall and 17 feet in diameter (25.9 × 5.2 m).

A dome-shaped roof supports the two double-bell and hopper charging units, and cold air is introduced between the two bell valves to prevent gas leakage. There are also four equispaced top air inlets located in the furnace sides above the charge level. Preheated air is bled from the tuyere bustle pipe through these inlets. This air burns the CO being produced by the furnace reaction and raises the temperature of the whole gaseous mixture at this location as much as 212°F (100°C). The resultant temperature, which can be as high as 1922°F (1050°C), is too high for zinc metal vapor to reoxidize, and it passes out through two offtakes located opposite each other just under the domed roof into the condensers.

Mixed sinter of lead oxide and zinc oxide is used in the blast furnace, in the frequent proportion of about two zinc to one lead, (42% zinc and 20% lead). However it is not necessary to have this ratio, and the process can treat a wide range of compositions equally well, including those where the lead content exceeds that of the zinc. It can also operate on charges containing as low as 20% zinc, whereas the retorting processes require much higher concentrations in the 50 to 60% zinc concentrate range for successful operation.

The reduction of zinc oxide to metal is endothermic, but the reduction of lead oxide is slightly exothermic:

$$PbO + CO = Pb + CO_2 \qquad \Delta H° = -15{,}540 \text{ calories per mole}$$

which means that the lead oxide is reduced without the need for more than a small amount of additional coke. The CO furnace gas from the zinc reduction is diluted with CO_2 gas from the lead reduction, but it is still high enough in CO content to use as a fuel to preheat the coke and tuyere air blast. The furnace gases have a composition of 20% CO and 12% CO_2. The liquid lead that is formed, along with precious metals, copper, other reduced trace metals, and slag, flows to the bottom of the furnace and is tapped every $1\frac{1}{2}$ hours through a water-cooled copper tapping block into a forehearth. The lead bullion settles to the bottom of the forehearth, from where it is drawn off for immediate refining or cast into ingots for refining elsewhere. The slag runs over the top into a granulation system and from there goes to the dump. Metal values in the slag run 5% zinc and 0.5% lead.

There are two lead splash condensers, one connected to the gas offtakes on either side of the furnace. In these, a curtain of fine lead droplets, put up by rotor blades, absorbs heat from the vapor and reduces the temperature to 1022°F (550°C). At this point the zinc vapor is quickly condensed to liquid metal and is absorbed in the liquid lead. A common type of condenser is divided into three stages, with four rotors in the first stage,

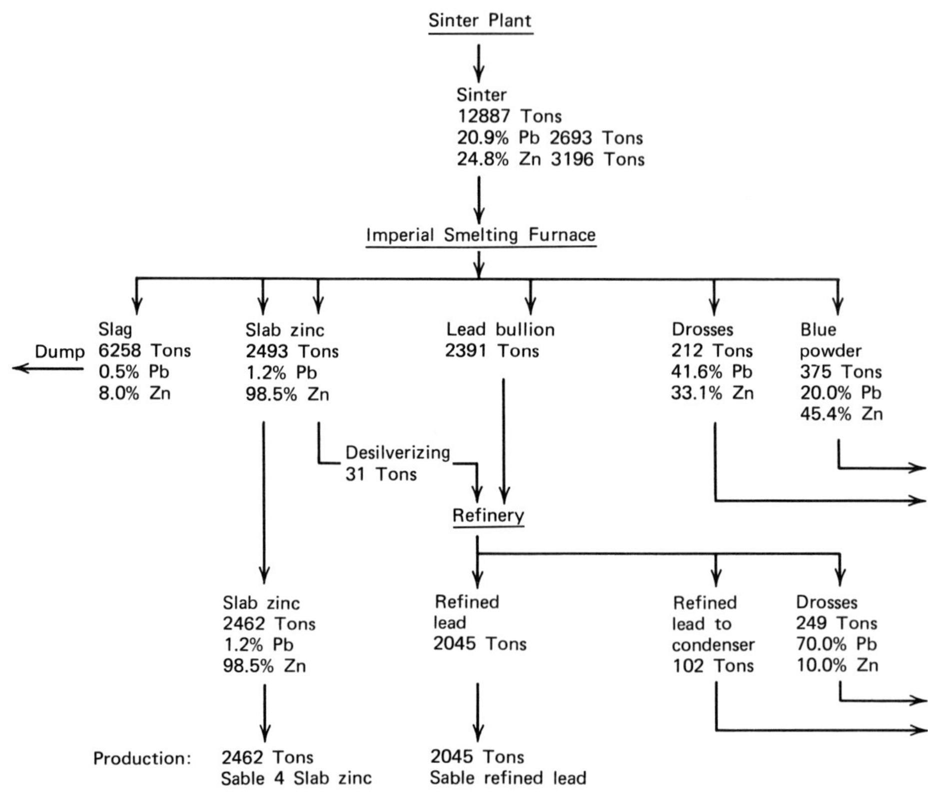

Figure 3.15. Zinc blast furnace treatment program for average month. *Source:* C. H. Cotterill and J. M. Cigan, Eds., AIME World Symposium of Lead and Zinc, Vol. 2, 1970, p. 680.

two rotors in the second stage, and two rotors in the third stage. The rotors are rotating vertical shafts equipped with blades which dip into the pool of lead in the bottom of the condenser and throw up the condensing and collecting spray of metal droplets. The droplets must be of sufficiently small size to provide maximum liquid to vapor contact, and this droplet size will depend on the shape of the rotor blades, the immersion depth of the blades, and its speed of rotation.

The hot lead with zinc in solution is run at 1022°F (550°C) from the condenser through an underflow baffle forming a gas seal into a cooling launder. These refractory-bottomed launders are made up of mild steel water jackets enclosed by refractory and steel casings. The hot lead as it leaves the condenser contains about 2.5% zinc and is un-saturated; but as it is cooled down to 824°F (440°C) in the launder, the lead has less solubility for zinc and is saturated at roughly 2.25% zinc. This excess zinc, containing some 1.2% lead, rises as a liquid layer above the lead in a mild steel quiescent separation bath, because of the difference in densities, and overflows continuously from here into a heated zinc holding bath. It is then cast into ingots or sent on for further refining.

The lead from the separation bath runs by gravity under a weir, which holds back the separated zinc layer, and returns to the metal pool in the condenser. The cooling duty in the condenser is quite heavy, and this cooled lead is quickly reheated by the great amount of heat it absorbs from the entering hot furnace gases. For heat extraction requirements, 400 tons of lead must be circulated through the closed system—condenser, cooling launder, and separation bath—for every ton of zinc produced; or for a furnace output rate of 300 tons of zinc per day, 120,000 tons of lead must be circulated. Zinc metal produced by the blast furnace is Prime Western grade (1.2% Pb, 0.02% Fe), and recovery is 92%.

The furnace gases leaving the condenser, at the end opposite the blast furnace, are at 842°F (450°C) and must be as low in temperature as possible to carry off minimum amounts of uncondensed zinc. The gases pass through two stages of cooling and washing in spray towers and mechanical scrubbers, producing a clean gas at below 104°F (40°C). The solids removed are returned to the sinter plant, while the gas containing some 20% CO is used as required for fuel in the coke preheater and blast furnace stoves. Any gas not needed for fuel is vented to the atmosphere.

2a. Magnesium Chloride is produced as either a watery liquid or as granular crystals, which are transported to an electrolytic cell to dissolve in a molten salt electrolyte and then with the passage of current to break down and deposit as magnesium at the cathode and chlorine gas at the anode.

Liquid magnesium chloride is the product of reacting calcined magnesium oxide with chlorine in a chlorinator, while the granular crystals are precipitated from sea water. In this second operation, known as the *Dow process*, the magnesium that is in the sea water in solution is precipitated as magnesium hydroxide in large settling tanks by mixing the sea water with a slurry of calcium hydroxide. The precipitated magnesium hydroxide slurry, which is 17% magnesium hydroxide by weight, is filtered, and then neutralized with hydrochloric acid to give a 36% solution of magnesium chloride. This solution is dehydrated in fluosolids dryers to granules of composition 74% $MgCl_2$-20% H_2O and passed on in this form to the electrolytic cell. With both forms of $MgCl_2$ feed material the flow to the electrolytic cell is continuous, except when metal is being dipped in order to maintain the correct bath composition and level within the cell.

The cell itself is a steel-shelled pot 16 feet long (4.9 m), 5 feet wide (1.5 m), and 6 feet deep (1.8 m), with parallel sides and a rounded bottom. The electrolyte is composed of 20% magnesium chloride, 57% sodium chloride, 2% potassium chloride, 1% calcium fluoride, and 20% calcium chloride. It is kept at a temperature of 1292°F (700°C) partly by resistance to the passage of current and partly by external heating underneath the cell. Higher currents and voltages will give greater resistance heating and require less external heat to be applied.

The steel shell of the cell is the cathode, and two rows each with 11 equispaced artificial graphite electrodes 8 inches (20 cm) in diameter and 9 feet (2.7 m) long are suspended into the cell, through openings in a refractory cover, to serve as anodes. Each cell operates at about 6 to 7 volts and 60,000 amperes, with a current efficiency of 75 to 80%. Power requirements are on the order of 8 to 9 kW-hr per pound (454 g) of magnesium produced.

The passage of direct current through the electrolyte between the steel pot cathode and graphite anodes breaks down the magnesium chloride into chlorine and magnesium. The density of the bath is such that when the magnesium metal is produced at the cathode, it rises to the surface of the bath and is guided by inverted troughs to metal wells in the front of the cell. From here the molten metal, with a purity of 99.8% or better, is dipped two or three times daily and cast into pigs. Cell output will run to 1000 pounds (454 kg) magnesium per day and utilize as feed 4000 pounds (1816 kg) magnesium chloride.

The anode products from the cell are hot chlorine and hydrogen chloride, along with lesser amounts of carbon dioxide, and are diluted by water vapor and leaking air. The hot gases are collected under the tightly closed refractory cell cover and are piped away first in a brick flue; then, after cooling with a water quench, they flow off as a slurry in an acid-resistant pipe. These anode gases are converted to hydrochloric acid and recycled back into the process. Because of the corrosive and poisonous nature of the gases, great care must be taken both in collecting them and also in having the cell area extremely well ventilated. A suction of 0.25 inch of water (6.25 mm) is maintained on the cell to remove these product gases and the considerable amount of in-leak air.

The 20% water content of the granular magnesium chloride feed causes no problem, as most of the water flashes off immediately at the high cell temperature. However there is a consumption of graphite anode, 0.1 pound of anode per pound of magnesium produced (45.4 g per 454 g), from the reaction of the liberated oxygen in the water reacting with the anode to form CO and then passing off as CO_2 in the anode gases. There is no anode attack from the hot, dry chlorine gas also passing off.

There is always the possibility of some quantity of magnesium oxide passing into the cell along with the $MgCl_2$ feed material, this MgO being formed by the hydrolysis of $MgCl_2$. A limited amount of this magnesium oxide dissolves in the liquid bath and is electrolyzed, a considerable portion is rechlorinated to $MgCl_2$, and the remainder falls to the cell bottom where it forms a spongy sludge layer. This sludge layer is dipped out periodically, generally once a day. All United States production, which is about 50% of the world total, is by this electrolytic process.

2b. Magnesium Oxide produced by calcining dolomite,

$$MgCO_3 \cdot CaCO_3 = MgO \cdot CaO + 2CO_2$$

is reacted in the silicothermic process with 75% grade ferrosilicon in a ratio of 5:1 to produce metallic magnesium and dicalcium silicate:

$$2MgO \cdot CaO + FeSi = 2CaO \cdot SiO_2(Fe) + 2Mg$$

This is known as the *Pidgeon process* and is named for its developer. It is used extensively in Canada where it was developed, with Canada being one of the major magnesium producing countries.

The reaction is carried out in a small batch retort which produces 32 pounds (14.5 kg) 99.98% magnesium from a 240 pound (109 kg) charge in 8 hours. The charge to the retort is a 5:1 ratio of calcined dolomite (200 pounds, 90.8 kg) to ferrosilicon (40 pounds, 18 kg) which is ground, mixed, briquetted, and preheated to 1292°F (700°C) before being charged to the retort. As the reaction is endothermic, this preheating carries sensible heat to the retort as well as baking the briquettes for strength.

Figure 3.16. Magnesium retort furnaces, Pidgeon process, 24 retorts per furnace. *Source:* Courtesy of Leeds and Northrup.

The retorts are of nichrome steel, 10 feet long (3 m) and 10 inches (25 cm) inside diameter, with a 1 inch (2.5 cm) thick wall, supported horizontally in banks of up to 24, and projecting 8 feet (2.44 m) into an oil- or gas-fired furnace. The end inserted into the furnace is rounded and closed, while the other end, the 2 foot (61 cm) projection out of the furnace, is open with a gasketed flat steel disc bolted on as an end seal. This end seal is removed to charge the retort with fresh, hot briquettes and also to remove the spent briquettes and condensed magnesium metal after the reaction is completed.

The 2 foot (61 cm) projection of the retort from the furnace serves as a condenser to solidify the magnesium metal vapor which is released from the briquettes, with the hot end of the retort being heated to 2100°F (1150°C) and magnesium having a vaporization temperature of approximately 2012°F (1100°C) and a solidification point of 1202°F (651°C). Cooling water jackets are placed around the projecting end of the retort, and a vacuum connection (0.025 mm Hg) at the top outside end draws the magnesium vapor into this cooled zone to condense as crystals of magnesium on a cylindrical steel sleeve. The sleeve is removeable and is inserted for ease in extracting the condensed magnesium ring. This ring is referred to as a magnesium crown or muff and is melted and ladled into cast forms.

Calcined dolomite is preferred over magnesium oxide for this silicothermic reaction process for the reason that CaO will displace MgO from any $MgSiO_2$ formed. But if no CaO is present then the silicate will tie up a large proportion of the available magnesium. Aluminum can be substituted as a reducing agent in place of ferrosilicon, but generally it is more expensive, and the cheaper ferrosilicon is used.

The nichrome retorts undergo severe stresses during the cyclic heating in the batch processing, so that they begin to scale and crack and have a life of only about eight months. The spent briquettes, consisting of dicalcium silicate along with the iron from the ferrosilicon, retain their shape in the retort without softening or melting. At the end of the 8 hour reaction period, after the vacuum has been shut off and the retort opened, they are withdrawn and discarded.

3. Titanium Tetrachloride in the *Kroll process* is reduced to titanium metal by either liquid magnesium or sodium in a closed steel reaction vessel with an inert helium or argon atmosphere at a temperature range in practice of 1314 to 1688°F (712 to 920°C):

$$TiCl_4 + 2Mg = Ti + 2MgCl_2 + Mg \text{ excess}$$

$$TiCl_4 + 4Na = Ti + 4NaCl$$

These are both batch operations.

The magnesium reduction reaction is carried out in a cylindrical-shaped steel vessel, often with a rounded bottom, which is used in a variety of sizes ranging from small units 3 feet high (0.91 m) by $2\frac{1}{2}$ feet in diameter (0.76 m), producing 250 pounds (113.4 kg) of titanium metal, to quite large units 9 feet high (2.74 m) by 5 feet in diameter (1.52 m), producing 3000 pounds (1362 kg) of metal. There is negligible corrosion or reaction at the operating temperature of the retort between the mild steel of the vessel and the helium or argon inert gas, the titanium tetrachloride and magnesium reactants, or the titanium and magnesium chloride products. Consequently any pickup of iron is minute. However titanium will react strongly with, and be oxidized by, any iron oxide rust or scale, and this surface oxide is reduced to metal by filling the vessel with hydrogen and heating to above red heat, then flushing out the hydrogen with helium or argon.

The magnesium bars for the reaction, with 15 to 20% excess over that theoretically required, are cleaned, pickled in hydrochloric acid, dried, and put in the reactor vessel in a "dry room." A shallow dome-shaped lid with two vent holes to which are connected inlet pipes is welded to the open top of the vessel, which is then filled with helium or argon and moved to, and lowered upright into, a gas- or oil-fired furnace. The lid is welded in place for each individual run, with the weld ground away to open the retort at the end of the reaction.

The vessel after positioning in the furnace is heated to about 1560°F (850°C), at which temperature the magnesium (mp 1202°F, 650°C) has melted. Liquid $TiCl_4$ is now run in at a controlled rate through one of the inlet pipes and reacts with the molten magnesium to produce titanium and magnesium chloride.

The reaction is exothermic, with the temperature in the reactor being controlled by the rate of addition of the $TiCl_4$. Temperature range in practice runs from 1314 to 1688°F (712 to 920°C), with the minimum temperature of 1314°F (712°C) being the melting point of $MgCl_2$. At any lower temperature the molten magnesium tends to crust over with solidified $MgCl_2$ and not react properly with the $TiCl_4$. The maximum temperature of 1688°F (920°C) is run up to for the last hour of the reaction, with the temperature over the most part of the processing period kept at about 1560°F (850°C). The maximum temperature is restricted because titanium will form an alloy with the steel reactor above 1832°F (1000°C), and the exothermic reaction also becomes too rapid and uncontrollable above 2012°F (1100°C).

The second inlet pipe through the lid is used to maintain the pressure of the inert gas in the vessel at slightly above atmospheric during the complete period of the run.

A 3 foot (0.91 m) by $2\frac{1}{2}$ foot (0.76 m) retort which is charged with 330 pounds (150 kg) of magnesium and 1070 pounds (486 kg) of liquid titanium tetrachloride over a 6 hour period will produce 250 pounds (113.5 kg) of titanium metal.

The titanium chloride reacts almost completely, and 80 to 85% of the magnesium is used. Attempts to add additional $TiCl_4$ to react with this excess magnesium have not been successful, as it has been found that some of the $TiCl_4$ reacts with the titanium instead, to give titanium trichloride and dichloride:

$$3TiCl_4 + Ti = 4TiCl_3$$

$$2TiCl_3 + Ti = 3TiCl_2$$

The least complicated method of treating the reaction products is to remove the reactor from the furnace and allow it to cool to room temperature, still under a small positive pressure of inert gas. The weld on the lid is then ground off, the lid removed, and the reactor tipped on its side. In this position the cool, solidified charge is either bored out

Figure 3.17. Titanium pots in gas-fired furnaces during the titanium tetrachloride reduction stage. *Source:* Courtesy of RMI Company.

Figure 3.18. Red hot pots, after several hours of heating, removed from the furnace and sent to a cooling room. *Source:* Courtesy of RMI Company.

on a large lathe or chipped out with a pneumatic hammer, with the product falling into a rolls crusher where it is broken to $\frac{1}{4}$ to 1 inch size (6 to 25 mm). The broken mixture of titanium, magnesium chloride, and excess magnesium passes to a rotary leacher where dilute hydrochloric acid dissolves the magnesium chloride and magnesium, leaving the titanium metal, which is in the form of sponge. This metal is water washed, dried, and compressed into elongated briquettes. The briquettes are welded together to form an electrode which is melted in a vacuum arc-melting consumable electrode electric furnace to produce a homogeneous metal ingot.

There is another alternate method of treating the reactant products. In this method, a tap hole at the bottom of the reactor is opened twice during the reaction and once at the end to draw off most of the magnesium chloride that is being produced. The tap hole is sealed each time by localized cooling with a water jacket freezing a plug of magnesium chloride in the tap hole. The remaining reactant products are cooled to room temperature

Figure 3.19. After cooling, the pot is opened and titanium sponge and salt is chipped out by pneumatic hammer and fed into crushers for proper size reduction. *Source:* Courtesy of RMI Company.

and removed as before by boring or chipping, after which they are placed in stainless steel baskets and inserted into a vacuum distillation retort, which can be evacuated to 1 mm Hg by a mechanical pump. Distillation is carried out at 1610 to 1688°F (875 to 920°C) for 35 hours. Then the retort is cooled, still under vacuum, to 1505°F (815°C) and then further cooled to room temperature under an inert gas. Finally dry air is sucked through for 6 hours to remove any volatiles, giving a total treatment time of about 50 hours.

Titanium sponge is left in the basket after this treatment and is broken loose with a compressed air hammer to be briquetted and vacuum melted as before. Both the remaining magnesium chloride and excess magnesium are distilled off.

In some other instances the vacuum distillation is done directly in the retort, without removing the solidified charge, and here again the magnesium and magnesium chloride are distilled off leaving titanium sponge to be broken out, briquetted, and vacuum melted.

The titanium metal produced by any of these methods will have a purity of about 99%, with the major impurities being magnesium, iron, chlorine, and manganese.

The reduction of titanium tetrachloride with sodium is in general very similar to that done with magnesium. Purified liquid sodium at a temperature a bit above its melting point of 207°F (97°C) is run into the argon-filled reactor and heated to 1292°F (700°C). Liquid titanium tetrachloride is then dripped in at a rate to maintain an operating temperature of between 1562 and 1652°F (850 and 900°C), this temperature range being governed on the low side by the 1472°F (801°C) melting point of sodium chloride and on the high side by the 1832°F (1000°C) temperature where alloying takes place between titanium and the steel retort. At the end of the reaction the vessel is sometimes first drained of molten sodium chloride, and in other instances the whole charge is solidified before being bored or chipped out. As vacuum distillation is more difficult, requiring a higher temperature with sodium chloride than with magnesium chloride, leaching in dilute hydrochloric or 2% nitric acid is the process used to remove sodium chloride from the titanium sponge. After this the sponge is water washed, dried, briquetted, and melted in a vacuum to homogeneous metal.

The major difference in the magnesium and sodium processes is that while an excess of magnesium was used, a deficiency of sodium is put in the reactor along with a small excess of titanium tetrachloride. This is to avoid any excess of sodium at the end of the reaction which would be a fire hazard during the leaching procedure, when the metallic sodium could react with water and burn.

A second method of sodium reduction has accurate charges of molten sodium put into large argon-filled and sealed pots, which are then cooled and moved to the reduction area. In this area a continuous reactor is charged with accurate amounts of $TiCl_4$ and metallic sodium, giving a mixture of titanium dichloride and sodium chloride which is run into the first sodium-charged pots:

$$3TiCl_4 + 6Na = 3TiCl_2 + 6NaCl$$

These pots are now transported to a furnace where at 1562 to 1652°F (850 to 900°C) the second reduction stage takes place of titanium dichloride reacting with sodium to produce titanium sponge and sodium chloride:

$$TiCl_2 + 2Na = Ti + 2NaCl$$

The red-hot pots are removed from the furnace to cool after the reaction is completed. Then, as before, the solidified product is chipped out, crushed, and leached in dilute HCl to dissolve out the sodium chloride; and after water washing, the titanium sponge is dried, briquetted, and vacuum melted to solid ingot.

4. Aluminum Oxide, with required purity of greater than 99.5% Al_2O_3 on a dry basis, which is produced by the pressure leaching and precipitation of Al_2O_3 from bauxite, is electrolyzed in a molten bath of cryolite to give as products aluminum metal of 99.5% purity at the cathode and oxygen gas at the anode.

The electrolytic reduction cell or pot, known as the *Hall cell*, consists of a rectangular, open-topped sheet steel box with typical dimensions of 20 feet long (6.1 m), 10 feet wide (3 m), and 3 feet deep (0.91 m). For ease of working, these cells are often positioned below the floor with only about one third of the cell height projecting up above floor level. A carbon lining 6 to 10 inches thick (15 to 25 cm), backed by a layer of refractory insulation, covers the bottom and sides of the sheet steel shell. This carbon lining is the cathode, and steel pins are inserted into the bottom section to carry current to the bus bars.

The carbon lining can be provided in two ways, either of which should last for three to four years before having to be replaced. It can be built up of carbon blocks joined by carbon cement; or a monolithic lining can be constructed by ramming a hot plastic paste of mixed coke, tar, and pitch between a cast iron form and the steel shell, then baking this in place by heating the entire cell to 1382°F (750°C).

There are three types of carbon anodes in general use, and the type of anode configuration used distinguishes the three types of pots, namely, *prebaked (PB)*, *horizontal-stud Söderberg (HSS)*, and *vertical-stud Söderberg (VSS)*. The prebaked cells produce the greatest amount of aluminum in the United States, with 61.9% of total production (1970), followed by the horizontal stud Söderberg with 25.5% and the vertical stud Söderberg with 12.6%.

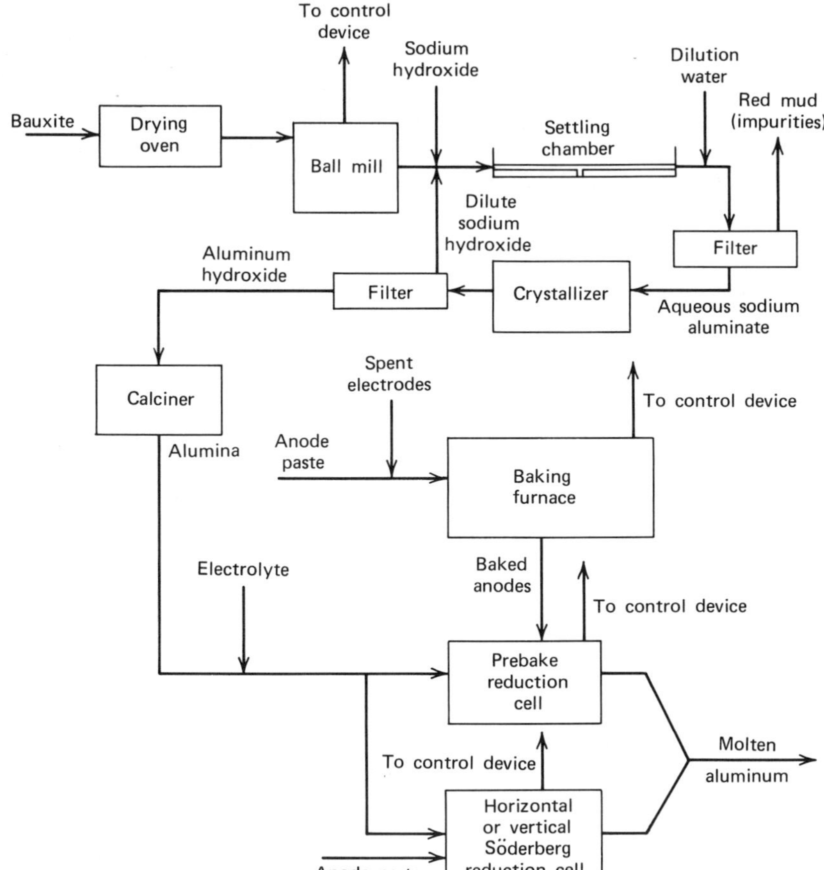

Figure 3.20. Primary aluminum production process flow chart. *Source:* Compilation of Air Pollution Emission Factors, U.S. EPA AP-42, 1973.

The Prebaked Anodes **(PB)** consist of blocks that are formed from a carbon paste and baked in an oven prior to their use in the cell. The blocks can be up to 400 pounds in weight (182 kg) with dimensions of 12 inches high, 20 inches wide, and 31 inches long (30 × 60 × 77.5 cm) and are individually supported in the bath in two equal rows down its length, typically 14 to 24 per cell. Steel studs are inserted into a hole on top of each finished baked block and are anchored in place by pouring molten iron around the stud. These studs conduct current to the anode through electrical connections bolted to them and also provide the method of support and positioning of the anodes in the bath. Each block is raised or lowered individually for proper positioning and is replaced after it has been gradually consumed, at a rate of about 1 inch (2.5 cm) per day, by reacting with the oxygen given off at the anode and forming CO.

Prebaked anodes give the best electrical efficiency, but are more expensive to make than the continuous Söderberg types, requiring fabrication and rodding facilities not used in the Söderberg system. While the Söderberg system requires less labor and is a continuous method of feeding the anode carbon, it is more difficult to collect the off-gases for subsequent treatment for fluorine recovery, necessitating that a much larger volume of gases be collected and treated than for a prebaked cell.

The Söderberg Anode consists of one large block of carbon contained inside a rectangular sheet aluminum casing, held together by perforated steel channels, leaving a gap of about $1\frac{1}{2}$ feet (0.45 m) at each end and 3 feet (0.9 m) on each side between the anode and the carbon lining of the cell. Briquettes of mixed petroleum coke and pitch, or carbon paste, are dumped into this casing and enough heat rises from the cell to melt the briquettes, distil off any volatiles, and finally bake the mixture into a hard monolithic carbon mass by the time it has descended to a point about 20 inches (51 cm) above the bath.

On the order of 0.5 pound (0.23 kg) of carbon anode is consumed as CO per 1 pound (0.454 kg) of aluminum metal produced, and as the anode is consumed during electrolysis, the Söderberg anode is lowered into the bath at a rate to compensate for this consumption. Continuous anode source is achieved by fastening an additional aluminum casing to the top of the anode shell and filling this with more briquettes to be melted and baked in turn. One disadvantage of baking the paste in place is that heavy organic mate-

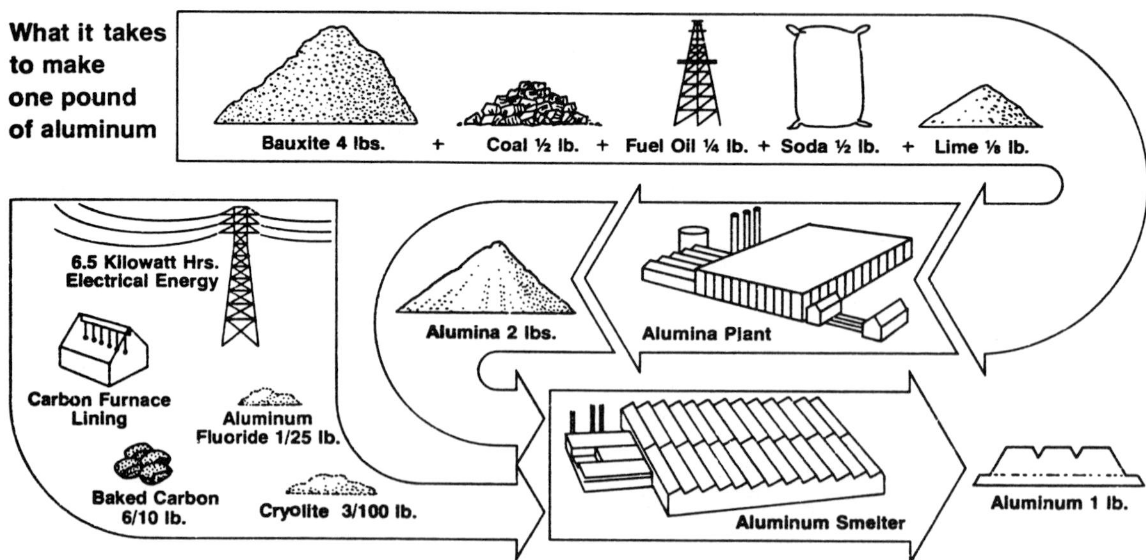

Figure 3.21. Materials for aluminum metal production. *Source:* Courtesy of The Aluminum Association, Inc.

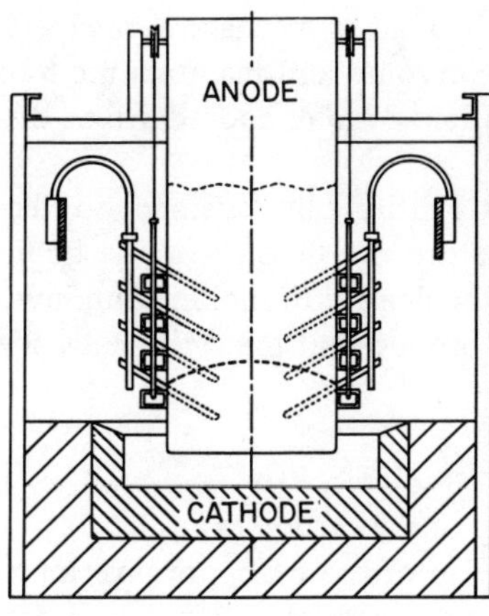

Figure 3.22. Aluminum electrolytic cell horizontal stud Söderberg anode. *Source:* S. W. Martin and H. W. Nelson, *J. Metall.*, Vol. 7, No. 4, 1955, p. 542.

rials (tars) are added to the effluent stream. These often cause plugging of the ducts, fans, and control equipment, which seriously limits the choice of air cleaning equipment that can be used.

Current is supplied to the Söderberg electrodes by steel studs which are inserted into the anode paste. Three or four rows of pins are inserted horizontally down each side of the *horizontal-stud Söderberg anode (HSS)*, and two to four rows of pins vertically down into the *vertical-stud anode (VSS)*. These pins are connected to the bus bars at their free ends; and, as well as carrying current, they act as supports for the anode.

As the Söderberg anode is consumed and lowered, the lowest pins, as well as the steel support channels on the casing, must be removed and these pins are moved to a higher row in the anode box. No pin is ever positioned to be closer than 8 inches (20 cm) to the bottom of the anode.

The *Hall cell* in operation has a bath of molten electrolyte as deep as 14 inches (35 cm) and is essentially a solution of liquid cyrolite (Na_3AlF_6) into which is dissolved 6 to 10% by weight 99.5% pure aluminum oxide. Small quantities, 2 to 5%, of aluminum fluoride (AlF_3) are also used in the electrolyte to combine with Na_2O, an impurity in the Al_2O_3, and these combine to form artificial cryolite. An addition of small quantities, 4 to 10%, of fluorspar (CaF_2) lowers the melting point of the bath but increases its density, while MgF_2 increases the bath conductivity and NaF may increase the solubility of Al_2O_3. Consumption of the bath materials is about 5 pounds (2.3 kg) per 100 pounds (45.4 kg) aluminum produced.

A crust of frozen electrolyte mixed with aluminum oxide freezes at the top of the bath where it contacts the air and gives the bath a protective covering. Bins positioned above and on either side of the length of the cell hold 200 mesh alumina, and periodically some

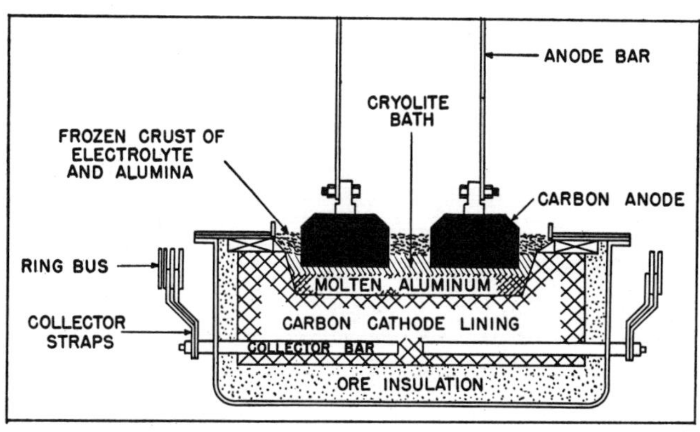

Figure 3.23. Aluminum electrolytic cell prebaked carbon anodes. *Source:* S. W. Martin and H. W. Nelson, *J. Metall.*, Vol. 7, No. 4, 1955, p. 541.

of this is dumped down on the crust, the crust is broken by hand or mechanical breakers, and the Al_2O_3 powder is stirred in to be dissolved in the cryolite and maintain the 6 to 10% aluminum oxide content. The Al_2O_3 dissolution is fairly rapid, and additions dissolve in less than 10 minutes.

The bath temperature is maintained at 1832°F (1000°C) entirely by resistance to electrical current flow through the cell, and at this temperature the density of the liquid cryolite, with its dissolved alumina, is about 2.10, while the density of molten aluminum is about 2.28. This makes it possible for aluminum being produced at the cathode by the dissociation of the Al_2O_3

$$2Al_2O_3 = 4Al + 3O_2$$

to collect as a liquid layer at the bottom of the cell, on top of the carbon cathode.

The anodes, which project down through the crust on top of the bath, are positioned quite deeply into the electrolyte so that there is usually only a distance of 2 inches (5 cm) between the bottom of the anode and the top of the layer of molten aluminum. Alumina is consumed in direct proportion to the weight of metal produced, with 2 pounds (0.90 kg) of Al_2O_3 needed for 1 pound (0.45 kg) of aluminum.

The other product from the cell is oxygen, which is given off at the anode. This has the beneficial effect of continuous agitation in the bath from the rising bubbles and greatly assists in the speed of dissolution of Al_2O_3 in the cryolite. However the oxygen also attacks the carbon anode, forming CO and CO_2 in approximately equal ratios and consumes the anode at the rate of one half the weight of aluminum metal produced.

A cell of the size discussed will produce on the order of 600 pounds (273 kg) of aluminum per day. This is removed every one to three days by a vacuum technique, in which insulated cast iron pots with air-tight lids and downward sloping spouts have the spout inserted down into the liquid metal layer and then siphon molten aluminum through the spout up into the pot. The liquid metal can either be cast into molds or poured into a blending and holding furnace.

A problem known as "anode effect" can occur in a cell if the alumina concentration drops from a normal 6 to 10% down to 2% or below. When the alumina concentration drops to this lower level, the electrical resistance of the cell increases sharply from 5 to 30 or 40 volts, due mainly to a gas film which envelopes the anode causing the current passage between electrolyte and anode to be a multitude of tiny arcs as the current jumps

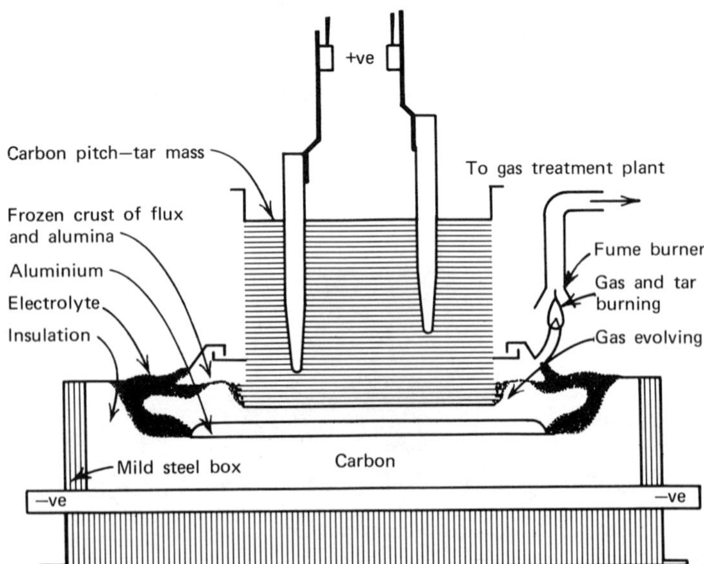

Figure 3.24. Aluminum electrolytic cell vertical stud Söderberg anode. *Source:* W. Alexander and A. Street, *Metals in the Service of Man*, 6th ed., Pelican Books, London, 1976, p. 56. Reprinted by permission of Penguin Books, Ltd.

Table 3.7. Raw Material and Energy Requirements for Aluminum Production

Cell operating temperature	1740°F (950°C)
Current through pot line	60,000 to 125,000 amp
Voltage drop per cell	4.3 to 5.2
Current efficiency	85 to 90%
Energy required	6.0 to 8.5 kW-hr/lb aluminum (13.2 to 18.7 kW-hr/kg aluminum)
Weight alumina consumed	1.89 to 1.92 lb Al_2O_3/lb aluminum (1.89 to 1.92 kg Al_2O_3/kg aluminum)
Weight electrolyte fluoride consumed	0.03 to 0.10 lb fluoride/lb aluminum (0.03 to 0.10 kg fluoride/kg aluminum)
Weight carbon electrode consumed	0.45 to 0.55 lb electrode/lb aluminum (0.45 to 0.55 kg electrode/kg aluminum)

Source: Compilation of Air Pollution Emission Factors, U.S. Environmental Protection Agency, AP-42, 1973.

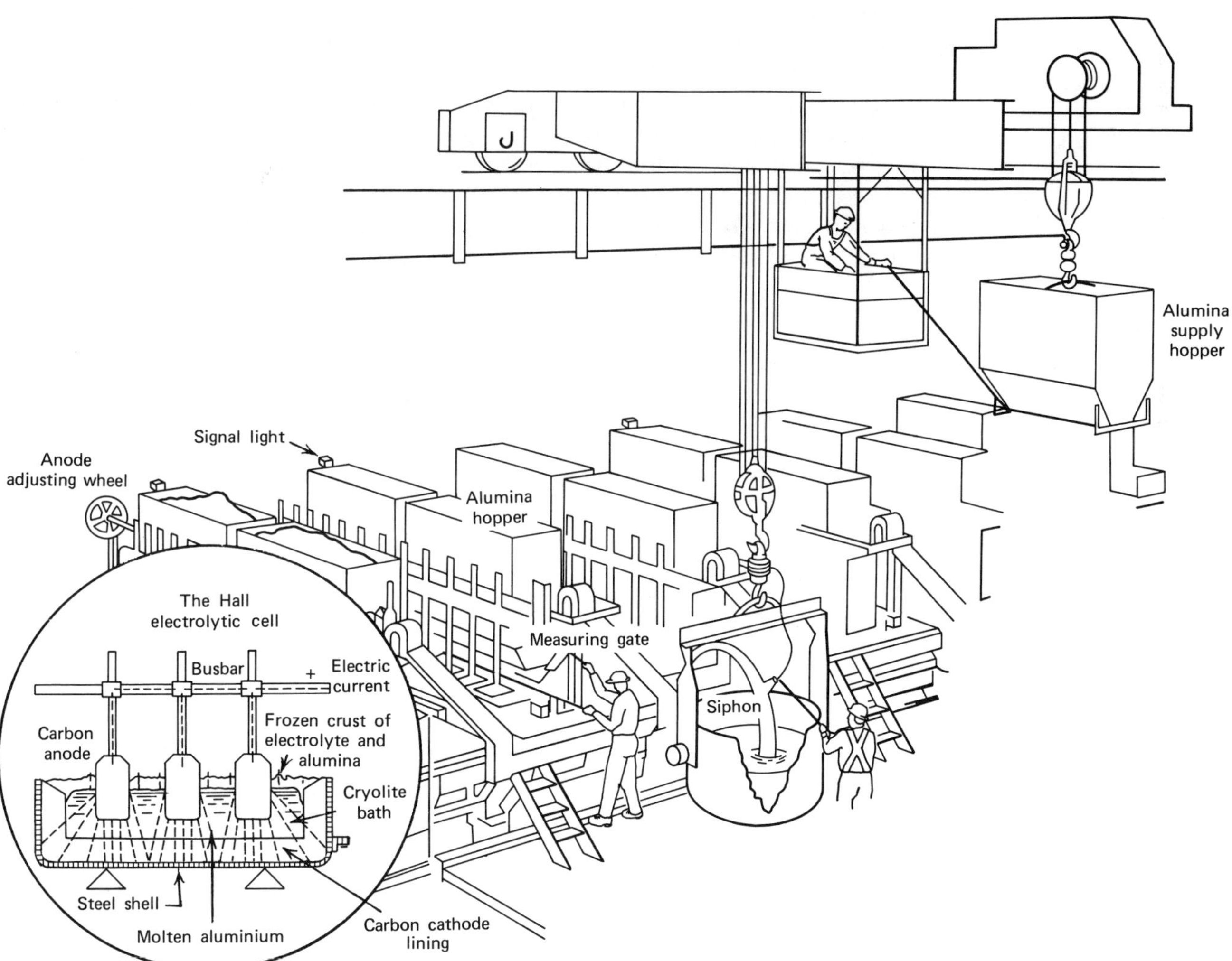

Figure 3.25. Tapping of molten aluminum. *Source:* Courtesy of The Aluminum Association, Inc.

the gas gap. As long as the anode effect continues, the gas evolved at the anode consists almost exclusively of carbon monoxide, which causes much greater than normal anode consumption. The sudden increase in cell voltage can be noted by an operator watching the change in the glow of an incandescent light bulb connected across the electrical leads to the cell.

As soon as "anode effect" is detected, the crust of frozen cryolite on top of the bath is broken, and more Al_2O_3 is added to bring the alumina content up to the required 6 to 10%, and the cell returns to normal operation.

The tendency is toward larger cells, with the majority of plants operating with cells in the 80,000 to 100,000 ampere range and some going as high as 150,000 amperes. The voltage drop across a single cell is 4.5 to 6 volts; and across a potline, or row of 50 to 200 cells connected in series, it is 600 to 800 volts. Anode current densities are 4 to 9 amperes per square inch (0.62 to 1.40 amperes per cm^2). Current efficiencies of 85 to 90% and 6.5 to 8 kW-hr are required to produce 1 pound (0.454 kg) of aluminum.

To prevent any atmospheric pollution, all gases leaving the cells are passed through a collection system which removes aluminum oxide, carbon monoxide, and fluorides. The fluorides are a particular concern, and great care is taken to scrub them out of the effluent gases.

Alcoa Smelting Process The latest development in the production of primary aluminum metal is the Alcoa smelting process, for which the applications for patents were announced in January 1973.

As in the Hall process, the Alcoa method uses bauxite as the raw material, and this too is refined by the Bayer process to purified Al_2O_3. This alumina is combined with chlorine in a reactor unit at 1290 to 1650°F (700 to 900°C) to make an aluminum trichloride powder, which is then dissolved in a liquid electrolyte of sodium and lithium chlorides and electrolytically processed in a completely enclosed cell (1290°F, 700°C) that separates it into molten aluminum and chlorine gas. The chlorine is caught and continuously

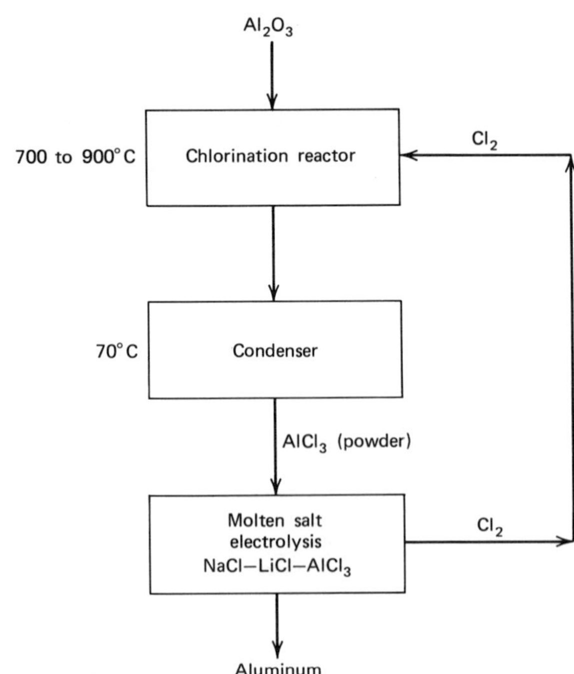

Figure 3.26. Electrolysis of aluminum metal from aluminum chloride, Alcoa process.

recycled back to the chloridizing reactor in a closed loop, while the liquid aluminum metal is withdrawn from the cell and cast or sent to a holding furnace.

Several advantages are apparent for this type of cell. These are, first, that the cell temperature is considerably lower at 1290°F (700°C) than the 1832°F (1000°C) used in the Hall cell, so the electrical energy can be reduced by as much as 30%, giving a new, lower range of 4.55 to 5.6 kW-hr required to produce 1 pound of aluminum (0.454 kg). Another, second advantage is a closed-loop system which recycles the chlorine gas produced; and, third, it disposes of the need for scarce cryolite and the expense of containing fluoride cell emissions. A last advantage has to do with labor costs, with the new process being a closed and continuous operation which will require only a fraction of the workers needed for the several operations in running a line of Hall cells.

Other New Smelting Processes With the Hall Process in use since 1900, and with some obvious shortcomings to this process, several other attempts in addition to the new Alcoa process have been made to improve the aluminum extraction operation. While none of these has been commercially successful so far, active investigation is continuing and may eventually produce a revised aluminum recovery process.

One of these studies, the *Alcan process* (Aluminum Company of Canada), reached pilot plant stage, but after 12 years of study and $40 million expenditure, the work was suspended in 1968. Bauxite and coke were smelted in an electric furnace to produce a crude alloy of aluminum, silicon, and iron, which was then reacted with aluminum trichloride vapor at 2372°F (1300°C) to form aluminum monochloride:

$$2Al + AlCl_3 \overset{1300°C}{=} 3AlCl$$

The AlCl gas was rapidly cooled to about 1290°F (700°C) in a spray of molten aluminum, and at this temperature the aluminum monochloride disproportionated to form liquid aluminum and aluminum trichloride vapor:

$$3AlCl \overset{700°C}{=} 2Al + AlCl_3$$

The $AlCl_3$ vapor was recovered and recycled to use in the first stage of the reaction.

Another process, the *Toth process*, reacts clay and other alumina material with coke and chlorine at high temperature to produce volatilized anhydrous aluminum trichloride:

$$Al_2O_3 + 3C + 6Cl = 2AlCl_3 + 3CO$$

The aluminum trichloride is separated from impurities by fractional distillation and reacted with manganese to form manganese dichloride and aluminum.

REFINING

Refining is carried out as the final operation in the purification of the metals. In one instance it is done to recover a valuable metal by-product as well as to produce an extremely high-purity metal. In the other instance refining is used only to produce a high-purity metal without regard for conserving the impurity elements. In the first case fire refining is used, and in the second case electrolytic refining is used.

Because of the treatment processing during the normal pyrometallurgical treatments of the reactive metals, the grade of metal produced from the smelting step is sufficiently high for most commercial use, and only a relatively small tonnage of these metals are further refined.

FIRE REFINING

Fire refining of reactive metals is done on a somewhat limited scale in order both to improve the grade of the metal by removing impurities and to collect these impurity metals as valued by-products of the operation.

The smelting operation preceding fire refining is retorting, and metals produced by retorting will have as impurities other metals in the retort charge that have vapor pressures fairly close to that of the major metal being recovered and will be in the vapor state at a temperature below that at which the retort is operating. These metals will all be reduced in the retort at the same time and pass off as combined metal vapors to be cooled and deposited together as a liquid alloy in the retort condenser.

As most retorting gives a comparatively pure product, the amounts of impurities to be removed are relatively small, and the impure metal is often of sufficiently high grade that it can be used commercially in many high-tonnage applications without going to the added expense of fire refining.

The method used in fire refining of reactive metals is very different from that used for the nonreactive metals. With all the nonreactive metals the process of selective oxidation was the chief operation used, and in this it was necessary that the impurity metals be more reactive with oxygen than the metal being fire refined. Therefore, when air was blown into an impure metal liquid bath, the impurity metals oxidized and collected as an oxide slag, or passed off as a volatile oxide gas, leaving a bath of purified, fire-refined metal.

However it is not possible to attempt selective oxidation with reactive metals, as they would react with air or an oxidizing agent and immediately be converted back into a metal oxide. One process that can be used is fractional distillation, where impurity metals with even a small range of boiling points can be individually selectively removed, each in a quite pure form, and leave behind a highly purified metal bath. Preferably the impurity metals would have a boiling point lower than that of the metal being fire refined, so that refining could take place by vaporizing only the small amount of impurity rather than the large amount of the metal being treated, and also this could be done at a low temperature to save on heating costs.

Liquidation is also used, taking into account the percent solubility of one metal in another at varying temperatures, so that if the temperature is lowered the solubility of the impurity in the host metal is decreased, and the impurity metal will partially separate as a separate liquid layer.

Fire refining by fractional distillation is a continuous process, while liquation is carried out as a batch process.

TYPES OF FIRE REFINING

The two types of processes used for fire refining differ very considerably both in the equipment used and in type of operation that is being done, though in both cases the object is the same—to remove impurity metals from the host metal.

In one process the impure liquid metal from retorting is added in the liquid state to a vertical fractional distillation column where the temperature is maintained at a level that the boiling point of one or more of the metals in the mixture is exceeded and will pass off as metal vapor, leaving behind a high-purity residual metal portion that did not reach its boiling point and remained as a liquid. This can be repeated through several stages, each at different temperatures, until all the impurity metals are removed.

Another process depends on holding the impure liquid metal at a temperature close to the freezing point so that impurities soluble to a greater percentage at higher temperatures will separate out as a separate liquid layer at the holding temperature.

Vertical Fractional Distillation is carried out on a continuous basis in two vertical columns, which can convert 25 tons per day of an impure metal at 98% purity to one of +99.995% purity.

Each of the two columns is made up of a stack of superimposed, rectangular-shaped silicon carbide trays, with each tray holding a pool of metal, and the height of the pool being regulated by a dam around an overflow hole. The overflow from one tray is caught on the tray immediately below it, and the descending flow of metal follows a zigzag traverse from tray to tray.

The impure liquid metal from the retorting operation is added to a tray in the first distillation column. Then as the metal zigzags downward the increment that has a boiling point below the temperature of the column will be vaporized and pass upward. Liquid metal is run off at the bottom of the column to be collected as one product, while metal vapor passing upward is condensed as a second product.

This condensed liquid metal now passes into the second distillation column, and a second similar distillation process takes place.

Liquation is done as a batch operation in a two-compartmented, coal-fired reverberatory furnace with a reducing atmosphere. This atmosphere discourages oxidation of any of the metal charge.

Slabs of impure metal are added to the first, larger compartment of the furnace and are melted, so that the metal and any impurities are liquid and homogeneous.

The furnace is now cooled close to the eutectic temperature of the metal and its major impurity. At this temperature the impurity metal is very much less soluble than at the higher temperature used to melt the slabs and will separate out as an individual liquid layer. This separated layer settles to the bottom of the furnace and is tapped off from there.

The purified metal that is left is only slightly above its freezing point, and it is allowed to overflow into a second small, heated compartment in the furnace to permit the temperature to be raised to where the metal is more fluid for easy tapping.

Purification by liquation will increase the metal purity from 98% to 99%.

FIRE REFINING PROCESSES

1a. Zinc Vertical Fractional Distillation, also developed by the New Jersey Zinc Company, is used to remove the lead and cadmium impurities from liquid zinc produced by the New Jersey Zinc's vertical retort and St. Joe Minerals' electrothermic furnace processes, upgrading it from 98% zinc after retorting to 99.995% plus purity after fractional distillation. Cadmium in particular is recovered as a by-product, and its association with zinc is the major source of cadmium supply.

Separation is possible in this way because of the widely divergent boiling points of lead (2948°F, 1620°C), zinc (1668°F, 907°C), and cadmium (1432°F, 778°C). By utilizing a two-column vertical unit the zinc and cadmium can be vaporized in the first column at 2228°F (1220°C), while the lead remains liquid. Then in the second column at 1562°F (850°C), the zinc remains as a liquid while the cadmium vaporizes.

Liquid zinc runs from the vertical retort condenser, or electrothermic furnace condenser, through a launder to a holding furnace, and from there through a heated trough to the first distillation column. About 25 tons per day of retort zinc is fed into the circuit, and to this is added 15 tons of recycled zinc from the first column, giving a total of 40 tons per day being treated. The recycled 15 tons is contained in the mixture of lead, iron, and other high boiling-point impurities that flow out the bottom of the column into a two-compartment holding pot, where it is separated by liquation. The first compartment of the pot is cooled to settle out a high-density lead layer and a zinc–iron dross, with the liquated zinc now reasonably free of lead and iron overflowing into the second compartment and back to the initial holding furnace for refeeding into the distillation column.

Each of the two columns is made up of 50 rectangular silicon carbide trays, 2 feet by 4 feet (0.61 by 1.22 m) and 6 inches deep (15 cm). These trays are superimposed to fit tightly together and prevent vapor leakage. They are adapted to hold a pool of metal, the depth of which is set by an overflow dam around a 3 inch diameter opening (7.5 cm) near the end of the tray. The trays are stacked with the openings on successive trays alternating from side to side, so that the metal overflow from one tray is caught on the tray below and the descending flow of metal follows a zigzag path downward from tray to tray. This design facilitates heat transfer and provides an extremely large surface from which the vapor can escape as the whole tray is covered with molten metal, but only to a shallow depth as each tray is an independent unit.

The ascending metal vapors follow the same path upward as the liquid takes downward, so that there is good liquid–vapor contact and liquid scrubbing of the vapor as it passed up through the column.

The lower two fifths of each column is inside a gas-fired furnace supplied with recuperators to heat the combustion air, and these supply the heat required in the columns for distillation. Uniform heating is achieved by supplying burners in the furnace at three different levels.

The liquid zinc from the holding furnace runs into the first, so-called "lead column" in a steady stream, at a height of two or three trays above the top of the heating furnace, and this flow is regulated to ensure that all the trays are full of metal and overflowing steadily into the trays below, in order that no tray may run dry or become overheated. The design of the trays differs somewhat inside the furnace in that these "distilling trays" have a trough extending around the outer periphery and a raised center, to present the maximum amount of heated tray wall to the liquid zinc, and at the same time to have a greater holding capacity for the liquid feed. The "refluxing trays," positioned above the furnace, are flat, with no peripheral trough, and hold a much smaller quantity of liquid metal. Controlled cooling (refluxing) takes place in these upper trays, with some

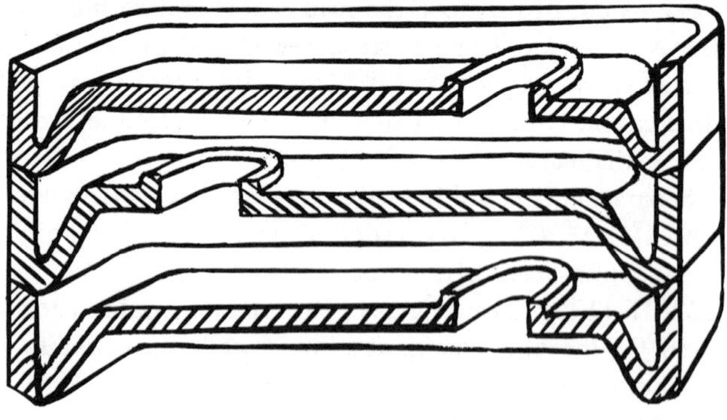

Figure 3.27. Distillation trays for fractional distillation zinc refining. *Source:* The Carborundum Co.

condensed liquid metal from the rising vapor phase produced in the furnace zone collecting and overflowing to run down from tray to tray.

The vapor which passes off at the top of the lead column consists of two thirds of the zinc fed in and all of the cadmium, and this travels into a condenser where the zinc and cadmium are converted to liquids and pass into a second, so-called "cadmium column." Two "lead columns" can operate in series with one "cadmium column," as the greater amount of impurities are removed in the first processing stage and a much smaller amount of material is treated in the second stage.

The design and operation of the "cadmium column" is essentially the same as that of the "lead column," except for the lower furnace temperature required and the fact that all the trays are the flat type, similar to those used in the upper "refluxing" section of the "lead column." This is because there is only a comparatively small amount of low boiling-point cadmium to be removed, and much less boiling capacity is required than in the "lead column" where most of the zinc fed in had to be volatilized. To ensure adequate refluxing, a condenser is set on top of the column of trays, and the vapor passing through this condenser goes into a second, smaller condenser to give a zinc–cadmium alloy containing 15% cadmium. About 160 pounds (72.6 kg) of this alloy is produced daily.

The zinc metal running out of the bottom of the cadmium column is special high-grade zinc, 99.995% pure, and is used for applications such as die casting alloys, which require this type of purity.

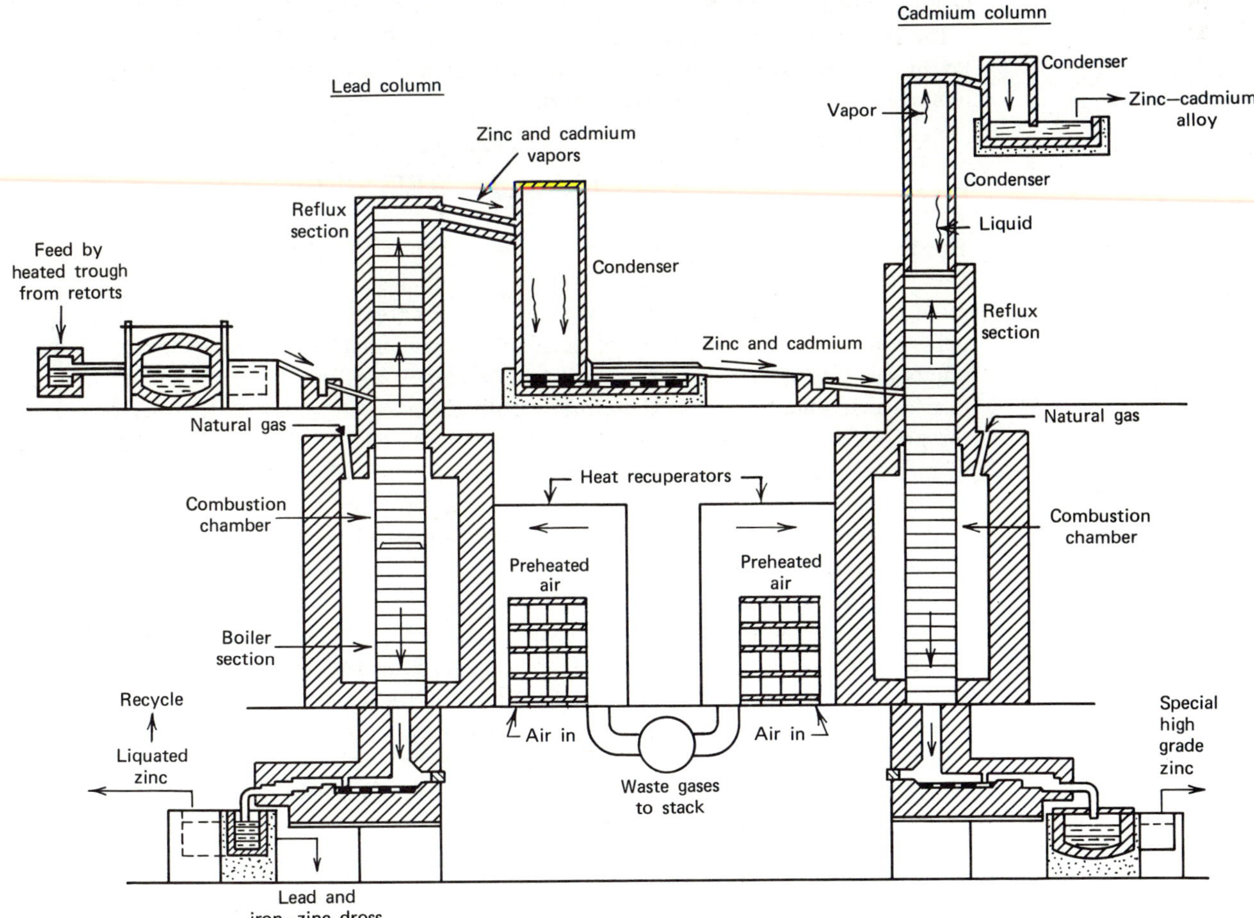

Figure 3.28. Zinc fire refining distillation columns. *Source:* C. H. Cotterill and J. M. Cigan, Eds., AIME World Symposium of Lead and Zinc, Vol. 2, 1970, p. 547.

1b. Zinc Liquation is used to separate lead and iron from retort zinc but has no effect on cadmium. The principle of this refining operation is to reheat or melt the zinc in a reverberatory furnace and then allow the temperature to drop to near the freezing point. The lead in excess of that soluble in the zinc at the lower temperature will now settle out as a liquid layer on the bottom of the furnace and can be withdrawn. Zinc purity can be increased from 98% to 99%, with lead in the zinc being decreased from 2% to 1% and iron decreased from 0.10% to 0.025%.

A two-compartmented reverberatory furnace is used, fired with coal and kept short of air to have a reducing atmosphere above the liquid metal bath to prevent zinc from oxidizing. Slab zinc is added to the large compartment of the furnace and melted at a temperature above 1472°F (800°C), at which temperature the lead and zinc are completely miscible.

The furnace is then cooled to slightly above 786°F (418°C), at which point the eutectic temperature is reached and zinc will dissolve only 0.85% lead. The excess lead over this eutectic quantity comes out of solution and settles as a bottom layer in the furnace. The lead is still quite fluid at this temperature because of its low melting point (327°F, 164°C) and is easily tapped from the furnace.

However the refined zinc–lead eutectic mixture remaining is close to its freezing point and would solidify if run out of the furnace directly into the open air. In order to prevent this from happening and prevent the metal from freezing in the tapping spout, it is allowed to overflow into a second small, independently heated furnace compartment, where the temperature is raised enough to allow easy tapping.

The iron impurity collects as an iron–zinc alloy, of about 0.5% Fe, 97.5% Zn, and 2% Pb, in a mushy layer between the fluid upper layer of refined zinc and the lower layer of liquid lead. From here it is removed periodically.

ELECTROLYTIC REFINING

This treatment method produces metal of extremely high purity for a reactive metal that does not lend itself to easy treatment by the other common methods of refining. Fire refining is not possible because of the high rate of oxidation that would occur if selective oxidation was used, and fractional distillation is not practical because of the high boiling points involved. Electrolysis from an aqueous bath is also very difficult and is not practical as a refining method. Therefore a method has to be used that to a degree repeats the process by which the reactive metal was first produced, namely, electrolysis of a bath of molten mixed fluorides at high temperature. In this case impure metal from a liquid anodic layer at the bottom of the cell is electrolyzed and transferred through an intermediate fluoride layer to a top cathodic layer of superhigh-purity refined metal.

The degree of purification that is accomplished in this electrolytic refining is rather small, as the metal being refined is already of comparatively high grade as produced by careful quality control in the initial extraction process; the degree of refining is only to increase the overall purity by some 0.45% in the refined metal. This is initially considerably higher than the purity of several of the nonreactive metals, where a metal with as much as a few percent impurities will be electrolytically refined to higher purity.

In this case also there is no special attempt to recover the impurity metals as highly valuable by-products, as is frequently done for the nonreactive metals, but it is refining chiefly to upgrade the purity of the host metal.

TYPES OF ELECTROLYTIC REFINING

Only one method of electrolytic refining is used, and this is to refine commercially pure metal to a higher degree of superpurity when it is required in this state for special applications.

The construction of the cell is an open-topped, mild steel box with a carbon anode which makes up the floor of the cell. Three carbon cathodes, equispaced down the longer dimension of the cell, are attached to steel stubs which both support and position them in the cell and carry current.

The electrolyte is made up of three separate liquid layers which do not intermix and take their relative positions depending on their different densities. The heaviest and bottom layer is the impure metal to be refined. Above this is a mixed fluoride electrolyte. At the top is the layer of high-purity refined metal.

ELECTROLYTIC REFINING PROCESSES

1. Aluminum metal, commercial-grade, of 99.5 to 99.8% purity as produced in the Hall cell can be upgraded to 99.95% in the *Hoope cell* to produce a metal for special applications, with increased softness and superior chemical resistance, in a continuous operation.

The high boiling point of aluminum, 4523°F (2500°C), and its high rate of oxidation make it unsuitable for the conventional methods of refining by fractional distillation or selective oxidation, but it is possible to refine it by electrolysis in a bath of fluorides, with aluminum metal ions from the anode collecting as refined metal at the cathode.

The fluorides in the bath can be mixtures of different proportions. Cryolite, Na_3AlF_6, to which has been added sufficient barium fluoride, BaF_2, to give the correct density is one that is used. Other baths have aluminum fluoride, AlF_3, calcium fluoride, CaF_2, sodium fluoride, NaF, or strontium fluoride, SrF_2, replacing some of the barium fluoride. Typical compositions of these salt mixtures are 30 to 38% BaF_2, 25 to 30% NaF, and 30 to 38% AlF_3, with the remainder being cryolite, or in another bath 18% BaF_2, 16% CaF_2, 36% AlF_3, and the remaining 30% as cryolite.

The working temperature of the cell is between 1652°F (900°C) and 1832°F (1000°C). This is entirely maintained by the heat of the internal resistance of the cell bath to the passage of current through it. The components in the cell are all liquid and fluid at this temperature, and the cell operates at 5 to 7 volts and 20,000 amperes.

The rectangular, water-cooled steel shell is 13 by 10 feet (3.96 × 3.05 m) and 5 feet high (1.52 m), lined with a refractory heat- and electrical-insulating layer of solidified electrolyte which freezes on the water cooled side walls. In this container are placed the three separate, different-density layers of materials that make up the working cell.

A carbon anode is the cell bottom, backed up by a layer of heat-insulating refractory brick. And on top of this carbon block, which is connected by steel bars to the current-carrying bus bars, is an anode layer of copper–aluminum alloy containing 25 to 30% copper. Molten aluminum is added to the copper–aluminum alloy at regular intervals to maintain the required density of at least 2.9 and replenish the supply of aluminum to be refined. The aluminum is added through a submerged well on one side of the cell or through a graphite tube projecting down through the bath to this bottom layer. The mixture of molten fluorides with a density of 2.5 is positioned as the middle layer in the bath, and above this is a layer of high-purity aluminum with a density of 2.3, into which the three equispaced carbon cathodes dip. The cathodes are inserted 2 inches (5 cm) into this aluminum layer. Cathode consumption is approximately 6% of the weight of refined metal produced.

A frozen crust of aluminum covers the bath where the liquid aluminum metal comes in contact with air, and this is periodically broken to ladle off in graphite ladles about 200 pounds (91 kg) of refined metal per day.

Inasmuch as the density differences between the three liquid layers are quite small (2.9, 2.5, and 2.3), and the density of the fluoride electrolyte increases or decreases with rising or falling temperature and the density of the aluminum–copper anode layer shifts as aluminum ions are removed and its composition changes, there is always the danger of mixing of the three layers, with the resultant contamination of the refined aluminum

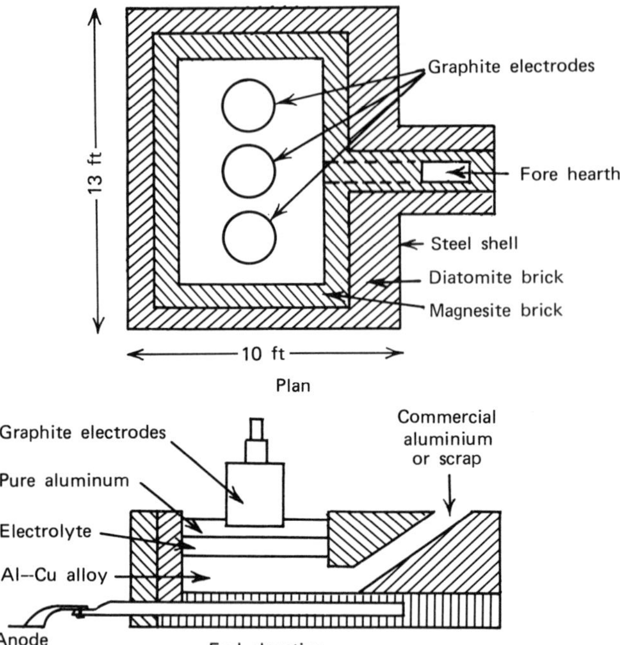

Figure 3.29. Hoope cell aluminum refining process. *Source:* W. H. Dennis, *Metallurgy of the Non-Ferrous Metals*, Pitman, London, 1961, p. 305.

layer. These factors limit the ultimate purity of the product that can be obtained, with copper from the anode alloy being the chief impurity found in the refined aluminum. Iron and silicon also remain in the anode layer and eventually reach concentrations so high that the layer must be completely replaced.

A modification of the Hoope process that uses an electrolyte of 60% barium chloride, $BaCl_2$, 23% aluminum fluoride, AlF_3, and 17% sodium fluoride, NaF, or a mixture of cryolite, Na_3AlF_6, along with barium chloride and aluminum fluoride operates very much like the Hoope cell but at the lower temperature of 1382°F (750°C). This lower temperature is possible because of the lower melting point of the barium chloride.

The chloride–fluoride electrolyte is not as stable as the all-fluoride electrolyte, and the composition has to be kept adjusted by regular additions of the different salts. However this is not a serious problem and the required additions are easily made. The lower bath temperature does have the advantage that it provides a less chemically reactive medium, so that it is possible to line the steel shell of the cell with magnesite brick rather than having to resort to cooling water jackets to freeze part of the electrolyte as a cell lining.

The three layers which separate in this case are a 33% copper in aluminum alloy of density 3.0 at the bottom of the cell on top of the carbon anode, the mixed chloride–fluoride electrolyte of density 2.7 in the middle, and in the upper zone an almost pure aluminum layer of density 2.3 at the cathodes. The three layers are each 5 to 8 inches deep (12.5 to 20 cm).

The purity of the metal produced is normally 99.994 to 99.996% aluminum, with small residual impurities of iron, silicon, and copper. The cell operates at 5 to 7 volts and 16,000 amperes, with a current density of 0.3 to 0.4 amperes per square centimeter. It requires 8 to 10 kW-hr to refine 1 pound (454 g) aluminum.

REACTIVE METALS,
PYROMETALLURGICAL EXTRACTIVE PROCESSES

The similarity in treatment of the reactive metals is that in all their processing, with the exception of some of the primary roasting and calcining, the stages are isolated from air by a variety of methods so that reoxidation of the metal being produced does not occur.

Table 3.8. Reactive Metals, Pryometallurgical Extractive Processes

Zinc

Sulfide

A. Beneficiate
Oxidizing roast
Batch horizontal retort reduction
Condense metal vapor
Liquation refining

B. Beneficiate
Oxidizing roast
Vertical retort reduction
Condense metal vapor
Fractional distillation refining

C. Beneficiate
Oxidizing roast
electrothermic furnace reduction
Condense metal vapor
Fractional distillation refining

D. Beneficiate
Oxidizing roast
Blast furnace reduction
Condense metal vapor

Magnesium

Carbonate

A. Calcine
Chlorinate
Molten salt electrolysis to metal

B. Calcine
Batch vacuum retort reduction
Condense metal vapor

Chloride

Precipitate as magnesium hydroxide
Neutralize to magnesium chloride
Dry to granules
Molten salt electrolysis to metal

Titanium

Oxide

Chlorinate
Batch retort reduction to metal
Slow cooling
Removal by boring
Leaching of residues
Briquetting
Vacuum melting

Aluminum

Oxide

A. Molten salt electrolysis to metal
Molten salt electrolysis refining

B. Chlorinate
Molten salt electrolysis to metal

This considerably complicates the different treatment steps compared with those used for the nonreactive metals where reoxidation was not a major problem and necessitates that treatment be carried out so that the operating procedures can be carefully controlled to the acceptable conditions required.

The purity of metal produced from the smelting stage is generally of higher grade than that produced for the nonreactive metals and of a grade that is suitable, for the most part, for commercial application without further refining. However refining is done on relatively small amounts of a few reactive metals that are required in highest purity for special applications.

The combinations of processes using pyrometallurgical methods to extract and refine zinc, magnesium, titanium, and aluminum are given in Table 3.8.

REFERENCES

Advances in Extractive Metallurgy, Symposium, The Institution of Mining and Metallurgy, London, Elsevier, New York, 1968.

Bunshah, R. F., Ed., *Techniques of Metal Research*, Vol. 1, Pt. 2, Techniques of Materials Preparation and Handling, Wiley, New York, 1968.

Compilation of Air Pollution Emission Factors, 2nd ed., U.S. EPA, Office of Air and Water Programs, #AP-42, 1973.

Cotterill, C. H. and J. M. Cigan, Eds., *AIME World Symposium on Mining and Metallurgy of Lead and Zinc*, Vol. 2, Extractive Metallurgy of Lead and Zinc, Port City Press, Baltimore, 1970.

Coudurier, L., D. W. Hopkins, and I. Wilkomirsky, *Fundamentals of Metallurgical Processes*, Pergamon, New York, 1978.

Dennis, W. H., *Metallurgy of the Non-ferrous Metals*, Pitman, New York, 1961.

Fitterer, G. R., Ed., *Applications of Fundamental Thermodynamics to Metallurgical Processes*, Gordon and Breach, New York, 1967.

Gerard, G., Ed., International Symposium on the *Extractive Metallurgy of Aluminum*, Vol. 2, Aluminum, Interscience, New York, 1963.

Habashi, F., *Principles of Extractive Metallurgy*, Vols. 1 and 2, Gordon and Breach, New York, 1969, 1970.

Imperial Smelting Progress, Imperial Smelting Processes, Ltd., W   .am Press, England, 1963.

Jamrack, W. D., *Rare Metal Extraction by Chemical Engineering Techniques*, Vol. 2, Pergamon, New York, 1963.

Annual Review Issue, *J. Metall.*, Vol. 28, No. 3, 1976.

Annual Review Issue, *J. Metall.*, Vol. 29, No. 3, 1977.

Annual Review Issue, *J. Metall.*, Vol. 30, No. 4, 1978.

McMahon, A. D. et al., *The U.S. Zinc Industry: An Historical Perspective*, U.S. Bureau of Mines Information Circular 8629, Washington, D.C., 1974.

Mantell, C. L., *Electro-Chemical Engineering*, McGraw-Hill, New York, 1960.

Milazzo, G., *Electrochemistry*, Elsevier, New York, 1963.

Mineral Facts and Problems, U.S. Bureau of Mines, Bulletin 667, Washington, D.C., 1975.

Modern Mineral Processing Flowsheets, Denver Equipment Co., Denver, 1962.

Newton, J., *Extractive Metallurgy*, Wiley, New York, 1959.

Patterson, S. H., Aluminum from bauxite, *Am. Scientist*, Vol. 65, No. 3, 1977.

Pehlke, R. D., *Unit Processes of Extractive Metallurgy*, Elsevier, New York, 1973.

Powell, C. F., J. H. Oxley, and J. M. Blocher, Jr., Eds., *Vapor Deposition*, Wiley, New York, 1966.

Powell, R. L., *Chem. Eng. Prog.*, Vol. 50, No. 11, 1954.

Ryan, W., Ed., *Non-ferrous Extractive Metallurgy in the United Kingdom*, The Institution of Mining and Metallurgy, London, 1968.

Rosenqvist, T., *Principles of Extractive Metallurgy*, McGraw-Hill, New York, 1974.

St. Joe Minerals, Lead and zinc, *Eng. Min. J.*, Vol. 177, No. 11, 1976.

Thomas, R., Ed., *Engineering and Mining Journal Operating Handbook of Mineral Processing*, Vol. 1, McGraw-Hill, New York, 1977.

4. Reactive Metals, Hydrometallurgical Treatments

Leaching processes have been used for many years as parallel treatment methods to pyrometallurgy in treating some reactive metals such as zinc. They are also the only methods used on a large scale to treat other reactive metals, such as aluminum and uranium. Alumina processing is an old, well-established operation, while uranium processing is comparatively recent.

The general processing varies to a considerable degree in the treatment of these metals, with some being more easily taken into solution than others, and some being of extremely low grade that requires concentrating the very dilute leach solution before precipitation can take place. This preliminary concentration then becomes a major part of the whole overall operation.

The precipitate in the case of zinc is relatively pure metal which needs only melting and casting to commercial slabs, while with alumina the precipitate is a pure chemical compound in hydrated powder form that must only be calcined. However with uranium several additional chemical and pyrometallurgical extraction and refining steps must be carried through in order to recover the final product in metal form from a crude crystalline precipitate.

These processing treatments are grouped together in their major categories in Table 4.1.

ROASTING

Roasting of concentrates prior to leaching is done with two different purposes in mind. In one case an oxidizing roast is carried out on a sulfide concentrate to have the resulting oxide and sulfate calcines more soluble in the leach solution than is the initial sulfide. An oxidizing roast can also convert leach-soluble impurity metals present in the concentrate to insoluble oxides. In the other case a chloridizing roast is used as a preliminary step to convert troublesome impurities in an oxide ore into volatile chlorides which will go off in the roaster gases, or it can produce insoluble chlorides which remain as residue

Table 4.1.   Reactive Metals, Hydrometallurgical Treatments

Roasting Methods

1. Zinc sulfide concentrate–flash, fluid-bed, hearth
2. Uranium oxide concentrate–hearth

Leaching Methods

1. Zinc oxide concentrate–tank percolation, tank agitation
2. Uranium oxide concentrate–tank percolation, tank agitation, heap, in place
3. Aluminum oxide–pressure

Precipitation Methods

1. Zinc sulfate solution–electrolytic
2. Uranyl sulfate solution–chemical
3. Sodium aluminate solution–chemical

Refining Methods

1. Zinc–melt and dross
2. Ammonium or sodium diurante–chemical then fire
3. Aluminum trihydrate–fire

in the leaching process to follow. A chloridizing roast can also produce water-soluble compounds which are removable by a simple water wash before the leaching stage begins.

The oxidation roasting reactions for the sulfides are the same as those discussed previously for the roasting of reactive and nonreactive metal concentrates by pyrometallurgical methods, in that metal sulfide combines with oxygen to form oxide and metal sulfate, with SO_2 gas being produced:

$$MS + 3O = MO + SO_2$$

$$2SO_2 + O_2 = 2SO_3$$

$$MO + SO_2 = MSO_4$$

With the chloridizing roast, the chloridizing compounds added to the charge will react in most cases to form a metal chloride and a sulfate:

$$2NaCl + MS + 2O_2 = Na_2SO_4 + MCl_2$$

$$4NaCl + 2MO + S_2 + O_2 = 2Na_2SO_4 + 2MCl_2$$

TYPES OF ROASTING

The types of roasters used for preliminary treatment of reactive ores and concentrates before leaching are *multiple-hearth, flash*, and *fluid-bed roasters.*

In the case where a sulfide concentrate is being roasted to an oxide and sulfate, the flash and fluid-bed roasters are preferred, though some use is also made of the hearth roaster. All of these roasters have large capacities, with the flash and fluid-bed types producing the largest throughput tonnage by their quicker reaction times.

The chloridizing roast is done in a multiple-hearth roaster, as here a chloride salt is blended into the charge to the roaster and it is most important that this salt decompose into its component parts and recombine with new elements in the roaster charge, a situation that can be carried out best in the slower, carefully controlled hearth roaster.

This roaster is also best for removing volatile chlorides and for oxidizing certain impurity metals in the feed to oxides, which are then insoluble in the leaching to follow and will be left as residue.

These three types of roasters are discussed in detail in the sections dealing with the pyrometallurgical treatments of nonreactive and reactive metals.

ROASTING PROCESSES

1. Zinc Sulfide Concentrate containing from 50 to 55% zinc as zinc sulfide and about 30% sulfur is roasted to give a zinc oxide calcine that will contain about 65% zinc and below 2.0% total sulfur, with practically all of the sulfur being as sulfate. Roasting is necessary as zinc sulfide has only limited solubility in the dilute sulfuric acid leach solution used in the process to follow, whereas both the zinc oxide and zinc sulfate in the roasted calcines are very soluble. The amount of sulfate sulfur remaining is of value in the leaching process as a makeup to replace sulfuric acid that has been lost by spillage, evaporation, and entrainment in the insoluble leach residue.

The roasting reaction is exothermic,

$$2ZnS + 3O_2 = 2ZnO + 2SO_2 \qquad \Delta H° = -111,000 \text{ calories per mole}$$

and no additional fuel is required. The desired roasting temperature is approximately 1652°F (900°C); at this temperature the required conversion of zinc sulfide to zinc oxide

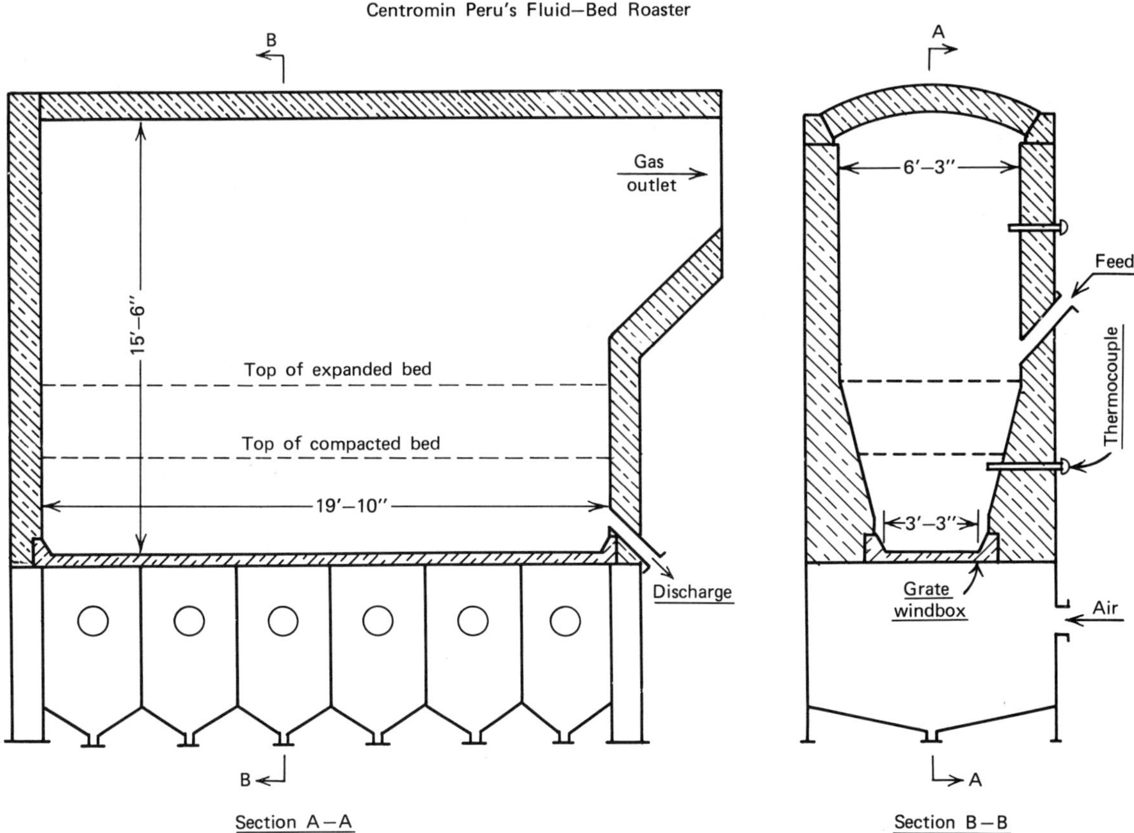

Figure 4.1. Fluid-bed roaster for high-temperature roasting of zinc concentrates, Oroya, Peru. *Source:* Carlos E. Roggero, High temperature fluid bed roasting of zinc concentrates, Presented at AIME Annual Meeting, New York, 1962.

will take place. However in many instances the roasting temperature has to be lowered to 1202°F (650°C) because of the increased tendency at high temperatures to form zinc ferrites, and these ferrites are insoluble in sulfiric acid:

$$Fe_2O_3 + ZnO = ZnO \cdot Fe_2O_3$$

Consequently a balance has to be arranged between roasting temperature and the particular concentrate being treated, to have both a minimum of zinc remaining in the calcines as insoluble zinc sulfide, where one unit weight of sulfide sulfur will tie up two unit weights of zinc and a minimum of zinc remaining as insoluble zinc ferrites. The amount of zinc sulfate produced is not as critical, as the zinc in this condition is recoverable in the leaching step, and the sulfate radical is useful makeup for acid loss during processing.

The conventional 25 foot diameter multiple-hearth roaster (7.62 m), the 25 foot diameter (7.62 m) flash roaster, and the fluid-bed roaster are all used for zinc oxide roasting. In each case fine, dry feed is required, and this is usually ball milled to minus 200 mesh. A combination of scrubbers, cyclones, and an electrostatic precipitator is used to catch the roaster dust and recycle it back in to the operation.

2. Uranium Oxide Ores Some uranium ores, particularly carnotite, $K_2O \cdot 2UO_3 \cdot V_2O_5 \cdot 3H_2O$, which contains sufficient V_2O_5 to be a major source of vanadium, are given a preliminary chloridizing roast with sodium chloride before a subsequent acid leach to extract the uranium. Five to 10% of the ore weight of NaCl is mixed in with crushed and ground ore, and the mixture is roasted on a hearth roaster at 1472 to 1562°F (800 to 850°C) for 1 to 2 hours. The multiple-hearth roasters are the conventional type with eight to ten hearths each 18 to 20 feet (5.5 to 6.1 m) in diameter.

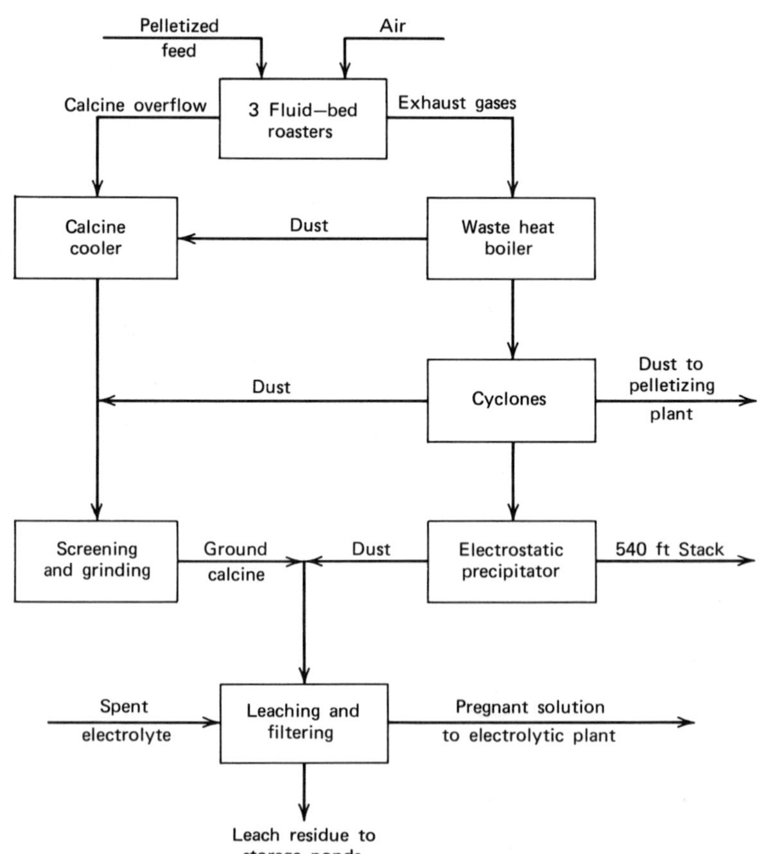

Figure 4.2. Zinc roasting plant flow sheet, Oroya, Peru. *Source:* Carlos E. Roggero, High temperature fluid bed roasting of zinc concentrates, Presented at AIME Annual Meeting, New York, 1962.

Decomposition and recombination of compounds take place to have the various metal impurities in an easily removed form or one in which they are inactive and do not have any effect in the further processing of the uranium.

The vanadium in the carnotite combines with the sodium of the sodium chloride to form water-soluble sodium vanadate, and 75% of the vanadium content can be removed from the roasted calcines by a water wash. The sodium vanadate contains up to 66% V_2O_5 and is treated to recover this vanadium. The uranium values are unaffected and left in the insoluble residue to be extracted by a following sulfuric acid leach.

Other impurities in the ore are removed as volatile chlorides in the roaster gases, and these are sulfur, arsenic, and antimony. Native silver converts to insoluble silver chloride, which will remain as a residue in the acid leaching process, while carbonaceous and organic matter will decompose at the roasting temperature used. Any iron present in the ore is oxidized to ferric oxide Fe_2O_3, which is only slightly soluble in the acid leach solution and will also remain as a residue.

The operating details of the multiple-hearth roasters used have been discussed at length in the sections covering roasting in the pyrometallurgical treatment of the nonreactive and reactive metals.

LEACHING

Leaching of the reactive metals as a method of extraction is carried out for certain specific reasons, in addition to the fact that the solid to be selectively dissolved is in a form highly soluble in the leaching solution. The conditions in every case make it impractical or too costly to consider treatment by pyrometallurgical methods and include the following four types of situations.

The first of these concerns a roasted concentrate with a large number of harmful impurities, each in small amounts, which can all be taken into the pregnant solution along with the major metal; then these impurities are selectively precipitated and removed to leave a purified pregnant solution. A second condition has to do with ores or concentrates that are very low grade and contain only a fraction of a percent of the major metal. With these, leaching is a much less costly process than smelting, and the low-value pregnant solution that results can be concentrated to much less dilute solutions by solvent extraction or ion exchange methods. A third situation has to do with a metallic oxide which is relatively high grade with a high metallic content but does not lend itself to furnace reduction of the oxide to metal using carbon as the reduction agent. This metal oxide will be mainly converted to carbide by the carbon instead of being reduced to metal. Then fourth and finally, the metal oxide under consideration may have much too high a melting point for furnace processing to be feasible.

The solids to be leached are prepared in all cases by crushing and grinding to have them in a fine state, in some cases as fine as minus 200 mesh. This gives very good solid–liquid contact during the leaching stage, and a maximum surface area of the metallic compound to be dissolved is exposed to the leach solution for subsequent fast dissolution.

Leaching solvents are both acidic and basic, with sulfuric acid being the most common acid used. The basic solutions are usually sodium hydroxide or sodium carbonate. Sulfuric acid is a strong, nonselective solvent that is good for most metal oxides and sulfates but is ineffective on sulfides and makes it necessary that these first be roasted to oxide or sulfate. The solid compounds go into solution as sulfates, and such impurities as silica and most of the iron oxide is left as insoluble residue. It is an efficient leach solvent for complex concentrates as well as both high- and low-grade ores.

The basic leaching solvents are used preferentially in cases where a metallic oxide is associated with basic gangue, and it would be prohibitive in cost to use an acid leach liquor that would have to be used in sufficient quantity to neutralize all the basic gangue as well as dissolve the metallic oxide. They are also used in situations where the gangue material is much less soluble in a basic solution than in an acid solution, so that the insoluble residue after leaching will contain a much higher percentage of the waste material in the feed, and leaching has been more selective. In general, the basic solvents are rather slower acting than are acid solvents, but this can be speeded up by having the basic leach done under relatively mild pressure.

Some low-grade oxidized ores can be leached with the assistance of iron- and sulfur-oxidizing bacteria. In this situation there is no direct attack of the bacteria on the oxide, but rather acid ferric sulfate, which is formed with the assistance of the bacteria providing the oxidant to ferrous sulfate, leaches the minerals in the ore:

$$4FeSO_4 + O_2 + 2H_2SO_4 = 2Fe_2(SO_4)_3 + 2H_2O$$

$$MO_2 + Fe_2(SO_4)_3 = MO_2SO_4 + 2FeSO_4$$

The leaching equipment used is relatively simple and is for the most part a batch operation that can be carefully controlled. Agitation tanks are used with atmospheric pressure leaching for both acid and basic solutions, and agitation is carried out by rotating impellers, compressed air injections, or compressed air rising up a center "standpipe" in the tank. Pressure leaching is done in mild steel autoclaves equipped with agitators and is generally used with a basic solvent. Time of leaching varies and will depend to a considerable degree on the strength of the solution, the difficulty of taking the solid into solution, and whether or not pressure is used. The range of time is from 1 hour using a pressure autoclave to 18 hours with atmospheric pressure in an open tank.

Oxidizing agents are added in some cases to have all the desired metallic oxide in the valance state where it will be most soluble. In other cases an added oxidizing compound will precipitate a major impurity as an oxide, which then becomes part of the insoluble residue that is left.

After the leaching operation is completed, the insoluble residue is separated from the pregnant solution and discarded, while the solution goes on to concentration, purification, and precipitation.

TYPES OF LEACHING

Oxides, containing some sulfates from roasted calcines, are leached in either batch type or continuous agitation tanks with a sulfuric acid solvent. Low-grade oxides in ores and concentrates are leached in batch-type agitation tanks with either sulfuric acid or sodium carbonate solutions, whereas high-grade oxides containing impurities soluble in H_2SO_4 are pressure leached in horizontal autoclaves in a sodium hydroxide solution. Heap leaching and leaching in place are also now being applied to low-grade ores and to deposits that are too deep and too low grade to be treated by bringing them to the surface for processing.

Continuous leaching is more frequently used than batch leaching for sulfuric acid treatment of combined oxides and sulfates, with the same types of leaching equipment being used in each case. Pachuca tanks are again a common type of agitation tank, and these are usually open-topped, cylindrical wooden tanks with conical bottoms and with an air lift which draws material from the bottom of the cone and raises it to the top of the tank. Wood staves are a common construction material for Pachuca tanks and stand up fairly well to the acid leach liquor. However the steel hoops around the staves should be encased in plastic for protection as they are quite susceptible to attack by any slops, spills, or drips of acid on them. Stainless steel tanks with stainless steel mechanical agitators are also used, as are concrete tanks lined with sheet lead and acid proof brick, also mechanically agitated.

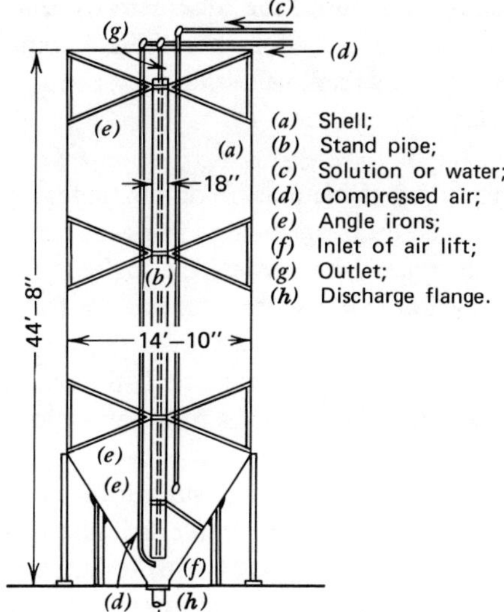

Figure 4.3. Pachuca tank air lift agitation. *Source:* J. L. Bray, *Nonferrous Production Metallurgy*, Wiley, New York, 1947, p. 506.

The Continuous Process is a two-step, counter-current leach, with the treatment consisting of two stages, a neutral process and an acid process.

The neutral leaching step is first. In this operation a large excess of metal oxide is added to acidic spent electrolyte. The metal oxide is dissolved as impure metal sulfate, while insoluble elements and unwanted impurities remain as residue. As the acid is consumed, the pH rises, at which time the leach tanks are discharged into so-called "neutral thickeners" to separate the remaining solids from the pregnant solution. Only part of the metallic content of the oxide feed is dissolved in this first leach, so that the residue still contains much of the desired metal content as well as most of the unwanted impurities.

The clear solution overflow from the neutral thickeners goes to a purification step to remove trace elements which will be harmful during electrolysis and then is pumped to the tank house for electrolytic precipitation of the metal content.

The neutral thickener underflow residue goes to the second stage, the *acid leach process*, where, in mechanically or air agitated tanks, acidic spent electrolyte is added to take the remaining metal oxide into solution. The acid content of the solution must be kept low so that only a minimum of the impurity compounds will be taken into solution along with the desired metal oxide. These second leach tanks are also discharged into thickeners, with the clear acid thickener overflow being pumped back to add to the tanks used for the first stage neutral leach. This overflow product utilizes its remaining free acid for further leaching, as well as adding to the volume of pregnant solution going on to purification and precipitation.

The acid thickener underflow is filtered and washed to recover any contained leach solution, then dried and shipped to smelters to recover the metal content remaining.

Batch Leaching of combined oxides and sulfates is done on a smaller scale than with the continuous process, but with the same general processing equipment. With the batch process, roasted oxide calcines are added to high-acid, spent electrolyte solution. The acid in the mixture is neutralized by the oxide addition, and the mixture is then reacidified by the addition of more spent electrolyte, which in turn is used up in taking more metal oxide into solution. The final remaining acid can be neutralized by the addition of either lime or more of the oxide calcine.

The finished products in the leach tank are separated by filtration, with the pregnant solution going to purification and then to electrolytic precipitation, while the insoluble residue is sent to the smelter to recover the remaining metallic contents.

The batch system ensures that each leach tank can be held until the treatment for that particular lot of material is completed under close control. It is useful when there is some deviation in the analysis of the roasted calcines coming in as feed, necessitating variations in the leaching procedures for optimum metal recovery.

Low-Grade Oxide Ores and Concentrates can be treated by either an acid or a basic leach with the common solvents used being sulfuric acid or sodium carbonate. The acid leach is more rapid and is usually preferred for this reason, unless the waste rock in the ore is basic, which would make an acid leach economicallly impractical. In these cases a basic leach must be used.

With either leaching agent the oxide is most soluble in the hexavalent state. If the oxide is found in the tetravalent state, an oxidizing agent is added to convert it from tetravalent to hexavalent.

Thiobacillus ferrooxidans bacteria, which are able to oxidize ferrous sulfate to ferric sulfate and which will then combine with a metallic oxide to form soluble metallic sulfate, are also useful in leaching low grade-oxide ores and concentrates. This leaching is carried out in conventional air or mechanical agitation tanks, with protection needed for corrosion in sulfuric acid leaching but not in sodium carbonate leaching. For acid leaching a common arrangement is a series of wood stave tanks, each with a rubber-covered impeller for agitation and with the feed ground to 35 mesh and leached for times ranging from 8 to 18 hours.

When leaching is completed, the pregnant solution is separated from the insoluble residue by filtering, thickening, or cyclone classifying. The sands are discarded to a tailings pond, while the leach solution, which is very dilute in values (1.5 gpl) due to the low grade of this type of feed material, is concentrated to 70 to 75 gpl by either ion exchange or solvent extraction processes, and the metallic values are then chemically precipitated.

Pressure Leaching of High-Grade Oxides, which cannot be leached in sulfuric acid solutions because the impurities in the feed material are also soluble, is done in a basic sodium hydroxide solution which will dissolve the metal oxide but not the impurities.

The oxides are crushed and then ground in a fine grinding mill. Sodium hydroxide and hot water are added to the grinding mill so that the product comes out as a slurry of controlled consistency. This slurry is pumped to a horizontal, mild steel digester tank heated by steam under pressure and constantly agitated. Two main combinations of leaching conditions are carried out, depending on the difficulty of taking the metallic oxide values into solution while leaving the impurities as an insoluble residue. The reaction in both cases is the same:

$$M_2O_3 + 2NaOH = 2NaMO_2 + H_2O$$

The pregnant solution is separated from the insoluble residue by decantation and filtration, after which the liquid is cooled, the metal is precipitated as the trihydrate, $M_2O_3 \cdot 3H_2O$, and then calcined to form a pure oxide M_2O_3.

Heap Leaching of Ore that is too low grade for tank or pressure leaching to be economical is done in an area adjacent to where the leaching of higher-grade material is being carried out. This leaching consists of piling the low-grade ore in heaps, flooding it with leach solution, and then draining the pregnant solution off into sumps. The solution is recirculated from the sumps to the ore heaps until it has reached the desired strength and is then pumped to the precipitation circuit to remove its metal content.

Leaching In-Place of Ores that are too deep and too low grade for treatment by conventional underground mining and surface processing can be successfully carried out by underground leaching methods which have low capital and operating costs, along with minimal environmental disruption from surface workings and tailings disposal.

As with other leaching operations of this type, previously described, a leach solution is circulated down through a fractured ore body to dissolve the values and is then pumped to the surface and the values are extracted.

LEACHING PROCESSES

1. Zinc Oxide Calcines, containing a few percent of zinc sulfate from the roasting of zinc sulfide concentrates, are leached with spent electrolyte solution from the electrolytic precipitation cells which follow. The acidity (H_2SO_4) of the spent solution has been increased as a consequence of zinc deposition from it and therefore is a much more effective solvent of zinc oxide at the leaching stage. A series of purification procedures after leaching leads to the production of zinc sulfate solution of a high degree of purity, which is necessary in the successful electrodeposition of high-grade zinc.

Leaching as a treatment method was initially evolved to treat large tonnages of complex lead–zinc–silver sulfide ores which were impossible to concentrate with the beneficiation technology of the time to the 60% zinc content needed for retorting, and only 30 to 40% zinc concentrates were obtainable. Improved technology, especially in froth flotation, has now increased the concentrates from these complex ores to 50 to 55% zinc, which is improved to 65% zinc by roasting, with the leaching process being established as one of the major extraction methods, and giving about 78% of the total world zinc tonnage produced. This is an increase from 56% in 1968 and is partly due to the requirements for less fuel energy. It is environmentally cleaner than the pyrometallurgical retort and smelting methods.

Continuous Process Leaching There are two different leaching methods used, one of which is a *continuous process* while the other is a *batch process*. The continuous process is much the more commonly used of the two and is in general a two- or three-stage, cyclic procedure. In this process the zinc oxide calcine, sometimes first finely ground to minus 325 mesh in a wet. closed-circuit ball mill–centrifugal classifier arrangement, is added to an agitation-type leach tank. These tanks can be a variety of styles—wood stave Pachucas with air lift agitation or stainless steel or lead-lined concrete tanks with stainless impellers. The leach solution in this first, so-called neutral leach is made up of spent electrolyte from the following electrolytic precipitation and zinc sulfate solution from the second leach, called the acid leach.

A predetermined quantity of calcine is automatically fed by preset controls from charge bins into the tank which is one third full of spent electrolyte and return acid leach solution. Here, with agitation over a 2 hour period, the free acid in the solution is used up in reacting with the ZnO in the calcines and taking it into solution as $ZnSO_4$. In some cases live steam is injected to increase the solution temperature to 208°F (98°C) in order to increase the reaction speed. The pH of the solution is carefully monitored throughout the operation.

The zinc oxide, added in excess to use up the free acid, raises the pH to a neutral 5 to 5.2 as the acidity decreases and the spent electrolyte which contained 10 to 11.5% H_2SO_4 drops to 0.5% H_2SO_4, or from 200 gpl H_2SO_4 to 5 gpl. From 50 to 75% of the soluble zinc present in the calcine is dissolved in this first step.

The final high pH of 5.2 also causes the precipitation as oxides and hydroxides of various impurities which have gone into solution along with the zinc oxide. To assist in this precipitation, air is injected under the impellers in those leach tanks using mechanical agitators, or an oxidizing agent such as MnO_2 is added. A good elimination of impurities is achieved, with ferric hydroxide precipitation removing most of the iron and insoluble oxides and hydroxide compounds forming to take out silica, alumina, arsenic, and antimony, as well as a considerable portion of the copper. Cadmium is not affected and is soluble with the zinc.

The neutral leach tanks are discharged into thickeners, with the clear overflow being run off to the purification circuit to remove the remaining major impurities of copper and cadmium, as well as lesser amounts of cobalt, nickel, germanium, arsenic, and antimony, which are all harmful during electrolytic deposition. The purification processes have to be carried out with great care as even small amounts of impurities will have a harmful effect. To accomplish this, it is fairly common to use both mechanically agitated batch tanks and continuous drum reactors.

Copper is removed first by adding zinc scrap to the impure solution in continuous drum reactors and precipitating the copper by replacement with zinc. This reduces the copper from 400 mgpl to less than 0.2 mgpl.

The solution from the copper circuit is clarified in a thickener, with the clear overflow going to the continuous cadmium circuit, where in mechanically agitated tanks maintained at 140°F (60°C) zinc dust is added to precipitate the cadmium. The cadmium content drops from 5 mgpl to less than 0.2 mgpl, and the solution is clarified in a filter press.

The final purification removes cobalt, nickel, germanium, tin, arsenic, and antimony and is carried out as a replacement batch process in mechanically agitated tanks. The solution from the cadmium circuit is heated to about 167°F (75°C), and 10 mgpl of antimony dust and 6 gpl of zinc dust are added. Agitation is continued until the cobalt content is lowered to 0.25 mgpl, at which time the analysis of each of nickel, germanium, arsenic, and antimony is less than 0.01 mgpl.

Purification has to be carried out to this extreme degree because of the harmful effects that even minute quantities of impurity metals can cause during the electroprecipitation

Table 4.2. Solution and Residue Analyses—Continuous Zinc Leaching, Balen Plant, Belgium

	Neutral Solution	Acid Solution	Final Residue
pH	5.2	2.8–3	—
Density	1380–1420	1375	—
Zinc	155–160 g/l	145 g/l	8–12%
Cadmium	450–500 mg/l	—	0.05–0.15%
Copper	100 mg/l	200 mg/l	0.1–0.2%
Ferrous iron	10–30 mg/l	—	—
Total iron	—	—	4–5%
Lead	—	—	30–40%
Silver	—	—	150–300 g/t
Total sulfur	—	—	9–11%
Silica	50 mg/l	200 mg/l	—
Antimony	3–6 mg/l	9–18 mg/l	0.08–0.12%
Arsenic	1–3 mg/l	—	0.5–0.8%
Tin	—	—	1–1.5%
Indium	—	—	0.25–0.35%
Germanium	1 mg/l	30 mg/l	—
Selenium	0.3 mg/l	0.5 mg/l	—

Source: C. H. Cotterill and J. M. Cigan, Eds., AIME World Symposium of Lead and Zinc, Vol. 2, 1970, p. 185.

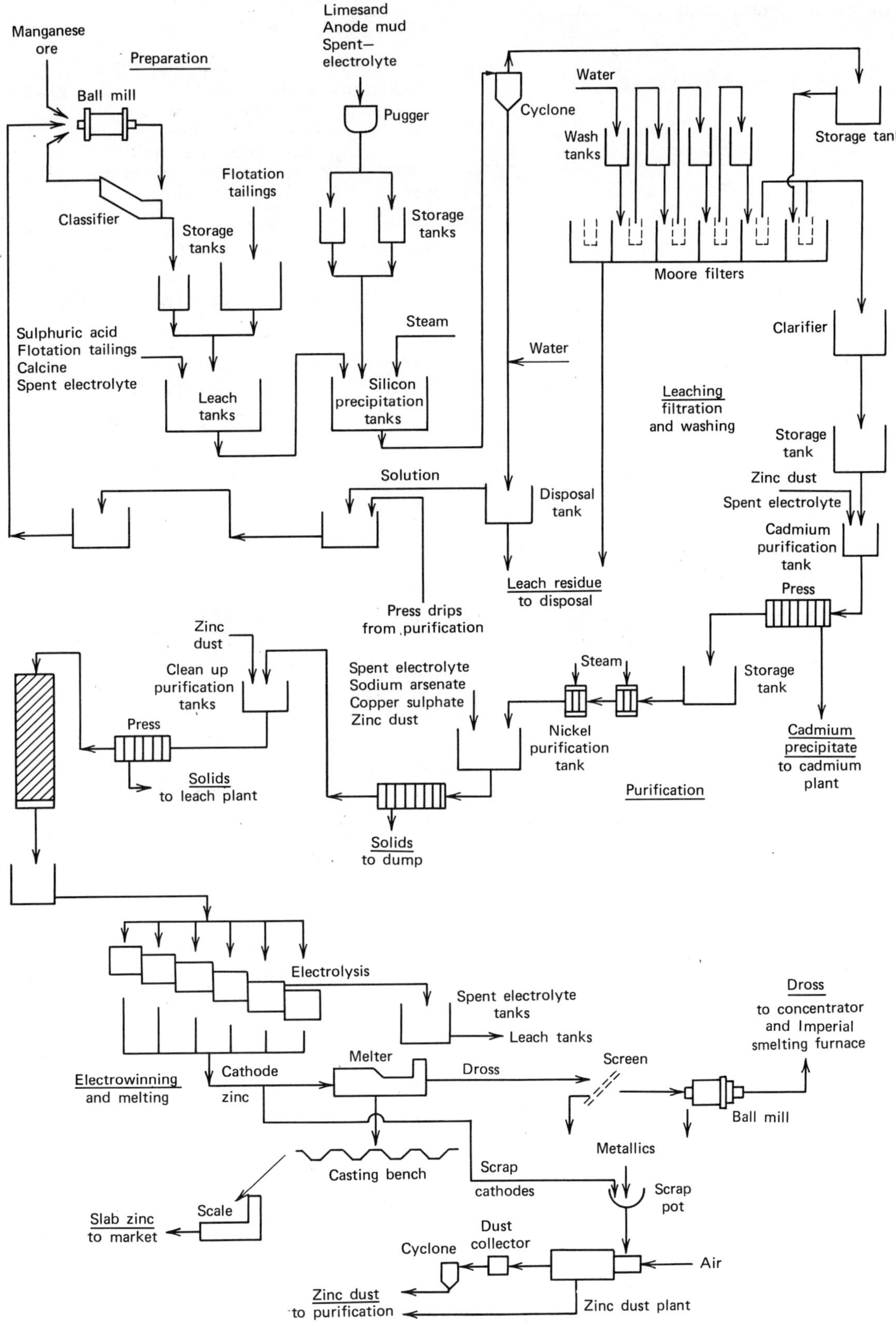

Figure 4.4. Zinc oxide calcines leached in sulfuric acid, purification of the leach solution, and electrolytic precipitation. *Source:* C. H. Cotterill and J. M. Cigan, Eds., AIME World Symposium of Lead and Zinc, Vol. 2, 1970, p. 685.

Table 4.3.   Analysis of Zinc Leach Solution
After Purification Circuit,
Balen Plant, Belgium

	Concentration (mg/l)	
Copper	Less than	0.2
Cadmium		5
Ferrous iron		15
Thallium		0.5–5
Rhenium	Less than	0.005
Gallium	Less than	0.005
Mercury	Less than	0.2
Selenium	Less than	0.002
Tin	Less than	0.02
Arsenic	Less than	0.01
Antimony	Approx.	0.01
Germanium	Less than	0.01
Aluminum		10
Cobalt		0.25
Nickel	Less than	0.01
Tellurium	Less than	0.001

Source: C. H. Cotterill and J. M. Cigan, Eds., AIME World Symposium of Lead and Zinc, Vol. 2, 1970, p. 189.

of zinc from the leach solution. Iron greater than 20 to 30 mgpl decreases the current efficiency, and copper above 10 mgpl assists cathodic corrosion and accentuates the effect of arsenic and antimony, which with 1 mgpl of arsenic causes rough cathode deposits and with 1 mgpl of antimony grows sprouts on the cathode and lowers current efficiency. As little as 1 mgpl of cobalt produces corrosion holes in the deposited zinc and lowers the current efficiency, while cadmium will deposit with the zinc and contaminate its purity.

The underflow from the neutral leach thickeners passes on to the acid leach cycle tanks, where the solids left from the first, neutral, leach are releached in equipment identical to that just used. Spent electrolyte is added to the leach tanks in sufficient quantity to dissolve as much as possible of the zinc oxide still present in the neutral leach solids, while at the same time the pH is kept high enough (finishing at pH 2.80) so that little of the impurities that have been precipitated during the neutral leach are redissolved. Total dissolution of calcine zinc in the combined neutral and acid leaches will be on the order of 85 to 90%.

The acid leach tanks are discharged into thickeners, with the clear overflow being recycled back to the first neutral leach tanks. This acid leach portion of the pregnant solution will now also pass from the neutral leach section into the purification stage preliminary to zinc precipitation, while the free acid that it still contains at pH 2.75 will be used up in dissolving zinc oxide in the calcine feed to the neutral leach tanks.

The solids settling out in the acid leach thickeners are filtered and washed, and the filtered residue is sent to the smelter to recover the still considerable metal values, which are made up of 8 to 12% zinc, 4 to 5% total iron, 150 to 300 gm per ton silver, and 30 to 40% lead, with small amounts of copper, cadmium, indium, tin, silica, antimony, arsenic, germanium, and selenium. The wash water is pumped back to the neutral leach tanks to recover whatever of the zinc solution that is contained in it.

This two-stage continuous leach emphasizes the importance of obtaining pure solutions to be sent on to electrolytic precipitation, while at the same time securing a high percentage of zinc extraction from the calcine feed. The neutral leach residues are treated with acid solution in the acid leach to remove the soluble zinc before these residues are permitted to leave the plant, while the neutral leach pregnant solution, to which is added the acid leach pregnant solution, is neutralized with an excess of calcine, which precipitates many impurity constituents. This neutral solution is then separated from the solid residue and pumped on to the purification circuit before electrolytic precipitation.

Batch Process Leaching of zinc calcine is not as commonly used as the continuous processes because of its smaller capacity, but it does have certain advantages. One of these is that zinc ferrites ($ZnO \cdot Fe_2O_3$) which are formed during roasting and are insoluble during normal leaching can be processed, while another advantage is that each lot of feed is individually treated and any deviation from a standard feed analysis can be accomodated. The types of leach tanks, Pachucas and mechanically agitated vats, are the same for both batch and continuous leaching.

Less close control is required during roasting as the formation of ferrites is not critical. As a consequence, the oxidizing roast can be carried out more efficiently as the roasting temperature does not have to be lowered to 1202°F (650°C) where ferrite formation is least. The ferrites that form are removed magnetically from the calcines and are treated separately from the zinc oxide portion.

High-acid (H_2SO_4) spent electrolyte of at least 60 gpl is heated to 140°F (60°C) and combined with a charge of ferrites in an agitated tank for 1 hour. The zinc ferrites decompose and MnO_2 is added to oxidize the iron as it is freed. The pH of the solution is raised by adding zinc oxide free of ferrites to use up the free acid by taking the soluble zinc into solution as $ZnSO_4$, and as the pH increases the oxidized iron and other metal impurities will precipitate. As much as 12 to 25 gpl iron can be in solution, with this reduced to less than 25 mgpl after precipitation. The final remaining acid can be neutralized to pH 5 to 5.3 by adding more zinc oxide calcine, or sometimes lime.

The heat of reaction of the decomposition of the ferrites raises the solution temperature to 223°F (106°C), and a considerable amount of water is evaporated as steam. Stacks are provided over the leach tanks to aid in the removal of this evolution.

When the leach charge has been neutralized and the iron fully precipitated, the leach tank pulp is dropped into a storage tank and from there goes to a filter to separate the insoluble residue from the pregnant solution.

The filter cake is water washed, repulped with water, and thickened, and the thickener underflow is filtered, dried, and finally shipped to the smelter to recover the valuable metallic contents, which are principally lead and zinc, with some silver.

After the leach-tank pregnant solution is initially filtered, the filtrate, which contains 215 gpl zinc, 600 mgpl cadmium, 300 mgpl copper, and 30 mgpl cobalt, is purified before going on to electrolytic precipitation of the zinc. This is carried out in the same manner as the purification treatments in the continuous process, after which the filtered, now purified, pregnant solution goes to the tank house for zinc removal, while the metal-bearing solids of the filter cake go to the smelter.

2. **Uranium Oxide** ores and calcines are leached in two different leach solutions, depending largely on the types and amounts of impurities present. High carbonate-content material is given a basic leach with a sodium carbonate solution, while the greater part of the uranium-bearing feed to leaching, which is not overly high in basic gangue content, goes through a sulfuric acid solution treatment.

Sodium Carbonate Leach Solutions have advantages over sulfuric acid leach solutions, in addition to their being required for high lime-content ores where the amount of acid necessary both to neutralize the basic gangue and leach the uranium values is prohibitive, in that corrosion problems are minimized and relatively simple and inexpensive equipment is adequate. Also, few ore components other than uranium, vanadium, and some silicates are soluble in the carbonate solution, which makes it relatively selective with a low reagent consumption and gives a comparatively pure pregnant solution. Uranium can be easily removed from the carbonate solution, which can then be regenerated for further leaching.

However there are also some disadvantages to leaching with sodium carbonate solutions. It is necessary to grind the ore finer, often to minus 200 mesh, with increased costs for this processing stage. The finer grind is necessary to liberate and expose the uranium mineral particles to the action of the leaching solution. This is required by the selectiveness of the sodium carbonate solution in attacking the uranium in preference to other materials in the ore. In addition, some uranium minerals are insoluble in carbonate solutions, as are many uranates that tend to form during roasting. Also, sulfates and arsenides tend to form compounds that reduce sodium carbonate concentration:

$$CaSO_4 + Na_2CO_3 = CaCO_3 + NaSO_4$$

These factors, along with the generally lower leaching rates of sodium carbonate in comparison to sulfuric acid leaching, make the acid leach process the most commonly used and preferred method.

Sodium carbonate leach solutions will readily dissolve simple oxide minerals of hexavalent uranium, while the quadrivalent oxides and mixed oxides can be dissolved only with the use of oxidants. The comparative slowness of the leaching operation, 96 to 100 hours in Pachuca tanks at atmospheric pressure, has brought about the use of high-pressure autoclaves which reduce the leaching time to 16 hours.

Pachuca tanks, used for atmospheric pressure leaching, are steel cylinders 18 feet (5.49 m) in diameter by 34 feet (10.37 m) high, with an 18 inch (45 cm) standpipe supported in the center of the tank for pulp agitation. Air for oxidation is blown into the tank through 1 inch (2.5 cm) pipes spaced evenly over the top of the tank area. Steam coils inserted in the tank keep the temperature at $170°F$ $(77°C)$. Leaching is carried out on a pulp containing 55 to 60% solids, and a solution strength of 30 grams sodium carbonate, Na_2CO_3, and 20 grams sodium bicarbonate, Na_2HCO_3, per liter.

The first step in high-pressure autoclave leaching is the fine grinding of the ore in a ball mill to which has been added hot leach solution. The liquid in the ball mill is a hot solution $(110°F, 43°C)$ of recycled carbonate–bicarbonate from the leaching circuit or a fresh hot solution of 5% sodium carbonate. The ground pulp, at a density in excess of 50% solids, is run from the grinding mills to a heat exchanger where it is further heated to $150°F$ $(66°C)$, and is then pumped into the autoclaves.

The autoclaves are horizontal, cylindrical steel tanks with dished-out ends and insulated with fiber glass to retain heat. The dimensions of the tanks are 8 feet (2.44 m) in diameter by 25 feet (7.62 m) long. They are connected in series, with six to nine to a bank. Steam coils are inserted in the first and third autoclaves in the line to raise the temperature to the 230 to $250°F$ $(110$ to $121°C)$ range required for dissolution. Leaching pressure varies from 30 to 90 psi (206.7 to 620 kPa). Agitators at either end of each tank keep the pulp in suspension, and the autoclaves are operated with about 80% of their volume occupied. Oxidation can be provided by adding $KMnO_4$ or by an aerator impeller with an air suction sleeve placed close to the surface of the pulp and located in the center of the tank. The hot discharge residue and pregnant solution pass from the last autoclave in the line to the heat exchanger, and the contained heat is used to preheat the fresh feed material coming into the leaching circuit from the grinding circuit.

The pregnant solution containing soluble sodium uranyl carbonate from both the Pachucas and the autoclaves is thickened and filtered to clarify and then passes to precipitation tanks to recover the dissolved values.

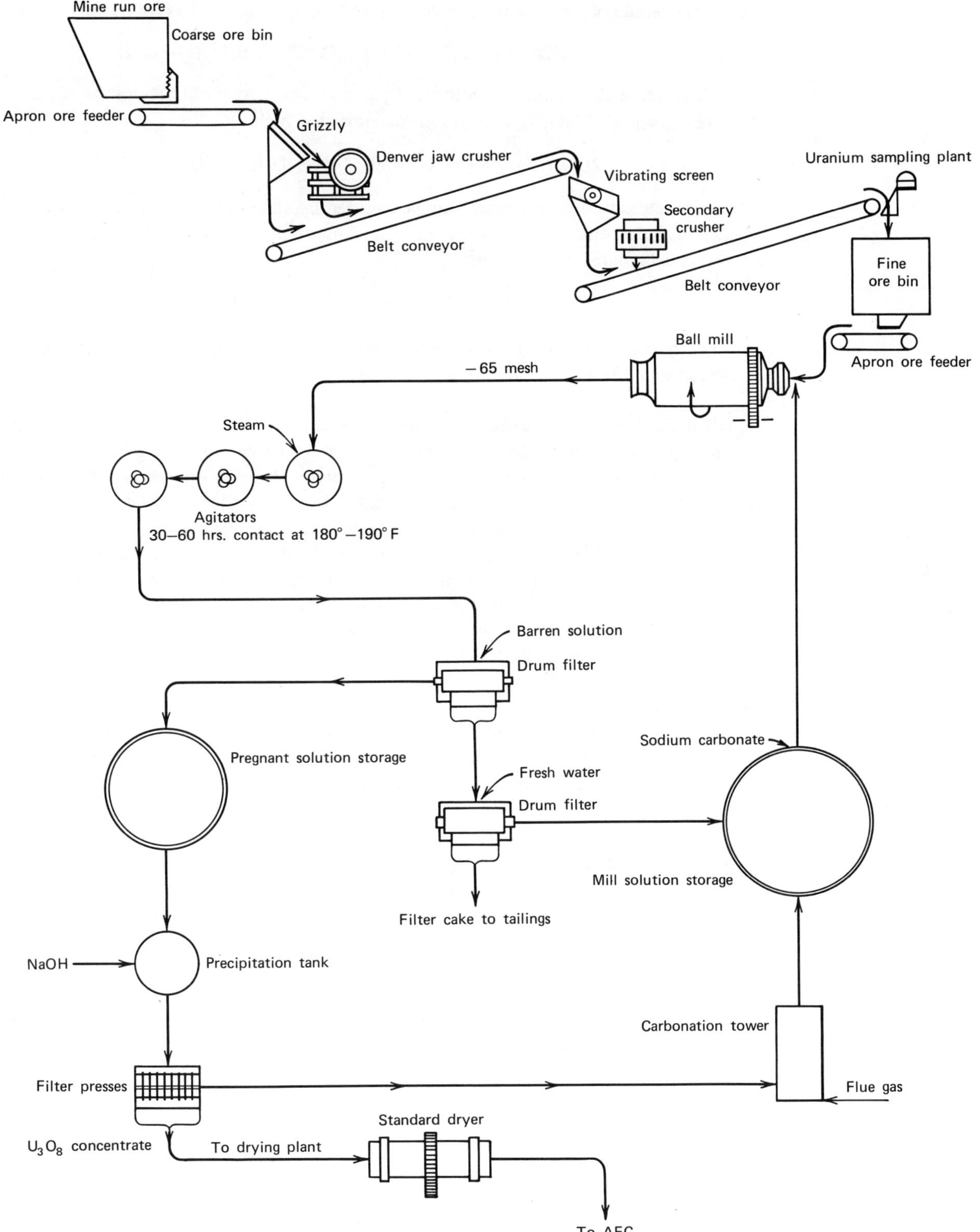

Figure 4.5. Carbonate leach of uranium ore flow sheet. *Source:* M. C. Irani, Denver Equipment Co. Deco Trefoil, Jan.–Feb. 1956, p. 18.

The overall reaction for the carbonate leach can be expressed as follows:

$$2UO_2^{2+} + 6CO_3^{2-} + O_2 + 2H_2O = 2UO_2(CO_3)_3^{4-} + 4OH^-$$

This reaction shows the formation of hydroxyl ions, which will precipitate uranates if the concentration of these hydroxyl ions becomes too high:

$$2UO_2(CO_3)_3^{4-} + 6OH^- + 2Na^+ = Na_2U_2O_7 + 6CO_3^{2-} + 3H_2O$$

The precipitation of uranates can be avoided by the addition of an acid cation such as sodium bicarbonate to the leach solution. This addition neutralizes the hydroxyl ion and prevents uranates from forming:

$$U_3O_8 + \tfrac{1}{2}O_2 + 3CO_3^{2-} + 6HCO_3^- = 3UO_2(CO_3)_3^{4-} + 3H_2O$$

Leach solution strength will vary from 2.5 to 10% sodium carbonate, with added amounts of 1.0 to 2.5% sodium bicarbonate.

Dilute Sulfuric Acid Leaching at atmospheric pressure and ambient temperature has been found to be the quickest and cheapest method of treating most uranium ores and calcines, with over 90% of all uranium-bearing materials being processed in this way. In particular, this is the accepted method of extraction from uraninite, UO_2, and pitchblend, U_3O_8, both of which are essentially the same ore.

The choice of leaching equipment is not especially critical, and a considerable variety is used successfully. Pachuca tanks, both of rubber-lined steel and wood stave construction, are common; they come in a general size range of 20 feet (6.1 m) in diameter and 45 feet (13.7 m) high. Center air lifts are used in some, and all have compressed air agitation at about 40 psi (275.6 kPa) to stir the pulp. The advantages of air agitation are the low first costs and low maintenance and power costs as compared to most mechanically agitated tanks.

However both slow-speed ($\tfrac{1}{4}$ rpm) and high-speed (65 rpm) mechanical agitation is also widely used and possibly gives improved extraction over air agitation alone. Impellers and shafts are protected from acid attack by rubber or neoprene coverings, and the leach tanks are commonly rubber lined with acid-proof brick bottoms. Slow-speed mechanical agitation is done in a thickener 32 by 30 feet (9.76 by 9.5 m) in dimension and with compressed air blown in to assist the slowly rotating rakes. High-speed agitation is done by several ship-type or turbine-type impellers inserted into a baffled tank, and is the best type of agitation to keep comparatively coarse feed, as coarse as 20 mesh, in suspension during leaching. A typical tank would be one 24 by 14 feet (7.32 by 4.27 m) equipped with three equispaced impellers, each 54 inches (1.35 m) in diameter. Ten tanks of this size can leach 1500 to 1800 tons of ore per day.

Leaching may be either batch or continuous, with continuous leaching now being used almost entirely. Continuous leaching lowers the cost of installation, operation, and maintenance, as well as improves the ease and consistency of control of the critical leaching variables.

Most uranium plants use a single-stage leach on ore ground to minus 35 mesh, and no attempt is made to recover unused acid from the leaching circuit. The circuit will have from four to 14 tanks in series, 16 feet in diameter by 16 feet deep (4.9 by 4.9 m) being a common size, and the pulp flows by gravity from one tank to the next. The pulp from the grinding circuit is pumped to the leaching tanks and varies from 45 to 65% solids, depending on the particular plant operation. Sufficient sulfuric acid is added to bring the pH to between 1.5 and 0.5, again depending on the individual plant requirements. The acid is often stage added over the first several tanks, and acid consumption will vary

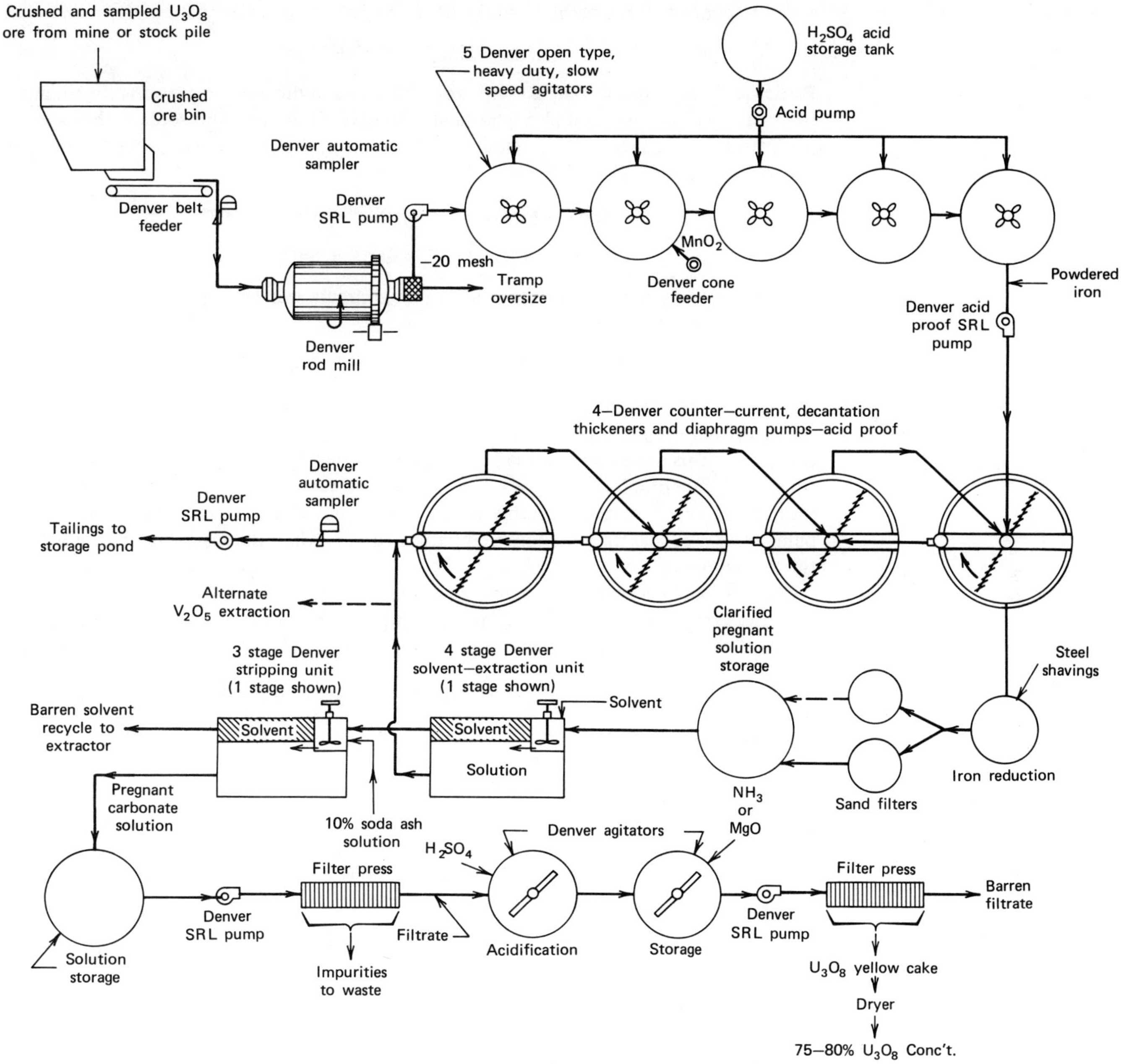

Figure 4.6. Uranium sulfuric acid leach followed by solvent extraction concentration. *Source:* Denver Equipment Co., *Modern Mineral Processing Flowsheets*, 1962, p. 208.

between 50 to 400 pounds (22.7 to 182 kg) of H$_2$SO$_4$ per ton of ore. The leaching period is between 8 and 18 hours, with the normal leaching time about 11 hours.

Acid leaching of highly basic ores is extremely expensive, with 1 pound sulfuric acid required to neutralize 1 pound lime, dolomite, or magnesite in the feed material. As a consequence, ores with more than 12% basic composition are not considered for acid leaching and are more economically treated by a calcium carbonate leach.

Uranium in its tetravalent state, U^{4+}, is only sparingly soluble in dilute sulfuric acid in the absence of an oxidant. Consequently an oxidizing agent such as ferric iron is usually added to convert all the uranium to the 6^+ hexavalent state as uranyl ion, UO_2^{2+}:

$$UO_2 + 2Fe^{3+} = UO_2^{2+} + 2Fe^{2+}$$

Ferric iron also forms a complex with any phosphate in the ore and prevents the uranium from precipitating as a complex phosphate. Manganese dioxide and sodium chlorate are also effective oxidants, and the choice of which one of the three to use generally depends only on local costs and availability:

$$2Fe^{2+} + MnO_2 + 4H^+ = 2Fe^{3+} + Mn^{2+} + 2H_2O$$

$$6Fe^{2+} + NaClO_3 + 6H^+ = 6Fe^{3+} + NaCl + 3H_2O$$

Recovery of uranium from the acid leaching circuit will be 95% or better, with the uranium in solution as uranyl sulfate, UO_2SO_4:

$$2U_3O_8 + 6H_2SO_4 + O_2 = 6UO_2SO_4 + 6H_2O$$

Because of the low grade of the ore being treated, 0.10 to 0.30% U_3O_8, the pregnant solution from the leach circuit is quite dilute and will contain only 0.8 to 1.5 gpl U_3O_8 equivalent. This solution is clarified by combinations of cyclones, thickeners, and vacuum filters and sometimes uses the help of a flocculating agent such as Separan to improve the settling properties of the insoluble residue. After clarification, the pregnant solution is pumped to a concentration and precipitation circuit, while the insoluble residue goes to the tailings pond.

Uranium can be recovered from low-grade ores such as brannerite, $(UCaFeThY)_3Ti_5O_{16}$, with the assistance of iron and sulfur oxidizing bacteria. There is no direct attack by the bacteria, which only oxidize the iron to ferric sulfate,

$$4FeSO_4 + O_2 + 2H_2SO_4 = 2Fe_2(SO_4)_3 + 2H_2O$$

with the acid ferric sulfate which is formed leaching out the values in the ore:

$$UO_- + Fe_2(SO_4)_3 = UO_2SO_4 + 2FeSO_4$$

Ferric iron from any pyrite associated with the ore will oxidize the tetravalent U^{4+} to hexavalent U^{6+}, which is soluble in the acidic ferric sulfate solution.

The uranium U_3O_8 equivalent in these solutions is very dilute and must be concentrated before precipitation.

Heap Leaching as done by the Petrotomics Company (Getty Oil) of Shirley Basin, Wyoming, is carried out on ore that is too low grade for conventional tank leaching by itself, or is in excess tonnage over what can be economically blended with higher-grade ores and leached with them.

Leach pads are prepared by placing six to eight mil thick PVC plastic sheeting, which will collect the pregnant solution and prevent seepage, over an area of about 150 by 150 feet (45 by 45 m). The pad is loaded with ore to a height of 15 feet (4.5 m), and paddies are arranged in squares of about 40 feet (10.25 m) on top of this heap. Raffinate from the plant, at about 25 gpl (11.25 lpm) flow rate, floods the paddies and trickles down through the heap. Perforated plastic pipe on about 36 inch centers (0.9 m) drains the leach liquor from the pad to two sumps. These sumps are 40 foot by 40 foot excavations (10.25 by 10.25 m) lined with 20 mil thick PVC plastic sheeting.

The solution is recirculated from the sumps back to the paddies on the ore heap until it reaches a strength of 0.5 grams per litre of U_3O_8, at which time it is pumped to the extraction circuit in the treatment plant processing the normal higher-grade ore.

Heap leaching recovers about 75% of the U_3O_8 contained in this low-grade material.

In-Place Leaching is an attractive extraction method for uranium deposits that are too low grade and too deep to be processed economically by conventional underground mining and surface treatment. Texas, in particular, has low-grade uranium-bearing sandstones at depths of 300 to 700 feet (90 to 210 m) which are particularly attractive to this type of recovery.

The first in-place leaching operation of this nature was begun in 1975 by the Atlantic-Richfield Company at George West, Texas, with planned production of 125 tons of U_3O_8 per year.

A leach solution of dilute sodium or ammonium carbonate and bicarbonate is circulated at the rate of 2000 gpm (9000 lpm) through a pattern of 66 injection wells and 46 extraction wells, over an area of three acres (1.2 hectare), bored into the ore zone.

To ensure a hydraulic gradient toward the extraction wells, more solution must be pumped out than is injected. Oxidation is necessary for satisfactory solution of the uraninite mineral in the ore, and this can be accomplished by adding hydrogen peroxide to the leach solution or injecting oxygen into the strata.

Pregnant leach solution from the extraction wells is first clarified by passing it through a column of activated carbon, and then the uranium is recovered from the clear solution by conventional ion exchange columns.

To yield the planned production of 125 tons U_3O_8, this leach solution must contain about 30 ppm U_3O_8, which is only one twentieth of the uranium content of the normal surface leaching of higher-grade ores. But ion exchange concentration processing works very well on dilute alkaline solutions and there are no problems in this circuit. After the uranium is removed, the depleted solution is recycled back to the injection wells. Ground waters in the area were studied and classified before leaching began, and a ring of monitor wells around the leaching area warns if any leach solution is escaping out into the ground water.

Several other similar types and sizes of operations were begun in Texas in 1976. Among these were the Westinghouse Corporation at Bruni, the Union Carbide Corporation in Duval County, and the Mobil Oil Corporation in Webb County.

3. Aluminum Oxide is found as bauxite, an impure form of monohydrates or trihydrates of alumina, containing 45 to 65% Al_2O_3, and with major impurities of SiO_2 and Fe_2O_3, which can go as high as 12% SiO_2 and 25% Fe_2O_3. Titanium oxide, TiO_2, is also present in amounts up to 3%, and the combined water will vary from 14 to 36%.

Bauxite cannot be leached in sulfuric acid because some of the iron impurities contained in it are also soluble, so a more selective sodium hydroxide leach is used which takes the alumina into solution as sodium aluminate:

$$Al_2O_3 + NaOH = 2NaAlO_2 + H_2O$$

The monohydrates of alumina, $Al_2O_3 \cdot H_2O$, are less soluble in sodium hydroxide solutions than are the trihydrates, $Al_2O_3 \cdot 3H_2O$, which necessitates more severe leaching conditions for the monohydrates. In both cases mild pressure leaching is used after preliminary crushing and grinding the ore to minus 70 mesh. Sodium hydroxide and hot water are added to the ball mills in which the fine grinding takes place so that the ground product is taken out as a slurry of controlled consistency and pumped on to the digester tanks. These tanks, which are constructed of mild steel, are positioned horizontally, are constantly agitated, and are heated by steam jackets or the injection of live pressurized steam.

The two principle variations of the leaching operation are either a batch leach with duration times of 2 to 8 hours in a digester maintained at 390°F (199°C) and 210 psi (1447 kPa), in a 400 gpl caustic soda solution for the monohydrate ore, or a continuous process with $\frac{1}{2}$ to 1 hour leaching time in a solution at 290°F (143°C) and 60 psi (413 kPa), of 170 gpl strength, for the trihydrate ore.

In both processes the pregnant solution is separated from the insoluble residue, known as red mud, by countercurrent decantation in thickeners and clarification in pressure filters, after which the pregnant liquor is cooled until it becomes supersaturated and is then treated to precipitate the values.

The red mud contains Fe_2O_3, TiO_2, and a complex sodium aluminum silicate compound which represents a loss of both caustic soda and alumina. The quantity of this compound that is discarded is related to the silica content of the bauxite, with approximately 1.1 weight unit of alumina and 1.2 weight unit of sodium hydroxide being lost for each 1.0 weight unit of silica present in the bauxite.

This qualifies only bauxite ores that have less than 8% silica for treatment by the foregoing leaching processes. Other ores of an upper limit of 15% silica are treated by a different, modified process. In practice, some bauxites combined with clays containing as much as 25% silica are mined, and this can be blended with low-silica ores to give a combined feed material of 12% silica that can then be successfully treated by the modified process.

This process subjects the high, 12 to 15% silica bauxite to the conventional leach with sodium hydroxide, after which the separated red mud containing the complex sodium aluminum silicate compound is first ground in a ball mill to minus 200 mesh and then sintered in a rotary kiln with limestone and sodium carbonate. The calcine from the kiln is cooled and reground with water in a ball mill, and then the slurry from grinding is passed on to a vacuum filter. The grinding with water leaches out soluble sodium aluminate which is recycled back to the digesters after clarification by vacuum filtering. The solids from the filtering, so-called brown mud, contain the silica as dicalcium silicate, which is a composition somewhat similar to portland cement, and it is discarded.

PRECIPITATION

The processing operation following that of leaching is precipitation, in which the metal values which were taken into solution during leaching are now precipitated in a solid form from the pregnant solution.

Three general methods of precipitation are used, one of these being electrolytic, in which the pregnant solution must first be carefully purified of all contaminant metals which can either upset the delicate balance of the cell in which deposition is taking place or can deposit along with the desired metal and lower its purity. Electrolytic precipitates are usually quite pure and are frequently used commercially in this form with only a minor amount of further refining. The solvent solution is regenerated so that it can be reused as leaching solution in the leaching circuit, after the deposition by electrolysis of part of its metal content.

The second method of precipitation is by chemical methods, in which compounds of the desired metal can be formed by reagent additions; and then, by changing the pH, new conditions are established in which these compounds are no longer soluble in the pregnant solution. For the most part these leach solutions are very dilute, due to the low grade of the ore which is being treated. As a consequence they must first go through concentration processes to improve the strength of the solution before precipitation is carried out. The chemical purity is only moderately high and needs considerable further refining and conversion to obtain the refined metal state.

The final method of precipitation comes about by decomposition of the pregnant solution during slow cooling, so that if the solution is seeded with crystals of the compound to be formed during precipitation, this seeding helps to initiate further crystal growth and separation of the compound out of the solution. The precipitate recovered from this type of separation is quite high in purity but also needs conversion and refining to obtain refined metal finally from it.

TYPES OF PRECIPITATION

Electrolytic Precipitation is used on sulfate leach solutions, after purification, which have been obtained by dissolving oxide or sulfate roasted calcines in sulfuric acid. The process is continuous, with only a part of the metal content being removed by deposition at the cathode electrode. The removal of this portion of the metal regenerates the acid solution, which is then recycled back to the leach tanks to be used as fresh leach solution:

$$MSO_4 + H_2O + e = M + H_2SO_4 + \tfrac{1}{2}O_2$$

Purification of the leach solution is necessary before electrolytic deposition in order to have as pure a product as possible come out of solution and deposit at the cathode when the decomposition voltage is applied to the cell. The decomposition voltage is a direct current voltage, which is a definite value for each metal and varies according to the position of the metal in the electromotive series, so that a sufficiently high decomposition voltage will deposit not only the desired metal but also all other metals which lie below it in the electromotive series, and in this manner will contaminate the cathode product. Other impurity metals in the electrolyte leach solution, while not depositing at the cathode, have the unfortunate effect of causing decreases in cell current efficiency, resolution of the cathode deposit, and rough, sprouted deposits.

The anodes and cathodes are both inert metals, with the deposited metal being stripped from the cathode at regular intervals. The cell temperature is not allowed much of a rise, as this higher temperature intensifies the effect of impurities in the electrolyte solution. The cells are usually of concrete and lined inside and covered outside with an acid-resisting material.

The electrical connections to the cells are by the multiple system circuit, with one more anode than cathode. In general their overall operation is similar to that used for electrolytic precipitation of nonreactive metals.

Chemical Precipitation is used on both sulfuric acid and sodium carbonate leach solutions obtained from the leaching of low-grade oxide ores. The chemical precipitate formed in both these cases is brought about by adding a reagent to the leach solution which combines to form an insoluble compound at the pH of the solution. Recovery is very good, and extraction of the metal value as a precipitate is close to 100%. However the chemical compound formed contains only 72 to 85% metallic value and must be further processed to have a final product of refined metal.

The leach solutions are quite dilute due to the low metal content in the ore being treated. Because of this the acid pregnant solutions are concentrated before precipitation to increase their metal content from 1.0 gpl to 10 to 40 gpl. Carbonate leach solutions are not as dilute as the acid solutions, 2.5 gpl, and are not usually concentrated before precipitation.

Concentration and Purification of sulfuric acid leach solutions are carried out by one of two methods, *solvent extraction* or *ion exchange*, with solvent extraction the more widely used process.

Solvent extraction in its simplest form involves two steps; the first is extraction and the second is stripping. In the extraction step an organic solvent is brought into contact with the solution to be treated, and the metal content is transferred from the pregnant leach solution to the solvent. In the stripping operation, the loaded solvent is brought into contact with a suitable aqueous stripping solution and the metal is transferred back to an aqueous phase. As the volumes of the organic solvent and aqueous stripping solutions are much less than the volume of the low metal-concentration leach solution, the metal values are now held in a much smaller volume than was originally the case, and their concentration has correspondingly increased. The overall operation can be made to approach a continuous system, with close to 100% recovery of the metal values.

The ion exchange process brings solid resin beads into contact with the pregnant leach solution, with the resin beads then exchanging ions on their surfaces for other metal ions in the solution and in this manner removing the metal value. The loaded resin beads are then eluted with a strong negative ion solution, such as brine. This reverses the ion exchange, removes the metal value from the resin surface to the eluting solution, and regenerates the resin beads so that they can be reused in the circuit. The volume of the eluting solution is much less than the volume of the pregnant solution, and as the metal values are now being held in a small volume of solution, their concentration is greatly increased. As with solvent extraction, the overall process can be operated as a continuous circuit with close to 100% metal recovery.

While it is possible to treat carbonate leach solutions by ion exchange, they are generally only separated from the solid leach residues before precipitating out the metal values. This liquid–solid separation is carried out in combinations of washing thickeners and vacuum filters, and then the clarified solution is passed to the precipitation circuit for recovery of the contained values.

The final method of precipitation is by the *decomposition* of the clarified leach solution, which is the product of a sodium hydroxide pressure leach of an oxide ore. The clear solution is agitated in tanks in the presence of seed crystals of the precipitate, which induces the formation of its own species:

$$2NaMO_2 + 4H_2O = 2NaOH + 2M(OH)_3$$

The trihydrate of the metal is the precipitate formed and is quite pure; while the sodium hydroxide, which is separated from the precipitate by filtering, is recycled back to the leaching circuit of the process. The temperature during precipitation is held at 77 to 122°F (25 to 50°C), in which range a coarse grained precipitate is formed that is a desirable size for filtering.

PRECIPITATION PROCESSES

1. Zinc Sulfate leach solution after purification is pumped to storage tanks in the cell house and flows from there to the electrolytic cells for precipitation, where it becomes the electrolyte in these cells:

$$ZnSO_4 + H_2O + \underset{(3.4\ V)}{e} = Zn + H_2SO_4 + \tfrac{1}{2}O_2$$

Leach solution is added to the electrolytic cells in controlled amounts, either continuously or in batches, with a batch being added once every 24 hours if this system is used. The object is to keep the zinc content and acid content of the electrolyte at a uniform level. As electrolysis proceeds and the acid strength rises to 150 to 200 gpl, some of this high-acid solution is withdrawn from the cells and sent to the acid storage tanks, from where it will be recycled to the leach circuit for reuse. This withdrawn solution is replaced in the cells by fresh neutral leach solution, which reduces the cell acid strength to about 115 gpl, the normal range for electrolysis. The fresh leach solution, in addition to being low in H_2SO_4, carries from 100 to 160 gpl zinc as zinc sulfate. Electrolysis removes from 50 to 70% of this by deposition of zinc metal on the cathodes, depleting the solution to about 50 gpl, and the remaining zinc sulfate content is withdrawn in the high-acid solution recycled to the leaching circuit and forms a consistent circulating load.

The cells are constructed of lead-lined concrete and have typical dimensions of 15 feet long (4.57 m) by 2.8 feet wide (0.85 m) by 5.5 feet deep (1.68 m). These cells can be arranged in cascades of six to 12 cells each so situated that solution will flow by gravity from the highest "head" cell of each cascade to the second, from the second to the third, from the third to the fourth, and so on, down the series, with the last cell discharging into a launder. A cell room will have 300 to 400 cells divided into two to four units, with each unit having its own electrical circuit for flexibility. Pipes and launders are also of acid-proof construction and are lead-lined, glass fiber-reinforced polyester or polyethylene.

Cell temperature is important and is held at 95 to 113°F (35 to 45°C) by circulating cooling water through lead coils placed in each cell. High temperature intensifies the bad effect of impurities in the cell; and as heat is generated during electrolysis, the cooling coils are needed to keep this heat within reasonable limits.

The anodes are fabricated from 99% lead–1.0% silver sheet $\frac{1}{2}$ inch thick (1.25 cm) and are inert to the electrolyte solution. The cathodes are also inert and are rolled aluminum sheet $\frac{1}{4}$ inch thick (0.6 cm), and either the cathodes or anodes are equipped with rubber or porcelain buttons on the submerged portion to serve as electrode spacers. This distance is kept at about $3\frac{1}{2}$ inches (9 cm) center to center of anode to cathode. There is one more anode than cathode in each cell. A cell of the typical dimensions previously given will have 46 anodes and 45 cathodes, whereas a smaller-sized cell, also in common use, has 28 anodes and 27 cathodes.

Most plants operate with a current density in the range of 20 to 40 amperes per square foot (0.09 m^2) of cathode area, with variations at different plants ranging from 20 to over 100 amperes. The higher current density means that the operating conditions are adjusted for higher voltage per cell, higher acid strength needed, larger volume of cooling water required, faster zinc deposition, and less solution to handle. The choice of the current density chosen for each individual plant operation will then depend on the balance required for all these various factors.

The voltage required for the electrolysis of a zinc sulfate solution is theoretically 2.35 volts, but in actual practice 3.25 to 3.5 volts is required due to current loss and leakage throughout the electrical circuitry. The decomposition voltage of zinc sulfate is above that of hydrogen, and normally it would be expected that hydrogen would evolve instead of zinc. However the hydrogen overvoltage with respect to zinc in an acid solution is high enough to let the zinc plate out of the zinc sulfate solution without the evolution of a great amount of hydrogen at the cathode.

Ampere efficiency of zinc electrolysis, which is the ratio of metal actually recovered to that theoretically recoverable by the amount of current used according to Faraday's law, is maintained at 87 to 94% over a long-term average. The chief factors in lowering ampere efficiency are the impurities in the electrolyte, which have a devastating effect in even minute quantities by causing resolution of deposited zinc or improper zinc plating. These impurities are removed in the purification circuit between the leaching and electrolytic precipitation operations.

Most plants operate on a basis of pulling the cathodes from the cells every 24 to 48 hours to strip off the layer of deposited zinc, which is very pure and analyzes 99.995% zinc. This is done by removing the cathodes from a cell by an air hoist and then moving the load by monorail to a stripping rack on the main floor of the cell house, where the actual stripping or pulling the zinc from the aluminum cathode is done with a hand tool. The stripped zinc is weighed and moved to a cathode storage area.

The cells must be cleaned frequently because of the accumulated slimes, mostly manganese dioxide added to oxidize iron in the leaching circuit and brought in with the electrolyte. This MnO_2 both forms on the anode as a slime coating and collects as a sluge in the bottom of the cell. To keep this under control, a cascade of cells will be cleaned out about every one and one half weeks, at which time the anodes are also washed off with high-pressure water. Calcium sulfate (gypsum) is also a problem at some plants by depositing and building up in the solution lines. This is removed by high-pressure water cleaning at intervals, as required, and by frequent reaming of the cell feed orifices to keep them open and running.

2a.   Uranyl Sulfate solution resulting from the sulfuric acid leaching of low-grade uranium oxide ores is purified and concentrated before precipitation by one of two methods, *solvent extraction* or *ion exchange*. The more widely used method currently in vogue is solvent extraction, which is popular because of its low capital cost, its speed, simplicity, selectivity, and high efficiency which gives over 99% extraction.

Solvent Extraction The solution to be treated by solvent extraction is first clarified by thickening, usually followed by filtering, after which it is known as the pregnant aqueous and contains from 0.8 to 1.5 gpl U_3O_8. The pregnant aqueous solution is pumped to the extraction circuit, where it comes in contact with an organic–kerosene mixture called the organic. The organic has greater affinity for the uranium than does the water solution and combines with it. Then, due to the immiscibility of the organic in the water solution, the organic being lighter than water rises to the top and is separated from the uranium-depleted aqueous phase, which is termed the raffinate. Since the volume ratio of aqueous to organic solutions in the extraction circuit runs from $2\frac{1}{2}:1$, to $5:1$, depending on the type of organic used, the concentration of the uranium in the extraction liquor will be in this same proportion.

The extraction is carried out in four or five stages in compartmental tanks built of wood staves or fiberglass-lined concrete. The tanks are built both round and rectangular, with a round tank 27 feet (8.2 m) in diameter and 9 feet (2.74 m) deep partitioned off with three mixing chambers and three settling chambers, while a round tank 22 feet (6.7 m) in diameter and 9 feet (2.74 m) deep is partitioned into two mixing and two settling chambers. The mixing chambers are stirred by motor-driven stainless steel impellers. The flow of aqueous and organic solutions through the circuit is countercurrent, with the pregnant aqueous flowing in one direction and the organic in the opposite. The aqueous flows from introductory mixer to settler, to second mixer, and so on, on through the tank, with the flow induced either by gravity or acid proof pumps. The organic flow recirculates back to each mixer from the adjacent settler, using either a combination of air lifts and pumps or a vertical pumping turbine. The vertical pumping turbine is superceding older recirculating equipment and can be set to give a type of flow so that 150 gpm (682 lpm) of the aqueous and 22 gpm (100 lpm) of the organic are advanced, while 150 gpm (682 lpm) of the organic is recirculated back.

A number of compositions of organic solution are used, such as 3.5% diethylhexyl-phosphoric acid, mixed with 2% isodecanol and 94.5% high flash-point kerosene. Another mixture used is 5% Alamine 336, 2.5% isodecanol, and 92.5% kerosene. The organic loss in the circuit is negligible, amounting to only 0.3 gallon (1.4 litre) per 1000 gallons (4550 litres) of pregnant aqueous treated:

$$2R_2HPO_4(\text{organic}) + UO_2(\text{aqueous}) = UO_2(R_2PO_4)_2(\text{organic}) + 2H^+(\text{aqueous})$$

where R = diethylhexyl group.

The barren leach solution, the raffinate with 0% U_3O_8, is discarded as waste material. Frequently there is no preliminary treatment before discarding, though in some cases it is first put through an organic scavenging tank.

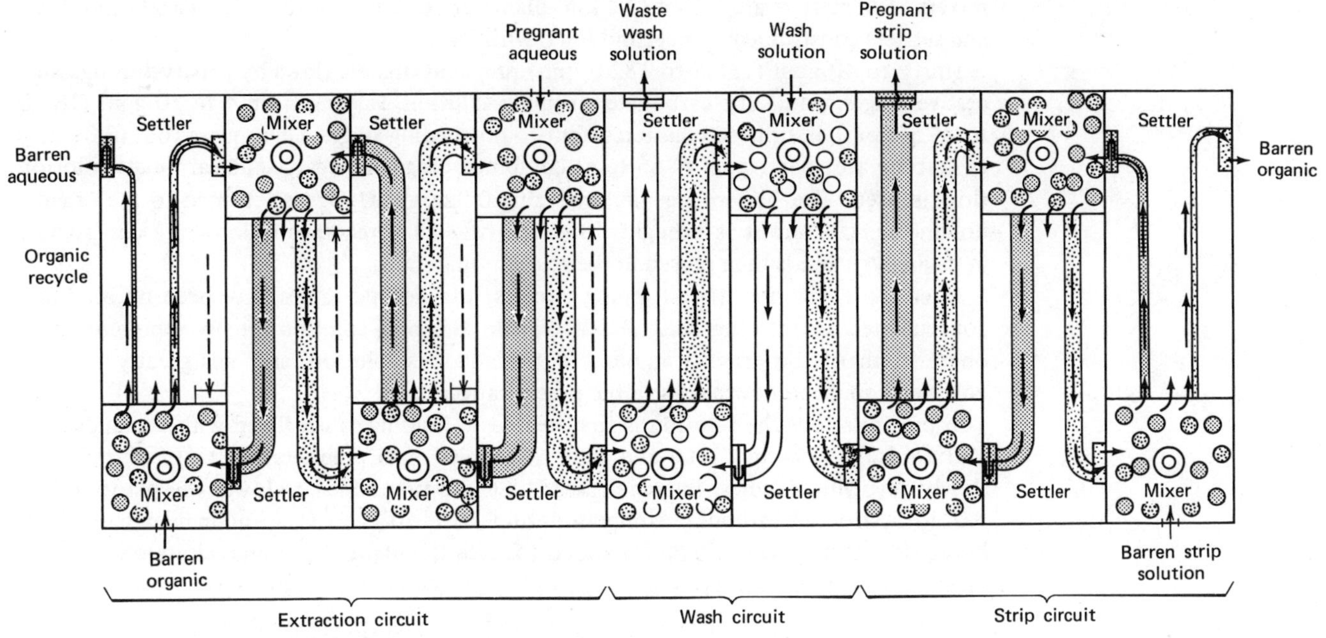

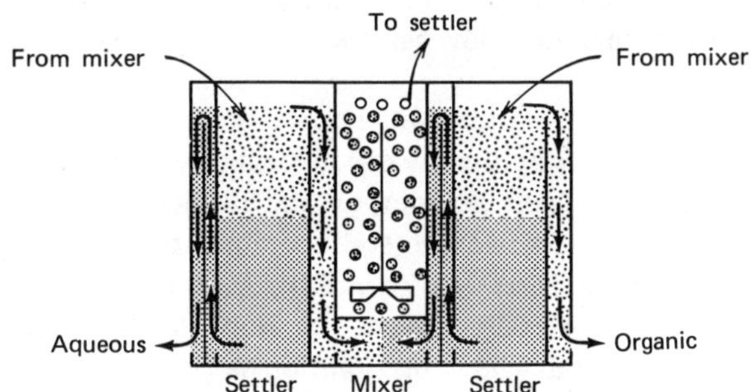

Figure 4.7. Typical mixer–settler solvent extraction system. *Source:* W. C. Hazen, Denver Equipment Co. Deco Trefoil, Aug.–Sept.–Oct. 1963, p. 17.

The organic, which now contains the uranium values in a concentration of some 2.5 to 6 gpl U_3O_8, is mixed with a 10% solution of sodium carbonate. The sodium carbonate has a greater affinity for the uranium than does the organic and picks it out. The solution mixture then flows to a settling tank where the barren organic, being lighter again, rises to the top and can be separated from the uranium carbonate solution layer below. This is known as the stripping circuit. The barren organic solution is returned to the extraction circuit for reuse, while the uranium carbonate solution containing 30 to 40 gpl U_3O_8 is sent to precipitation. The same upgrading of solutions again takes place in the stripping circuit as the amount of stripping solution utilized to remove the mineral values from the pregnant organic is usually only one tenth to one fifteenth of the volume of the organic:

$$UO_2(R_2PO_4)_2 + 3Na_2CO_3 = Na_4UO_2(CO_3)_3 + 2R_2NaPO_4$$

The stripping section consists of three stages of mixers and settlers, with the mixer tanks $3\frac{1}{2}$ feet (1.07 m) in diameter by 7 feet (2.1 m) deep, agitated by vertical turbine mixers. The settlers are 8 foot (2.44 m) diameter cone-bottomed tanks, with both mixers and settlers constructed of steel and lined with PVC.

Thirty to 40 gpm (136.5 to 182.0 lpm) pregnant organic flows by gravity through successive stages, while the carbonate stripping solution, varying from 4 to 10 gpm (18.20 to 45.50 lpm), is airlifted countercurrent to proceeding stages. One square foot (0.09 m^2) of settling area per 1 gpm (4.55 lpm) combined pregnant organic and carbonate solution flow is used. The barren organic, analyzing 0.025 gpl U_3O_8, is returned to the organic holding tank where it is brought up to the original percentage ratio of the compounds in its analysis and is then reused in the extraction circuit.

After the extraction and stripping circuits, the desired values have been purified and concentrated so that the volume of pregnant stripping solution usually represents only one fiftieth of the original amount of the aqueous solution, and this greatly reduced volume of solution now passes to the precipitation stage.

In precipitation, the solution is first heated in holding tanks fitted with steam coils to 130 to 160°F (54 to 71°C) and then discharged into precipitation tanks stirred with slowly revolving paddles. In some cases H_2SO_4 is first added to lower the pH to 2.5 to 3.0, to destroy all carbonate compounds and drive off the CO_2. And in all cases sodium hydroxide or magnesium oxide is added to raise the pH to 7.0 and precipitate sodium or magnesium diuranate ($Na_2U_2O_7$), yellow cake, which is 75 to 85% U_3O_8:

$$2Na_4UO_2(CO_3)_3 + 6NaOH = Na_2U_2O_7 + 6Na_2CO_3 + 3H_2O$$

The slurry containing the precipitated uranium is pumped to a plate-and-frame press or precoated vacuum filter for filtration and washing to remove the soluble impurities. After this treatment the yellow cake is dried and packaged for shipment to the refinery.

Ion Exchange   The other widely used method of recovery of uranium from a sulfuric acid leach solution is by the use of ion exchange columns. This may be used on clear solutions after countercurrent decantation and filtration, or with unfiltered pulps which have only been cyclone classified and still contain 4 to 8% of minus 325 mesh solids. In this latter case the process is referred to as RIP, or resin in pulp process.

Uranium is largely present in sulfuric acid leach solutions as a complex anion $UO_2(SO_4)_3^{4-}$, and when this solution is brought into contact with the active surface of the cation resin, RX^-, then an exchange of ions between the two will take place until equilibrium is reached:

$$4RX^-(\text{resin}) + UO_2(SO_4)_3^{4-}(\text{feed}) = R_4UO_2(SO_4)_3^{4-} + 4X^-$$

where R represents the resin cation and X^- represents anions such as Cl^- or NO_3^-.

Ion exchange processes usually have a fairly high capital cost per ton of product, partly from the fact that the process depends on the rate of diffusion of ions into the solid resin beads, which is a relatively slow operation. This then necessitates a fairly large amount of in-process materials and a correspondingly large plant, with this alleviated to some degree by having the overall processing as fully continuous as possible. The treatment is an efficient one, however, and ion exchange methods will extract up to 99% or better of the uranium in the solution being processed.

Leach solutions going to ion exchange usually contain 1 to 2 gpl U_3O_8, but in certain cases it is economically justified to extract uranium from leach solutions containing as little as 0.2 gpl. These very low-grade solutions are recovered as a by-product of some gold mining operations, and the fact that gold is also being extracted from the same ore and the cost of the mining and milling operations can be shared between the gold and the uranium makes the recovery of uranium economically worthwhile.

A conventional ion exchange column being fed with a clarified leach solution is contained in a rubber-lined, mild steel cylinder with outward dished ends, 7 feet (2.13 m)

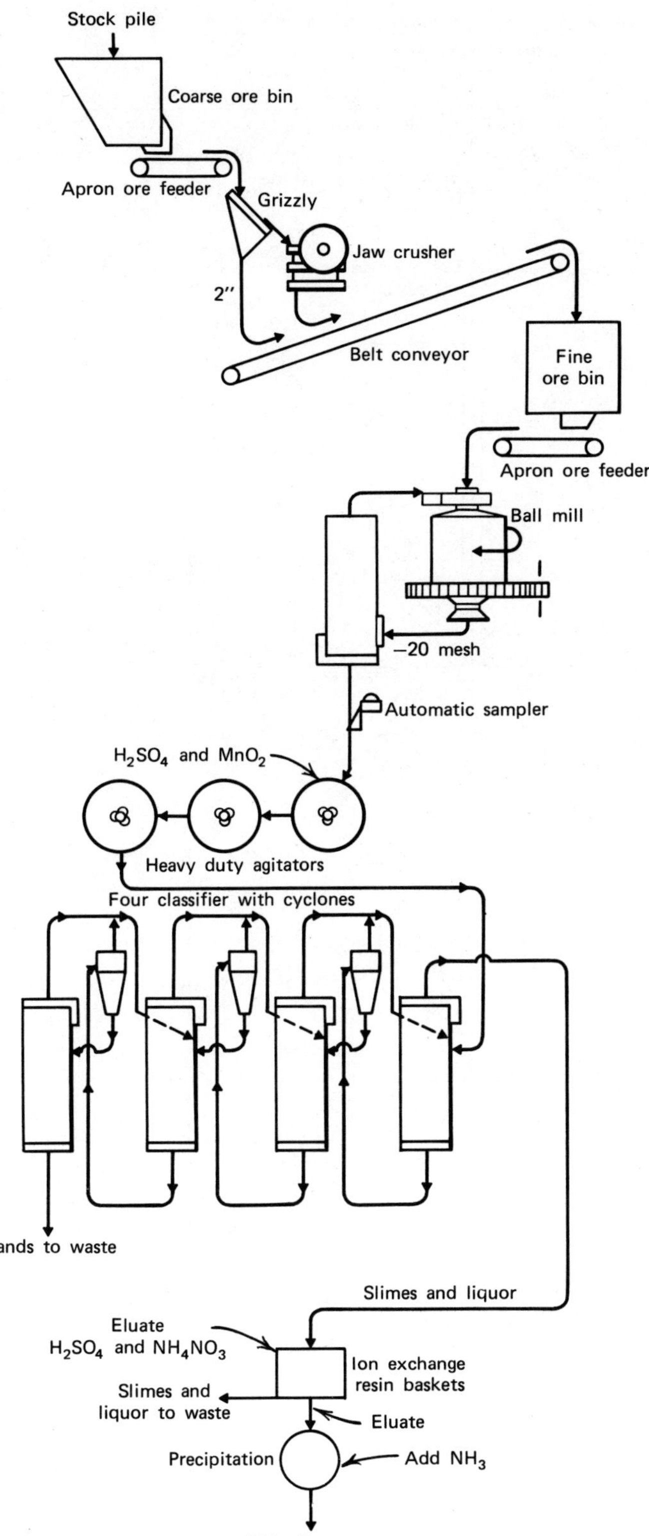

Figure 4.8. Acid leach of uranium ore with classifiers preceding ion exchange. *Source:* M. C. Irani, Denver Equipment Co. Deco Trefoil, Jan.–Feb. 1956, p. 11.

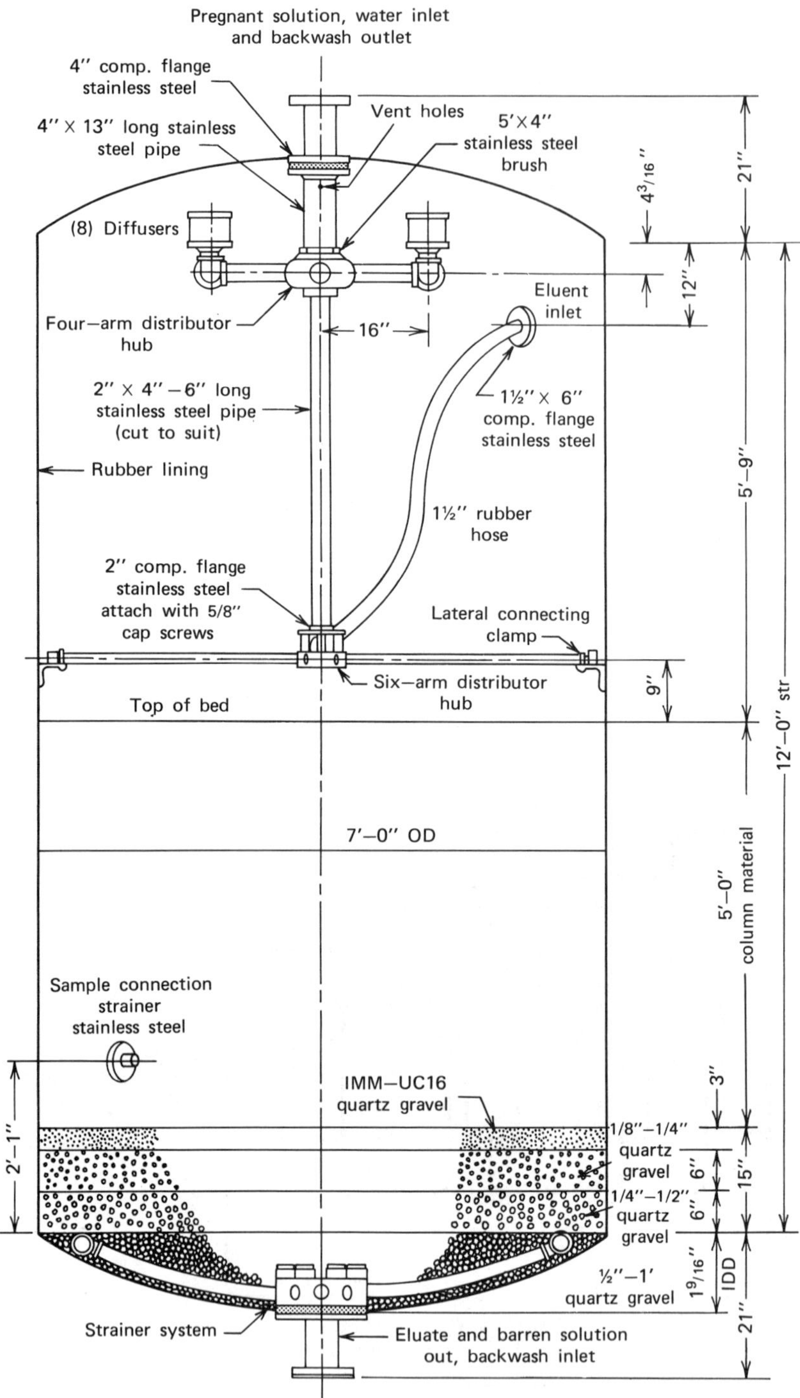

Figure 4.9. Ion exchange column for uranium extraction. *Source:* W. Q. Hull and E. T. Pinkney, *Ind. Eng. Chem.*, Vol. 49, No. 1, 1957, p. 7.

in diameter and 12 feet (3.6 m) high. The dished base and bottom 15 inches (37.5 cm) of the cylinder are packed with a gravel bed, graded from a maximum of 1 inch (2.5 cm) at the bottom to coarse sand at the top. A strainer is located under the gravel bed at the base of the steel cylinder and is attached to the single bottom outlet. A 4 foot (1.22 m) deep layer of resin beads is positioned on top of the gravel bed, and in the open space in the column above the beads is a stainless steel diffuser and distribution system for the incoming solutions, which is attached to a single inlet pipe. Spent liquor and pregnant eluate leave the tank through the bottom outlet. In this case both ion exchange and elution are carried out with a downward flow of solutions, while the flow is reversed from bottom to top for backwashing operations.

The columns are used in sets of three or four in series, with one column being eluted while the others in the series are used for resin absorption and are being fed with acid leach solution. The column being eluted is kept in reserve after the resin beads have been stripped and rejuvenated and will be returned to the circuit as a replacement when the first column in the series becomes saturated with uranium and must in turn be removed to be eluted. This saturation of the number 1 column is determined to have occurred when the barren effluent leaving the series of columns either contains 1% of the uranium content of the feed solution or exceeds 0.001 gpl. At this time the number 2 column becomes the number 1 column in the series, the number 3 becomes number 2, number 4 becomes number 3, and the freshly stripped and rejuvenated column is put back in the circuit as the last, number 4, column. Usually several lines of columns are run in parallel, so that a plant will be operating with a dozen or more columns overall, which necessitates a great maze of tanks, pipes, and pumps for a continuous operation.

Absorption is carried out by the resin beads in the ion exchange column until loaded to slightly over 3 pounds (1362 g) U_3O_8 per cubic foot (0.0283 m^3), which would take place over a period of 5 hours with a flow of solution to the columns of 60 gpm (273.0 lpm). The number 1 column is now taken out of the circuit for eluting, and a forward displacement wash of one bed volume of water, or 0.1N sulfuric acid, is given to displace the residual pregnant liquor it contains to columns 2 and 3. Column 1 is then backwashed, with two bed volumes of water, or 0.1N sulfuric acid, to wash off any physically entrained solid impurities on the surface of the beads, after which the column is now ready to be stripped.

Two- or three-stage elution is commonly used and builds up the uranium content of the eluate solution to 10 to 20 gpl. With a typical three-stage operation, return recycled barren eluate from the precipitation circuit is placed in a 5600 gallon (25,480 litre) propeller-agitated tank, and sodium chloride (or ammonium nitrate) is added to give a 1 molar solution strength. This tank is designated as tank 3. Saturated column 1 is stripped with solution from tank 1, and this solution is sent to the precipitation circuit. The column is now stripped twice more with eluate from storage tanks 2 and 3, which solutions are then advanced to tanks 1 and 2, respectively.

The chloride (or nitrate) eluate solution, in addition to removing the absorbed uranium ions from the resin beads, rejuvenates the resin by replacing the negative ions Cl^- or NO_3^- removed during absorption. After a final water wash to remove residual chloride or nitrate ions, the resin in column 1 is again ready to return to the absorption cycle and will be added, when required, as the last ion exchange column in the series.

Nitrate elution is less corrosive and two to three times more efficient than chloride elution, but it is a considerably more expensive reagent and must be recycled and reused.

Precipitation of the uranium concentrate from the eluant is accomplished by raising the pH to 6.5 to 7.0 with ammonia (sodium hydroxide, magnesia), at which point diuranate, $(NH_4)_2U_2O_7$, 50 to 80% U_3O_8, yellow cake, is precipitated. The yellow cake is filtered and dried before being packing in drums for shipment to the refinery, while the barren eluate is recycled to the stripping circuit to conserve chemical reagent costs.

The Resin in Pulp process is a simplification of the conventional ion exchange process in that the expensive solids separation after acid leaching can be omitted and pregnant solutions containing as high as 8% of minus 325 mesh solids from cycloning can be successfully processed. To accomplish this, the resin is held in stainless steel baskets and gently agitated in the leach pulp. The mesh size of the basket retains the resin beads but allows the fine ore particles to pass on through, away from the resin beads. After absorption of

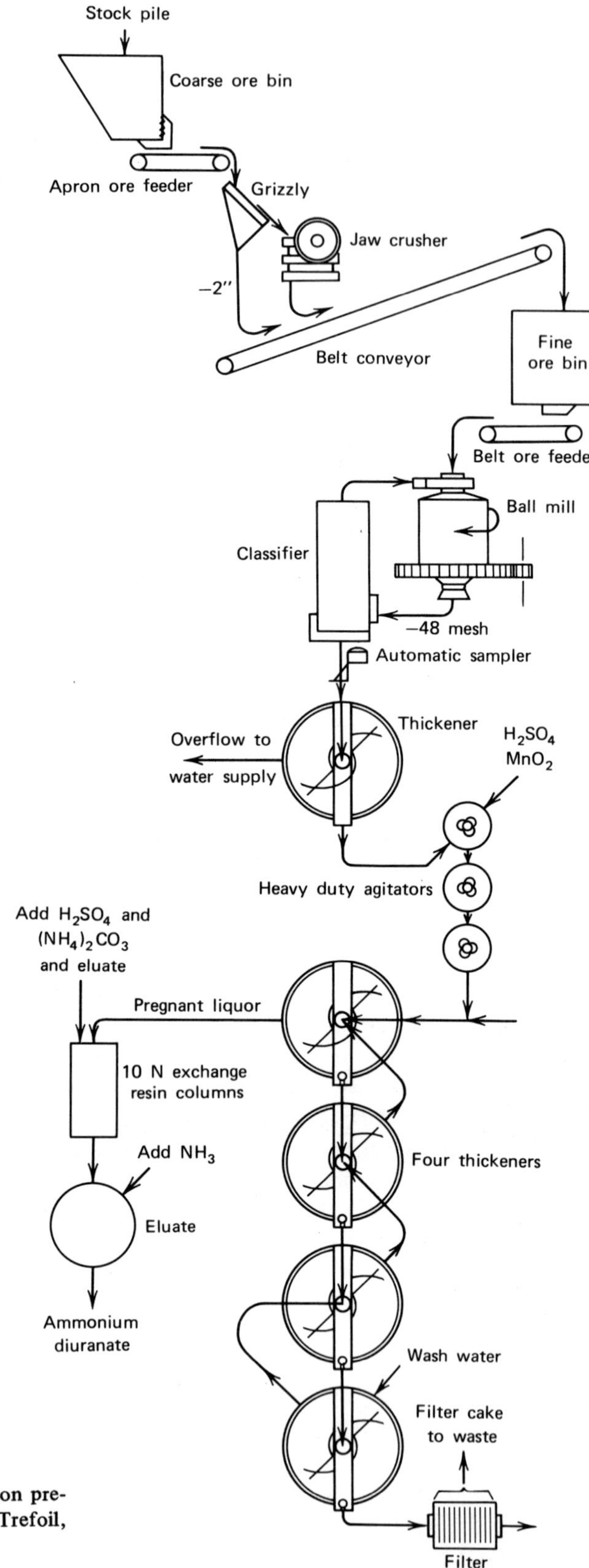

Figure 4.10. Acid leach of uranium ore with countercurrent decantation preceding ion exchange. *Source:* M. C. Irani, Denver Equipment Co. Deco Trefoil, Jan.–Feb. 1956, p. 11.

the uranium by the resin, elution and then precipitation is carried out, as was done in the columnar ion exchange process.

From six to ten rubber-lined steel cells, each containing a cubical 20 mesh stainless steel wire cloth basket, are used in an absorption bank, with 12 to 14 banks of these cells used in one circuit. The cubical baskets will be from $4\frac{1}{2}$ to 6 feet (1.37 to 1.83 m) to a side, and each is loaded with 30 to 50 cubic feet (0.85 to 1.415 m^3) of plus 20 mesh anionic resin beads to a depth of about 18 inches (45 cm) in the bottom of the basket. The baskets are moved up and down slowly in the pregnant leach pulp (five to ten times per minute) and are sometimes arranged so that adjacent baskets are counterbalanced. On the downstroke, the resin is circulated within the basket allowing good contact with the uranium leach solution, while on the upstroke, the basket is drained of solution to allow new solution to enter during the next down stroke.

Flow through the banks of cells is in series, with the flow from bank to bank either by gravity due to an 8 inch (20 cm) differential level between banks, or with airlifts being used to transfer solution and pulp from one bank to the next. Normally, at one time, nine banks of a 14 bank operation will be on absorption while five banks will be on elution. Retention time of solution through the 14 banks is about 100 minutes.

The relative position of each bank is to advance it through the absorption cycle until it is fully loaded, then to remove it from absorption and pass it to elution. The average resin loading will be about 3 pounds (1362 g) U_3O_8 per cubic foot (0.0283 m^3), and it takes some $1\frac{1}{4}$ to 2 hours for the number 1 bank of cells in the series to reach this limit. Thereupon it is removed from the absorption circuit and the uranium content stripped by elution. When the number 1 absorption bank is changed to the elution cycle, the bank which has been on the elution cycle longest is switched over to the number 9, or last, bank on the absorption cycle. This allows the fresh, rejuvenated resin beads in the number 9 bank to contact the weakest uranium pulp feed liquor and strip out the last traces of uranium before the barren solution is discarded as tailings.

When the number 1 absorption bank is taken out of the absorption circuit, it is drained of solution which does to a holding tank to be metered gradually back through the rest of the absorption circuit. This prevents any surge and spilling of solution that might occur if the bank was emptied too quickly into the rest of the circuit.

Before elution, wash water is passed through the bank to remove any remaining acid solution and to clean the surface of the resin beads of any solid particle picked up from the leach pulp. The washed bank is then placed as the number 5, or last, bank on the elution cycle. In this way the elution solution contacts the fully loaded resin last before going on to precipitation and gives the highest possible uranium content in the stripping solution for precipitation.

The uranium is eluted from the resin using an acidified molar nitrate solution of ammonium or sodium nitrate with nitric acid. Fresh eluant, consisting of recycled barren eluant solution to which makeup nitrate and nitric acid have been added to give a $1M$ solution, is put in the first, or oldest, bank on elution. This allows the freshest solution to contact the bank that has been on elution the longest to strip all traces of uranium before the bank is washed with water and placed back in the absorption circuit. The eluant is advanced through successive banks on elution, with the final solution from the number 5, or last, bank containing 11 gpl U_3O_8 when it is pumped to precipitation. The baskets of resin beads are gently agitated vertically during elution at the same rate as was used for loading during absorption.

Precipitation is accomplished by first heating the eluate to about 140°F (60°C) in a steam jacketed heat exchanger and then adding ammonia in three stages to a final pH of 7. The resultant slurry is thickened and filtered to recover the yellow cake, ammonium diuranate, $(NH_4)_2U_2O_7$, which is dried for packaging before shipping it to the refinery. The thickener overflow is reused as recycled eluant in the elution circuit. The resulting yellow cake averages 75 to 85% U_3O_8, if drying is done at a sufficiently high temperature of above 500°F (260°C) to upgrade the salt by driving off water of crystallization and ammonia.

The overall process is quite efficient, despite the comparative complexity of its flow sheet, with recovery in both the ion exchange and precipitation circuits being close to 99%.

2b. Uranyl Carbonate solutions, resulting from the sodium carbonate leaching of uranium oxide ores high in lime, have their uranium values precipitated by several chemical methods. A standard method is to raise the pH to above 11 with sodium hydroxide, precipitating the uranium as sodium diuranate:

$$2Na_4UO_2(CO_3)_3 + 6NaOH = Na_2U_2O_7 + 6Na_2CO_3 + 3H_2O$$

Four or five 16 by 18 foot tanks (4.9 by 5.5 m) in series will provide a residence time of 6 hours, and to them is added the pregnant carbonate solution plus a 15 to 18% sodium hydroxide solution. Mechanical agitation is provided and precipitation is continuous. The precipitate is removed by a filter press and dried for shipment to the refinery. The average composition will be 72 to 77% U_3O_8.

Another standard method is to acidify the solution to pH 3 with sulfuric acid which destroys the carbonate. Carbon dioxide formed is removed either by boiling or pulling a vacuum over the solution. Ammonia (or sodium hydroxide) is added to raise the solution to pH 7, and the uranium is precipitates as $(NH_4)_2U_2O_7$ or $Na_2U_2O_7$.

In cases where uranium and vanadium are both present in the carbonate leach solution, the destruction of the carbonate by adding sulfuric acid will completely precipitate both metals as sodium uranyl vanadate, $Na_2O \cdot 2UO_3 \cdot V_2O_5 \cdot nH_2O$, yellow cake.

3. Aluminum Oxide, dissolved from bauxite ore by pressure leaching with a sodium hydroxide solution, goes into solution as sodium aluminate, $NaAlO_2$. This solution is taken to precipitation tanks where the aluminum is removed from the pregnant solution as crystals of aluminum trihydrate, $Al(OH)_3$:

$$2NaAlO_2 + 4H_2O = 2NaOH + 2Al(OH)_3$$

The pregnant solution is first clarified in a two-stage operation. The first stage consists of settling and countercurrent decantation in thickeners which remove most of the suspended solids and give as a product a cloudy liquid. This liquid is given a final clarification in plate-and-frame presses, with the resulting clear liquor diluted to a specific gravity of 1.23 and cooled to improve the supersaturation. After this it is pumped to large, 25 feet in diameter by 25 feet high (7.62 by 7.62 m) mild steel precipitation tanks equipped with agitators, to separate out crystals of aluminum trihydrate.

This precipitation is carried out in the presence of freshly prepared, fine seed crystals of aluminum trihydrate, which are added to the tanks. Then, with the sodium aluminate solution slowly hydrolyzing on the surface of the seed, larger crystals of aluminum trihydrate build up and grow to a size of over 100 microns in diameter that can be easily filtered.

The rate of the precipitation reaction is proportional to the total surface area of the aluminum trihydrate particles present at any given time and seems to be favored by small seed particles with their proportionately larger surface area. Solution temperature also has an effect on rate, with the best results when the temperature is kept in the range of 77 to 122°F (25 to 50°C).

The precipitation process is rather slow, with the liquid in the tanks being stirred for 30 hours at the end of which time the precipitated hydrate flows to a hydroseparator where a separation is made between fine and coarse particles. The coarse-sized underflow

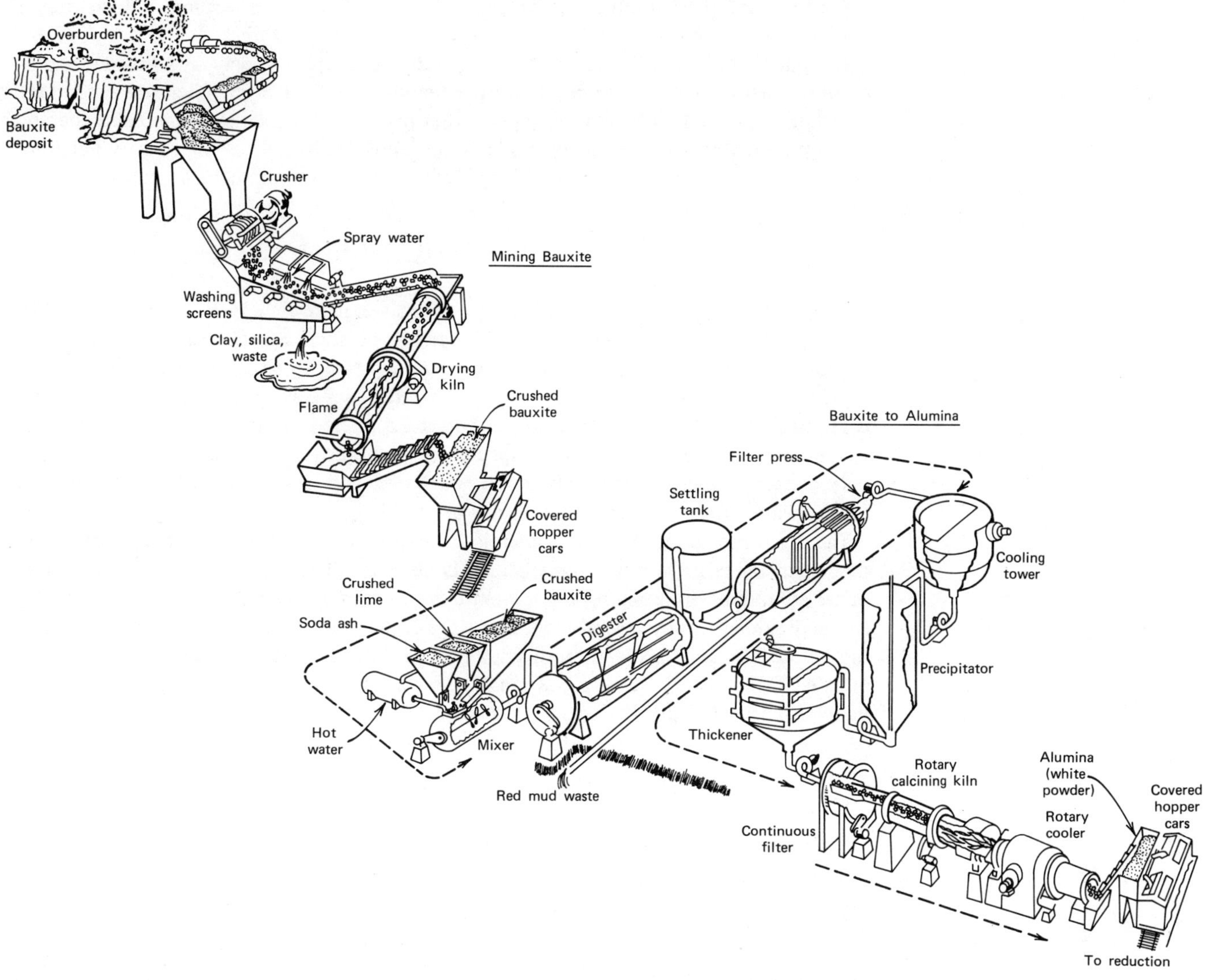

Figure 4.11. Mining bauxite and processing bauxite to alumina. *Source:* Courtesy of The Aluminum Association, Inc.

from the hydroseparator is classified and filtered and is the final $Al(OH)_3$ product of the process.

The overflow from the hydroseparator goes to a thickener where the dilute caustic liquor is separated from the fine aluminum trihydrate crystals. These crystals, taken off in the thickener underflow, are returned to the precipitation tanks as seed, while the overflow is fed into evaporators that concentrate the dilute sodium hydroxide solution for reuse in the leaching circuit. Some fresh NaOH makeup is also added to this recycled solution to bring it up to full strength and to compensate for that portion of the hydroxide lost in the red mud.

REFINING

Refining concludes the processing treatment given to the reactive metals and metal compounds that have been recovered from their ores and concentrates by leaching and precipitation, with these products being given whatever treatment is required to have them in commercially pure form.

The purity and composition of the precipitates varies quite widely, and this dictates the refining treatment necessary to obtain the metal in its final form. Chemical compounds with fairly low metal content will need rather involved treatments of combined chemical and pyrometallurgical processing, while compounds and metals of high purity need only a single stage of simple fire refining to produce a product of commercial quality.

TYPES OF REFINING

Precipitates from hydrometallurgical processing are recovered in three different forms, with the refining steps that are taken to give a purified product varying considerably depending upon the form and purity of the precipitate. In situations where the precipitate is an electrolytically deposited high-grade cathode metal, only a minimum of drossing is required in the furnace where the cathodes are melted to cast into commercial shapes. A minimum of treatment is also required for high-purity chemical compounds that are only contaminated by chemically combined water, and in this case calcination will drive off the water and give the required purity of product.

The third precipitate form is a chemical compound containing only 50 to 85% of the desired metal value in oxide form, and it is necessary in this case first to concentrate the metal compound by leaching and solvent extraction and then reduce the stripped powder to the dioxide form by several reduction stages. The metal dioxide powder is converted to the trifluoride, and this is reduced with a reactive metal in a thermite reaction to its final purified form, using overall a combination of chemical and fire refining operations ultimately to achieve a purified metal product.

MELTING AND REFINING PROCESSES

1. Zinc Metal is deposited at the cathode in the electrolytic precipitation cell in very pure form analyzing about 99.995% zinc, with the chief impurities being lead, iron, cadmium, copper, and silver. Lead is the major impurity and is only on the order of 20 to 25 grams per ton of zinc.

Most electrolytic zinc is sold as rectangular slabs weighing approximately 55 pounds (25 kg). To obtain these the zinc cathode deposition, stripped from the aluminum cathodes of the cell, is melted and cast into these slabs.

Induction furnaces, or small reverberatory furnaces 12 feet wide by 21 feet long with a 2 foot bath depth ($3.66 \times 6.40 \times 0.61$ m), are used to melt the zinc cathode sheets, and the liquid zinc is ladled by hand or pumped by a centrifugal graphite pump to the slab molds.

The purity of the cathode zinc is only minimally improved by this melting operation, but considerable amounts of zinc oxide can form (up to 2.5% of the zinc melted), and this collects as a dross on the surface of the furnace bath, which must be skimmed off and returned to the leaching stage. There is also zinc oxide fume in the furnace gases; these are retrieved in a bag house and recycled with the dross.

2. Ammonium or Sodium Diuranate, $(NH_4)_2U_2O_7$ or $Na_2U_2O_7$, only contains 50 to 85% U_3O_8 in this form of yellow cake and must first be concentrated before the stages of separation and refining begin. This is accomplished by dissolving the various forms of the crude uranium diuranate in nitric acid and subjecting the resulting leach solution to concentration by solvent extraction.

The yellow cake is delivered by screw feeders from enclosed hoppers to stainless steel, steam coil-heated, stirred dissolver tanks into which nitric acid has been metered. The tanks are 7 feet deep by 10 feet in diameter (2.13 × 3.05 m) and are partitioned into quadrant compartments by vertical dividers. The nitric acid feed slurry, which is kept at 212°F (100°C), is successively passed from compartment to compartment, with the fourth quadrant compartment discharging the slurry to a cooler. From here at a temperature of about 104°F (40°C) it is passed along to a bank of rotary, precoat-type vacuum filters, where the silica and small amounts of other insoluble impurities are removed and discarded. Washing of the filter cake is done on the filter drum, and the washings are added to the main filtrate of uranyl nitrate. The combined filtrate and washings provide a clear liquid for solvent extraction recovery at a uranium concentration of 300 gpl and a free acidity of up to $3N$.

The conditions under which solvent extraction is carried out for a very selective extraction of uranium in the next stage is of the utmost importance, as it is at this point that a highly purified uranium product is obtained, and the operations which follow it do not significantly influence any changes in the degree of purity.

The preferred solvent phase is 20% TBP (tri-*n*-butyl phosphate) solution in odorless kerosene. Extraction takes place at ambient temperature in an assembly of stainless steel mixer–settler units where flow is controlled by gravity. A common arrangement has the aqueous feed solution put into mixer 8 in a series of 16 mixer-settler units, and this flows countercurrent to the solvent phase which is fed into the assembly at one extreme. The extraction of uranium takes place by complex formation in the solvent phase in the first eight units of the mixer-settler assembly. The loaded solvent then flows through the remaining eight units where it is scrubbed with pure uranyl nitrate to remove impurities:

$$UO_2 + 2NO_3 + 2TBP = UO_2(NO_3)_2 \cdot 2TBP$$

The loaded solvent is heated to 122°F (50°C) and stripped with a very dilute solution of nitric acid in water in a 12 unit mixer-settler assembly in which the acidified water flows countercurrent to the loaded solvent. The overall extraction efficiency is about 99.99%, with the uranyl nitrate produced being of very high purity.

The pure uranyl nitrate solution derived from the foregoing solvent extraction is then concentrated before being converted to UO_3 by thermal decomposition. This concentration by evaporation at 158°F (70°C) produces a solution of about 70% uranyl nitrate with a uranium concentration of 800 gpl.

Evaporation results in the uranium being in the form of uranyl nitrate hexahydrate, $UO_2(NO_3)_2 \cdot 6H_2O$, which is converted to UO_3 (orange oxide) powder by atomizing the hexahydrate through spray nozzles into a fluid bed reactor held at 572 to 662°F (300 to 350°C). The reactor is circular, 10 inches in diameter (25 cm), and constructed of stainless steel. Fluidizing air is preheated and passed upward through a distributor plate to a fluid bed of UO_3, 2 or 3 feet deep (60 to 90 cm), with the total air usage being as high as 2 tons per ton of uranyl nitrate hexahydrate processed. The continuously formed uranium trioxide leaves the fluid bed reactor through a product discharge line and goes to a storage hopper for the next process:

$$UO_2(NO_3)_2 \cdot 6H_2O = UO_3 + 2NO_2 + \tfrac{1}{2}O_2 + 6H_2O$$

The UO_3 (orange oxide) is next converted to UO_2 (brown oxide) by reduction with hydrogen in a second cylindrical, stainless steel fluid-bed reactor, operating at 932 to 1112°F (500 to 600°C):

$$UO_3 + H_2 = UO_2 + H_2O \qquad \Delta H° = -24 \text{ kcal per mole}$$

The reactors are about 7 feet high by 14 inches in diameter (2.1 m × 35 cm), with a bed depth of approximately 4 feet (1.22 m). The reducing gas is a mixture of hydrogen and nitrogen in ratios varying between 3:1 and 9:1, with the composition adjusted to a suitable concentration for reduction. Uranium dioxide (brown oxide) flows from the discharge opening in the reactor and is collected in the product hopper.

While the foregoing processes can be run satisfactorily on a continuous or semicontinuous basis, the operation of the conversion of UO_2 to UF_4 by hydrofluorination with HF gas is best carried out as a batch process, because of the fairly long 4 to 8 hour residence time required for the reaction to progress to the state where only 1 ot 2% of the UO_2 remains unconverted.

A fluid-bed reactor is used here also. It is very similar in design to that used previously for the reduction of UO_3 to UO_2, with the exception that Inconel (Ni 75%–Cr 15%–Fe 10%) is used as the construction material instead of stainless steel, for the reason that Inconel is better able to withstand the action of hydrogen fluoride at the 842 to 1112°F (450 to 600°C) operating temperature. The solid product from the reactor is UF_4 (green salt), while the gaseous products are hydrofluoric acid and steam. HF is recovered in a distillation plant and is recycled to the process.

$$UO_2 + 4HF = UF_4 + 2H_2O \quad \Delta H° = 43 \text{ kcal per mole}$$

The final operation is the reduction of uranium tetrafluoride with magnesium to produce uranium metal and a magnesium fluoride slag:

$$UF_4 + 2Mg = U + 2MgF_2 \quad \Delta H° = -89 \text{ kcal per mole}$$

Both calcium and magnesium can be used for this reduction reaction, with the advantage going to magnesium because of its lower cost, purity, and convenient size and form.

The reaction is carried out on a batch basis in a steel reaction vessel with a flanged lid which can be fastened down to seal the vessel and retain the charge during reaction. These reaction vessels are used in a variety of sizes. A common size has a 13 inch inner diameter and 45 inch height (32.5 × 112.5 cm). This size will accommodate a charge of 318 pounds (144 kg) UF_4 to which has been mixed in a 4 to 5% excess of high-purity magnesium chips of minus 10 plus 50 mesh particle size.

A liner must be provided in the steel reaction vessel, and this is usually done by vibrating or jolt packing powder around a mandrel positioned in the empty reactor. Finely ground MgF_2 slag, CaF_2 slag, or fused dolmitic oxide are all satisfactory lining materials.

The reduction reaction is highly exothermic and abrupt. The charged reaction vessel is lowered into a resistance-heated furnace and slowed heated over several hours to approximately 1200°F (650°C), where ignition takes place and the temperature rapidly rises to 2372°F (1300°C). This temperature is sufficiently above the melting point of uranium (2066°F, 1130°C) for good slag–metal separation to take place, with the liquid uranium collecting in the bottom of the lined reactor and the layer of slag above it. After the reaction is complete, the vessel is removed from the furnace and first cooled in air then in water, after which the reactor is opened and inverted for the slag and uranium billet to fall out. The uranium, referred to as a derby or biscuit, is chipped free of any adhering slag and is remelted in a vacuum furnace to be cast into fuel element bars or ingots suitable for rolling. The MgF_2 slag is retained to be finely ground and used for lining material in the reaction vessels.

Recovery is better than 98%, with the uranium derbies being produced in sizes varying from 350 pounds (158 kg) to as large as 570 pounds (260 kg).

3. Aluminum Trihydrate ($Al_2O_3 \cdot 3H_2O$) is the coarse crystalline material separated as underflow from the hydroseparator which follows the precipitation operation in the precipitation circuit.

The chemically combined water is removed by calcining at about 1832°F (1000°C) to form anhydrous Al_2O_3 of greater than 99.5% purity on a dry basis, as is required for feed material to the electrolytic aluminum reduction cell. The calcining is carried out in rotary kilns, some of which are fitted with electrostatic precipitators to collect the escaping dust:

$$2Al(OH)_3 + heat = Al_2O_3 + 3H_2O$$

REACTIVE METALS,
HYDROMETALLURGICAL EXTRACTIVE PROCESSES

The reactive metals are treated by hydrometallurgical methods for a variety of reasons. For instance, some zinc is processed in this manner because of the complexity of the ore and the difficulty in beneficiating it to obtain a concentrate that is of sufficiently high grade or purity to be treated by one of the several pyrometallurgical retorting methods. However these shortcomings can be accommodated in a leaching process and very high-purity cathode zinc can be precipitated from leach solutions.

Uranium ores are leached because of the extremely low-grade ores and concentrates that are processed. These can be recovered and concentrated on a large-scale basis with very high recovery efficiencies, and this without the problem of the oxidation of the products, which would be a great problem if a high-temperature pyrometallurgical process was used.

The extraction of alumina from bauxite is also a process that makes use of a nonoxidizing condition to remove selectively an oxide associated with waste material difficult to separate by pyrometallurgical treatment.

The combinations of processes using hydrometallurgical methods to separate and refine zinc, uranium, and alumina are given in Table 4.4.

Table 4.4. Reactive Metals, Hydrometallurgical Extractive Processes

Zinc	Ion exchange or solvent extraction concentration
Sulfide	Precipitation of yellow cake
Roast to oxide	Nitric acid leach
H_2SO_4 tank leach, single or double	Solvent extraction
Pregnant solution purification	Air reduction to UO_3
Electrolytic precipitation	Hydrogen reduction to UO_2
Melt and cast	Hydrofluorination to UF_4
Uranium	Thermic reduction UF_4 by Mg
	Vacuum melt and cast
Oxide	
Beneficiate	***Alumina***
Chloridizing roast, purifying leach	*Oxide*
Tank leach with H_2SO_4 or sodium carbonate	NaOH pressure leach
Heap leach with raffinate	$Al(OH)_3$ seeding, precipitate
In-place leaching with sodium carbonate	Classify, filter
	Calcine to Al_2O_3

REFERENCES

Bellamy, R. G. and N. A. Hill, *Extraction and Metallurgy of Uranium, Thorium, and Beryllium*, Pergamon Press, New York, 1963.

Burkin, A. R., *The Chemistry of Hydrometallurgical Processes*, D. Van Nostrand, New York, 1966.

Clegg, J. W. and D. D. Foley, Eds., *Uranium Ore Processing*, Addison–Wesley, Reading, Mass., 1958.

Cotterill, C. H. and J. M. Cigan, Eds., *AIME World Symposium on Mining and Metallurgy of Lead and Zinc*, Vol. 2, Extractive Metallurgy of Lead and Zinc, Port City Press, Baltimore, 1970.

Dennis, W. H., *Metallurgy of the Non-ferrous Metals*, Pitman, New York, 1961.

Garbella, E. J., Joy Manufacturing Company, Denver Equipment Company, Denver, Bulletin M4-B136, Deco Trefoil, Winter Issue, 1968–1969.

Gerard, G. and P. T. Stroup, Eds., *Extractive Metallurgy of Aluminum*, Vol. 1, Interscience, New York, 1963.

Habashi, F., *Principles of Extractive Metallurgy*, Vol. 2, Gordon and Breach, New York, 1970.

Hazen, W. C., Joy Manufacturing Company, Denver Equipment Company, Denver, Bulletin, T4-B32, Deco Trefoil, Aug.–Sept.–Oct. 1963.

Henrie, T. A. and D. H. Baker, Jr., Eds., Electrometallurgy, *Extractive Metallurgy Division Symposium on Electrometallurgy*, AIME, New York, 1968.

Irani, M. C., Joy Manufacturing Company, Denver Equipment Company, Denver, Bulletin M4-B84, Deco Trefoil, Jan.–Feb. 1956.

Jamrack, W. D., *Rare Metal Extraction by Chemical Engineering Techniques*, Vol. 2, Pergamon, New York, 1963.

Annual Review Issue, *J. Metall.*, Vol. 28, No. 3, 1976.

Annual Review Issue, *J. Metall.*, Vol. 29, No. 3, 1977.

Annual Review Issue, *J. Metall.*, Vol. 30, No. 4, 1978.

Mantell, C. L., *Electrochemical Engineering*, McGraw-Hill, New York, 1960.

Miller, J. D. A., Ed., *Microbial Aspects of Metallurgy*, Elsevier, New York, 1970.

Mineral Facts and Problems, U.S. Bureau of Mines Bul. 667, U.S. Department of the Interior, Washington, D.C., 1975.

Mines Branch Research Report R33, Department of Mines and Technical Surveys, Ottawa, 1959.

Mines Branch Technical Bulletin TB30, Department of Mines and Technical Surveys, Ottawa, 1961.

Oroya Metallurgical Operations, Cerro De Pasco Corporation, (now Centromin Peru), 1972.

Pehlke, R. D., *Unit Processes of Extractive Metallurgy*, Elsevier, New York, 1972.

Quinn, J. E., *The Mines Magazine*, July 1960, p. 36.

Quinn, J. E., Joy Manufacturing Company, Denver Equipment Company, Denver, Bulletin M4-B99, Deco Trefoil, Jan.–Feb. 1959.

Quinn, J. E., Joy Manufacturing Co., Denver Equipment Co., Denver, Bulletin M4-B101, Deco Trefoil, May–June–July 1959.

Quinn, J. E., Joy Manufacturing Co., Denver Equipment Co., Denver, Bulletin M4-B104, Deco Trefoil, Jan.–Feb. 1960.

Ryan, W., Ed., *Non-ferrous Extractive Metallurgy in the United Kingdom*, The Institute of Mining and Metallurgy, London, 1968.

Rosenbaum, J. B., Minerals Extraction and Processing: New Developments, in *Materials: Renewable and Nonrenewable Resources*, Abelson, P. H. and A. L. Hammon, Eds., American Association for the Advancement of Science, Washington, D.C., 1976.

Rosenqvist, T., *Principles of Extractive Metallurgy*, McGraw-Hill, New York, 1974.

Seeton, F. A., Joy Manufacturing Co., Denver Equipment Co., Denver, Bulletin M4-B111, Deco Trefoil, May–June–July 1961.

Seeton, F. A., Joy Manufacturing Co., Denver Equipment Co., Denver, Bulletin M4-B115, Deco Trefoil, Aug.–Sept.–Oct. 1962.

Solvent Extraction in Metallurgical Processes, *Proc. International Symposium, Technologisch Instituut K. VIV*, Antwerp, 1972.

Uranium bacterial leaching, *J. Metall.*, Vol. 18, No. 11, 1966.

Van Arsdale, G. D., *Hydrometallurgy of Base Metals*, McGraw-Hill, New York, 1953.

Wadsworth, M. E., and F. T. Davis, Eds., Unit Processes in Hydrometallurgy, *Metallurgical Society Conference AIME*, Vol. 24, Gordon and Breach, New York, 1964.

Wilkinson, W. D., *Uranium Metallurgy*, Vol. 1, Uranium Process Metallurgy, Wiley, New York, 1962.

5. Environmental Considerations:
Gas Recovery and Cleaning of Particulate Matter, Heat Exchangers, and Water Cleaning of Solids

The control of pollutants from metallurgical processing has long been a troublesome and expensive problem and one that has been increasing in importance over recent years. Greater emphasis has been put on minimizing contamination, and higher standards of control on permissible air and water pollutants are now being set by strict legislation in many countries.

The awareness of this problem in the metallurgical industry is not new. As long ago as 1821, at the beginning of the Industrial Revolution, there was concern over the SO_2 furnace emissions from the copper smelters at Swansea, Wales, where these emissions were damaging crops in the surrounding area and had to be controlled. By the 1860s these smelters were managing their SO_2 gas by treating a part to produce sulfuric acid (with about 40% of the total sulfur going to the smelters recovered in this way) to the extent of the marketable consumption of the acid produced, and the remaining SO_2 gas was sent up tall stacks to be dissipated before coming back to the ground.

In general, process gas to be cleaned will contain solid particulate matter, which can be removed by various types of dust collectors, and gaseous chemical compounds of which sulfur dioxide is by far the most common, although other gases such as chlorine, hydrochloric acid gas, and fluorides are also the by-products of some operations. Chemical methods of combination remove these gaseous products.

Water to be cleaned also contains considerable amounts of solid matter which can be removed by clarifying steps, and it also contains dissolved heavy-metal compound contaminants which can be removed by precipitation. It also has a high acid content which can be neutralized by the addition of a basic material to lower the pH to acceptable levels.

There are several important reasons for cleaning gases and water from metallurgical processing. The most important of these are as follows:

1. To recover value-bearing particulate material which can be returned to the plant for reprocessing. A copper smelter treating minus 200 mesh flotation concentrate might lose as high as 8% of the input tonnage as flue dust. If this is not recovered and recycled it would greatly affect the economics of the overall process.

2. Environmental pollution, both from the viewpoint of workers exposed to harmful gas emission in dangerous concentrations (U.S. EPA standard of 5.0 ppm for workers exposed to SO_2) and such gases as SO_2 which cause agricultural damage by combining with water vapor in the atmosphere to form sulfurous and sulfuric acid and burn leaf crops. The contamination of water supplies by heavy metal compounds and high acid content has also come under strict governmental regulation for health reasons. (Canadian Ministry of the Environment rates 1 ppm as maximum safe trace metals concentration.)

3. To remove value-bearing gaseous by-products such as SO_2, which can be used as the feed material for the production of marketable sulfuric acid or elemental sulfur.

4. To clean high-calorific product gas containing high percentages of combustible CO that having been freed of contained particulate matter can be piped and fed through burners, without danger of plugging, and be used wherever required as a fuel gas throughout the plant.

5. Many metallurgical gases are the product of high-temperature furnacing operations and leave the process carrying a high content of sensible heat. This heat can be recovered in such devices as waste heat boilers and recuperative systems before the cooled gas is permitted to escape to the atmosphere.

6. To clarify plant water of solid matter so that it can be recycled as clean process water or be discharged to nearby streams without danger of contamination.

DUST COLLECTORS–REMOVAL OF PARTICULATE MATTER FROM GASES

Dust collecting is done by a variety of methods, with the method chosen taking into account (1) the size of particles to be removed, (2) the temperature of the gas, (3) the gas volume, (4) the gas velocity, (5) whether the collector is to be used alone or as part of a series, and (5) whether the sensible heat in the gas is to be recovered or not.

Particle sizes and types are broken into four categories (1 micron = 1/1,000,000 meter = 1/25,400 inch = 3.937×10^{-5} inch):

Fume   0.05 to 1.0 micron, fine, solid particles formed by the condensation of metal vapors.

Dust   1.0 to 50 microns, small, solid particles formed from the breaking of larger particles.

Mist   0.5 to 10 microns, liquid droplets generated by condensation.

Smoke   0.05 to 1.0 micron, fine, solid particles resulting from the incomplete combustion of organic materials.

Dust loadings are given in grains per cubic foot or grams per cubic meter (1.0 grain per cubic foot = 2.3 grams per m^3; 1 grain = 0.065 gram).

There are four designations of dust loadings:

Light $\frac{1}{2}$ to 2 grains per cubic foot (1.15 to 4.6 grams per m^3)

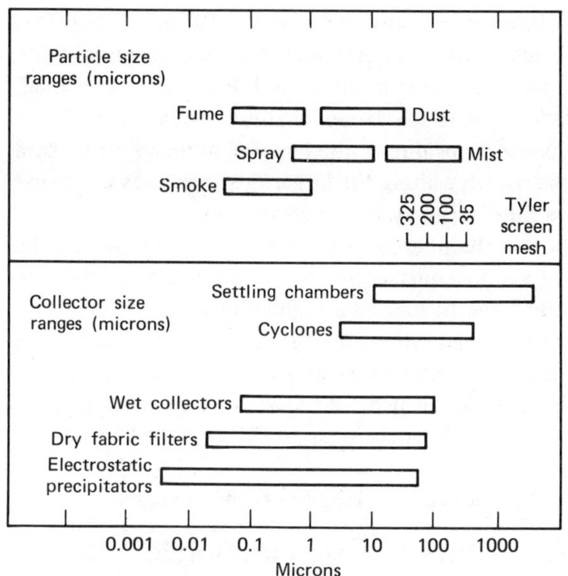

Figure 5.1. Comparative particle size ranges and corresponding applicable collecting equipment.

Medium   2 to 3 grains per cubic foot (4.6 to 6.9 grams per m^3)
Moderate   3 to 5 grains per cubic foot (6.9 to 11.5 grams per m^3)
Heavy   5 or more grains per cubic foot (11.5 or more grams per m^3)

There are eight types of dust collectors in common usage, each of which has its best collector efficiency within a certain particle size range. These collector types and their efficiency ranges are as follows:

Settling Chamber   90% efficient above 50 microns.
Cyclone   70 microns 20% efficiency, 100 microns 92% efficiency.
Multiclone   3 microns 20% efficiency, 70 microns 99% efficiency.
Bag House   range 0.5 to 100 microns 99% efficient.
Spray Tower   10 microns 88% efficiency, 90 microns 98% efficiency.
Venturi Scrubber   0.2 micron 30% efficiency, 5 microns +99% efficiency.
Electrostatic Precipitator   0.1 micron 82% efficiency, 2 microns +99% efficiency.
Wet Scrubber   0.3 micron 20% efficiency, 9 microns 99% efficiency.

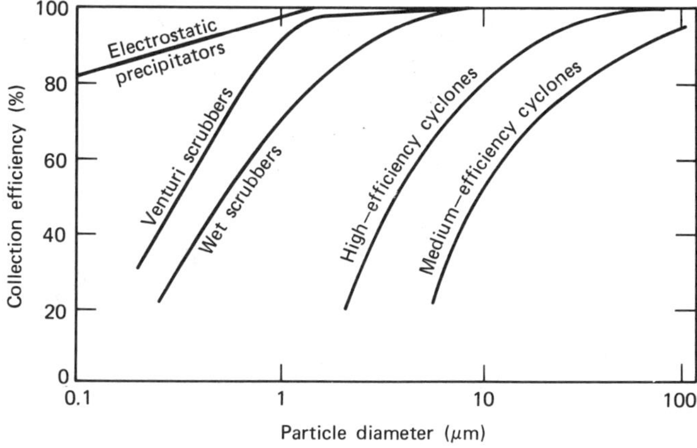

Figure 5.2. Comparative particle size collector efficiencies. *Source:* A. T. Rossano, Ed., *Air Pollution Control Guidebook for Management*, Environmental Services Division, ERA Inc., 1969, p. 143. Copyright assigned to McGraw-Hill, New York.

1. Settling Chamber (Gravity Settling Chamber, Expansion Chamber, Balloon Flue) This is one of the oldest and simplest methods of dust collection and is used to remove the coarser particles above 50 microns. The settling chamber works on the principle of a small flue carrying gas at a high velocity entering a much larger chamber with a greatly increased cross-sectional area. The gas velocity drops due to the larger chamber volume, and as the reduced velocity is no longer able to carry along the larger dust particles in the gas stream, they drop out. A small discharge flue carries away the cleaned gas.

Gas temperature is not important as the chamber is of steel construction and can be lined with heat-resistant firebrick. It is of simple construction with no mechanical parts to wear out rapidly and can be made quite large to handle a high-volume gas throughput. The design may have the settling chamber placed either vertical or horizontal, and as the gas is put through the unit dry, there is no cooling or loss of sensible heat. The settling chamber is a primary unit which can be used alone if the dust particles are all large or as the first unit in a series if there is a variety of particle sizes to be recovered by several collectors.

The efficiency of a horizontal settling chamber can be obtained from the equation

$$\text{weight efficiency} = \frac{(\text{particle terminal velocity})\,(\text{chamber length})}{(\text{chamber height})\,(\text{gas velocity})}$$

or

$$N_w = \frac{(U_t)\,(L)}{(H)\,(v_g)}$$

where U_t is measured in ft/sec, H in ft, L in ft, and v_g in ft/sec.

The equation for calculating the theoretical minimum size particle that can be collected at 100% efficiency for a horizontal chamber is

$$\text{minimum particle diameter} = \left[\frac{18\left(\begin{matrix}\text{gas}\\\text{viscosity}\end{matrix}\right)\left(\begin{matrix}\text{chamber}\\\text{height}\end{matrix}\right)\left(\begin{matrix}\text{chamber}\\\text{gas}\\\text{velocity}\end{matrix}\right)}{\left(\begin{matrix}\text{chamber}\\\text{length}\end{matrix}\right)\left(\begin{matrix}\text{particle}\\\text{specific}\\\text{weight}\end{matrix} - \begin{matrix}\text{gas}\\\text{specific}\\\text{weight}\end{matrix}\right)} \right]^{1/2}$$

or

$$D_p\ \min = \left[\frac{18(\mu_g)\,(H)\,(v_g)}{(L)\,(\gamma_p - \gamma_g)} \right]^{1/2}$$

with D_p in ft, v_g in ft/sec, γ_p in lb/cu ft, μ_g in lb-sec/sq ft, L in ft, γ_g in lb/cu ft, and H in ft.

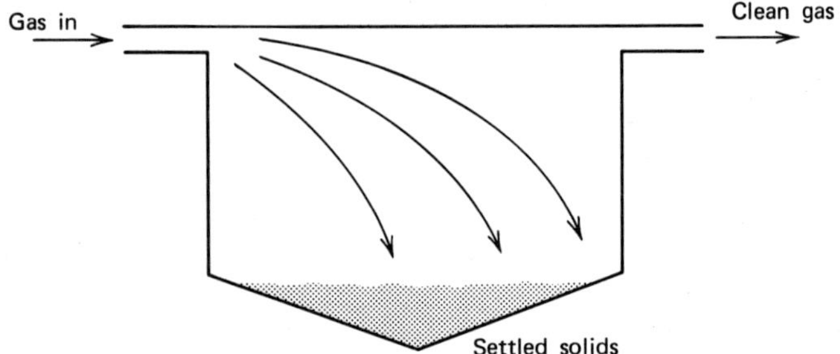

Figure 5.3. Settling chamber.

The relationship for the cross-sectional areas of the flue and chamber relative to the velocity of the gas in each is given by the expression

$$\text{velocity of gas in chamber} = \left(\frac{\begin{array}{c}\text{cross-sectional} \\ \text{area flue}\end{array}}{\begin{array}{c}\text{cross-sectional} \\ \text{area chamber}\end{array}} \right) (\text{velocity of gas in flue})$$

or

$$v_{gc} = \left(\frac{A_f}{A_c} \right) (V_{gf})$$

with v_{gc} in ft/sec, A_f in sq ft, v_{gf} in ft/sec, and A_c in sq ft.

2. **Cyclones** are also types of collectors with no moving parts that are used dry and conserve the sensible heat in hot gases passing through them. The cyclone is best at treating medium-sized particles, having an efficiency of only 20% recovery on small 7 micron particles and increasing this to 92% with larger, 100 micron sizes.

The cyclone is a low-cost collector, inexpensive to operate, and relatively small in size, and in its variety of manufactured forms is the most commonly used type of collector, with small body diameters (under 9 inches, 22.5 cm) being the most efficient.

The principle of operation is that a high-velocity gas stream enters near the top cylindrical section of the cyclone and impinges tangentially on the curved surface. Centrifugal force resulting from the tangential velocity of the gas carries the dust particles to the wall of the cyclone, where they slide downward due to gravity and are led out through a cone-shaped bottom section. A discharge opening at the top of the collector carries off the cleaned gas.

If there is excessive scouring and wear on the steel cyclone shell from the sand blasting effect of fine dust particles entering the unit at a high velocity, this wear can be con-

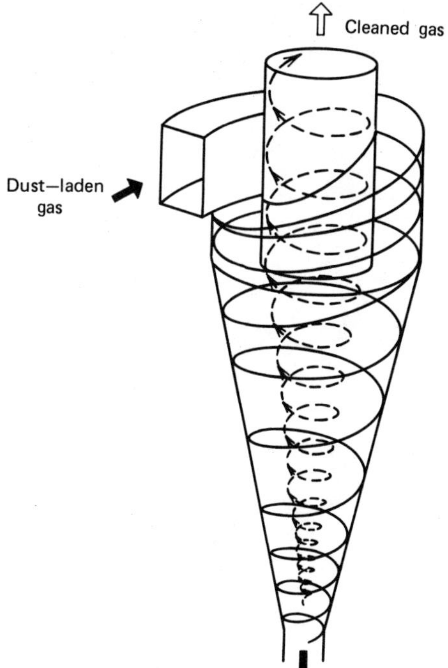

Figure 5.4. Cyclone dust collector. *Source:* J. R. Boldt, Jr., *The Winning of Nickel*, Methuen and Company, Ltd., London, 1967. Copyright Inco Limited, 1967, p. 253.

trolled by inserting porcelain linings in the cylindrical feeding section of the cyclone. The cyclone can be used alone as a primary unit, or as an intermediate collector if used as one unit in a series.

The centrifugal force which separates particles of particular mass is given by the equation

$$\text{centrifugal force} = \frac{(\text{particle mass}) \, (\text{particle tangential velocity})^2}{(\text{cyclone radius}) \, (\text{gravitational acceleration})}$$

or

$$F_c = \frac{(m_p) \, (V_{pt})^2}{gr_c}$$

with F_c in lb, V_{pt} in ft/sec, g in ft/sec^2, m_p in lb, and r_c in ft.

According to the foregoing equation the magnitude of centrifugal force that can be exerted on a given particle in a certain size of cyclone is a function of the particle mass and the tangential velocity, with tangential velocity being dependent on the velocity of the gas entering the cyclone.

Theoretically the efficiency of particle removal for a given cyclone design increases as the ratio of centrifugal force to drag force increases:

$$\text{ratio} = \frac{\text{centrifugal force}}{(\text{drag force}) \, (\text{friction of particle})}$$

$$\text{ratio} = \frac{(\text{particle diameter})^2 \, (\text{particle density}) \, (\text{particle tangential velocity})}{(\text{gas viscosity}) \, (\text{cyclone diameter})}$$

or

$$\text{ratio} = \frac{(D_p)^2 \, (\rho_p) \, (v_{pt})}{(\mu_g) \, (D_c)}$$

with D_p in ft, ρ_p in lb/cu ft, v_{pt} in ft/sec, μ_g in lb-sec/sq ft, and D_c in ft.

For a given set of conditions the operating efficiency (the ratio) will increase with a decrease in cyclone diameter for cyclones of the same design proportions.

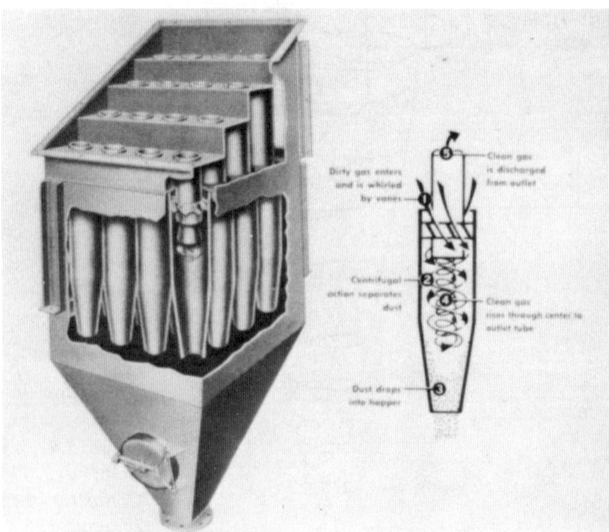

Figure 5.5. Multiclone dust collector. *Source:* Courtesy of the Western Precipitation Division of Joy Manufacturing Company.

3. Multiclones are a form of cyclone collectors, also used dry, where several small-diameter cyclones are combined together inside a single container shell and operated collectively as one unit. This takes advantage of the greater operating efficiency of small-diameter cyclones, and because the small cyclones are splitting the gas dust load and operating in parallel, the several cyclones in a multiclone can accommodate an overall large gas volume that has a heavy dust loading with quite high overall cleaning efficiency.

The multiclone also works best at collecting and removing medium-sized particles and has an efficiency of only 20% with 3 micron particles, but 99% with 70 micron particles.

4. Bag Houses are filter-type collectors and are one of the oldest, most reliable types still in widespread general use. The gas containing particulate matter is blown into the interior of fabric bags, whereupon the gas passes on through the fabric and out of the collector and leaves behind the solids, dust, or fume that are retained by the fabric. After the first brief instant the filtering action is actually being done by the dust cake that has deposited on the fabric filter, rather than by the size of weave in the fabric bag itself.

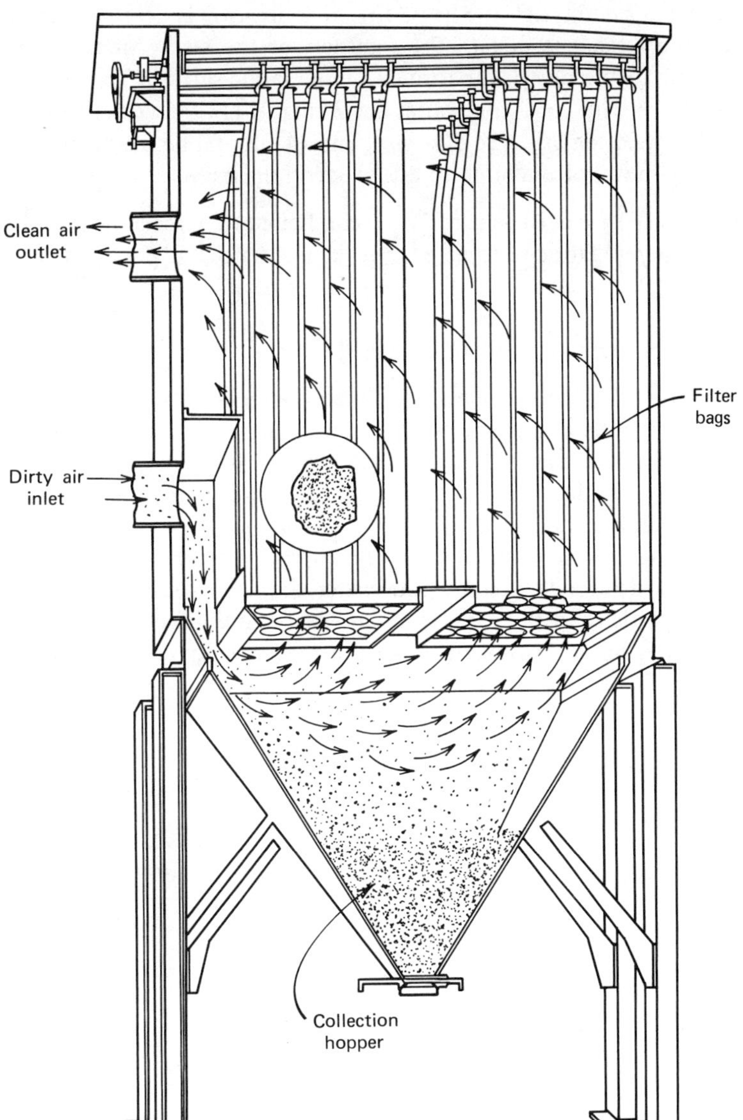

Figure 5.6. Bag house fabric filter for dry particles. *Source: Control Techniques for Particulate Air Pollutants*, U.S. Public Health Service Publ. AP-51, 1969, p. 103.

Periodically, the bags, which are used in large groups in a bag house, are shaken, or the flow of gas is momentarily reversed to shake or blow off the particle layer that has collected inside the bags. This dust falls down into a hopper at the bottom of the bag house and is removed from here. Bag life is from six months to two years, and the bags will range in size from 5 to 20 inches in diameter (12.5 to 50 cm) and up to 44 feet high (13.2 m). A torn bag must be replaced immediately as it will leak and is not filtering.

The bag house can be used over a wide range of particle sizes, 0.05 to 100 microns, with efficiency to 99%, but is generally considered best as a collector for smaller-sized particles and gases with a light to medium dust loading. Otherwise the installation of a group of bags to handle large gas volumes with heavy dust loadings would be enormously large and less practical.

The fabric of the bag and the closeness of the weave are matched to the operating conditions, which will be (1) particle size, (2) gas temperature, (3) resistance to abrasion, and (4) resistance to acid or alkali attack.

The most common filter bag materials and their properties are shown in Table 5.1. As the thickness of the dust cake which builds up on the fabric increases, so too does the pressure required to push the gas through the bags. This pressure drop is expressed as the sum of two resistances—the resistance due to the tightness of the weave of the fabric itself and the resistance due to the dust layer formed on the fabric surface. The pressure drop will continue to increase with time, and varies according to (1) the velocity of the gas through the fabric, (2) dust loading, (3) dust size, and (4) the fabric used, and it is controlled by the periodic removal of the built-up dust cake on the bags:

$$\text{pressure drop} = \begin{pmatrix} \text{superficial} \\ \text{velocity} \end{pmatrix} \left[\begin{pmatrix} \text{coefficient of} \\ \text{fabric resistance} \end{pmatrix} + \begin{pmatrix} \text{coefficient of} \\ \text{cake resistance} \end{pmatrix} \begin{pmatrix} \text{cake weight} \\ \text{per areas} \end{pmatrix} \right]$$

or

$$\Delta P = (v_s) \left[K_f + (K_c)(W_c) \right]$$

with ΔP in lb/sq ft, v_s in ft/sec, K_f dimensionless, K_c dimensionless, and W_c in lb/sq ft.

The bag house can be used alone as a single unit for conditions where particle sizes are small and dust loadings or gas volumes are not excessive, or it can be used as the final collector in a series to remove the last remaining fine dust particles. In this case it would probably follow a settling chamber and a cyclone.

5. Spray Towers utilize particle collection by fine water droplets, with the drops of water contacting the solid particles in the gas stream and carrying them out of it. In its simplest form the spray tower has a downward flow of atomized water droplets sprayed from nozzles to completely fill the interior of the tower, and rising through this spray is

Table 5.1. Filter Bag Materialsa

	Maximum Long Operating Temperature	Tensile Strength	Abrasion Resistance	Acid Resistance	Alkali Resistance
Cotton	180°F (82°C)	B	C	P	E
Wool	200°F (93°C)	C	C	F	P
Fiberglass	500°F (260°C)	A	D	F	P
Nylon	200°F (93°C)	A	A	P	E
Dacron	275°F (135°C)	A	B	G	F
Felt (Nomex)	425°F (218°C)	A	B	F	E

*a*E—excellent, G—good, F—fair, P—poor, A—highest, B—second, C—third, D—lowest.

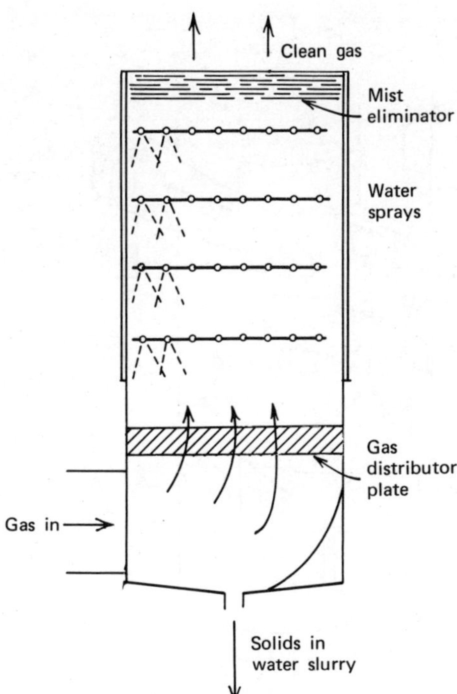

Figure 5.7. Spray tower. *Source: Control Techniques for Particulate Air Pollutants*, U.S. Public Health Service Publ. AP-51, 1969, p. 55.

an upward flow of dirty gas to be cleaned. The solids are carried to the bottom of the spray tower by the water droplets and are drained from here to be dewatered and recovered.

Spray jets are available for a variety of spray shapes to fill the tower interior to give the required fineness of water droplets and to supply the volume of water required. For maximum cleaning efficiency these spray jets will match the dust particle size, the gas volume, and the gas velocity through the tower.

Particles of all sizes can be removed, but finer sizes below 1 micron are more difficult, and as a consequence the spray tower is considered a primary or secondary unit where it has efficiencies of 88% on 10 micron material and up to 98% on 90 micron particles.

The spray tower is a comparatively simple, inexpensive type of collector that can handle large gas volumes with all degrees of dust loadings, but is used to best advantage when gas cleaning conditions are not extreme. As it is a wet-type collector the gas passing through the tower is cooled and any sensible heat is lost.

6. Venturi Scrubbers are a wet-type gas cleaning device in which the dirty gas and water for cleaning are atomized in a moving gas stream. The relative velocity between gas and water droplets is very high, with gas velocities in the range of 200 to 400 feet per second

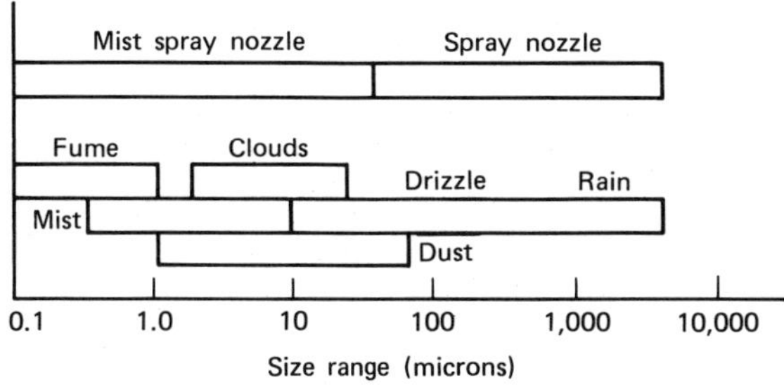

Figure 5.8. Particle size comparison—spray tower water droplets in relation to dust and fume particle sizes.

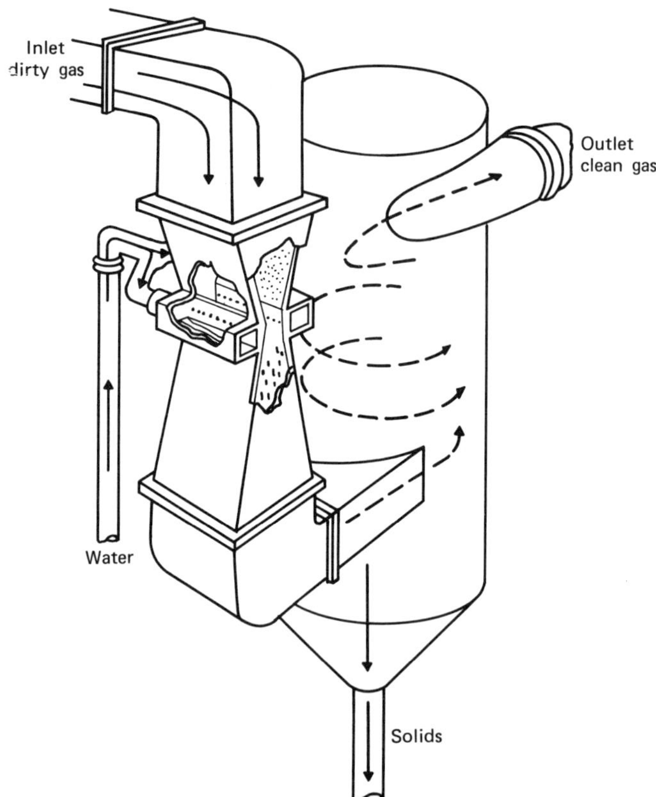

Figure 5.9. Venturi scrubber. *Source: Control Techniques for Particulate Air Pollutants*, U.S. Public Health Service Publ. AP-51, 1969, p. 59.

(3660 to 7320 m per minute), and this high velocity promotes large treatment capacity. The scrubber has a converging–diverging Venturi throat in its design, and it is at this location that the water droplets are added to give maximum liquid–gas contact.

Collection efficiency is proportional to and increases with the pressure drop. Therefore high pressure-drop Venturi scrubbers, by adjustments of their pressure drop, the water spray flow rate, and the input gas velocity, are of sufficiently high efficiency to collect even submicron size particles.

The liquid droplets which have contacted the dust particles collect on the walls of the lower diverging section of the scrubber, below the throat constriction, and flow downward in rivulets and sheets into the bottom of the collector. From here they are drained off, and the solid particles are separated from the liquid. The cleaned gas is taken off at a discharge duct in the lower section of the unit.

The Venturi scrubber is a cheap, simple type of dust collector that is very efficient for small-sized particles, +99% for 5 micron size, and can handle large gas capacities in a quite small unit because of its rapid rate of treatment. Its usefulness is as a single primary unit if all the particles to be removed are small, or as the final treatment for the smallest particles if used in a series of collectors.

As this too is a wet collector, a hot gas is cooled when contacting the water droplets and any sensible heat content is lost.

7. Electrostatic Precipitators make use of electricity to remove wet or dry particles in a flowing stream of gas. While it is an expensive type of collector, it is generally considered to be the best for removal of very small particles, with an efficiency of 82% for 0.1 micron sizes and +99% for 2 micron solids.

The precipitator uses two electrodes, a discharge electrode of negative potential and a collector electrode of positive potential. A rectifier converts the voltage between these

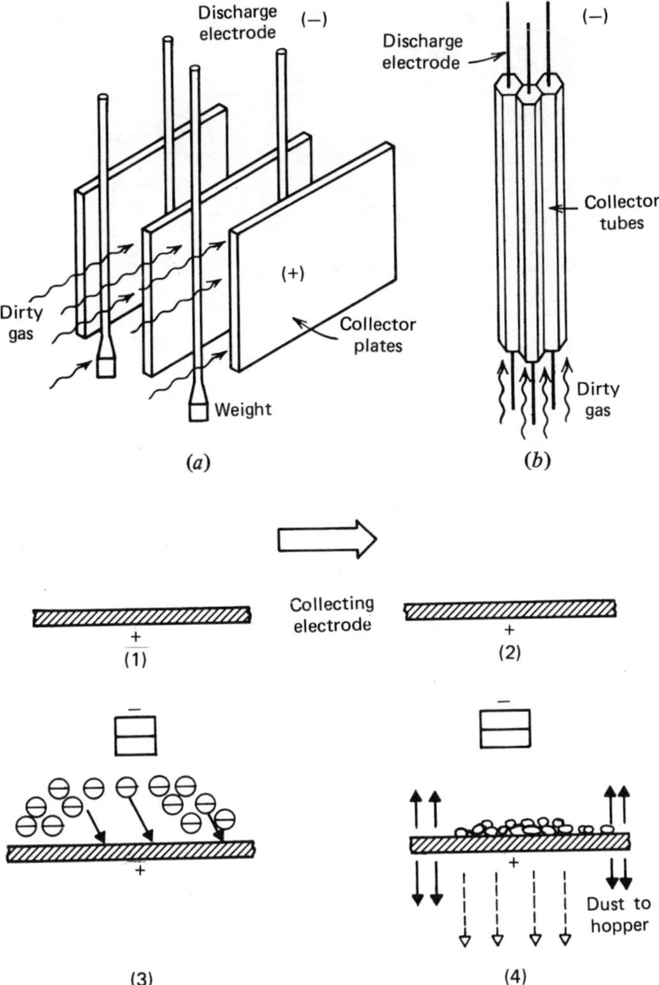

Figure 5.10. Electrostatic precipitator operational details. (*a*) Wire-in-plate type; (*b*) wire-in-tube type. Sequence of events in an electrostatic precipitator: (1) generating a strong electrical field between electrodes; (2) passing the suspended solids to be collected through the field whereupon they are electrically charged by means of ionization; (3) the charged particles are then transported to a collecting surface by means of the force exerted on them by the electric field; (4) the electrically charged particles precipitated on the collecting surface are neutralized and removed, usually by rapping or shaking the collecting electrode. *Source:* Reproduced, with permission, from *Engineering Digest*, Volume 21, No. 10, 1975.

two electrodes to direct current and steps it up to 80,000 volts. There are different design types of precipitators. One common type has the two electrodes as plates positioned next to each other (plate type), while another type has the discharge electrode as a wire suspended inside a tube which is the collector electrode (tube type).

At the rectified high voltage in the precipitator, a corona of ionized gas molecules forms around the negative discharge electrode, and the negative and positive gas ions are attracted to the surface of the electrode of opposite polarity. The negatively charged gas ions have a greater distance to travel to the collector electrode, and on their way they contact and pass on their negative charge to most of the dust particles in the gas stream passing through the precipitator. These negatively charged particles are then also attracted to the positively charged collector electrode, where they build up a discharged layer of dust particles.

In a similar fashion, any dust particles that become positively charged will travel the short distance to the negatively charged discharge electrode and collect there as a discharged layer.

Periodically, in a dry-type precipitator, rappers dislodge the built-up layers of particles to fall into a bin for removal. In a wet-type precipitator, the moist solids can be washed off and down with a flowing film of water across the electrodes.

Because of its high efficiency for collection of small particles and ability to handle large gas volumes, the electrostatic precipitator can be used as a single unit if only fine particles

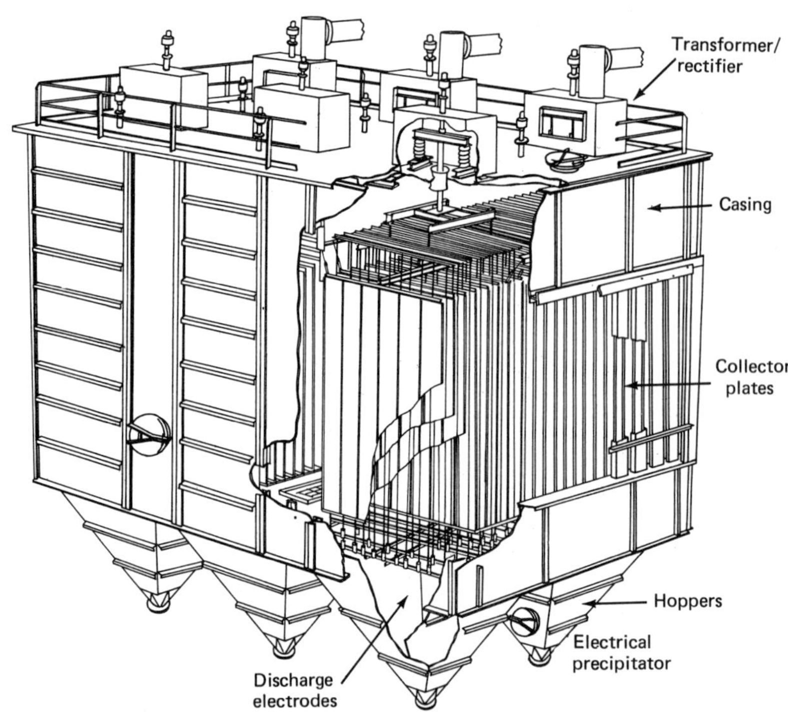

Figure 5.11. Electrostatic precipitator. *Source:* Courtesy of the Western Precipitation Division of Joy Manufacturing Company.

are to be removed, or it can also be the final stage in a series of several collectors removing a variety of particle sizes.

The theoretical efficiency of an electrostatic precipitator is given by the expression

$$\frac{\text{precipitator}}{\text{efficiency}} = 1 - e^{-\left(\frac{\text{collection area}}{\text{gas flow rate through collector}}\right)(\text{drift velocity of particles to collector electrode})}$$

or

$$\text{precipitator efficiency} = 1 - e^{-(A/Q)(w)}$$

where Q is measured in ft/sec, A in sq ft, and w in ft/sec. Highest efficiencies occur for a low gas flow rate (Q), for a high drift velocity (w), and for a large collection area (A).

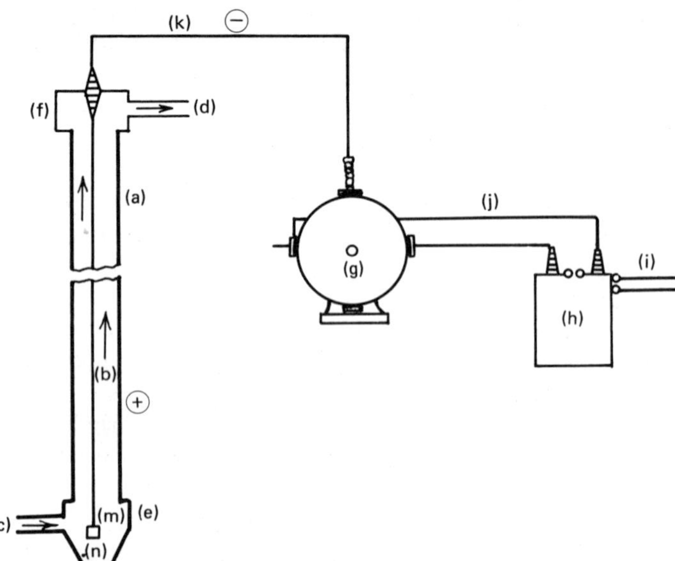

Figure 5.12. Operating components wire-in-tube-type electrostatic precipitator: (a) collecting electrode; (b) discharge electrode; (c) gas inlet; (d) gas outlet; (e) lower header; (f) upper header; (g) rectifier; (h) transformer; (i) low-tension lines; (j) high-tension lines; (k) high-tension rectified current; (m) weight; (n) hopper. *Source:* J. L. Bray, *Nonferrous Production Metallurgy*, Wiley, New York, 1947, p. 194.

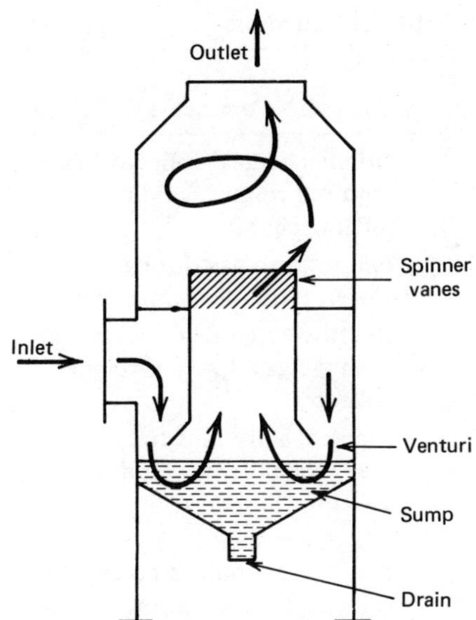

Figure 5.13. Simple scrubber, Turbulaire type. *Source:* Courtesy of the Western Precipitation Division of Joy Manufacturing Company.

Higher efficiencies can be obtained by increasing the collection area, though to increase efficiency from 90% to 99% requires doubling this area, and from 90% to 99.9% requires tripling the area.

An increase in flow rate reduces the efficiency, and even a small increase means a large increase in the number of particles escaping the collector. A drop from 99% efficiency to 97% efficiency triples the emissions, and it is these emissions that are the important variable in all gas cleaning operations.

8. Wet Scrubbers There are many types of wet scrubbers which operate on the principle of immersing the dirty gas directly into a water bath, where the contact between water and particulant matter removes the solids from the gas stream and the gas passes out of the unit cleaned.

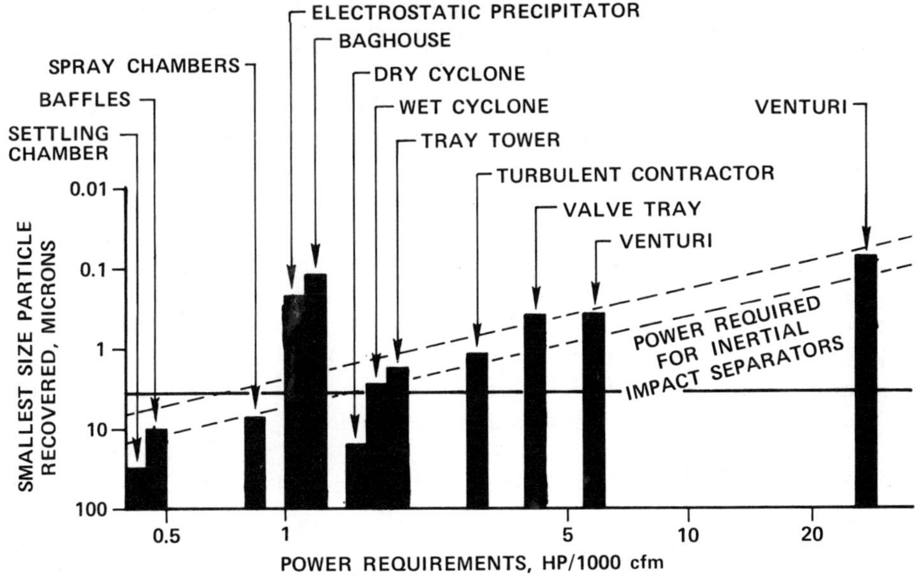

Figure 5.14. Power requirements versus particle sizes for the various collectors. *Source:* J. C. Agarwal, H. W. Flood and R. A. Giberti, *J. Metall.*, Vol. 26, No. 12, 1974, p. 9.

Table 5.2. Power Requirements of Dust Collectors

Type of Collector	Power Requirements (hp/1000 cfm gas)	Comments
Settling chamber	0.2	Nonmechanical collector, fans required to push gas into settling chamber
Spray tower	0.75	Fan to push gas through tower, pump for water spray
Electrostatic precipitator	1.0	Rectified electrical energy, fan to move gas slowly through collector
Bag house	1.5	High-capacity fans to move gas through large volumes of collector bags
Cyclone	2.0	Gas blown in at high velocity to achieve required tangential velocity for separation
Venturi scrubber	6.0	Gas and water spray blown and pumped at high velocity.

Capacities are somewhat limited because of the size of the water bath that would be required for any large-volume gas treatment. This type of scrubber is most useful as a single unit for small- to medium-sized gas cleaning applications. Efficiency drops to 30% with submicron sizes (0.2 micron) but increases to 99% with larger 9 micron sizes.

Energy Consumption used by the different dust collectors varies considerably. The simpler, less mechanical designs have far lower power requirements than the more complicated, sophisticated pieces of equipment, and those operating at high gas velocity and treating larger volumes need more and larger blowers and fans (see Table 5.2).

GAS CLEANING—REMOVAL OF CONTAMINATING COMPOUNDS FROM GASES

1. Sulfur Dioxide is the most common compound found in product gases from metallurgical treatments due to the fact that sulfide ores are the form in which many of the common nonferrous metals (copper, lead, zinc, nickel, cobalt) are found, and their processing includes roasting, smelting, and converting, all of which are oxidizing operations that produce large quantities of SO_2 gas.

Consequently, as SO_2 has long been considered an undesirable pollutant in the atmosphere because of human respiratory and agricultural damage and because stricter clean air standards are being legislated by many countries to control smelter emissions, the removal of SO_2 from metallurgical process gases is of great importance.

For example, the legislative controls recently enacted on SO_2 by the U.S. Environmental Protection Agency have established the following limits:

1. Not more than 10% of the sulfur entering a smelter can be emitted to the atmosphere.
2. Workers must not be exposed to an atmosphere containing more than 5.0 ppm SO_2.
3. Yearly ambient air standards at ground level must not exceed 0.03 ppm SO_2.
4. The average SO_2 content for any one day must not exceed 0.14 ppm.

In Belgium, legislation states that SO_2 discharges must not exceed 0.1% by volume of discharged gas, while in Sweden provisional guidelines have been laid down at the following SO_2 limits: $\frac{1}{2}$ hour average, 0.25 ppm; one week average, 0.10 ppm; one month average, 0.02 ppm.

There are three methods of treating gases containing SO_2 to remove the sulfur. The first, and by far the most widely used, is to convert the SO_2 to sulfuric acid by the *contact process*. Secondly, sulfur can be *extracted in the form of elemental sulfur*, and thirdly, the SO_2 can be *combined with calcium oxide* to form calcium sulfite and calcium sulfate.

Contact Acid Process This process works best and most economically on gases containing at least 3.5% SO_2 and preferably 10 to 14% SO_2. However it has been made to operate, with the inclusion of additional cycles and at considerably increased expense, on gases containing down to 2.2% SO_2. The strength of acid produced varies somewhat from plant to plant and usually is in the range of 98 to 99.6% H_2SO_4, which then is diluted with water to 93% H_2SO_4, the usual commercial market strength.

The hot smelter gas must first be cooled and cleaned to prevent contamination during processing, and it is particularly important to eliminate any impurities such as arsenic trioxide, which forms during the roasting of arsenic containing material.

The hot metallurgical process gases are first passed through a waste heat boiler to recover their sensible heat and then to a dry electrostatic precipitator to remove particulate matter. If the dust content is very high, a cyclone collector can be installed ahead of the electrostatic precipitator to divide the particulate removal into two stages.

The cleaned gas is now put through a water scrubbing tower to wash it, followed by a refrigeration cooler to lower the temperature below the dew point. A wet electrostatic

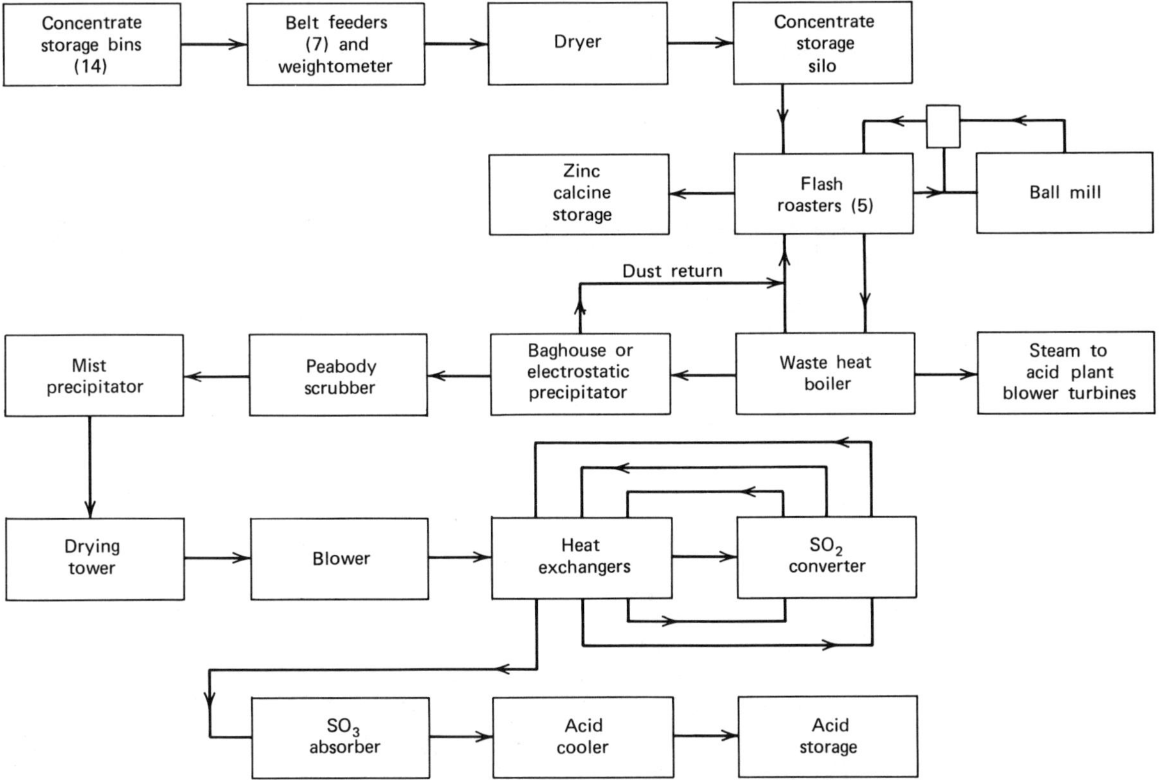

Figure 5.15. Contact acid process, Bunker Hill Company. *Source:* C. H. Cotterill and J. M. Cigan, Eds., AIME World Symposium of Lead and Zinc, Vol. 2, 1970, p. 116.

Table 5.3.   Contact Acid Plant Economics[a]

Gas Stream SO$_2$ Concentration (%)	Capital Costs ($M)	Operating Costs (dollars per ton of acid produced)
2	15.0	14.00
4	8.5	7.20
8	5.0	4.20
12	4.5	4.00

Source: Oxygen in Copper Smelting, Air Products and Chemicals, Inc., Allentown, Pennsylvania, 1972, p. 6.

[a]Basis: Feed gas stream of 300 tons of sulfur equivalent per day.

By increasing furnace gas SO$_2$ content from 2% to 4%, capital investment for acid plant facilities can be reduced by about $7M and operating expenses cut in half.

precipitator removes the mist formed in the cooling cycle and any dust which is still remaining after the initial dust removal treatment.

When cooling roaster gases, a sulfuric acid mist forms which contains any As$_2$O$_3$ that is present, and the removal of this mist in the wet precipitator also eliminates any problem from As$_2$O$_3$ in the steps to follow. In particular it eliminates the problem of the contamination of the vanadium pentoxide catalyst during the contact conversion operation.

The clean, cool gas next passes through a tower where it is dried by spraying with 93% H$_2$SO$_4$ and then enters the contact process system. This system consists of a converter and a heat exchanger, followed by a gas cooler and an absorption tower.

The converter is a brick-lined cylindrical steel shell which usually contains four trays of vanadium pentoxide catalyst. Sulfur dioxide is catalytically oxidized to the trioxide by atmospheric oxygen,

$$2SO_2 + O_2 = 2SO_3$$

Figure 5.16. Sulfuric acid plant, catalyst towers (Hoboken). *Source:* Courtesy of Metallurgie Hoboken–Overpelt.

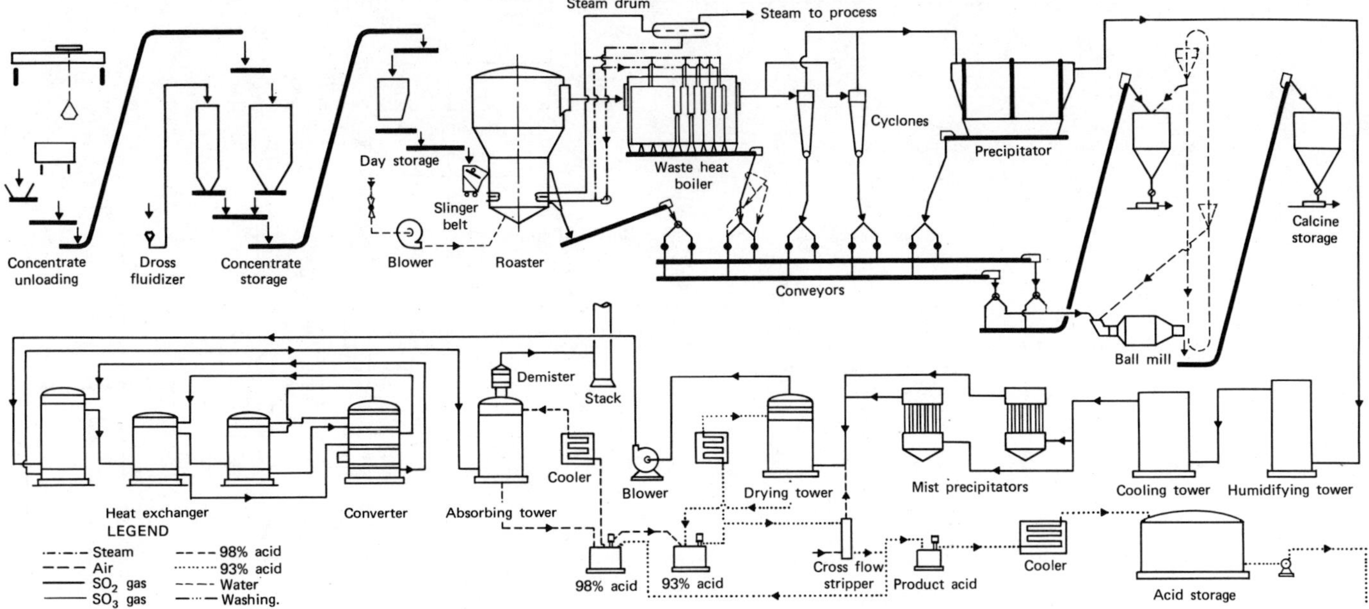

Figure 5.17. Contact acid plant, Canadian Electrolytic Zinc Ltd. Sulfuric acid produced from sulfur dioxide in zinc roaster gas. *Source:* C. H. Cotterill and J. M. Cigan, Eds., AIME World Symposium of Lead and Zinc, Vol. 2, 1970, p. 146.

with oxidation taking place on the surface of the catalyst at atmospheric pressure. Conversion is 95 to 98% and contact time 2 to 4 seconds. The reaction is exothermic and is controlled at 790 to 1100°F (420 to 600°C) to obtain the greatest conversion rate. The heat content of the gases leaving the contact chamber can be used through a heat exchanger to heat the cold entering SO_2 gas to the necessary initial temperature; or, if a heat exchanger is not used this initial heating can be done by a fuel-fired preheater.

The SO_3 gas, cooled to approximately 400°F (200°C), is sent to an absorption tower where the SO_3 is absorbed in circulating 98.5% H_2SO_4. The SO_3 combines with the water in the acid to produce 99.6% H_2SO_4:

$$SO_3 + H_2O \longrightarrow H_2SO_4$$

This 98.5% strength H_2SO_4 gives optimum absorption efficiency. Cast iron and alloy steels are not attacked by these high acid strengths, and this greatly simplifies the choice of equipment materials needed for pumping and recirculation.

Part of the circulating acid is continuously removed from the absorption system and diluted with water to the 93% H_2SO_4, which is the usual commercial market strength. The tail gases leaving the absorption tower contain sulfuric acid mists and also still can hold as much as 0.1 to 0.2% SO_2. Scrubbing these gases with lime solution precipitates the sulfur as calcium sulfite or calcium sulfate, and a wet electrostatic precipitator will remove both these and any H_2SO_4 mist before the gas is finally released from a stack. This final released gas has less than 0.01% SO_2, and over 99% of the SO_2 in the acid plant input gas has been removed.

Elemental Sulfur manufacture is a second process that can be used to treat high SO_2 smelter gases, particularly those containing over 10% SO_2. Elemental sulfur has fine, long-term storage characteristics and is adaptable to other different uses than is sulfuric acid.

In this process the gas is first cooled, cleaned, and dried in a similar manner to that used for the contact acid process. It is then reduced with carbon (natural gas or coal) or

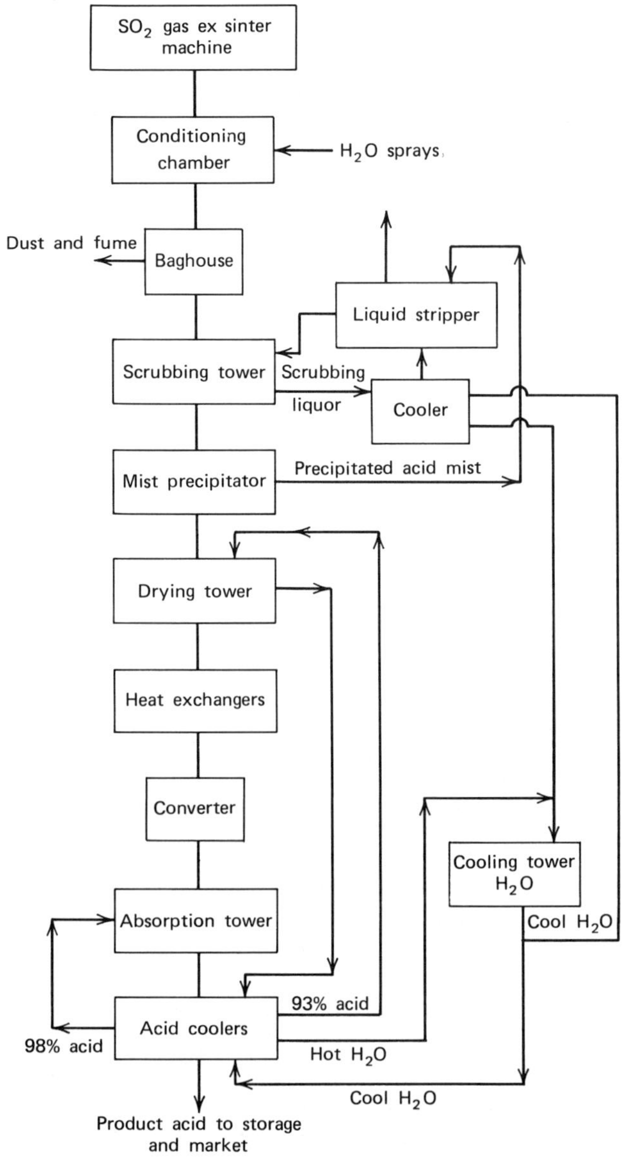

Figure 5.18. Acid plant flow sheet, St. Joe Minerals Corp. *Source:* C. H. Cotterill and J. M. Cigan, Eds., AIME World Symposium of Lead and Zinc, Vol. 2, 1970, p. 733.

hydrogen, at temperatures of 932°F (500°C):

$$SO_2 + C \longrightarrow CO_2 + S(g)$$

$$SO_2 + 2H_2 \longrightarrow 2H_2O + S(g)$$

A mixture of SO_2 and air is passed continuously through a furnace, and the resulting gases are condensed to 340°F (170°C) to give liquid elemental sulfur, which is tapped from the condenser and cast into blocks.

The gas passing from this first condenser still contains some SO_2, so it is reheated to 465°F (240°C) and reacted with hydrogen sulfide in a one- or two-stage reduction to give sulfur gas,

$$2H_2S + SO_2 = 2H_2O + \tfrac{3}{2}S_2(g)$$

which is again condensed to liquid, tapped, and cast. The tail gas contains about 1% SO_2 and is now released from a stack to the atmosphere.

Wet Scrubbing by Fine Calcium Oxide or limestone of off-gases containing small amounts of SO_2 has been carried out where the SO_2 concentration is too low for reasonable treatment for either H_2SO_4 or elemental sulfur production, and also to further clean tail gas from contact acid plants prior to its final discharge to the atmosphere.

Spray towers and Venturi scrubbers are used with water–solid slurries of both fine (-200 to -325 mesh) CaO and $CaCO_3$. The reactions occurring with the SO_2 in the gas are

$$CaO + SO_2 = CaSO_3$$

and to some extent some

$$CaSO_3 + \tfrac{1}{2}O_2 = CaSO_4$$

The solids are removed by settling tanks, Venturi scrubbers, and centrifuges, to be discharged to a disposal site, while the cleaned gas is released from a stack.

The process is more efficient on lower concentrations of SO_2 and decreases in percent removal as the amount of SO_2 in the feed gas increases, as shown below:

Inlet SO_2 % Concentration	% SO_2 Removed
0.3 to 0.6	85
1.0	75
1.5	72

2. Fluorides, both gaseous and particulate, constitute the most serious pollutants from aluminum electrolytic reduction plants and are kept under control by primary and secondary systems of gas removal. The primary system consists of hooding for each cell connecting into a common duct which discharges to a cleaning operation, while the secondary system would be located on the roof to capture any fluorides that escaped the cell hoods.

The gases are cleaned first by multiclones or dry electrostatic precipitators. Then these are followed by cyclonic towers, spray towers, or Venturi scrubbers. In some instances alumina has been used as a fluid-bed or bag filter coating to absorb gaseous fluoride; it appears to be very efficient, although it is not applicable to cells that produce emissions containing condensible organic matter as this fouls the systems.

3. Chlorides and hydrochloric acid gas are released from magnesium electrolytic cells and must be caught and processed. They are collected under a tightly closed refractory cell cover and piped to a plant to separate the chlorine and hydrochloric acid for recycling.

GAS DISSIPATION FROM TALL STACKS

Even the most efficient dust collecting or chemical treatment processing does not remove all of the pollutants in a gas. There is always a small amount of contaminant that has to be released finally to the atmosphere. In addition, there are conditions where SO_2 concentrations are too low to be efficiently or economically removed as H_2SO_4 or elemental sulfur. There are also conditions where because of the location of a smelter, remote from any industrial manufacturing area, there is no market within shipping range for all the H_2SO_4 that can be produced from available SO_2 gas, and some smelter SO_2 is in excess. Finally there are the smelters in areas where there are low population densities and little agriculture to be damaged by SO_2 discharge.

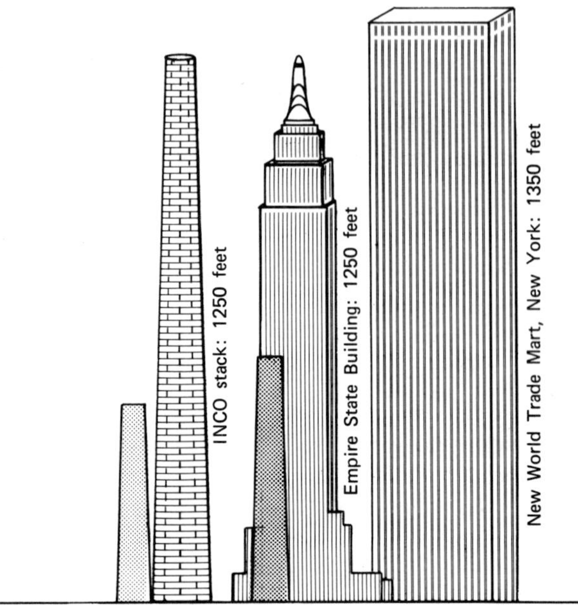

Figure 5.19.   INCO stack, tallest in the world, Sudbury, Ontario, Canada.

In all of these cases the solution has been to disperse the pollutants from tall stacks so that both particulant matter and SO_2 are widely scattered and diluted before they reach the ground and will no longer be at the level of harmful concentrations. To this end there has been a tendency to build taller and taller stacks to put the gases with their pollutants out into the upper air currents and increase the probability of dispersion over a larger area to ensure the greatest dilution of contaminants. The largest stack at the present time is that of the International Nickel Company, which is 1250 feet high (375 m), has a base diameter of 110 feet (33 m), and tapers to a top diameter of 45 feet (13.5 m).

While stacks for pollution dispersion are an old and continuing method, they are not ideal in all situations. Some of their drawbacks are that (1) temperature inversions hold down gas stack plumes and limit dilution and dispersion; (2) a decrease in wind velocity decreases dispersion; (3) there is no effect on emission standards that limit the percent of total sulfur entering a smelter that can be emitted to the atmosphere (U.S. EPA, 10% of total sulfur); (4) there is a strong possibility that prevailing winds carry smelter gases long distances, with SO_2 returning to earth hundreds of miles away as an acid rain, which lowers the pH of lakes and stunts forest growth up to as much as 60%. This acid rain seems to be most harmful in springtime, during the early growing season.

The maximum pollution concentration that a particular stack can handle under specific conditions of emission rate and wind velocity is given by the expression

$$\begin{array}{c}\text{maximum}\\ \text{pollution}\\ \text{concentration}\end{array} = 2.15 \times 10^5 \left(\dfrac{\text{emission rate}}{\text{wind velocity}}\right)\left(\dfrac{1}{\text{stack height}}\right)^2 \left(\dfrac{\text{diffusion coefficient}}{\text{diffusion coefficient}}\right)$$

or

$$C_{\max} = 2.15 \times 10^5 \left(\frac{Q}{v_w}\right)\left(\frac{1}{H}\right)^2 \left(\frac{K_p}{Kg}\right)$$

with $C_{\max}$ in ppm, Q in cu ft/sec, v_w in ft/sec, and H in ft; K_g and K_p are 0.08 and 0.05, respectively, for normal atmospheric conditions.

The maximum distance from the stack at which C_{max} occurs is given by the expression

$$\text{maximum distance} = \frac{1}{2}\left(\frac{\text{stack height}}{\text{diffusion coefficient}}\right)$$

or

$$X_{max} = \frac{1}{2}\left(\frac{H}{K_p}\right)$$

with X_{max} in ft, H in ft, and $K_p = 0.05$.

HEAT RECOVERY

Heat Exchangers The gases from pyrometallurgical processes are at elevated temperatures, some quite high, and contain an appreciable amount of sensible heat which is commonly recovered and used before the gas is allowed to pass on to pollution control treatment. In some gas cleaning operations where temperature is important and limited, as in bag houses, where the fabric bags have a definite temperature limit, and in electrostatic precipitators, where the electrode plates might warp and short circuit, the cooling of the gas is an integral part of the overall processing and an additional reason for removal of the sensible heat.

This is generally done in heat exchangers or waste heat boilers, which are similar in general design. In the heat exchanger, cold incoming air or gas passes around ducts through which is passing the hot furnace gases, and heat is transferred by convection through the duct walls and heats the cold air. The same general principle occurs in waste heat boilers, where cold water is passed through tubes around which flows hot furnace gases and heat is transferred to the water to heat it, or if sufficiently hot, to turn it to steam. Hot gases can also be used for drying, and removal of moisture from wet material by passing over or through it.

Gases With High Calorific Value Some metallurgical processes, such as the reduction of zinc oxide by carbon in a retort or blast furnace, produce carbon monoxide as the gaseous product of the reaction:

$$ZnO + C = Zn + CO$$

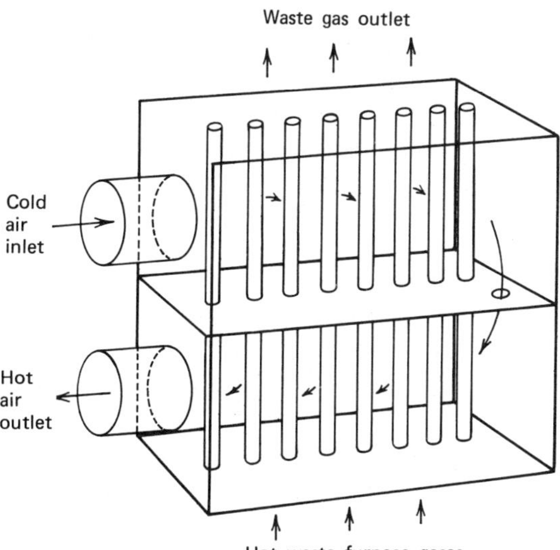

Figure 5.20. Hot gas heat recovery system. *Source:* W. H. Dennis, *Metallurgy in the Service of Man*, MacDonald and Jane's Publishers Ltd., London, 1961, p. 56.

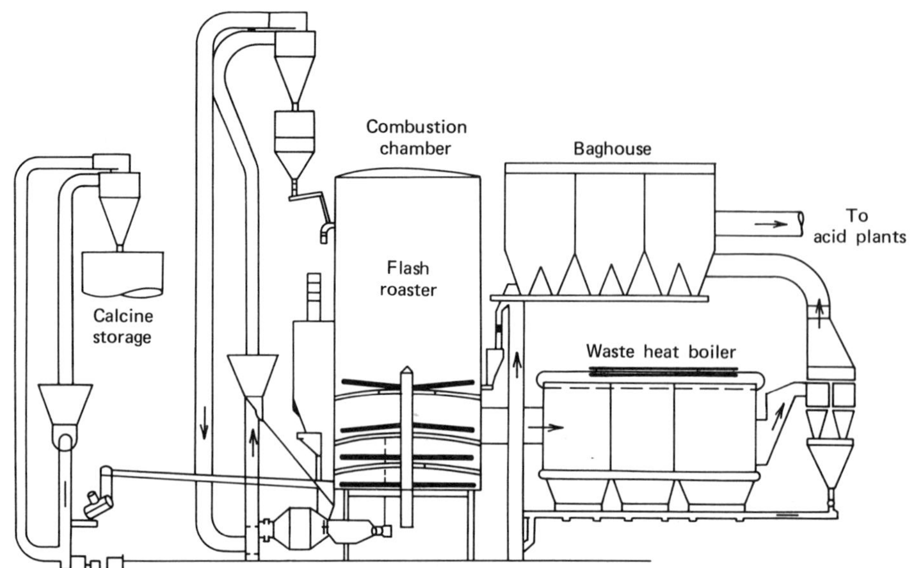

Figure 5.21. Waste heat boilers in conjunction with flash roaster. *Source:* C. H. Cotterill and J. M. Cigan, Eds., AIME World Symposium of Lead and Zinc, Vol. 2, 1970, p. 119.

This CO has a high calorific value and can be recovered and used as a fuel gas, rather than being wasted by allowing it to escape to the atmosphere. In the New Jersey Zinc Company's vertical retort, this recovered CO gas is used as fuel to heat the retort charge to its reaction temperature, and makes up 30% of the total fuel used. Natural gas is added to the CO to supply the other 70% of the total fuel requirements.

If particulate matter is swept out in the high-calorific gas passing from a process, it may have to be cleaned in a dust collection system before being used as a fuel, both to recover the particulate matter and to ensure that the solid particles in the gas are not going to plug the fuel gas burner in any way.

WATER CLEANING

Particulates As several of the dust collecting types such as the spray tower, scrubber, Venturi scrubber, and the wet electrostatic precipitator are wet-type cleaners, the particulate matter that is removed from the gas stream in each case is not dry, easily managed dust but is a very thin slurry of the collected solids in water. To recover these solids for recycling or disposal, they must first be separated from the water.

This separation can be accomplished in a two-stage operation, the first of which is thickening, and the second is filtering. After this processing is complete, the solids in the slurry are removed and recovered and the water portion is clarified so that it too can be reused and recycled through the dust collector, used as plant process water, or returned to nearby streams.

Thickening is essentially settling, with the particles taking hours to gravitate slowly downward and settle out of the liquid in a large tank. The operation is continuous, with feed slurry being added to the center upper portion of the tank and giving as products a clear solution overflow at the top and a thick mud underflow at the bottom, both being steadily removed.

The settling rate of the particles is increased by flocculation, whereby aggregates of the suspended solid particles form and have a faster settling rate than do individual particles alone. Flocculation is accomplished by the addition of electrolytes (lime) which neutralize the electrical charges on the particles and allow them to collect together into clumps.

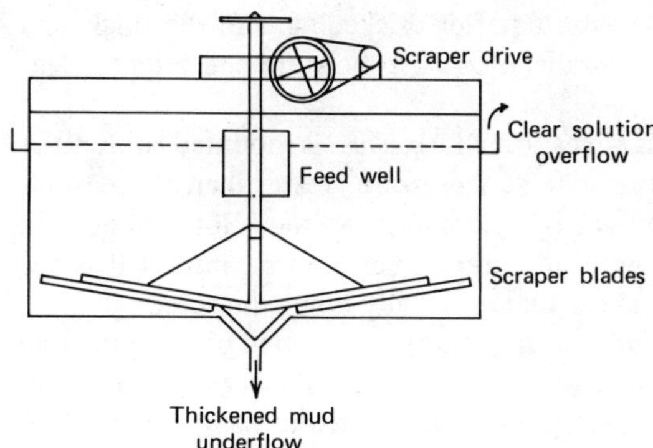

Figure 5.22. Thickener, cross section view, for separation of fine particulants from water.

A very slowly turning rake with scraper blades at the bottom of the thickener tank pushes the settled thickened mud to a central discharge opening, from where it is drawn away by a diaphragm pump. The mud underflow discharge will be on the order of a liquid-to-solid ratio of 2:1 or 1:1 and is sent on to vacuum filtering for further dewatering.

Thickeners are available in a great variety of sizes, with the common standard dimensions being,

Diameter	Depth
5 to 7 ft (1.5 to 2.1 m)	4 ft (1.2 m)
8 to 11 ft (2.4 to 3.3 m)	6 ft (1.8 m)
12 to 25 ft (3.6 to 7.5 m)	8 ft (2.4 m)
26 to 40 ft (7.8 to 12 m)	10 ft (3 m)

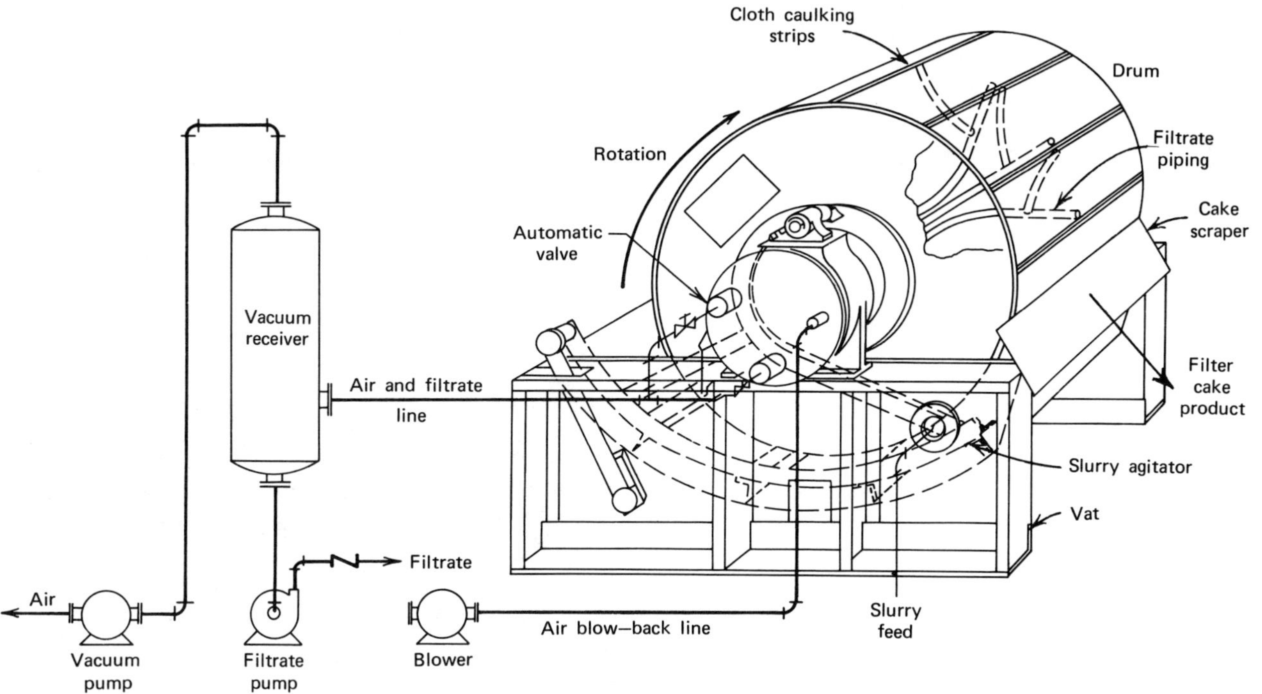

Figure 5.23. Drum-type vacuum filter. *Source:* Courtesy of Komline–Sanderson Engineering Corporation.

Vacuum Filtering is the second stage of dewatering after thickening, and the thickened mud gives a product after filtering which contains 6 to 10% moisture and a quite clear water discharge.

The principle of filtering is that a porous sheet, usually canvas, is mounted so that the material to be filtered can be brought to one side at a pressure greater than at the other side. Then if the pore openings in the sheet are of a suitable size, the solids will be held back while the water passes through. As with all filters, after the first instant it is the deposited solid cake rather than the porous sheet that is actually doing the filtering.

The filters are continuous in operation and are frequently constructed on the principle of a drum divided into segments and covered by a filter cloth. The segments are connected separately so that either vacuum or pressure can be applied to each one individually. The drum turns slowly down through a tank filled with thickened slurry, and vacuum is applied to that segment immersed in the tank, drawing a cake of solids against the filter cloth and sucking water through it into the drum interior, from where it is discharged. As the drum section rotates out of the tank, vacuum suction is continued, drawing still more liquid from the cake, and continues to suck through over 180° of the rotation. Then vacuum is cut off and pressure is applied to pop out the filter cloth and crack and break off the nearly dry cake, which falls off into a bin or onto a conveyor belt for removal. The drum segment from which the cake has just been removed is now turning under the slurry in the tank, and vacuum is again applied to complete the cycle and continue the operation.

If the filter cake is desired to be completely dry, it can be put through a kiln-type drier, and hot gas from a furnacing operation can sometimes be used as the drying medium.

Combined Heavy Metals In treating industrial waste waters, chemical, physical, or biological methods are used, with chemical treatment generally being the one most commonly used. This includes pH neutralization and the precipitation of metals from unwanted dissolved compounds.

As an example, water containing hexavalent chromium as Na_2CrO_4 can be reduced with sodium hydrosulfite in a pH range of 6.5 to 9 and precipitated as insoluble chro-

Figure 5.24. Vacuum filtering of wet dust (Hoboken). *Source:* Courtesy of Metallurgie Hoboken–Overpelt.

mium hydroxide:

$$2Na_2CrO_4 + 3Na_2S_2O_4 + 2NaOH + 2H_2O = 6Na_2SO_3 + 2Cr(OH)_3$$

Other heavy metals such as copper, zinc, and iron also precipitate as insoluble metal hydroxides or carbonates. These precipitates are removed by carefully controlled filters to permit the water at 0.05 ppm trace metals concentration now to meet drinking water specifications of 1 ppm and to be discharged into nearby municipal water sources.

Polyelectrolytic systems are also a new treatment method beginning to have an application in such industries as uranium mining and treatment to settle mine water sludge, sludge dewatering, and metal recovery.

The mine water, at a low pH of 2.5, usually contains iron and sulfate ions in addition to uranium. Lime is added to neutralize the acid content and raise the pH to about 11, and polyelectrolyte is also added at this stage to improve the settlability of any solids. The sludge, or settled solids, combined with iron and uranium which precipitate, are removed and the clear liquid is recycled for use as plant process water.

Acid Mine Drainage An unwanted by-product of some mining activities where the ore deposit contains iron pyrite, FeS_2, is acid mine drainage. This sulfide compound oxidizes when exposed to air and water and forms ferrous sulfate and sulfuric acid:

$$2FeS_2 + 7O_2 + 2H_2O = 2FeSO_4 + 2H_2SO_4$$

The high-acid mine waters then flow into nearby lakes and streams through the natural drainage process to damage municipal and industrial water supplies, destroy marine life, and attack metal and concrete bridges, turbines, and dams.

The water can become still more acidic from the further oxidation of the ferrous sulfate, $FeSO_4$, into ferric sulfate, $Fe_2(SO_4)_3$, and the ferric sulfate reacting with water to form more sulfuric acid and increase its acidic concentration level still further:

$$Fe_2(SO_4)_3 + 6H_2O = 2Fe(OH)_3 + 3H_2SO_4$$

This is a serious, widespread pollution problem for which there is no immediate easy and inexpensive solution. A secondary problem of additional concern is that the sulfuric acid formed can also dissolve other metals in the rocks that it contacts, with the result that a large variety of toxic metals may enter the water as contaminants.

The most common treatment method is to neutralize the sulfuric acid with a basic material and raise the pH to 7. Lime, CaO, and limestone, $CaCO_3$, have been used for this purpose, as has magnesium carbonate, $MgCO_3$:

$$CaO + H_2SO_4 = CaSO_4 + H_2O$$

$$CaCO_3 + H_2SO_4 = CaSO_4 + H_2O + CO_2$$

$$MgCO_3 + H_2SO_4 = MgSO_4 + H_2O + CO_2$$

SPECIFIC METAL TREATMENTS

1. Copper Copper is produced primarily (over 90%) from low-grade sulfide ores, with much of the processing treatment being pyrometallurgical. These pyrometallurgical steps include roasting, smelting, converting, and refining, and the high temperatures attained cause volatilization of a number of the trace elements present. The product gases also contain dust, sometimes in considerable amounts, as well as sulfur dioxide, since all the operations are oxidizing to a greater or lesser degree.

The values of the dust in the process gases, along with the additional value of the volatilized elements, dictates for economic reasons the efficient collection of dust and fume

Table 5.4.   Uncontrolled Emissions of Particulates from Copper Smelting

Type of Operation	Particulates (g/m^3)	Gas Volume (m^3/ton concentrate)
Roasting	15	1300
Smelting (reverberatory furnace)	4	2000
Converting	12	10,000

for recycling or recovery. To accomplish this, a combination of cyclones and electrostatic precipitators is most often used.

The uncontrolled emissions of particulates are summarized in Table 5.4. The solid particles are caught in the dust collectors.

Many of the gases pass off at quite elevated temperatures ($2550°F$, $1400°C$) from furnaces and converters. This heat can be recovered for plant use in waste heat boilers before the gas is finally released.

Sulfur dioxide is one of the main pollutants in the gaseous emissions from a copper smelter. The variations of SO_2 concentration in the gas from the different processes is a major sulfur treatment problem, as the contact acid process works best at SO_2 concentrations above 3.5% and the elemental sulfur process at concentrations above 10% SO_2.

Roasting is the first treatment step, and multiple-hearth roaster gases average 5%, while fluid-bed roasters average 10 to 15% SO_2. Both of these are above the minimum for contact acid treatment, but only the fluid-bed roaster is of high enough concentration for elemental sulfur production. For about half the smelters, the roasting step has now been eliminated in order to supply a greater percent of sulfur in the feed to the following furnace treatment, with a correspondingly higher percent of SO_2 in the furnace gases.

Furnace Smelting follows roasting, and the variety of furnaces used produce gases with considerable differences in SO_2 content. Reverberatory furnaces, because of the burning of a large amount of fuel with air inside the furnace, give a large volume of gas with a low SO_2 content, from a few tenths up to 1.5%, which is very expensive to remove. This gas is customarily passed out a tall stack after particulate matter has been removed. However smelting of unroasted concentrates to increase the percent SO_2, combustion with oxygen or air–oxygen mixtures to reduce the total volume of off-gas, and judicious blending of reverberatory gas with other higher SO_2 smelter gases have all been used to provide adequate SO_2 concentration for acid plant feed from reverberatory furnaces.

Experimental work is also underway to treat low SO_2 concentration reverberatory gas by wet limestone scrubbing, and this has given efficiencies in test runs averaging 74% removal. The reverberatory gas containing 0.2 to 1.5% SO_2, after cooling to about $150°F$ ($66°C$) in water sprays which also remove some of the particulate matter, contacts a 5 to 15% solids slurry of minus 200 mesh limestone in a Venturi scrubber. The slurry runs off into a holding tank from where most of it, with the addition of some fresh limestone slurry, is recycled back to the Venturi. A portion from the holding tank, in an amount equal to the fresh slurry added, is removed and centrifuged to dewater the solids, and the solids are sent to waste.

The reaction mechanism is believed to be two-stage, with sodium bicarbonate, $Ca(HCO_3)_2$, forming first from the limestone and then reacting with SO_2 to form calcium sulfite, $CaSO_3$. The presence of oxygen oxidizes some of the sulfite to calcium sulfate, $CaSO_4$:

$$CaCO_3 + CO_2 + H_2O = Ca(HCO_3)_2$$

$$Ca(HCO_3)_2 + SO_2 = CaSO_3 \cdot \tfrac{1}{2}H_2O + 2CO_2 + \tfrac{1}{2}H_2O$$

$$CaSO_3 + \tfrac{1}{2}O_2 = CaSO_4$$

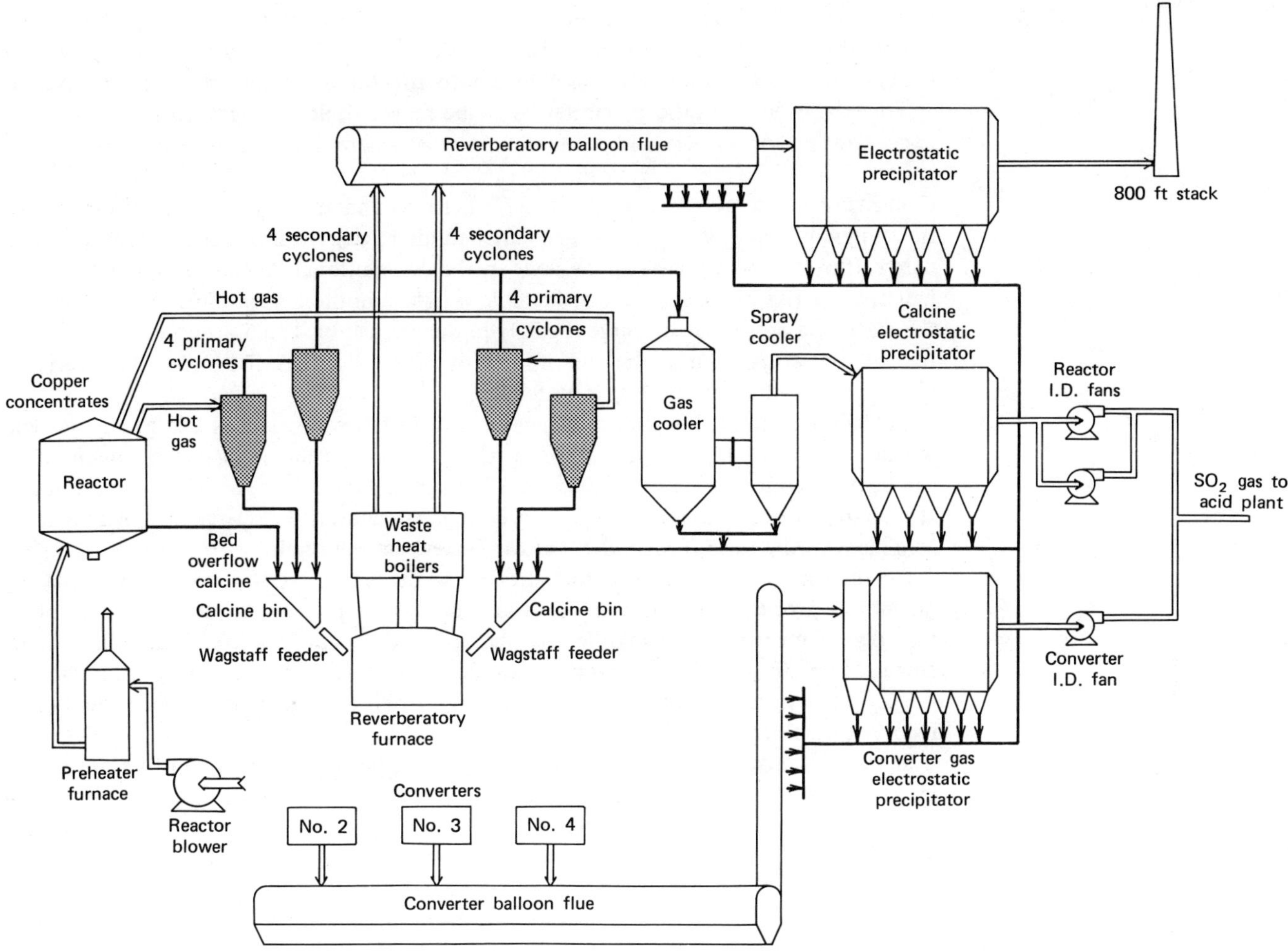

Figure 5.25. Hayden smelter. Gas collection and cleaning system. *Source:* R. P. Ehrlich, Ed., *Copper Metallurgy*, AIME Extractive Met. Div., 1970, p. 145.

The electric matte smelting furnace is very similar in design to the reverberatory furnace expect that it uses electrical resistance heating instead of burning fossil fuel. As a consequence no larger volumes of diluting air are needed for fuel combustion, which means the off-gas volume is drastically reduced and the SO$_2$ concentration is raised up to 5%. This is high enough to be used for acid plant feed.

The two flash smelting furnaces, while similar in their autogenous smelting of unroasted concentrate, produce quite different gaseous products. The International Nickel furnace injects dried copper concentrate entrained in 95% oxygen through lances at opposite ends of a rectangular reverberatory-shaped furnace, with the reaction heat of sulfide oxidation sufficient for smelting. The off-gas contains 70 to 80% SO$_2$, and this is cooled, cleaned, and treated to recover liquid SO$_2$ for marketing. The second type of flash smelter, the Outokumpu furnace, injects dried copper concentrate in oxygen-enriched (35%) air into a vertical cylindrical smelting chamber, and the off-gas, from the autogenous smelting that takes place is 18% SO$_2$. This concentration of SO$_2$ is sufficiently high for either elemental sulfur or sulfuric acid production.

Blast furnace smelting is still used in a few smelters in Africa and Japan, with finely ground flotation concentrate being agglomerated by sintering or pelletizing before being fed to the furnace. The product gases contain about 5% SO_2 along with a heavy dust loading due to the volume of air blown into the furnace. The solids are removed by cyclones and electrostatic precipitators to be recycled, and the gas is of sufficient SO_2 concentration for sulfuric acid production in areas where there is a market for it.

Converting is the third step in copper smelting, and the copper matte from furnace smelting is blown with air or oxygen-enriched air to slag off iron as iron oxide and eliminate sulfur as SO_2. Converting, as done in the conventional Peirce–Smith converter, is a batch, stagewise process, with considerable variation in the percent SO_2 in the off-gas. At the start of a fresh matte charge the SO_2 in the gas will be 4 to 5%, approaching the content of roaster gas. But at the end of the blow when the charge is almost all converted to blister copper, the percent SO_2 is quite lean.

Provisions to treat this gas can be made in different ways. If it is a large smelter with several converters, then the off-gas from the many converters at different blowing stages can be blended to have always at least 4% SO_2 and a gas suitable for H_2SO_4 manufacture. If blending is not practical, then the early, high-concentration SO_2 gas can be sent to the acid plant, while the late, lean SO_2 gas can be sent up the stack.

The continuous copper smelting processes, the commercially proved Noranda and Mitsubishi process, and the still experimental Worcra process all take advantage of the fact that converting is an exothermic operation while smelting is endothermic. The excess exothermic heat from converting is used for smelting. In addition to improving the fuel efficiency, they produce a uniform flow of rich SO_2 gas suitable for processing to recover the sulfur.

In the Noranda process, pellets of combined copper concentrate and flux are thrown into a cylindrical, horizontally aligned furnace and are melted by a fuel-fired burner. Air injected through submerged tuyeres does the converting, and the gas passing out contains 4% SO_2. When the burner air is enriched to 50% O_2 and the tuyere air to 35% O_2, the SO_2 content of the off-gas increases to 13%.

The second commercial process, the Mitsubishi process, utilizes three stationary hearth furnaces. In the first fuel-heated furnace, copper concentrate and flux are injected through lances with air containing 25% O_2 and smelted to matte and slag. These flow to a second furnace, which is electrically heated, to clean and remove the slag, and the matte then flows on to a third furnace for converting to blister copper. Air containing 25% O_2 is blown into the converting furnace. The combined off-gas from the three furnaces averages 10% SO_2.

The experimental Worcra process injects copper concentrate and flux with air through lances into a stationary furnace, where it is smelted to matte and slag. Then the matte is blown to blister copper with air put in through side lances. The furnace gas in this case is also about 10% SO_2.

Fire Refining is the last stage in which SO_2 is removed, and the concentrations are not sufficiently high to make sulfur recovery economical. Blister copper contains about 0.05% dissolved S from conventional converting and about 0.5 to 2% S from the continuous processes. Air is blown in to the bath of liquid blister copper in the fire refining furnace, the S is oxidized to SO_2, and the sulfur content is lowered to a level of 0.001 to 0.003% S.

Removal of sulfur during refining, even though a quite small amount is present, is necessary in order to cast strong, thin, smooth-surfaced anodes for the electrolytic refining step to follow. If the sulfur is not removed, some will combine with the 0.5 to 0.8% dissolved oxygen in the blister copper to form SO_2 gas, which will then leave blisters and bubbles in and on the cast anodes.

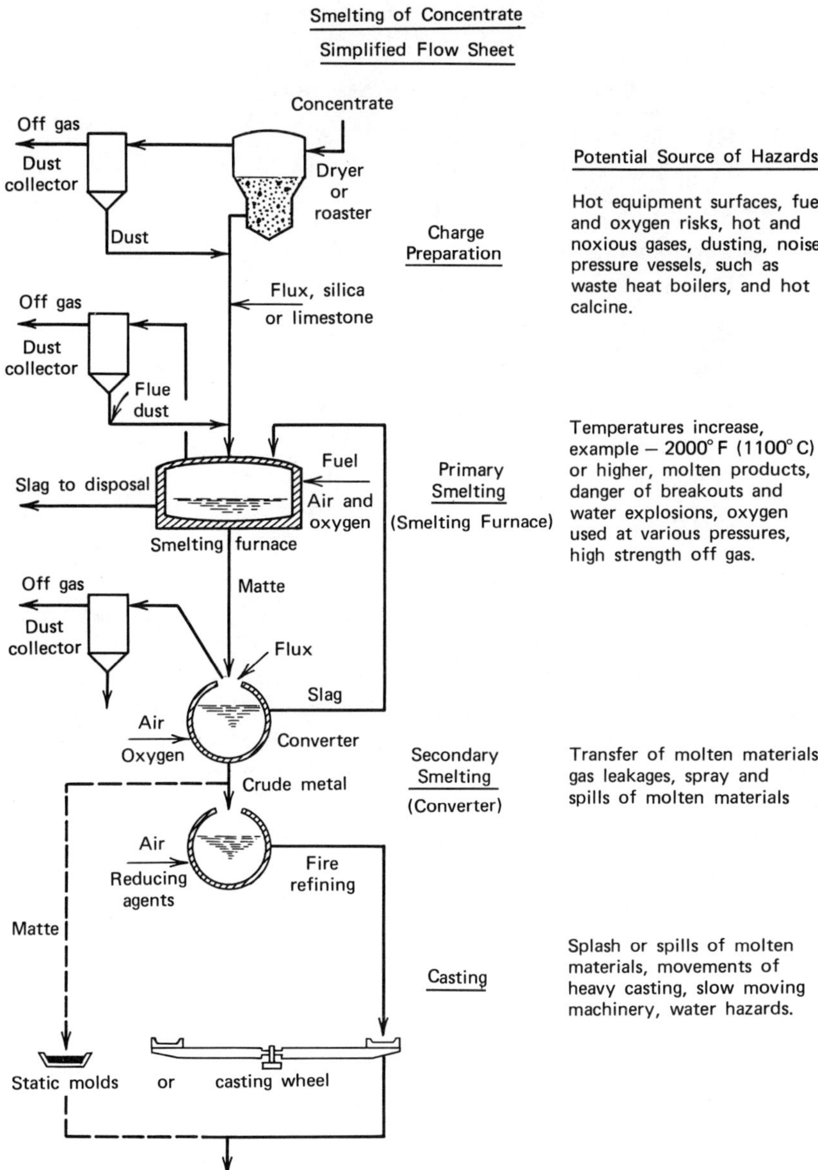

Figure 5.26. Copper smelter waste products for treatment from roasting, smelting, converting, and refining. *Source:* John C. Taylor, Approaches to dust and fume control in copper and nickel smelters, Presented at the AIME Annual Meeting, 1979.

Roast Leaching Process. The leaching of roasted copper sulfide concentrate is being carried out in several locations—Zaire, Zambia, and most recently the Hecla Mining Company's operation at Lake Shore, Arizona. The advantage of this system is that it is environmentally cleaner than full pyrometallurgical treatment and uses its own SO_2 products. Sulfide concentrates are roasted to sulfate or oxide and leached in sulfuric acid produced from the SO_2 gas from the roasters. Combined oxide–sulfide ores can also be efficiently processed in this type of operation, with the only limitation being the balance of sulfide supplied which will produce sulfuric acid against sulfuric acid consumed in leaching the oxide concentrate.

Electrorefining and Electrowinning. Purification of impure metal anodes or deposition of copper from leach solutions are carried out in a tankhouse in generally quite similar pro-

Table 5.5.   Percent SO$_2$ in Gas from Copper Roasting, Smelting, and Converting

Process	% SO$_2$ in Gas
Multiple-hearth roaster	5
Fluid-bed roaster	10–15
Blast furnace	5
Reverberatory furnace	0.2–1.5
Electric furnace	5
INCO flash furnace	70–80
Outokumpu flash furnace	18
Peirce–Smith converter	4
Noranda	4–13
Mitsubishi	10
Worcra	10

cesses, in that high-purity copper metal is deposited at the cathode, a gas is given off at the anode or cathode, and the electrolyte contains a high concentration of sulfuric acid.

The tankhouse is a fully enclosed building. To prevent the accumulation of acid mist carried off by the escaping gas, with its accompanying high building and equipment corrosion and difficult working conditions, certain design modifications have been made in these structures. A ventilation system using a wind tunnel concept with a continuous sweep of fresh air drawn through the tankhouse by large fans is one of these. Structural steelwork should be of the wide flange type with no lattice girders to avoid dead air spaces where acidic moisture can accumulate and cause a high corrosion rate. Thin gauge stainless steel siding is more corrosion resistant than other types of siding. Finally protection against cell mist can be provided by a blanket of several layers of hollow, $\frac{3}{4}$ inch (1.875 cm) polypropylene balls covering the electrolytic cells.

2. Nickel Nickel processing very closely parallels the treatment for copper, with copper and nickel sulfides found together in most of their ores. Common processing is carried out in some cases through the converting operation before a separation is made of the nickel from the copper. Some of the processes also now used for copper were initially developed for nickel, such as flash smelting and oxygen converting, and in these cases the treatment and recovery of dust, SO$_2$, and heat in the smelter gases is carried out in the same way for both metals.

The conventional processing for nickel is to roast nickel sulfide concentrate in multiple-hearth or fluid-bed roasters and then smelt the roasted calcines; or in some instances, as is done with copper, to smelt unroasted concentrate, in reverberatory, electric, or flash smelting furnaces. Continuous combined smelting and converting has not yet been applied to nickel, and the last major blast furnace operation at Falconbridge Nickel in Canada was replaced by an electric furnace smelter completed in 1980.

Matte from the smelting furnaces is blown to nickel oxide in Peirce–Smith converters or to crude nickel metal, equivalent to blister copper, in the newer top-blown rotary converter using an oxygen lance.

3. Lead Lead is another of the major sulfide ores which is processed almost exclusively by pyrometallurgical treatment. The sulfide concentrate is usually roasted to oxide and agglomerated on a traveling-grate sintering machine. The resulting lead oxide sinter is smelted in a blast furnace with coke to reduce the oxide to lead metal.

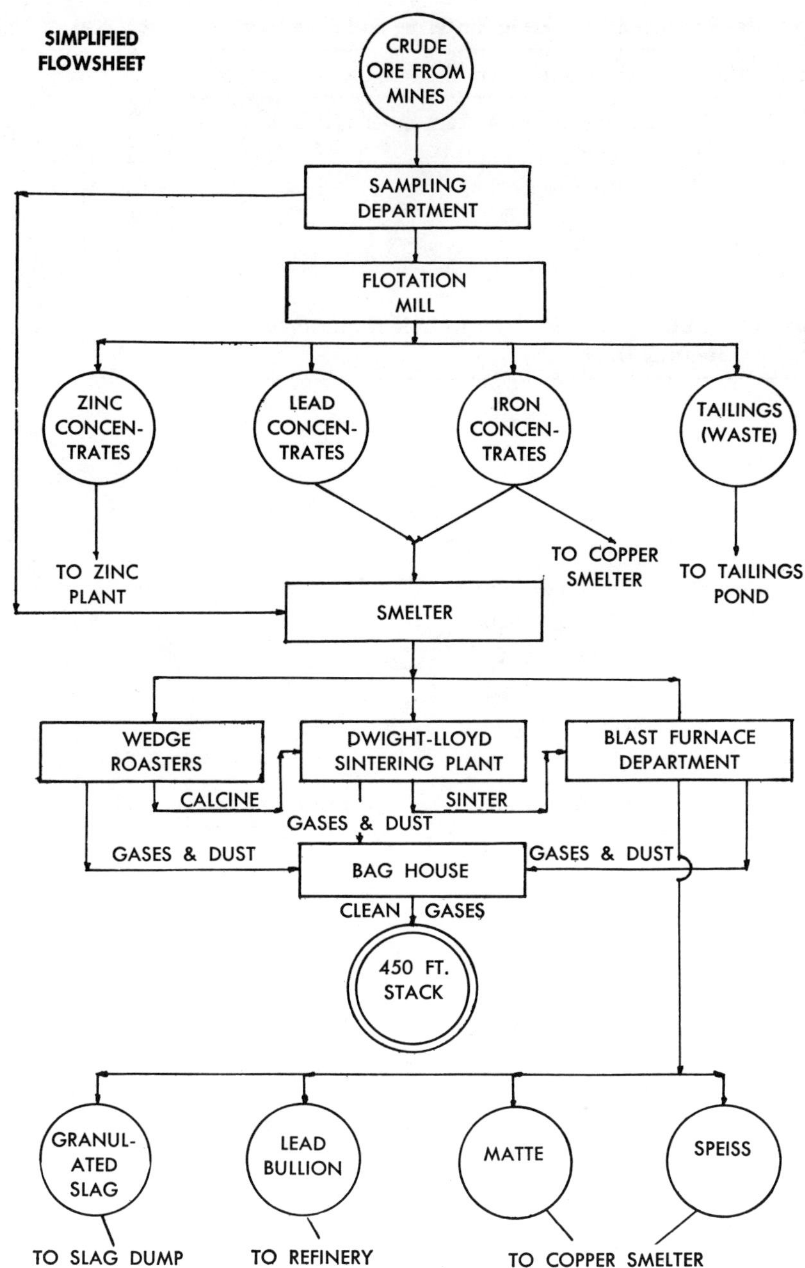

Figure 5.27. Lead smelter gas cleaning by bag house and tall stack. *Source:* United States Smelting, Mining and Refining Company.

The effluent gases from the roast sintering and smelting contain considerable particulate matter because of the high gas volume nature of the operations, and these are collected and recovered in combinations of settling chambers for the coarse dust and either electrostatic precipitators or bag houses for the fine particles.

The solid emissions and gas volume are summarized in Table 5.6. Approximately 2 unit weights of concentrated ore are required to produce 1 unit weight of lead metal.

Sulfur dioxide is also a component of the roaster gases, as lead sulfide concentrate analyzing 15% is roasted to sinter of 1.5% S, with the oxidized sulfur forming SO_2. Overall plant emissions from combined sintering and blast furnace operations will average 660 pounds SO_2 per ton concentrated ore treated (330 kg per MT), and this gas has generally in the past been put up the stack after dust removal, as it is usually too low for economic H_2SO_4 production.

Table 5.6. Particulates and Gases Produced from Lead Sintering and Smelting

Type of Operation	Particulates (g/m^3)	Gas Volume (m^3)	SO$_2$ Content (%)
Sintering	2–15	3000 (per ton sinter)	1.5–5
Blast furnace	5–15	15,000–50,000 (per ton coke)	—

Table 5.7. Bag House Operations, Buick Lead Smelter, AMAX-Homestake:
Operating Data

1. *Main Bag House*

Make	Wheelabrator
No. of compartments	14
No. of bags, total	5824
Bag size, D $\times$ L	20.3 cm $\times$ 6.1 m
Bag material	Filtron 260-50
Maximum volume (m^3/min)	12,750 at 110°C
Normal volume (m^3/min)	8300 at 90°C
Air-to-cloth ratio (m^3/m^2)	0.59:1
Operating pressure (mm H$_2$O)	100–150
Normal dust loading (g/m^3/min)	1.26

2. *Hygiene Bag Houses*

	Sinter Crushing	Furnace Charge
Make	Pangborn type	CT
No. of compartments	3	3
No. of bags, total	312	456
Bag size, D $\times$ L	12.7 cm $\times$ 3.4 m	
Bag material	Wool	
Maximum volume (m^3/min)	380 at 45°C	580 at 45°C
Air-to-cloth ratio (m^3/m^2)	0.90:1	
Operating pressure (mm H$_2$O)	250	
Normal dust loading (g/m^3/min)	—	

3. *Main Bag House Fans*

Type	Sturtevant double inlet
Size, D $\times$ W, two fans	2 m $\times$ 4 m each
Rated capacity at temp. for each	6650 m^3/min at 110°C
Static pressure (mm H$_2$O)	355
Horsepower	800
Revolutions per minute	900

4. *Operating Labor*

Three-shift operation includes dust pug mill

Operator	3
Labor	1
Total	4 per day

Source: C. H. Cotterill and J. M. Cigan, Eds., AIME World Symposium of Lead and Zinc, Vol. 2, 1970, p. 776.

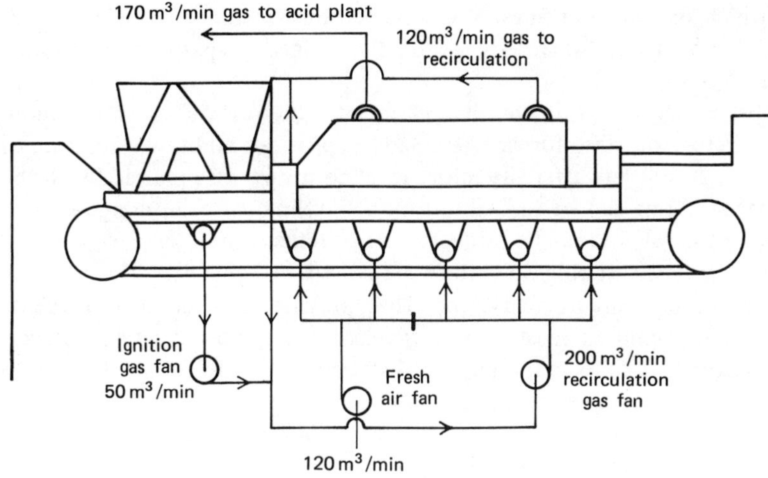

Figure 5.28. Sintering machine gas flow, Naoshima lead smelter. *Source:* C. H. Cotterill and J. M. Cigan, Eds., AIME World Symposium of Lead and Zinc, Vol. 2, 1970, p. 858.

4. Zinc Most zinc comes from sulfide lead and zinc ores, which are separated by beneficiation for further individual processing and metal extraction. The zinc sulfide concentrates are treated by pyrometallurgical or combined pyrometallurgical–hydrometallurgical processing to convert the zinc sulfide to metal.

There are three steps in the pyrometallurgical processing:

1. Roasting the zinc sulfide concentrate to remove most of the sulfur and give a zinc oxide calcine.

2. Sinter the calcine to oxidize the remaining sulfide and agglomerate the fine calcine into a dense, permeable furnace feed.

3. Reduce the ZnO sinter (cake or briquettes) with carbon in a vertical, electrothermic, or horizontal retort, or in a blast furnace, to produce metallic zinc.

Roasting is carried out to a high degree of oxidizing efficiency, as only ZnO can be reduced by carbon in the retorting and any ZnS left unoxidized will be lost as unreduced retort residue. As a consequence, 93 to 97% of the input sulfur in the roaster feed is converted to SO_2, and zinc plants use several types of roasters and combinations of types to achieve the best roasting. These types include multiple-hearth roasters, flash roasters, fluid-bed roasters, and sintering machines.

The emissions of particulates and SO_2, along with gas temperatures from the various roasters are shown in Table 5.8. The gas from the multiple-hearth roaster has a sufficiently high SO_2 content to be sent to a contact acid process for H_2SO_4, while the gas from flash or fluid-bed roasting can be used for either acid or elemental sulfur. The high particulate content, especially with the flash and fluid-bed roasters, necessitates having efficient dust removal, and the high gas temperature also provides a recoverable value in heat energy. The exhaust gas first passes through a waste heat boiler to recover some of the contained heat, followed by cyclones to remove the coarser particles and an electro-

Table 5.8.   Zinc Roaster Emissions

Type of Roaster	Particulates (% of feed)	SO₂ Content (%)	Gas Temperature
Multiple-hearth	5–15	4.5–6.5	1300°F (704°C)
Flash	50	8–12	1880°F (982°C)
Fluid-bed	35–50	9–12	1700°F (927°C)

static precipitator which recovers the fines. The collected dust is added to the main calcine stream going to the next stage of sintering, while the cleaned gas passes on to a sulfur removing operation.

Sintering accomplishes two things, in that it agglomerates the fine calcines into a more satisfactory form of retort or blast furnace feed, and it provides additional roasting to remove whatever sulfur is still left after the initial roast to insure that practically all the zinc is as zinc oxide. Sintering can be carried out on a traveling grate sintering machine to make sinter cake, or the calcines can be briquetted or pelletized and then baked.

The emissions from typical sintering operations are shown in Table 5.9. The resulting gases are low in SO_2 and difficult to treat alone. They present the alternative of blending in with the higher SO_2 content roaster gases or releasing from a tall stack. The particulate content is caught and collected in an electrostatic precipitator or Venturi scrubber to be recycled back through the sintering step.

Reduction of the zinc oxide to metal with carbon,

$$ZnO + C = Zn + CO$$

is the final treatment process and can be carried out by one of the several methods of retorting or in a zinc blast furnace.

Comparative results for the retorting processes are,

Type of Operation	Particulates
Vertical retort	1.25 to 3.75 g/m^3
Horizontal retort	0.1 to 0.3 g/m^3

These retorting operations are quiet processes, with heat being provided to the outer wall of the retort. The only gases produced inside the retort are zinc metal vapor and carbon monoxide, and as a consequence the dust loading is also quite light.

The zinc metal vapor is condensed to liquid metal, and the gas high in CO is recovered from some of the retorting methods for use as an auxilliary fuel. In the vertical retort the zinc-free gas is scrubbed and returned to make up 30% of the fuel used to heat the retorts to their operating temperature. Natural gas makes up the remaining 70% of the fuel requirement. The electrothermic furnace is very similar, with the gas going through a washer for cleaning and then piped to a gas burner in a rotary preheater which is used to preheat the feed material going to the retort. There is no way to recover the CO gas from the open-ended horizontal retort, and it burns to CO_2 as it escapes from the hot zinc condenser out into the air.

The zinc blast furnace also produces gas containing CO, but because of the nature of the process with large volumes of air being blown in through the tuyeres, the off-gas is diluted and has a low calorific value with only 15 to 20% carbon monoxide content. After the zinc vapor is condensed out, the gas is sent through a gas-washing tower and a Thiessen disintegrator, which cools the gas to below 104°F (40°C) and cleans it to less than 40 milligrams per cubic meter of solids.

Table 5.9. Zinc Sintering Emissions

Feed Material	S in Feed (%)	Particulates (% of feed)	SO_2 in Gas (%)
Calcine	8	5	1.5–2
Calcine	2	5–7	0.1
Concentrate (80% recycled)	31	5–10	1.7–2.4

The clean gas goes to a booster fan to increase its pressure and then is burnt as fuel in hot blast stoves which preheat the tuyere air going to the blast furnace.

Roast Leaching Process This is a combined pyrometallurgical–hydrometallurgical operation where the zinc sulfide concentrate is first roasted, as in the simple pyrometallurgical treatment, and then leached in a sulfuric acid solution. The roaster gases are used to manufacture the H_2SO_4 used for leaching by the contact acid processes, and the gases are cleaned of dust in the same manner as previously described.

5. Aluminum The ore for aluminum production is bauxite, which is a hydrated oxide of aluminum associated with silicon, titanium, and iron. Bauxite is upgraded and purified by the Bayer process to produce pure Al_2O_3 for feed material to the aluminum electrolytic reduction cell. To accomplish this, it is dried, ground in ball mills, and pressure leached in a sodium hydroxide solution. Iron oxide, silica, and other impurities are insoluble, while the soluble aluminum is precipitated from the pregnant leach solution as aluminum hydroxide and finally calcined to pure alumina:

$$2Al(OH)_3 + \text{heat} = Al_2O_3 + 3H_2O$$

In the bauxite grinding, and particularly in aluminum hydroxide calcining, as well as material handling operations, considerable dust is produced. This is usually caught by various dry dust-collecting devices such as cyclones, multiclones, or dry electrostatic precipitators. However there are some cases where wet scrubbers are used, or even sometimes a combination of both dry and wet collectors, to remove the solid particles and recycle them back into the process.

Operation	Particulates
Bauxite grinding	6 lb/ton (3 kg/metric ton) of bauxite processed
Materials handling	10 lb/ton (5 kg/metric ton) of bauxite processed
Calcining	300 to 400 g/m^3 of hot gas

Aluminum metal is made in the Hall cell, which involves the electrolytic reduction of alumina dissolved in a molten salt bath of cryolite, Na_3AlF_6, and various other salt additives:

$$2Al_2O_3 + \text{electrolysis} = 4Al + 3O_2$$

There are three types of Hall cells, differing only in the form of the carbon anode which is inserted down into the liquid bath from above. These three types of anode configuration are (1) prebaked (PB), (2) horizontal-stud Söderberg (HSS), and (3) vertical-stud Söderberg (VSS). The prebaked cells are the most common, with the horizontal-stud Söderberg being second and the vertical-stud Söderberg the least used.

Emissions from the reduction cell consists of both gaseous and particulant matter, with the gases being mostly hydrogen fluoride and carbon monoxide. The solid emissions consist of alumina and carbon from anode dusting, cryolite, aluminum fluoride, calcium fluoride, shiolite ($Na_5Al_3F_{14}$), and ferric oxide.

The source of fluoride emissions is the fluoride electrolyte which contains cryolite (Na_3AlF_6), aluminum fluoride (AlF_3), and fluorspar (CaF_2). For normal operating conditions the weight ratio of NaF to AlF_3 is maintained between 1.36 to 1.43 by the additions of Na_2CO_3, NaF, and AlF_3, and it has been found that increasing this ratio will decrease the total fluoride effluents. Cell fluoride emissions are also decreased by lowering the operating temperature and increasing the bath content of Al_2O_3. The ratio of

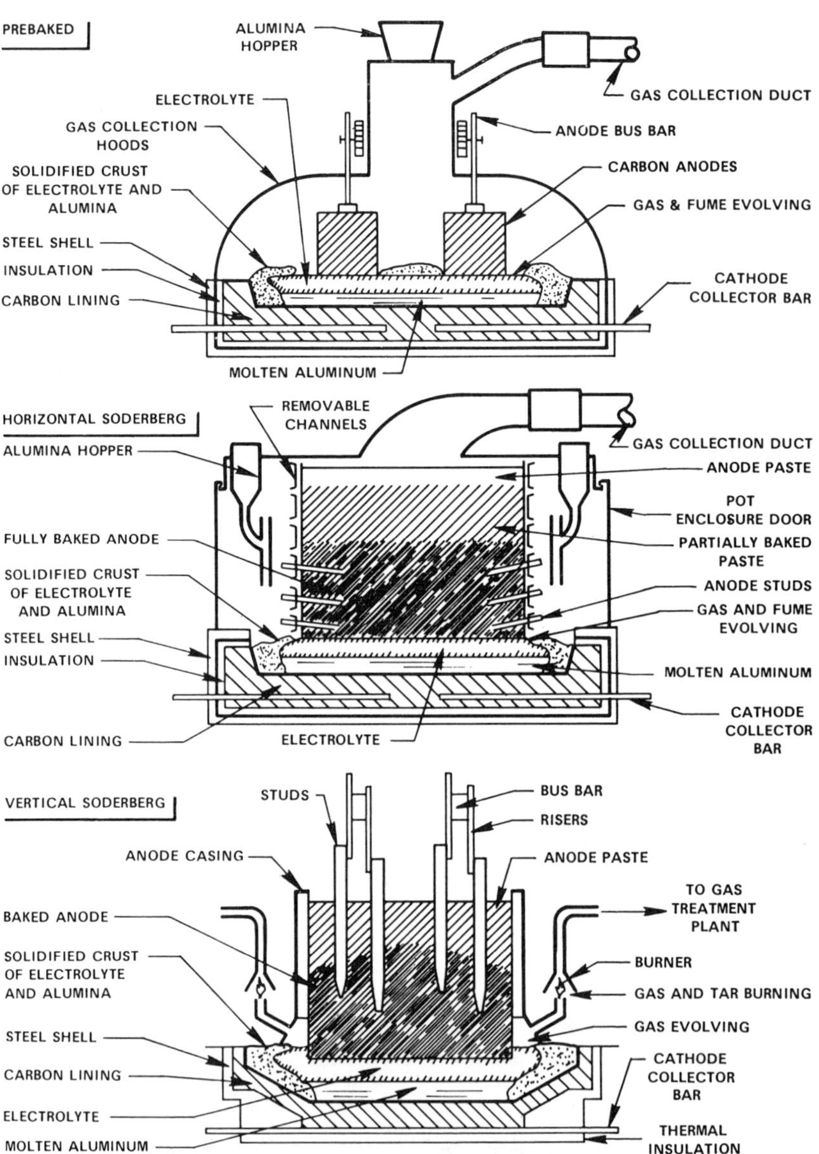

Figure 5.29. Systems of gas removal from the three types of aluminum reduction cells. *Source:* R. E. Iverson, *J. Metall.*, Vol. 25, No. 1, 1973.

gaseous fluorides (mostly HF) to particulate fluorides varies considerably, with prebaked the lowest, horizontal-stud Söderberg next, and vertical-stud Söderberg cells the highest.

Fluoride emissions from the three types of cells are shown in Table 5.10.

The solid particle sizes in emission from the cells are quite small, with the greatest single proportion less than 1 micron in diameter for the prebaked and horizontal-stud Söderberg cells (Table 5.11).

Because of the variety of gases and particulants coming from the cells, many different kinds of control devices have been used. Aluminum plants using prebaked carbon anodes may have dry dust collectors to remove solids and scrubbers to remove gaseous fluorides, while plants using Söderberg electrodes may omit dry collecting and scrub all effluent directly.

Dry alumina adsorption is also a technique that has been used at several prebaked and vertical-stud Söderberg installations. In this type of process both gaseous and particulate

Table 5.10. Aluminum Cell Fluoride Emissions[a]

Type of Cell Operation	Gaseous Fluorides	Particulate Fluorides
Prebaked anode	81.3 lb/ton (40.65 kg/metric ton)	20.4 lb/ton (10.2 kg/metric ton)
Horizontal-stud Söderberg anode	98.4 lb/ton (49.2 kg/metric ton)	15.6 lb/ton (7.8 kg/metric ton)
Vertical-stud Söderberg anode	78.4 lb/ton (39.2 kg/metric ton)	10.6 lb/ton (5.3 kg/metric ton)

[a]Per ton of molten aluminum produced.

**Table 5.11. Comparison of Aluminum Cells
Particulate Size Distribution**

Particle Size (microns)	Weight %, Particle Size Distribution	
	Prebaked Anode	Horizontal-Stud Söderberg Anode
<1	35	44
1 to 5	25	26
5 to 10	8	8
10 to 20	5	6
20 to 40	5	4
>44	22	12

fluorides are controlled by passing the reduction cell gases through the entering alumina feed, and the fluorides are absorbed on the Al_2O_3. The overall control efficiency is 98%.

The types of control methods used with the three different cells are shown in Table 5.12.

Efficiencies of fluoride removal from the gases with these types of collectors and scrubbers are listed in Table 5.13.

In all cases, after treatment to remove the solids and contaminant gases, the cleaned process gas is released to the atmosphere from a stack.

6. Cobalt Cobalt is commonly found as a sulfide ore, in combination with other sulfides of nickel, copper, and iron. One of the more common methods of processing this ore is first to beneficiate it to a higher grade and then to roast the beneficiated concentrate to a sulfate. Cobalt and the other valuable metal sulfates are soluble in water or dilute acid solutions and can be extracted by leaching and precipitation.

Table 5.12. Pollution Control Methods for Aluminum Cells

Type of Cell Operation	Control Method
Prebaked anode reduction cell	Multiclone, fluid-bed dry scrubber, coated filter, dry electrostatic precipitator, spray tower, floating-bed scrubber, dry alumina adsorption
Horizontal-stud Söderberg cell	Spray tower, floating bed scrubber, wet electrostatic precipitator
Vertical-stud Söderberg cell	Spray tower, Venturi scrubber, wet electrostatic precipitator, multiclone, dry alumina adsorption

Table 5.13. Fluoride Removal Efficiency

Gas Treatment	HF Gas (%)	Particulate (%)	Total F (%)
Coated filter	90	98	94
Fluid-bed dry scrubber	99	98	99
Dry electrostatic + wet scrubber	98	90–95	95
Wet electrostatic + wet scrubber	99	99	+99
Floating-bed scrubber	98	87	95
Venturi scrubber	99	96	98
Dry alumina scrubber	98	96	98

The sulfatizing roast is carried out in a four-compartment fluid-bed roaster in one of the most modern treatment plants. And as with all fluid-bed operations where a fine-sized feed is used (50 to 60% minus 200 mesh), a large fraction of the total solids fed into the process (55 to 60%) is carried out with the roaster gases and must be recovered, as well as the off-gas carrying considerable amounts of SO_2.

The roaster gases pass from the top of the roaster to a battery of cyclone dust collectors. Then the recovered dust from this is immediately fed through screw feeders back into the beds of the second, third, and fourth compartments of the roaster.

From the cyclones the gases pass to a waste heat boiler, where their temperature is cooled from 1200°F (650°C) down to 625°F (330°C) and 140 Mcal is recovered as steam per ton of concentrate feed. A small amount of flue dust is also recovered from this operation.

From the waste heat boiler the gas passes to an electrostatic precipitator for final cleaning. This operates at 625°F (330°C) with two chambers in series and uses selenium rectifiers at 60 kV and 200 mA.

The efficiencies of the dust removal systems are 99.5% overall, with the individual efficiencies rated as follows: cyclones, 85 to 87%; boiler, 17 to 19%; electrostatic precipitation, 95 to 96%.

The recovered dust in each instance is fed back into the fluid bed roaster for reprocessing.

The composition of the clean gas is 4.8% SO_2, 0.8% SO_3, 13.0% O_2, and 81.4% N_2. This is transferred to a neighboring plant for sulfuric acid manufacture.

REFERENCES

ASARCO liquifies SO_2 from smelter gas. *Min. Eng.*, Vol. 26, No. 11, 1974.

Agarwal, J. C., H. Flood and R. Giberti, *J. Metall.*, Vol. 26, No. 12, 1974.

Biswas, A. D. and W. G. Davenport, *Extractive Metallurgy of Copper*, Pergamon, New York, 1976.

Boldt, J. R., *The Winning of Nickel*, Methuen, London, 1967.

Campbell, I. E., *The Removal of Sulphur Dioxide from Copper Reverberatory Furnace Gas by Wet Limestone Scrubbing*, Smelter Control Research Association, 1972, Manning, Selvage and Lee, Inc., New York.

Chanlett, E. T., *Environmental Protection*, McGraw-Hill, New York, 1973.

Cheremisinoff, P. N., Pollution Engineering Fundamentals Technical Update Seminar, *Eng. Dig.*, Vol. 21, No. 10, 1975.

Compilation of Air Pollution Emission Factors, U.S. EPA, Office of Air and Water Programs, Bulletin AP-42, 2nd ed., Washington, D.C., 1973.

Cotterill, C. H. and J. M. Cigan, Eds., *AIME World Symposium on Mining and Metallurgy of Lead and Zinc*, Vol. 2, Extractive Metallurgy of Lead and Zinc, Port City Press, Baltimore, 1970.

Deju, R. A., *Extraction of Minerals and Energy: Today's Dilemmas*, Ann Arbor Science Publishers, Ann Arbor, Michigan, 1974.

Ehrlich, R. P., Ed., *Copper Metallurgy*, Extractive Metallurgy Division Symposium on Copper Metallurgy, AIME, New York, 1970.

Environmental control, *Min. Eng. Ann. Rev.*, Vol. 25, No. 2, 1973.

Calspan Corp., Environmental Impacts of Virgin and Recycled Aluminum, prepared for U.S. EPA, U.S. Department of Commerce, National Technical Information Service, Washington, D.C., PB-253 487, 1976.

Iverson, R. E., Air pollution in the aluminum industry, *J. Metall.*, Vol. 25, No. 1, 1973.

Annual Review Issue, *J. Metall.*, Vol. 29, No. 3, 1977.

Lynn, D. A., *Air Pollution—Threat and Response*, Addison-Wesley, Reading, Mass., 1976.

Lurgi Geselleschaften, *Lurgi Manual*, Frankfurt am Main, 1961.

Marston, E. H., *The Dynamic Environment*, Xerox College Publishing, Lexington, Mass., 1975.

Mineral Facts and Problems, Bureau of Mines Bulletin 667, U.S. Department of Interior, Washington, D.C., 1975.

Mineral Resources and the Environment, Commission on Natural Resources, National Research Council, National Academy of Sciences, Washington, D.C., 1975.

Moriber, G., *Environmental Science*, Allyn and Bacon, Rockleigh, New Jersey, 1974.

Oroya Metallurgical Operations, Cerro De Pasco Corp. (now Centromin Peru), 1972.

Perkins, H. C., *Air Pollution*, McGraw-Hill, New York, 1974.

Pollution control systems, *J. Metall.*, Vol. 27, No. 2, 1975.

Priest, J., *Problems of Our Physical Environment*, Addison-Wesley, Reading, Mass., 1973.

Pryor, E. J., *Mineral Processing*, Elsevier, London, 1965.

Rosenbaum, J. B., Minerals Extraction and Processing: New Developments, in *Materials: Renewable and Nonrenewable Resources*, Abelson, P. H. and A. L. Hammond, Eds., American Association for the Advancement of Science, Washington, D.C., 1976, pp. 113–117.

Shaheen, E. I., *Environmental Pollution: Awareness and Control*, Engineering Technology Incorp., Mahomet, Michigan, 1973.

Shonle, J. I., *Environmental Applications of General Physics*, Addison-Wesley, Reading, Mass., 1975.

Solid Waste Processing, U.S. Dept. of Health, Education and Welfare, Environmental Control Administration, Bureau of Solid Waste Management, Bulletin SW-4c, Washington, D.C., 1969.

Stack gas cleaning equipment, *Pollution Equipment News*, Vol. 10, No. 1, 1977.

Stern, A. C., Ed., *Air Pollution*, Vol. 3, Sources of Air Pollution and Their Control, Academic Press, New York, 1968.

Sulphatizing roasting cobalt ores, *J. Metall.*, Vol. 23, No. 2, 1971.

Thomas, R., Ed., *Engineering and Mining Journal Operating Handbook of Mineral Processing*, McGraw-Hill, New York, 1977.

Trends in wastewater treatment, *Eng. Dig.*, Vol. 23, No. 1, 1977.

The U.S. Zinc Industry: A Historical Perspective, Bureau of Mines Information Circular 8629, U.S. Dept. of Interior, Washington, D.C., 1974.

Yannopoulos, J. C. and J. C. Agarwal, Eds., *Extractive Metallurgy of Copper*, Vol. 1, Pyrometallurgy and Electrolytic Refining, Metallurgical Society of AIME, New York, 1976.

Appendix

Appendix. Properties of the Elements

Element	Symbol	Atomic Number	Atomic† Weight	Density*	M.P. °C***	B.P. °C***	Chemical** Valence	First Ioniz. Pot. (ev.)
Actinium	Ac	89	(227)	–	1051	–	3	6.9
Aluminum	Al	13	26.98	2.70	660	2454	3	5.98
Americium	Am	95	(243)	11.7	996	–	3,4,5,6	6.0
Antimony	Sb	51	121.75	6.62	630.7	1382	3,5	8.64
Argon	Ar	18	39.95	1.78	−189	−186	0	15.75
Arsenic	As	33	74.92	5.72	–	613[2]	±3,5	9.81
Astatine	At	85	(210)	–	–	–	–	–
Barium	Ba	56	137.34	3.5	714	1643	2	5.21
Berkelium	Bk	97	(249)	–	–	–	3,4	–
Beryllium	Be	4	9.01	1.85	1279	2775	2	9.32
Bismuth	Bi	83	208.98	9.80	272	1562	3,5	7.29
Boron	B	5	10.81	2.34	2033	2555	3	8.30
Bromine	Br	35	79.91	3.12	−7	58	±1,5	11.81
Cadmium	Cd	48	112.40	8.65	321.1	765	2	8.99
Calcium	Ca	20	40.08	1.55	839	1442	2	6.11
Californium	Cf	98	(251)	–	–	–	3	–
Carbon	C	6	12.01	2.25[1]	3735	4830	±4,2	11.26
Cerium	Ce	58	140.12	6.77[3]	805	3264	3,4	5.60
Cesium	Cs	55	132.91	1.90	29	690	1	3.89
Chlorine	Cl	17	35.45	3.21	−101	−35	±1,7,5	13.01
Chromium	Cr	24	52.00	7.19	1878	2670	3,6,2	6.76
Cobalt	Co	27	58.93	8.85[4]	1497	2906	2,3	7.86
Copper	Cu	29	63.54	8.96	1084.5	2600	2,1	7.72
Curium	Cm	96	(247)	–	1342	–	3	–
Dysprosium	Dy	66	162.50	8.55	1409	2339	3	–
Einsteinium	Es	99	(254)	–	–	–	–	–
Erbium	Er	68	167.26	9.15	1499	2515	3	6.08
Europium	Eu	63	151.96	5.24	827	1599	3,2	5.67
Fermium	Fm	100	(253)	–	–	–	–	–
Fluorine	F	9	19.00	1.70	−220	−188	−1	17.42
Francium	Fr	87	(223)	–	–	–	1	–
Gadolinium	Gd	64	157.25	7.86	1314	3240	3	6.16
Gallium	Ga	31	69.72	5.91	30	2241	3	6.00
Germanium	Ge	32	72.59	5.32	938	2835	4	7.90
Gold	Au	79	196.97	19.32	1064.4	2976	3,1	9.22
Hafnium	Hf	72	178.49	13.09	2226	5400	4	6.8
Helium	He	2	4.00	0.18	−270	−269	0	24.58

Appendix.   Properties of the Elements (*Continued*)

Element	Symbol	Atomic Number	Atomic† Weight	Density*	M.P. °C***	B.P. °C***	Chemical** Valence	First Ioniz. Pot. (ev.)
Holmium	Ho	67	164.93	8.79[1]	1463	2577	3	–
Hydrogen	H	1	1.01	0.09	−259	−253	1	13.59
Indium	In	49	114.82	7.31	156	2003	3	5.78
Iodine	I	53	126.90	4.94	114	183	−1,5,7	10.45
Iridium	Ir	77	192.2	22.5	2458	5300	4,3,6	9.1
Iron	Fe	26	55.85	7.87	1538	3000	3,2	7.87
Krypton	Kr	36	83.80	3.74	−157	−152	0	14.00
Lanthanum	La	57	138.91	6.19[2]	921	3461	3	5.61
Lawrencium	Lw	103	(257)	–	–	–	–	–
Lead	Pb	82	207.19	11.36	327.5	1728	2,4	7.41
Lithium	Li	3	6.94	0.53	181	1338	1	5.39
Lutetium	Lu	71	174.97	9.85	1654	3322	3	6.15
Magnesium	Mg	12	24.31	1.74	650	1109	2	7.64
Manganese	Mn	25	54.94	7.43	1247	2154	2,7,4,6,3	7.43
Mendelevium	Md	101	(256)	–	–	–	–	–
Mercury	Hg	80	200.59	13.55	−38.4	357	2,1	10.43
Molybdenum	Mo	42	95.94	10.22	2615	5560	6,3,5	7.10
Neodymium	Nd	60	144.24	7.00[5]	1020	3133	3	5.46
Neon	Ne	10	20.18	0.90	−249	−246	0	21.56
Neptunium	Np	93	(237)	19.5	637	–	5,6,4,3	–
Nickel	Ni	28	58.71	8.90	1455	2735	2,3	7.63
Niobium	Nb	41	92.91	8.57	2472	4927	5,3	6.88
Nitrogen	N	7	14.01	1.25	−210	−196	−3,5,2	14.53
Nobelium	No	102	(254)	–	–	–	–	–
Osmium	Os	76	190.2	22.57	2700	5500	4,6,8	8.5
Oxygen	O	8	16.00	1.43	−219	−183	−2	13.61
Palladium	Pd	46	106.4	12.02	1554	3989	2,4	8.33
Phosphorus	P	15	30.97	1.83[6]	44	280	5,±3	10.48
Platinum	Pt	78	195.09	21.45	1772	4530	4,2	9.0
Plutonium	Pu	94	(242)	19.0	640	3242	4,6,5,3	5.8
Polonium	Po	84	(210)	(9.24)	254	–	2,4	8.43
Potassium	K	19	39.10	0.86	64	761	1	4.34
Praseodymium	Pr	59	140.91	6.77[5]	920	3219	3	5.48
Promethium	Pm	61	(145)	–	1028	–	3	–
Proactinium	Pa	91	(231)	15.4	1232	–	5,4	–
Radium	Ra	88	(226)	5.0	700	1142	2	5.28
Radon	Rn	86	(222)	9.96	−71	−62	0	10.75
Rhenium	Re	75	186.2	21.04	3186	5900	7,4,−1	7.87
Rhodium	Rh	45	102.90	12.44	1969	4500	3,4	7.46
Rubidium	Rb	37	85.47	1.53	39	688	1	4.18
Ruthenium	Ru	44	101.07	12.2	2500	4900	3,4,6,8	7.36
Samarium	Sm	62	150.35	7.49	1073	1755	–	–
Scandium	Sc	21	44.96	2.99[4]	1541	2837	3	6.54
Selenium	Se	34	78.96	4.79[5]	217[5]	685[5]	4,6,−2	9.75
Silicon	Si	14	28.09	2.33	1412	2685	4	8.51
Silver	Ag	47	107.87	10.49	961.9	2214	1	7.57
Sodium	Na	11	22.99	0.97	98	893	1	5.14
Strontium	Sr	38	87.62	2.60	769	1382	2	5.69
Sulfur	S	16	32.06	2.07[6]	113[7]	444.7	6,4,−2	10.36
Tantalum	Ta	73	180.95	16.6	2996	5425	5	7.88
Technetium	Tc	43	(99)	11.46	2130	–	7	7.28
Tellurium	Te	52	127.60	6.24[5]	450	991	4,6,−2	9.01
Terbium	Tb	65	158.92	8.25	1358	3047	3	5.98
Thallium	Tl	81	204.37	11.85	303	1459	1,3	6.11
Thorium	Th	90	232.04	11.66	1753	3850	4	6.95
Thulium	Tm	69	168.93	9.31	1547	1735	3	5.81
Tin	Sn	50	118.69	7.30[8]	231.9	2274	4,2	7.34
Titanium	Ti	22	47.90	4.51	1670	3267	4,3	6.82
Tungsten	W	74	183.85	19.3	3417	5930	6	7.98

Appendix.   Properties of the Elements (*Continued*)

Element	Symbol	Atomic Number	Atomic† Weight	Density*	M.P. °C***	B.P. °C***	Chemical** Valence	First Ioniz. Pot. (ev.)
Uranium	U	92	238.03	19.07	1132	3827	6,5,4,3	6.08
Vanadium	V	23	50.94	6.1	1903	3407	5,4,2	6.74
Xenon	Xe	54	131.30	5.90	−112	−108	0	12.13
Ytterbium	Yb	70	173.04	6.96	825	1195	3,2	6.22
Yttrium	Y	39	88.90	4.47^4	1511	3344	3	6.38
Zinc	Zn	30	65.37	7.13	419.6	907	2	9.39
Zirconium	Zr	40	91.22	6.49	1855	3588	4	6.84

Notes: $\dagger\,^{12}C = 12$
*Solids & Liquids: g/cm^3, 20°C. Gases: g/liter (STP).
**Useful valences, most stable stated first.
( )Integral mass of longest lived isotope.
***International Practical Temperature Scale−1968.
[1] Graphite. Diamond 3.51.
[2] Sublimes
[3] Face-centered cubic
[4] Close-packed hexagonal
[5] Hexagonal
[6] Orthorhombic, Monoclinic = 1.96
[7] Orthorhombic, Monoclinic = 119°C
[8] Normal tetragonal (β or white)

Index